An Evolutionary Study of Some Archaeologically Significant Avian Taxa in the Quaternary of the Western Palaearctic

John R. Stewart

BAR International Series 1653
2007

Published in 2016 by
BAR Publishing, Oxford

BAR International Series 1653

An Evolutionary Study of Some Archaeologically Significant Avian Taxa in the Quaternary of the Western Palaearctic

ISBN 978 1 4073 0089 4

COVER IMAGE *Drawing of a raven, by Tabitha Stewart Stacey (at 8 years of age).*

BAR Publishing is the trading name of British Archaeological Reports (Oxford) Ltd. British Archaeological Reports was first incorporated in 1974 to publish the BAR Series, International and British. In 1992 Hadrian Books Ltd became part of the BAR group. This volume was originally published by Archaeopress in conjunction with British Archaeological Reports (Oxford) Ltd / Hadrian Books Ltd, the Series principal publisher, in 2007. This present volume is published by BAR Publishing, 2016.

Printed in England

BAR
PUBLISHING

BAR titles are available from:

BAR Publishing
122 Banbury Rd, Oxford, OX2 7BP, UK
EMAIL info@barpublishing.com
PHONE +44 (0)1865 310431
FAX +44 (0)1865 316916
www.barpublishing.com

This work is dedicated to my mother and father, Joan Winifred Stewart (1932 – 2007) and Robert Octave Stewart (1937 – 1998). My father was a true naturalist and I can only hope to attain his level of understanding. He was my first and most influential teacher. My Mother provided all things lateral.

Preface

Palaeo-ornithological research has long been dominated either by the description and naming of isolated finds of fossil birds, or by the analyses of faunal remains from single localities with limited comparisons with other such faunas. Most practitioners have been unaware, or even ignorant, of the theoretical aspects of their work. Consequently, their activities seldom have been conducted within solid philosophical and methodological frameworks. It is probably fair to state that this characterises avian palaeontology to a higher extent than most other aspects of palaeontology.

Interactions between neo-ornitology and palaeo-ornithology also have been limited. The bulk of literature on ecomorphology, character displacement, theoretical aspects of speciation, intra-specific variation, etc., has seldom been incorporated in avian palaeontological publications. Syntheses of species evolution based on fossil specimens from several localities covering a wide geographic range, or time span, are exceptional.

In the present volume, the author discusses the consequences to avian palaeontology of the shortcoming mentioned above. The description of how shifting paradigms have influenced our conception of avian evolution is insightful and well written. For example, it is pointed out how the contrasting views of two influential palaeontologists on when most genera and species evolved (during the Miocene-Pliocene *versus* during the Pleistocene) is mirrored by their taxonomic decisions when describing fossil taxa. These conclusions are in turn important when extinction rates of Pleistocene birds are estimated.

What is a species? This is perhaps the most difficult biological question to answer. Yet the answer may affect the taxonomic conclusions about fossil birds dramatically. Currently several competing species concepts exist and one aim of the volume is to investigate the practical consequences of this.

The evolution of four bird taxa (*Grus*, *Lagopus*, *Corvus* and *Sturnus*) through the Quaternary in Europe have been investigated. The metric and morphological variation in their modern representatives were studied along a north-south transect from Scandinavia to Spain. The observed variation was compared to that found in fossils dating from the Middle Pleistocene until the Holocene. The author addresses several questions for each taxonomic group. Most of these concern the patterns of variation observed in the samples. The author obviously knows the subject very well and cites all the pertinent literature. Important and central research problems are addressed and studied by appropriate methods within a solid theoretical framework. The results and conclusions are presented in a clear way and are convincingly argued.

In summary, this is a significant and much welcomed contribution to avian palaeontology.

Per G. P. Ericson, Associate professor
Head of Department of Vertebrate Zoology
Swedish Museum of Natural History

Prologue

This short section serves to introduce this volume with particular reference to the archaeological significance of the work. It also aims to serve as a partial update of developments in European Quaternary avian studies since the contents was submitted as a PhD thesis (Stewart, 1999a), with particular reference to publications that have come out of the thesis or to those subjects related to the present volume.

This volume represents the published version of a thesis whose focus was mainly palaeontological, exploring the meaning of bird bones from a number of archaeological and geological sites in Europe. The emphasis was on evolution, both intraspecific and interspecific, and a biometrical survey of modern skeletons of four principle taxa was conducted to aid the interpretation of the Quaternary fossils of those same taxa. The taxa are cranes *Grus* sp., ptarmigan and grouse *Lagopus* sp., ravens *Corvus corax* and starlings *Sturnus* sp. These taxa were chosen for analysis because they are relatively common on archaeological sites and they do not possess too many identification problems to the genus level and do not have a great many similar species within the genus. In addition all four taxa have presented anomalously sized and proportioned fossils. These anomalous fossils have been variously interpreted as distinct extinct species like *Grus primigenia* (Milne-Edwards, 1867 - 71) and *Corvus antecorax* (Mourer-Chauviré, 1975), subspecies such as *Lagopus lagopus noaillensis* and *Lagopus mutus correzensis* (Mourer-Chauviré, 1975) or populations without further taxonomic decisions in the case of *S. vulgaris* (Harrison, 1980).

The general problems of identifying bird bones are discussed throughout, which will make this volume of interest to zooarchaeologists endeavouring to do so. The theoretical issues that exist in identifying animal bones are the same for birds as they are for other vertebrates so the contents should be of use to all those working on the bones of extant vertebrates.

The first section of this volume describes the historical background to the work by reviewing the literature on Quaternary birds with an emphasis on that covering Europe. The following chapter reviews the theoretical background to the study by considering such matters as species concepts and evolutionary theory which impact on the identification and interpretation of bird remains. This section of the volume is concluded with a list of questions that come out of the two preceding chapters. There follows a materials and methods section which gives details of the specimens analysed, both fossil and modern comparative, the measurements taken and the method for distinguishing starling bones from those of the thrush genus *Turdus*. What follows is a description of both the modern metrical data and that taken from the Quaternary fossils for each of the four taxa. Where possible data for each taxon was taken along a North-South transect through Europe. Therefore, institutions in Sweden, Denmark, Poland, Britain, Belgium, France and Spain were visited. The final section includes a discusion of the results with conclusions to the questions posed at the beginning of the thesis.

Publications on Quaternary birds that have appeared since the thesis was submitted are many and there is not the space here to document them exhaustively but the ones most pertinent to the present study should be mentioned here. In addition publications by the present author that expand on the conclusions in the thesis shall also be detailed.

Of the papers published by the present author, the most important is probably that on the timing of the speciation of continental birds and whether the fossil record reveals anything about it (Stewart, 2002a). The paper uses a meta-analysis of species and subspecies named from the Quaternary fossil record of Europe to illustrate that many of each are named on the same types of criteria. This analysis was used to support the suggestion that in general it is difficult to reliably identify birds to species and that often the genus is as far as is reasonably possible to go. This signifies that the timing of bird speciation will remain elusive from the point of view of most bird skeletal remains. The paper uses two examples (*Sturnus* and *Lagopus*), illustrated in greater debth in this volume, to show that even when taxonomy is obscure evolutionary pattern may be elucidated. A paper dealing with another perspective of the identification of birds examined the degree to which practitioners use the modern geographical ranges of birds to limit the taxa considered during identification (Stewart, 2005). This paper concluded that there were two factors which confounded this practise. First, if bird remains were old, and particularly if they are Pleistocene, then it is likely that distribution changes have occurred. Second, that more modern assemblages can include taxa from even further afield as humans have transported taxa as livestock, pets and curiosities from across the globe.

Perhaps the most important piece of work, certainly the most useful, not discussed in the historical background section (Chapter 1) is the 'Catalogue of Pleistocene birds of the Palaearctic' by Tyrberg (1998). This predates the date of the PhD thesis but was not available to the present author until after it was submitted. The catalogue is updated regularly on the web and is an exhaustive list of sites with faunas in the front part of the book and a list of taxa with sites at the back. This work lies behind many of Tyrberg's publications over the past decade or so with historical biogeographical treatments of various taxa (Tyrberg, 1991a, 1991b, 1995, 1999) as well as the paper on the maximum age of species (Tyrberg, 2002).

Zink with co-authors has meanwhile continued discussing the timing of bird diversification in the Quaternary from a biomolecular perspective (see Zink *et al.*, 2004 for a latest review). The latter paper by Zink and colleagues is in a volume representing the results of a meeting on the effects of the ice age on speciation of a variety of organisms including plants and animals (Willis *et al.*, 2004).

Of related interest to the study of birds and relating to the timing of the arrival of temperate woodland birds to Britain and northern Europe in general (Bramwell, 1984) was the debate about refugia during cold stages of the Pleistocene (Willis and Witaker, 2000; Hewitt, 1996, 1999, 2000; Tzedakis *et al*. 2002, Stewart and Lister, 2001; Stewart, 2003). The importance of this work lies both in the implications that exist for geographically based evolution (allopatric speciation) as well as to the implication that mixes of temperate and cold adapted organisms need not be stratigraphically mixed but may constitute genuine local biotas.

A review of marine birds in the Quaternary record has also been written with the aim of illustrating that unexpected finds on archaeological sites are not necessarily of human origin (Stewart, 2002). In the same volume were several papers on birds in archaeology including ones on taphonomy, domestic bird husbandry, site reports and birds illustrated on artefacts (Bochenski *et al.*, 2002). This was the latest proceedings of the ICAZ Bird Working Group which has existed since 1992 when the first meeting was held in Madrid.

Site reports continue to be written by the present author (Stewart, 1999c; Harrison and Stewart, 1999; Davis and Stewart, 2002; Stewart, in press a and b) and others (see Tyrberg, 1998 and http://web.telia.com /~u11502098/pleistocene.html). A review of the fossil record of wetland birds of North West Europe addressing the archaeological and ornithological audiences have also been written (Stewart, 2001; Stewart 2004). In the former, a description of the hypotheses explaining the identity of the large Western Palaearctic cranes is given. The same year that the thesis on which this volume is based came out von den Driesche published the results of a biometric survey of cranes (von den Driesche, 1999). The conclusion to both surveys (mine and that by von den Driesche) was similar suggesting that the common cranes of the earlier Holocene of were larger than they are today. Since that time, work has been undertaken to test the hypotheses using a combination biometric and ancient DNA techniques (Dobney *et al.*, in prep.).

Two particularly pertinent PhD theses have been completed since that on which this volume is based. They are those by Louchart (2002) dealing with the Pleistocene birds of Corsica and Sardinia and Cooper (1999) on the Pleistocene birds of Gibraltar. Interesting papers have come from those PhD's and those that are particularly notable are those on the new species of thrush from Mediterranean Europe (Louchart, 2004) and that on the first record of azure-winged magpie in the Pleistocene of Europe (Cooper and Voous, 1999 and Cooper, 2000).

The dating of sites in the Pleistocene, particularly the part before the time for which radiocarbon dating is useful, is continuing to be problematic. Recent U/Th dating of La Fage (Mourer Chauviré *et al.*, 2003) has confirmed the age of the site as Middle Pleistocene and not Late Pleistocene as had been feared by the present author. Recently Mourer-Chauviré (2004) returned to the subject of *Corvus antecorax* (Chapter 10) in a paper discussing some Pliocene bird remains. In this contribution she relegates the species to a subspecies level *Corvus corax antecorax* as suggested by Piers Brodkorb (1978). The paper by Mourer-Chauviré (2004) is important in relation to this volume as it shows large raven-like corvids existed well before the Middle Pleistocene, the age of *C. antecorax*. The latter was not known when the thesis was written. Furthermore, she discusses size change in ravens since the Pliocene which she speculates responds to the climate changes of the Pleistocene by increasing during the glacials and decreasing in interglacials. Finally, she sums up the paper's taxonomic section by saying that she had confirmed Tyrberg's calculation regarding species longevity as being *ca.* 6 Ma. The latter completely disregards the points made in Stewart (2002) that if species are difficult to identify it may not be realistic to make such conclusions. More recently, I considered this subject further (Stewart and Beech, 2006) when identifying Miocene birds remains from Abu Dhabi. It is not surprising that if species in genera today are hard to distinguish then Neospecies will be found which are as old as the oldest members of the genus in question. Clearly this is not a viewpoint that is popular as it is a negative take on the fossil record. However, I believe it to be realistic and not necessarily contradicting the claim by Olson and Rasmussen (2001) that modern species lineages date to the Pliocene.

Another issue to have seen developments since the thesis was submitted is that various taxonomic changes have taken place (Knox *et al.*, 2002; Sangster *et al.*, 2002) due to theoretical developments and technical advances in ornithology (Helbig *et al.*, 2002). These include ones where the names of the ptarmigan *Lagopus mutus* has been changed to *L. muta* and *L. lagopus scoticus* to *L. l. scotica* (Sangster *et al.*, 2002). The text has not been amended to account for this. There have also been changes whereby subspecies have been elevated to species. Of relevance here is the elevation of the carrion crow and hooded crow from different subspecies of *C. corone* (*C. c. corone* and *C. c. cornix*) to separate species (*C. corone* amd *C. cornix*) (Knox *et al.*, 2002). This has not been adopted throughout the volume as it is of debatable significance. It is also likely that *C. cornix* and *C. corone* are more similar to each other than each are to the raven *C. corax*. This makes the change recently adopted of little significance to the treatment of the raven in Chapter 10. It is also noteworthy that this is a reversal of a previous decision to subsume the two into *C. corone*. These decisions are a result of the kinds of philosophical changes discussed in the chapter on theory (Chapter 2; see Helbig *et al.*, 2002).

Abstract

This volume sets out to examine four European bird taxa in the Quaternary that exhibited interesting morphological anomalies in order to address whether these were the result of inter- or intraspecific processes. Modern skeletal material of these taxa from a wide geographical area was examined so as to make a more realistic assessment of the fossils than had previously been achieved. Similarly, fossils were studied from a wider geographical and temporal range than before. The study of the four chosen taxa was carried out with an acknowledgement of a variety of theoretical issues in biology, which affect the interpretation of such fossils.

Both European members of the genus *Lagopus* (Red/willow grouse *Lagopus lagopus* and ptarmigan *L. mutus*) were found to change in the robusticity of certain osteological elements through time. This change may have occurred at the Pleistocene / Holocene boundary, and has been interpreted as indicating a reduction in mean size, possibly due to the environmental transition taking place at this time. The study of cranes (*Grus*) revealed that the extinct species *G. primigenia*, which was named in the last century, should be synonymised with the common crane *G. grus*. This study found that there were few birds prior to the Roman period which corresponded in size to modern females of *G. grus*. As cranes are distinctly sexually dimorphic, it would seem that the large fossil bones which were described as a larger extinct species represent the males, and the fossils which were the size of modern males were probably the females. Ravens were also studied and were found to be highly variable in size and shape across their range in Europe. They appear to follow both Bergmann's and Allen's rules. A more thorough study of *Corvus antecorax*, the extinct middle Pleistocene raven, confirmed its size and metric-shape differences from all modern populations and subspecies available. However, the validity of the species' status is questionable because of the variability within the species today. Finally, the work on Quaternary fossil starlings (*Sturnus*), showed that a population of large starlings, probably with different wing bone length ratios, had existed in south-west Britain about 100,000 years ago. Furthermore, based on ecomorphological criteria, the fossil population of starlings may have been more sedentary than modern British starlings. Today, there are similarly large sedentary starlings in the south-west and west of Europe, some of which are distinct subspecies and one of which is a separate species. This led to the suggestion that the genus *Sturnus* has more than once given rise to large sedentary populations of starlings along its western and south-western limits. This is likely to be a climatically driven system, which may be a novel speciation mechanism for birds - the degree of migratory versus sedentary lifestyles of the various populations being the mechanism for isolation that enabled genetic differentiation.

The study has shown that it is necessary to examine as much modern and fossil material as is available before fossils can be interpreted realistically. This generally leads to a more conservative assessment of morphologically anomalous fossils because variation occurs both above and below the level of species. This in turn has led to a greater emphasis on ecomorphology than on taxonomy.

Acknowledgements

First of all I would like to thank Adrian Lister, my PhD supervisor, who showed faith in me by giving me the opportunity to study Quaternary birds at UCL on an NERC grant obtained to study size change in Quaternary vertebrates.

A number of institutions and personnel at those institutions throughout Europe should also be thanked for allowing access to both modern and fossil birds. The Collection managers included: Per Ericson at the Swedish Museum of Natural History; Jon Fjeldså at the Zoological Museum, University of Copenhagen; Zygmunt Bochenski and Adam Nadachowski at the Institute of Systematics and Evolution of Animals, Krakow; Storrs Olson of the Smithsonian Institute, Washington D.C.; the staff at the Musée d'Histoire Naturelles de Bruxelles, Belgium; Michel Phillipe at the Museum d'Histoire Naturelle, Lyon; Cécile Mourer-Chauviré at the Université Claude Bernard, Lyon; Fran Hernandez Carrasquilla and Arturo Morales at the Universidad Autonoma de Madrid and the staff at the Museo de Cienca Naturales de Madrid, Spain. In Britain Sandra Chapman at the Palaeontology Department, Natural History Museum; Robert Prŷs-Jones, Mark Adams and Don Smith at Sub-Department of Ornithology, Natural History Museum; John Nudds at Manchester Museum; Ruth Charles, Oxford University; Derek Yalden at the School of Biological Sciences at Manchester University; the staff at the Stoke on Trent Museum; Chris Hawkes Bristol University Speleological Society Museum; the staff of the Museum of Zoology and Sedgwick Museum at Cambridge University; Simon Davis and Sebastian Payne at the Ancient Monument Laboratory, English Heritage; Keith Dobney at the University of York, Dale Serjeanston at the University of Southampton University.

Invaluable discussions with friends and colleagues helped formulate my ideas about the data that I have studied. They include Simon Parfitt, Fred Owen, Mark Collard, Andy Currant, Fran Hernandez Carrasquilla, Jo Cooper, Colin Harrison, Cyril Walker, Peter Andrews, Zygmunt Bochenski, Zbigniew Bochenski, Steve Emslie, Cecil Mourer-Chauviré, Dale Serjeantson, Per Ericson, Andrew Chamberlain, Thijs van Kolfschoten, Tommy Tyberg, Sheila Hamilton-Dyer and Derek Yalden. Sadly, since this book was written Colin Harison and Fran Hernandez Carrasquilla have both died, the latter very prematurely.

Others who helped were simply there as my friends and without them I would probably not have finished. They are Tom Mbu, Susannah Ward, Jenny Heathcote, Andy Fairbairn, Karen Milek, Chris Stevens, Sandra Bond, Heather Binney, Adrian McKemey, Celia Lloyd, William Davies, Mick Kay and family, Rachel Massey, Keiko Tenaka and Marsha Rose.

Finally, I would like thank my family for their support including my mother and father; Jan and Tabitha; Jane, Neil and family; and my grandmother Winifred Few.

(March 1999)

Table of Contents

List of Figures

List of Tables

Section I

Aims and Background

The interpretation of Quaternary bird fossils has been the subject of much debate over the years (Amadon, 1963; Jehl, 1966; Harrison, 1979a, 1988a; Jánossy, 1987; Stewart, 1992 and others). On one level the argument is centered on identification and the question of the taxonomic level to which fossil material can be assigned. On another, there is the question whether macroevolution (speciation), or only microevolution has occurred during the Quaternary. The two problems differ, although they are strongly linked through the subjects of size and metric-shape, as size and metric-shape differences are often the justification for the identification of neospecies or the description of extinct species and subspecies in the fossil record (Table 1.1).

The aim of the present volume, therefore, is to identify the patterns of osteological size and shape in the Quaternary fossil record, and to compare them with the differing patterns in modern populations of the same and related species. The purpose of such an approach is to gain an insight into the processes which may be acting on taxa today and during the past, causing them to vary through time. Correlations between environmental parameters and the patterns of size and shape in the past, may be elucidated by a knowledge of similar correlations today. The greatest complication underpinning this study, is the overlap in such patterns, which often exist between similar taxa. Due to this, the taxa chosen for study are ones which present few such problems and are thus genera which, today, are relatively species poor.

This first section lays out the background to the approach taken, and draws on the past with an overview of previous research from an interpretative perspective (Chapter 1). The past work was also important when taxa were chosen for study, as it became clear that interpretations were in need of reinvestigation. Also dealt with in this section are the theoretical issues in biology which impact on such studies (Chapter 2), for which most authors do not explicitly state their beliefs. This is important, because these same authors have often interpreted the differences seen in fossils as indicative of various levels of taxonomic significance. The levels of taxonomic rank attributed appear to be arbitrary, although it may be that there are aspects guiding workers, such as the age of fossils, and whether the differences seen between the fossils and modern counterparts are in metric-shape or merely size related.

Finally, Chapter 3 attempts to refine the general aims, using a combination of the historical and theoretical perspectives, into a form which consists of questions or hypotheses, both general and specific. With this in mind, the taxa to be targeted shall be detailed and explained in terms of their relevance to these questions.

Table 1.1. Extinct Quaternary avian taxa described from the fossil record of France and Britain.

Taxon	Author	Age	Diagnosis
Anseriformes			
Somateria Gravipes	Harrison (1979a)	Cromerian *sensu stricto*	More robust tarsometatarsi than *S. mollissima.*
Accipitriformes			
Aquila chrysaetos bonifacti	Mourer-Chauviré (1975a)	M. Pleist	Slightly smaller than modern forms.
Buteo rufinus janossi	Mourer-Chauviré (1975a)	M. Pleist	Slightly smaller than modern forms.
Accipiter gentilis brevidactyla	Mourer-Chauviré (1975a)	M. Pleist	Short toes and varyingly different limb proportions than modern forms.
Falconiformes			
Falco antiquus	Mourer-Chauviré (1975a)	M. Pleist	Size between *F. rusticolus* and *F. cherrug.*

Table 1.1. (Cont.) Extinct Quaternary avian taxa described from the fossil record of France and Britain.

Galliformes			
Lagopus lagopus noaillensis	Mourer-Chauviré (1975a)	M. Pleist	More robust tarsometatarsi than modern forms.
Lagopus mutus correzensis	Mourer-Chauviré (1975a)	M. Pleist	More robust tarsometatarsi than modern forms.
Lyrurus tetrix longipes	Mourer-Chauviré (1975a)	M. Pleist	Females with longer

			tibiotarsi and tarsometatarsi.
Gallus europaeus	Harrison (1978b)	M. Pleist	Coracoid morphologically different to all modern species.
Alectoris sutcliffei	Harrison (1980)	M. Pleist	*Alectoris* species smaller than modern species.
Alectoris graeca mediterranea	Mourer-Chauviré (1975a)	M. Pleist	Tarsometatarsi with very well developed crest.
Alectoris graeca martelensis	Mourer-Chauviré (1975a)	M. Pleist	Shorter and more robust tarsometatarsi than modern forms.
Perdix palaeoperdix	Mourer-Chauviré (1975a)	M. Pleist	Small grey partridge.

Gruiformes

Grus primigenia	Milne-Edwards (1867 - 71)	Upper Pleistocene and Holocene	Distal tibiotarsus wider mediolaterally than *Grus antigone*.

Charadriiformes

Cepphus storeri	Harrison (1977b)	Early Pleist.	Carpometacarpus larger than *C. grylle* and *C. colomba* and smaller than *C. carbo*.
Stercorarius pomarinus philippi	Mourer-Chauviré (1975a)	M. Pleist.	Shorter and more robust tarsometatarsi and longer carpometacarpi than modern forms.

Columbiformes

Columba livia occitanica	Mourer-Chauviré (1975a)	M. Pleist	Different metric shape to Würmian birds*.
Columba livia minuta	Mourer-Chauviré (1975a)	M. Pleist	Smaller than Würmian birds*.
Columba livia lazarentensis	Mourer-Chauviré (1975a)	M. Pleist	More gracile limb bones than Würmian birds*.

* Fossils were compared with Würmian fossils because there are no truely wild modern skeleton of *C. livia* due to interbreeding with feral birds (Mourer-Chauviré, 1975a).

Strigiformes

Nyctea scandiaca gallica	Mourer-Chauviré (1975a)	M. - U. Pleist	Males, as today while females are larger than modern forms.
Bubo bubo davidi	Mourer-Chauviré (1975a)	M. Pleist	Some bones larger than modern forms.
Athene noctua lunellensis	Mourer-Chauviré (1975a)	M. Pleist	Larger than modern forms.

Apodiformes

Apus apus palapus	Mourer-Chauviré (1975a)	M. Pleist.	Larger than modern forms.

Table 1.1. (Cont.) Extinct Quaternary avian taxa described from the fossil record of France and Britain.

Passeriformes

Pyrrhocorax pyrrhocorax primigenius	Milne-Edwards (1875)	M. Pleist.	Smaller than modern forms.
Corvus pliocaenus janossyi	Mourer-Chauviré (1975a)	M. Pleist.	*Corvus* species between raven and crow/rook sizes.
Corvus antecorax	Mourer-Chauviré (1975a)	M. Pleist.	Corvus species between *C. praecorax* and *C. corax*.

Chapter 1

Historical Background

The first stage of any academic investigation is the review of past work to gain an insight into the questions which have prevailed since the subject's inception. It is only with such an historical perspective, that pressing issues can be identified so that research can be designed accordingly.

The purpose of this chapter is therefore to review the past literature dealing with Quaternary bird fossils, from the point of view of the interpretations made by previous authors. It is clear from the recent proceedings of the ICAZ Bird Remains Working Group, that different workers have different beliefs regarding the level to which certain taxa can be identified (Serjeantson, 1997). This is also very apparent when even the briefest survey of bird fossil site reports is undertaken. Taxa, such as the thrush species within the genus *Turdus*, may be given full specific identifications, left at the level of genus, or identified as a size category within the genus (Bramwell, 1984; Albarella and Davis, 1996; Wessie, 1988; Stewart, 1992). It is also apparent that certain authors are more willing than others to describe fossil bird bones which differ from modern counterparts as new taxa. The level of the taxonomic affinities ascribed to these fossils also varies; on some occasions new species are described, while at other times authors have preferred the use of subspecific rank. Related to these different approaches are beliefs, such as whether or not modern bird species evolved during the Pleistocene or during the Pliocene, and whether there may be intraspecific variation during the Quaternary over and above that seen in neospecies today. The different climatic and environmental circumstances of the Quaternary imply that the latter may well be the case, although the issue has largely been avoided because the existence of such additional variation would hamper identification and interpretation.

It is difficult to pinpoint the start of the discipline, and so the best known early event in the study of Quaternary birds may be an appropriate beginning. In 1839, a portion of bone was sent to Richard Owen from New Zealand. He, very perceptively, at once realized its avian affinities despite its enormous size. He named the bone *Dinornis*; it was the first flightless moa to be described (Owen, 1839). This identification was based on morphological characters shared with other ratite birds. This was important as the moas were at this point extinct, and the geographical range of other ratites at first made Owen question the deduction. This makes the rediscovery of moas (they had been known to the indigenous population of New Zealand), slightly unusual from a European perspective, as enigmatic extinct birds are far from the norm in our Quaternary record. This is in stark contrast to the fossil record of mammals of the area and helps explain the comparative lack of research into the Quaternary birds.

There were earlier references to Quaternary fossil birds, such as by Buckland (1822, 1823), although these were without detailed discussion. The sites in Britain known to have produced Quaternary bird fossils at that time were Lawford in Warwickshire, Berry Head and Kent's Hole in Devon, and Kirkdale in Yorkshire. Owen (1846) later referred to Buckland's work when describing the British fossil bird fauna as a whole. This work involved the pre-Quaternary fossil record in the main, although he did mention the remains of birds from "the fen and turbary deposits", which he said are "more easily referable to existing species".

Milne-Edwards of France was the first to dedicate much work to Quaternary avian palaeontology with his book 'Recherches Anatomiques et Paléontologiques pour servir a la Histoire des Oiseaux Fossils de la France' (1867-71). Among the many descriptions and observations, he recognized that a large crane had formely existed in the Palaearctic, which he named *Grus primigenia.* This crane has been the subject of controversy ever since, with suggestions that the remains represent the modern Asiatic sarus crane *Grus antigone*, or simply a large common crane *G. grus*. This is the first and most enduring of the controversies concerning the meaning of Quaternary avian fossils in Europe and illustrates many of the problems inherent in the subject (see later parts of this chapter). Milne-Edwards also commented on the size differences between the modern Scandinavian and temperate European nutcrackers *Nucifraga caryocatactes* and realized the implications this would have to the study of their bones. This is possibly the first reference addressing the importance of geographical size variation in birds and its relevance to their fossils. Ironically, it also implied, unwittingly, that his interpretation of the large extinct crane might have to be questioned.

E.T. Newton's paper on a red-throated diver from the Mundesley River Bed (Newton, 1883), a river channel deposit once exposed in Norfolk, used morphology of the fossil bones to allocate them to the genus *Colymbus* (now *Gavia*), as well as metrical comparisons with recent divers in the Palaearctic to indicate the species. This is the first work to imply that individual species within a genus may not differ in morphological detail, necessitating the use of metrical criteria.

Another interesting paper by Newton is that on the avian fauna of Merlin's Cave in Hertfordshire (Newton, 1924a), which details a large series of "ptarmigan" *Lagopus* bones. Amongst these, it was apparently clear that three "ptarmigans" were present, a large one which he referred to the red grouse *Lagopus scoticus* (now *L. lagopus scoticus*), a smaller one which agreed with the ptarmigan *L. mutus*, and a smaller one still which he referred to as "small ptarmigan *L. mutus* - small variety". This was all deduced from the "metacarpals and metatarsals", although the "upper and lower beaks" appeared too broad and robust for any present members of the *Lagopus* genus, let alone the smaller extinct "ptarmigan", unless there were considerable differences in skull to limb proportions.

Also present in Merlin's Cave, were a large series of "upper and lower beaks" of a large finch, which he assigned to the greenfinch *Chloris chloris.* He did, however, mention that they were large and robust in comparison with his recent greenfinch skeleton, and that he did not have access to a pine grossbeak *Pinicola enucleator* skeleton which needed to be eliminated. He attempted a comparison using a stuffed specimen, but this is wholly inadequate as considerable differences can exist between external bill shape and the underlying bones. This issue emphasizes another reccurring problem through the study of Quaternary birds: a lack of comprehensive comparative collections. This has hampered many workers since Newton's time and is still a problem since some species are either absent or represented by single individuals at major collections. The interesting assemblage of Merlin's Cave was thought to have been completely destroyed by bombing in W.W.II. (Harrison, 1987a). However, part of the collection has recently reappeared and is again available for study (Chapter 10). Never-the-less, the original work by Newton provides an interesting insight into possible evolutionary changes within the Quaternary and the presence of red grouse, if correct, would contradict Voous (1960) and others who considered the bird to have evolved in the Holocene from a relict population.

Other publications by Newton (Newton, 1894, 1916-17, 1922a, 1922b, 1923, 1924b) are mostly lists (some of which are illustrated) of species present in various deposits. The only mention of identification is when he regularly questions the reliability of his passeriform identifications due to their small size. Newton dominated the field of Quaternary birds at the turn of the century, studying in addition to British fossils, material from the Mediterranean Islands (Newton, 1922b) and Spain (unpublished notes in the Museo Nacional de Ciencias Naturales, Madrid). Unfortunately, many of his identifications were either never published, such as the Spanish example and his work on Pin Hole Cave in Derbyshire; or were published by others (Bell, 1915, 1922) and thus give no further insight into his methodology, or indeed the beliefs he held concerning these fossils.

A short publication by Barret-Hamilton (1889) merits mention, as he emphasized the difficulties in identifying fragmentary bird fossils. In this one page note, he discussed the endeavours by E.T. Newton, W.P.Pycraft (curator at the BM(NH)) and himself to identify a fragmentary mandible from Ballynamintra Cave, Co. Waterford. The specimen at first appeared to be that of a galliforme but eventually proved to be that of a kestrel *Falco tinnunculus.*

In 1891 Richard Lydekker (189la) published his 'Catalogue of Fossil Birds in the British Museum (Natural History)'. Lydekker's book was not merely a list, as there are descriptions of the osteological characters of the taxa. Examples include some differences between the corvid genera *Corvus* and *Pyrrhocorax,* and certain crane genera (*Grus, Anthropoides, Tertrapteryx* and *Balaearica*) which were considered "inseparable in palaeontology". There are, however, comments about size-related diagnostic criteria, such as the unsubstantiated statement that carrion crow *Corvus corone* tibiotarsi are larger than those of rooks *C. frugilegus.*

In the same year Lydekker (189lb) also published a review of British fossil birds. In this, he commented on the problems caused by the lack of reference material and was aware of the need to have "all the larger birds, not only of the British Isles; but also of the greater part of the Palaearctic region". This implies a realisation that distributions are likely to have changed and that all Palaearctic avian species should be considered when identifying Pleistocene fossils from Britain. He also says that "when we bear in mind the slight differences between the bones of many of the extinct species of birds, and also recollect that many of the mammals of Pleistocene period are extinct, the suggestion arises that if these fossil birds were before us in their flesh and feathers some of them would present differences from their living representatives which might entitle them to rank as distinct species. Still, however probable this idea may be, our only course is to take evidence as it is presented to us; and if this is to the effect that no differences of specific value can be detected between the bones of the British Pleistocene birds and those of the present day." This leaves no doubt that he saw the fossil record as potentially inadequate for the enlightenment of the past history of birds. The quote includes both a reference to the general conservative nature of avian post-cranial bones and its implications for identification, as well as the fact that bird species differ most from each other in aspects which will not enter the fossil record.

The next main British worker was D.M.A. Bate, who published on the faunas of La Cotte de St. Brelade in Jersey; Devil's Tower in Gibraltar; Camulodunum (Roman Colchester); and Har Dalam in Malta (Bate, 1916, 1915-16, 1928, 1947). Bate does not, however, discuss her means of species determination.

In 1933 Lambrecht published the 'Handbuch der Palaeornithologie' in which he gathered all the information available on a worldwide scale. Lambrecht also published a taxonomic study of birds based on postcranial osteology (Lambrecht, 1914). He investigated the differences between the carpometacarpi of bird taxa, and although this was not a palaeontological work, it has clear value for the study of fossil birds. Later in the century Bock (1962) and then Harrison (1969, 1976) used other skeletal elements for a similar taxonomic purpose. The work by Bock on passerine humeri was heavily influential on Jánossy, who published a key to the humeri of passerines in Central Europe (Jánossy, 1983).

During the 1930's and 1940's, the next centre for research into Quaternary birds arose in the United States, with authors such as H. Howard, A.H. Miller, A. Wetmore and P. Brodkorb. With them came a greater tendency to publish, and due to the number of individuals, debate and disagreement naturally followed. A subject which was prominent, was that of the age of living species. Wetmore (1951) expressed "a firm belief that our living kinds had their evolution both as genera and as species in the Miocene and Pliocene periods". He believed that most evolution in the Quaternary, in birds, was at a subspecific level (Wetmore, 1959). Brodkorb (1960) on the other hand, believed that most species had developed during the Pleistocene and notes that "almost no avian species are known to cross epochal lines". This last statement is notable for its disregard for the fact that "epochal lines" are artificial and themselves based largely on biostratigraphic divisions. This is further complicated because Wetmore did believe that extinctions of taxa occured (Wetmore, 1959), although the Pleistocene was clearly not seen as the significant period that Brodkorb believed.

It is likely that such beliefs influenced the various workers' interpretations of the fossils that they were studying, with Brodkorb being more prone to naming extinct species or referring specimens to extinct species named by others (Brodkorb, 1959; Wetmore, 1931). An example of this difference, is the species of pied-billed grebe *Podilymbus magnus* which Brodkorb revived after Wetmore had synonymised it with the modern *P. podiceps.* The extinct species was based on size, being consistently larger when compared with 8 modern specimens of *P. podiceps.*

Wetmore (1931), conversely, recognised the presence of two subspecies of gray crane *(Grus canadensis)* from Florida: a large one, which he believed could be either *Grus canadensis pratensis* or *G. c. tabida,* and a smaller one which was either *G. c. canadensis* or *G. c. nesiotes.* The two subspecies in each size group are apparently inseparable osteologically, and although Wetmore mentioned that *G. c. pratensis* is a resident in Florida today, he did not allow this to influence him into believing that the large bones were representative of that subspecies. This is despite the fact that the other large subspecies *G. c. tabida* is today confined to a more northerly and westerly range. He concluded that two races existed in the Pleistocene of Florida in an analagous situation to the present United States, where two subspecies are sympatric during migration, although it is not possible to say exactly what they were. This conclusion is to be praised because it acknowledges that bird distributions may change and that such criteria should not be used to influence the identification of fossils where no distinct morphological characters exist.

In Eastern Europe, Jánossy started work on bird faunas from a variety of countries including Hungary, Poland, Spain and Austria (Jánossy 1972, 1974a, 1974b, 1976a, 1976b, 1977, 1979a, 1979b, 1980, 1981, 1992). Jánossy is distinguished by being the author on European Quaternary birds who has recognised the greatest number of extinct species. An example of how new species diagnoses were achieved is the case of *Mergus connectens* (Jánossy, 1972). Here, on the basis of measurements on hundreds of bones of various elements, he deduced that a smaller species of *Mergus* was present at the Middle Pleistocene site of Stránská Skála, in Slovakia. The most interesting feature of this species is that, unlike modern species of the genus *Mergus*, the fossil form is not sexually dimorphic. This difference does not necessarily signify that the fossils were wrongly assigned to that genus, although it does merit an explanation or at least a mention, neither of which were supplied by Jánossy. More recently (Mlikovsky, 1982a, 1982b and Sanchez Marco, 1990), this species as well as others, e.g. *Buccephala angustipes*, *Anas percrecca*, *Anas submajor, Anser subanser* of Jánossy and *Somateria gravipes* of Harrison (1979a), have been questioned due a preconception that they should be considered conspecific with modern forms. Jánossy (1992) considers this "very dangerous for understanding of the evolution of palaeornithological research of our continent". A great deal of the argument has been of a personal nature, with authors accusing each other of "pseudoscientific" methods (Mlikowsky, 1982a, 1982b; Jánossy, 1987 and 1992). However, Jánossy commented that "the osteological differences are in such cases often very slight and hidden in several recent "sibling species", ecologically - etologically [sic] absolutely isolated". He also says that such closely resembling and related species indicate that "small differences also in fossil species indicate taxonomically - biologically quite different forms". He therefore feels justified in naming new species, some of which he describes as ancestors to modern species. It should be mentioned, however, that neither Jánossy nor his critics have considered the third alternative interpretation of the fossil specimens. That is, that due to the morphological similarities between certain modern species, it may not be possible to know whether specimens belong to modern or extinct species. The overlap in size between modern species can hinder Holocene fossil determination, let alone Pleistocene fossils which may have an even greater chance of demonstrating intraspecific variation (Stewart, 1992). Indeed, as shall be seen in Chapter 2, there is a growing body of literature on rapid morphological change in modern birds which would put into question Jánossy's reasoning. It seems that Jánossy was using false logic i.e. "All A are B" does not necessarily mean that "All B are A". He acknowledges the osteological similarity between closely related taxa and argues from this that small deviations from modern norms are taxonomically significant at the species level.

Already mentioned is Jánossy's key to humeri of Central European passerines (Jánossy, 1983). This uses a combination of morphological and metrical work to separate the various species, measurements being of importance to diagnose closely related species. However, the small number of specimens used in this endeavour, limits the use of his data.

The next British proponent in the field was Bramwell, who concentrated on faunas in Derbyshire, Staffordshire and Somerset caves (Bramwell, 1957, 1960a, 1960b, 1976, 1978, 1984; Bramwell and Yalden, 1988, Bramwell, *et al.*, 1987, 1990), as well as archaeological sites in Scotland and elsewhere (e.g. Bramwell, 1977). In his paper on bird distribution during the Late Glacial and Post Glacial of Britain (Bramwell, 1960a), he mentions the inherent difficulties in identifying bones "due to the very slight variation between members of the same genus, particularly with small waders, ducks, thrushes, finches and warblers, but the larger birds can usually be identified with some confidence". This is very enlightened, although the problem may be even more extensive than he suggested. Indeed, in his publication on Soldier's Hole (Bramwell, 1960b), he adds crows and small passerines in general to the list of problematic taxa.

Bramwell' s (1984) publication on Creswell Crags, which produced the largest fossil bird fauna in Britain in terms of both bone numbers and species, was an important contribution. Too many species may have been identified in certain taxa, such as the genus *Turdus*, so some may have to be abandoned (Stewart, 1992). What particularly concerned him was the timing of appearance of woodland birds in relation to the Pleistocene / Holocene boundary. He was aware (Bramwell, 1976) of the work on beetles by Coope, who showed that thermophilous insects were present in the last glacial stage during warmer interstadials. This may have influenced his suggestion that temperate birds, another mobile group of animals, may also have arrived in Britain during the Late Glacial (Bramwell, 1984).

His publication with Yalden in 1988 on Demen's Dale in Derbyshire, is notable in being the first British publication which made use of the Munich dissertations on the identification of bird bones (see below). It is notable that this was twenty one years after the first dissertation was published. This illustrates a general problem in this field where important publications may be poorly known and difficult to obtain. This inspired Stewart and Hernàndez Carrasquilla (1997) to publish a review of literature in which the means of bird identification were described.

It was in the late 1960's, that the Munich School under Boessneck and von den Driesch initiated the first of several studies of various modern taxa in the form of dissertations prepared by a succession of students (Woelfle, 1967; Bacher, 1967; Erbersdobler,1968; Kraft, 1972; Fick, 1974; Langer, 1980; Otto, 1981; Schmidt-Burger, 1982; Kelner, 1986; Gruber, 1990; Lorch, 1992). These were morphological and metrical comparative studies of modern species to enable archaeologists to identify bird remains. Kraft (1972), who deals with the smaller Central European Galliformes, is the only author in the series to have applied the criteria to fossils, albeit to measurements given in the literature. She does not, however, make any further conclusions regarding the fossils. It is noteworthy that she is also the only worker who considered geographic variation in skeletal elements.

Harrison was the next British worker to publish a great deal about Quaternary birds and over the 1970's and 1980's worked his way through most of the old, and a great deal of the more recent collections (Harrison, 1977a, 1977b, 1978a, 1978b, 1979a, 1979b, 1980a, 1980b, 1980c, 1985, 1986, 1987a, 1987b, 1988b, 1989a, 1989b; Harrison and Walker, 1977; Harrison and Cowles, 1977). The volume of work done by Harrison does not, unfortunately, include a great deal of comparative description of specimens in relation to their identification. Certain publications, such as that on the Quaternary birds of south-western England (Harrison, 1987a) deal briefly with the difficulties of separating carrion crow *Corvus corone* and rook *C. frugilegus.* He says that, excepting the skull, they cannot be separated, and that criteria formerly used may be invalid. In addition, he finds that blackbird *Turdus merula* and ring ousel *T. torquatus*; brambling *Fringilla montifringilla* and chaffinch *F. coelebs*; and stonechat *Saxicola torquata* and whinchat S. *rubetra*, cannot be split on osteological grounds. This is despite having earlier managed to separate *T. merula* and *T. torquatus* (Harrison, 1980a). Another apparent inconsistency, is the referral of a humerus and a distal coracoid from the Cromer Forest beds to a *T. merula / T. boulboul* (blackbird / grey-winged blackbird) superspecies in 1979, while later he only mentions the blackbird in relation to these deposits (Harrison, 1979a and 1985). The same is true of his use of a *Sturnus vulgaris / S. unicolor* (common starling / spotless starling) superspecies for starling bones from the same deposit. The impression given by the 1979 paper was that considerable problems existed in assigning some Quaternary bird fossils to modern species, as speciation may have taken place in the Pleistocene. Similarities in plumage and osteology, as well as the patterns of present distribution, imply a recent separation of species from a common ancestor. Harrison concluded that "in view of uncertainties about the status of the Cromer Forest Bed Series forms, and the lack of more certain views of the probably [sic] period of speciation in these genera, I prefer to use a modified superspecies concept rather than propose new names for putative ancestral forms". This realistic approach, which he did not pursue in further publications, may have been influenced by the great age for the Cromer Forest bed Formation material, as none of the later faunas he published were as old. However, the fact that he did not continue the use of superspecies in the paper summarising the birds from South-Eastern England, which included those from the Cromer Forest Beds (Harrison, 1985), implies that he abandoned this view for an unspecified reason. In naming a new species of eider *Somateria gravipes* (Harrison, 1979a), Harisson, like Jánossy, eventually came under criticism. In this case, the claim was that the thick-legged eider is in fact a pathologically deformed specimen and not taxonomically different from the present form (Mlikowski, 1982a and b).

Harrison's approach is best summed up by the quote: "in birds, size is usually consistent among individuals of the same species" (Harrison, 1988a). Here he implies very strongly that size can, in most instances, be reliably used to identify closely related species. In an earlier publication, Harrison (1980a) dismissed the importance of Bergmann's Rule in birds during the Pleistocene because he had found no evidence for larger birds in the past. It should be noted that rarely did Harrison make use of calipers when he compared specimens. Despite the fact that he reasserted this belief in 1988, there were a number of instances in his publications where he mentions size differences in birds. They include the eagle owl *Bubo bubo* from the Cromer Forest Beds (Harrison, 1985), and the ptarmigan *Lagopus mutus* and starlings *Sturnus vulgaris* at Tornewton Cave (Harrison, 1980b). However, none of these were considered to be taxonomically different from extant species.

Mourer-Chauviré, in publishing her Ph.D. thesis on the Middle and Late Pleistocene birds of France (Mourer-Chauviré, 1975a), provided a detailed analysis of a large number of species through time. Some taxa were found to include a series of specimens differently sized and proportioned from the modern forms, which were diagnosed as either extinct subspecies or species. She named a total of three new species and sixteen new subspecies. Using diagrams showing the relative lengths of certain skeletal elements of a variety of modern and fossil populations, Mourer-Chauviré (1975a) demonstrated metric-shape and size differences. She also made use of scatter plot diagrams to show absolute differences between modern and new species. These differences, while very apparent, are of debatable taxonomic significance. It is possible that the age of certain fossils influenced her taxonomic decisions; for instance the fossil raven *C. antecorax* is Middle Pleistocene, while the differences seen in various Late Pleistocene birds only receive subspecific status. However, she also names new subspecies such as *Lagopus lagopus noiallensis* and *L. mutus correzensis* from the same site where *C. antecorax* was recovered.

Mourer-Chauviré has not suffered the same kind of public criticism as has Jánossy for naming new taxa based on size and robusticity differences (Table 1.1). Her tendency to name subspecies rather than species is perhaps a more conservative reaction to the discovery of differently sized and proportioned specimens, although the criteria used to interpret one difference as subspecific, another at species level, are unclear. Indeed in certain instances, e.g. with the dipper *Cinclus cinclus*, she describes last glacial fossils as being larger than modern examples without attaching any taxonomic significance to them at all (Mourer-Chauviré, 1975a).

Other publications of Mourer-Chauviré include site reports, taphonomic studies, palaeoecological and palaeoclimatic studies, as well as general resumes (Mourer-Chauviré, 1975b, 1975c, 1979, 1983, 1993; Mourer-Chauviré and Antunes, 1991; Mourer-Chauviré *et al.*, 1977, 1980; Clot *et al.*, 1976 and Philippe *et al.* 1980). She has also contributed to studies on certain taxa such as the Balearic barn owl *Tyto balaearica* and most notably to the debate on the large Palaearctic crane species, as will now be discussed (Mourer-Chauviré *et al*, 1975; Northcote and Mourer-Chauviré, 1985, 1988).

As already mentioned, Milne-Edwards (1867 - 71) had already identified an extinct large crane *Grus primigenia* from French Pleistocene sites. This controversial topic has received constant attention and illustrates many of the problems inherent in the interpretation of Quaternary bird fossils. Certain authors, such as Lambrecht (1933) and Lydekker (1891c), viewed *G. primigenia* as conspecific with the sarus crane *Grus antigone*, while Driesch and Boessneck (1969) believed it to be the same as the common crane *Grus grus*. Driesch and Boessneck dismissed the species as they said that too little was known about the intraspecific variation within common cranes. There is another extinct Palaearctic crane in the form of *Grus melitensis*, which is claimed to be an island endemic restricted to Malta (Lydeker, 1890 and 1891c). This crane seems to be accepted by most authors, as morphological differences have been established between it and both *Grus grus* and *G. antigone* (Harrison and Cowles, 1977; Northcote, 1982a). *Grus primigenia*, meanwhile, having been identified in the Holocene of Germany and Britain as well as the Pleistocene of France and Britain, was thought to be a species living sympatrically with *G. grus* by Harrison and Cowles (1977). Northcote and Mourer-Chauviré (1985, 1988) tried to put the subject to rest by publishing morphological differences between the extinct crane and sarus crane, but did not eliminate the common crane in so doing. It seems to have been assumed that the size difference between the extinct form and the common crane was too great for them to be conspecific. This subject will be examined in Chapter 9.

Zygmunt Bochenski has contributed much to the subject of Quaternary birds and particularly to the application of measurements to the fossils. His publication on the Quaternary birds of Poland is noteworthy because of its treatment of the genus *Lagopus* and the use of geographically widely spread samples of the modern taxa to help expain the variation in the past samples

(Bochenski, 1974). Other publications by Bochenski and his son have dealt further with *Lagopus* evolution (Bochenski, 1984, 1985; 1991; Bochenski and Tomek, 1994), in addition to site reports (Bochenski, 1981, 1990) and a catalogue of Polish fossil birds (Bochenski, 1993). An interesting methodological approach has also been developed by Bochenski where sizes and proportions of skeletal elements, which were not part of the original type material, are estimated for extinct taxa (Bochenski, 1989). Later, the same principles were used to make use of the wealth of ornithological measurement data available to estimate bone dimensions of unrepresented species in skeleton collections (Bochenski and Bochenski, 1992). Zbigniew, Zygmunt's son, has also started to work in this field and completed a Ph.D. on the skeletal morphology of grebes throughout the world (Bochenski, 1994). Latterly, Zbigniew has concentrated on taphonomic studies and patterns of fragmentation and corrosion as a result of the attention of various birds of prey (Bochenski *et al.*, 1993, 1997).

A number of other authors are relevant here, although they have had less impact. They include Ballman, whose taxonomic work is very exacting, using a combination of morphological and metrical tools to diagnose species (Ballman, 1973 and 1978), his identifications being usually accompanied by justifications. Weesie's (1988) publication on the Quaternary birds of Crete, similarly dealt with the taxonomy of fossils in some detail. Eastham (1967), however, in describing the fauna of Gorham's Cave in Gibraltar, did not venture beyond the publication of a faunal list accompanied by ecological deductions.

Tyrberg has produced a number of review articles which have concentrated on Europe-wide temporal and spatial distributions of certain taxa (Tyrberg, 1985, 1991a, 1991b, 1995). His publication on the distribution through time and space, in relation to vegetation records, of the crossbills *Loxia* sp. is interesting and puts forward two hypotheses (Tyrberg, 1991b), which are more sophisticated than those originally proposed for the birds' diversification. However, Tyrberg does not question the identification of these fossils, which may well be erroneous and has a fundamental bearing on the use of such literature sources. Another paper reviewed all work on extinct island birds and their demise (Milberg and Tyrberg, 1993).

Island faunas have been a focus for work since the very beginning of the discipline, if the description of the moa is to be included (Owen, 1839). Later there was a significant amount of work on various Carribean islands by Wetmore, particularly during the 1920's and 1930's (see Milberg and Tyrberg, 1993). More recently attention has been diverted to the Pacific and a large number of extinct species have been discovered and named (see Milberg and Tyrberg, 1993). The extinctions, it seems, are mostly due to human colonisation, through over hunting and the damage done to the environment by introduced animals such as pigs, dogs, cats and rats.

A relatively new and healthy development, has been the growth in taxonomic studies of individual species, where all forms of variation are considered, including sexual, geographic and temporal (Northcote, 1979a, 1980b, 1981a, 1981b, 1982a, 1982b, 1982c, 1983, 1984, 1988; Fraguglione, 1982; Hufthammer, 1982; Ericson, 1986 and 1987a; Burness and Montevecchi, 1992; Ericson and Hernandez, 1997). Northcote published a number of papers on sources of intraspecific variation in mute swans *Cygnus olor* and whooper swans *C. cygnus*. The most interesting were those on size decrease over time in mute swans (Northcote 1981a and 1983). The reasons for this change in size are discussed and the possibility that domestication or microevolution may have caused the phenomenon are considered. Fraguglione (1982) reviewed the fossil record of the woodcock *Scolopax rusticola* and noted the tremendous degree of variation which exists within the species. In the same year Hufthammer published her Ph.D. on morphological variation in the great auk *Pinguinis impennis* in Scandinavia (Hufthammer, 1982). She found that the great auk decreased in size from south to north, contradicting the predicted relationship between body size and latitude of Bergmann's rule . This work was later extensively used when Burness and Montevecchi (1992) compared the northwestern and northeastern Atlantic great auk populations. They discovered that great auks breeding in the northwest were larger than those in the northeast. They discussed the cause of these variations and considered the following: "1) genotypically based size differences due to differing environmental pressures, e.g. sea surface temperatures; 2) phenotypically related size differences driven by inter-colony differences in prey selection; 3) founder effects generated by different sized establishers of different colonies". Finally, Ericson's study of eider (Ericson, 1986 and 1987a) used large samples from archaeological sites, as well as modern birds to assess the degrees of sexual, geographical and temporal variation within the species. He found a small amount of sexual dimorphism, which he tentatively suggests may be due to food competition, while geographic variation was similar to that in great auks, being larger in the south than in the north. Finally, the evidence for size change over time was assessed and he found that the eider has been decreasing in size since the medieval period. His conclusion was that food competition with other duck species had decreased over time, which has been shown to affect the mean size of eider populations, due to a decrease in selection against smaller body size. An important innovative aspect to Ericson's approach, is the emphasis he has placed on the ornithological literature dealing with recent morphological change (see chapter 2). This had received little or no attention until then. Together with Hernandez Carrasquilla, Ericson (Ericson and Hernandez Carrasquilla, 1997) took a similar holistic approach to a study of Holocene cormorants fom the Baltic. The conclusion drawn from this study was that the cormorant subspecies, *Phalacrocorax carbo sinensis*, which breeds there today, probably replaced the nominate subspecies *P. c. carbo* sometime after the mediaeval period.

Various strands of research into Quaternary birds have been undertaken during the 1980's and 1990's. While site reports still dominate the publications, there are new trends which are helpful to a better understanding of birds in the Quaternary. The taphonomic work by Zbigniew Bochenski and others is helping to lay down some of the fundamentals concerning what avian predators do to bird bones, and may hopefully better explain the damage found amongst fossils. There have also been a number of papers by Tyrberg, based on broad literature reviews, which are pointing to interesting patterns in past geographic distributions of birds, which are useful despite an unquestioning approach to fossil identifications. Meanwhile, there continues to be interest in island bird faunas, although islands are special situations with unusual faunas, which tell us little about what happened to birds during the Quaternary on the continents. Work on Quaternary fossil birds from continental areas other than Europe is still sparse. In North America there are few workers dealing with the subject and Emslie dominates the study of continental birds. Studies by him have included site reports (Emslie and Heaton, 1987; Emslie, 1985, 1986), and most importantly, considerations of the age of modern species (Emslie, 1992). In addition to this, there has been a description of a fossil scrub-jay from the Pliocene of Florida belonging to the neospecies *Aphelocoma coerulescens* (Emslie, 1996). The latter paper uses the morphology of fossil maxillae and mandibulae to indicate that a modern species was represented. Emslie believes that among Passeriformes, the bones of the bill are species-specific and hence can be reliably used to diagnose species. This belief appears to neglect the fact that the bill is highly influenced by natural selection (Grant, 1986), and that presumably the same shape could occur more than once, given similar environmemtal conditions. This is particularly the case with relatively simple shapes, such as those of the bones of the bill. Nevertheless, this example leads Emslie to believe that some species originate as far back as the Pliocene.

When viewed historically, the subject of Quaternary birds can clearly be seen to have undergone changes, while at the same time certain themes have continually been dominant. Those themes are centered around the questions of what, taxonomically, the fossils in the Quaternary represent. The emphasis is moving slightly, with taxonomic studies on single species or small groups of species becoming more prevalent. However, the geographic extent over which taxa are studied are still rather limited. This has clear consequences, if such mobile and potentially adaptable species are to be considered evolutionarily. The trend is, however, in the right direction, as it is only once individual taxa are studied in great detail over large geographic areas, that the significance of fossil birds and eventually their evolution may be ascertained. An ironic aspect to the history of the study of Quaternary birds, is that many enlightened views concerning the fossils were held very early on and were subsequently all but abandoned. Problems such as the implications of geographic variation and the need to consider species from the whole of the Palaearctic were expressed in the last century (Milne-Edwards, 1867 - 71; Lydekker, 189lb) and only recently are they coming to prominence. This has very much guided the approach taken during the present study and particularly the geographic and taxonomic scope of the project.

Chapter 2

Theoretical Background

2.1. Introduction

In the last chapter I demonstrated that different attitudes, usually unexpressed, have existed with respect to the significance of Quaternary bird bones and their variation from each other, and from their modern counterparts. In this chapter I will therefore explore the areas of biology and palaeontology which may have an impact on the interpretation of the taxonomy and evolution of Quaternary birds through their fossil remains. To this end all potential sources of osteological variation which may affect a modern or fossil sample shall be examined, including interspecific variation, the morphological overlap/separation between species, and intraspecific variation (sexual, geographic, temporal, pathological). In addition to these, I shall consider the implications of different species concepts and theories on modes and rates of species evolution, as well as the issues of species evolution versus immigration.

Many of the theoretical issues that need to be considered are linked, for example it may be difficult to consider a species concept without a speciation mode or a methodology for species identification. This makes it difficult to untangle the web of theory and to consider the interpretation of fossils in the light of individual theoretical concepts. It is important to do so, however, as theory always influences the conclusions arrived at during any analysis of data [(Popper, 1968)], as was certainly apparent in the overview of the literature on Quaternary birds. This has led to a proliferation of titles and subsections in this chapter, which in part serves as a glossary. Defining the terms could help confine the ambiguity which may befall these concepts when used later in the volume. After the definition and discussion of each aspect of theory, an attempt is made to explain how it affects the interpretation of Quaternary fossil birds.

2.2. Interspecific Variation and the Fossil species

Interspecific variation refers to the variety of morphologies among species and more inclusive taxa, and the degree to which they overlap or diverge. The nature of interspecific skeletal variation is important because it affects the identification and interpretation of bones of unknown taxonomic affinity, e.g. fossil birds. The taxa which warrant attention here are usually those which are most closely related, although phenomena such as morphological convergence can cause quite distantly-related taxa to overlap in morphology. The assumption is that phyletic distance is equivalent to morphological distance and this need not be the case. The problem is made all the worse when damaged or fragmentary fossils are to be identified and in such instances greater care should be taken.

Bochenski (1989) stated that the relative proportions of bone dimensions of birds remain constant within given systematic groups, which may imply that such characteristics can confidently be used to diagnose species. He mentions that the example of the bills of the "highly specialised Darwin's Finches" is an exception and that most species differ only in size. This is a dangerous stance, and while the skeletal proportions within groups are often similar, it is a matter of degree, and metric-shape changes can be superimposed onto a general constancy of form within a higher taxon. It is the more subtle metric-shape differences between or within species that can cause problems of identification. The degree of difference within species can even exceed that between species of the same genus (see later in this chapter). Clearly such phenomena will create problems both for the identification of neospecies (modern species) in the fossil record, and the recognition of extinct undescribed taxa. Metric-shape differences are often seen as more significant than mere size differences in this respect, but clearly are as potentially misleading as size.

There are three categories of species in the fossil record: the neospecies in the past, the 'ancestral species' and the truly extinct species. Neospecies are distinct from the last two categories since we possess knowledge about the biology and modern variation within such species in relation to other neospecies. This allows the uniformitarian principle to be used when they are identified in the fossil record. The 'ancestral species' is a concept out of favour in cladistics where the closest allowed possibility is the 'sister-group'. However, in the Quaternary, with finer stratigraphic coverage, it is possible to find forms at least very close to ancestors (Lister, 1992), and this approach will be followed in this volume. Such a species could be either a 'common ancestor' to two or more modern species, or may have only given rise to a single modern species. The latter are often called chronospecies and are not considered valid by some biologists and palaeontologists because no true speciation or cladogenetic event has occured. The true extinct species on the other hand are evolutionary dead ends, phyletic branches which have no modern descendents. Which of these three categories a Quaternary bird fossil falls into, determines the evolutionary interpretations of the group to which it belongs. Therefore, identification of an avian fossil is the single most important stage in the analysis of such material.

Another subject of concern in the identification of species concerns the perceptions carried by different workers that there may be theoretical grounds for expecting, *a priori*, a species to be morphologically distinct in certain ways. This has already been dealt with to some extent in Chapter 1. The significance of this is that the unexpressed beliefs of workers, regarding the processes and results of differentiation between species, influence their conclusions more than the practical comparative work they undertake.

2. 3. Species Concepts, Modes of Species Evolution and Rates of Species Evolution

The meaning of species has been discussed at length by a great number of authors, from before the time of Darwin through to the present day (e.g. Zink, 1997). Many of the arguments are about subtle differences of meaning, and consequently there is much discussion of semantics (Ereshefsky, 1992). However, what is important here are the main arguments and the degree to which they affect both our interpretation of avian fossils, and our perception of the macro-evolution of birds during the Quaternary. Some species concepts are described as 'process-related', and are linked to speciation mechanisms, whereas 'pattern-related' concepts are those which define species according to their morphology. It is, therefore, difficult to deal with the implications of these species concepts without considering their corresponding mechanism of species origination; so both will be dealt with in this section. The rates with which lineages evolve are also relevant, and will be dealt with here.

2.3.1. Species Concepts

Species concepts can be broadly divided into process-related and pattern-related concepts. Process-related species concepts are ones where the ways in which species are believed to arise are part of the definition of those species; while pattern-based concepts are those which rely more heavily on morphology. The former include the Biological Species Concept, the Recognition Species Concept, the Evolutionary Species Concept and the Ecological Species Concept. The pattern-based concepts include the Phylogenetic Species Concept, the Phenetic Species Concept and the Typological Species Concept.

Species concepts are fundamentally important in determining what is considered a species, and hence when a fossil should be considered as a distinct species, whether previously unrecognised, conspecific with a neospecies or a previously described extinct form. However, they are rarely discussed in taxonomic or other palaeontological works. Indeed, to my knowledge no publication dealing with avian fossils exists where the author states what species concept he or she favours. Until relatively recently this has not been a cause for concern as the vast majority of biologists and palaeontologists have accepted the 'Biological Species Concept' (BSC) of Theodosius Dobzhansky and Ernst Mayr (Dobzhansky, 1937; Mayr, 1942). Recently, however, the emergence and increasing acceptance of the 'Phylogenetic Species Concept' (PSC) of Eldredge and Cracraft (1980) and others, as well as the 'Recognition Concept' of Patterson (1981, 1985), have emphasised the need for a consideration of the different implications of these concepts.

In March 1996, the Linnean Society and the British Ornithologists' Club held a symposium entitled "Avian Taxonomy from Linnaeus to DNA". At this meeting support for the PSC appeared to be growing (Zink, 1997), although the BSC still has its supporters (e.g. Snow, 1997). However, the implications of the various concepts to avian palaeontology were not mentioned. Indeed there has not been any significant debate of the implications of the adoption of the Phylogenetic Species Concept in avian palaeontology.

In contrast, a recognition of the importance of species concepts has been made by palaeoanthropologists (e.g. Turner and Chamberlain, 1989; Kimbel and Martin, 1993) and other mammalian palaeontologists (e.g. Turner, 1993; Nadachowski, 1993), where the use of various species concepts has frequently been considered in relation to the recognition of species and speciation. Smith (1994) has reviewed a number of species concepts in relation to the problems of working with fossils, and concluded that only pattern-based concepts are practical in this instance. Smith was, however, using palaeontological examples that were no younger than Tertiary in age. This is significant as the palaeontology of the Quaternary differs from that of earlier periods in that uniformitarianism has greater prominence, which in turn increases the value of neontological species studies. As pointed out by Martin and Andrews (1993), the way modern taxa are divided taxonomically affects the analysis of modern species variation which we use as a reference for the interpretation of variation in the fossil record. Birds have received a considerable amount of relevant attention from neontologists (see below) and it is important therefore to consider the species concept which is most applicable to Quaternary fossil birds in the light of this work. There is a potential problem if individual species are not all defined on the same philosophical grounds. Agreement between the concepts used to define both the fossil species and their related modern taxa should ideally exist or the uniformatarian methodology becomes seriously flawed. Therefore, an approach to the species concept is proposed here for Quaternary palaeontology, which is uniformitarian in emphasis. I believe that it is worth articulating this approach so that what is ususally implicit becomes a coherent methodology linked to a theoretical framework. In this approach it matters less which particular species concept is followed in palaeontology, so much as there is consistency with neontological usage. The close temporal link between Quaternary fossils and their modern counterparts allows for a process-related concept to be adopted, although it remains to be seen which of the currently competing concepts will win the day, and in the present study, relevant situations will be appraised in the light of different concepts. The use of pattern-based species concepts, with respect to more temporally remote fossils, is a pragmatic approach and should not negate the consideration of the benefits of process-related species concepts to Quaternary fossils.

An important criticism which is often levelled at process-related species concepts is that it they are not operational (Sneath and Sokal, 1973). This signifies that a means of identifying a species does not follow from the definition of the species. Others have, however, dismissed this by saying that the nature of species cannot be defined operationally and that circularity will always creep into such arguments (Wiley, 1978).

Process-related Concepts

The Biological Species Concept

"Species are groups of actually or potentially interbreeding natural populations, which are reproductively isolated from other such groups" (Mayr, 1942).

"Species are groups of interbreeding natural populations that are reproductively isolated from other such groups" (Mayr, 1963).

"A species is a reproductive community of populations reproductively isolated from others that occupies a specific niche in nature" (Mayr, 1982).

The preceeding three quotes by Mayr are spread over 40 years and are essentially the same in their content in summing up the 'Biological Species Concept'(BSC), with the exception of the latest where ecology receives consideration. (Incidentally the quotes also serve to emphasize the semantic nature of certain arguments within the subject, as subtle differences in their wording can be taken to imply differences of emphasis.) The implications of this concept, in terms of emphasis on continual recombination of genotype, are that it is difficult to envisage a speciation event which does not involve a degree of allopatry. That is, for speciation to occur there must be geographical separation of part of a species' range. This mechanism has its roots in the modern synthesis of evolutionary biology. The modern synthesis was an attempt to bring together Darwinian evolution and Mendelian genetics. Therefore, the emphasis was on the potential interbreeding of populations rather than on the morphological characters of such organisms.

Although it is usually never stated, most avian palaeontologists, until recently, probably adhered to this concept of species with its associated speciation mode. This has probably contributed to certain aspects of their interpretation of the fossil record. Those palaeontologists working on island faunas, because of the allopatric situation, may have been more likely to name new species with the additional weight given by the BSC.

Continental faunas are more complex as areas of endemism are more difficult to perceive. Rand (1948) proposed a mechanism where floristic refugia during glacial episodes constituted such areas of endemism. This mechanism and similar ones to account for speciation at tropical latitudes (Mayr, 1969; Diamond and Hamond, 1980) have been used to examine the relative timing of various phylogenetic events (Selander, 1965).

The existence of a mechanism, involving Pleistocene climatic change, that could explain when and how the presently more closely-related taxa became isolated and subsequently split, allows for the possibility of other, unseen speciation events and so for further extinct taxa to have been present. (The extreme alternative is that most modern species arose before the Pleistocene.) If one accepts the possible existence of extinct species during the Quaternary, whether they be ancestral to modern species or truly extinct, one might be more inclined to give species-level taxonomic status to certain fossils. If for example, a differently sized bird of a particular genus appears to have existed in the Pleistocene, the fact that a worker believes that extinct species were present may influence him or her into naming a distinct taxon. This certainly appears to have been the case with the two American avian palaeontologists Alexander Wetmore and Pierce Brodkorb, whose opposing beliefs regarding timing of the appearance of modern species may have influenced their descisions when describing fossils from the Pleistocene (see Chapter 1). On the other hand, because the BSC allows for variation within species, it may influence one in the opposite way into considering the fossils as a subspecies or merely as differently-sized population of a modern species.

The Evolutionary Species Concept

"A species is an ancestral-descendant sequence of populations evolving separately from others and with its own evolutionary role and tendencies." (Simpson, 1961).

The Evolutionary Species Concept (ESC) was devised by the palaeontologist G.G. Simpson in an attempt to bring into the BSC an element of time and so make it more relevant to the study of fossil species. The ESC therefore incorporates the breeding elements of the BSC. The idea was problematic, however, because without access to information regarding whether the fossil populations, formed breeding populations the method behind the concept relied on morphological similarity as proof of breeding (Smith, 1994).

This concept raises no more theoretical implications to the understanding of bird fossils than are already put forward by the BSC. However, because the reliance on morphology means that in practice this concept becomes a pattern-based concept, the problems inherent in those concepts are likely to be encountered (see below). Although this could be argued for all process-related concepts as ultimately they all rely on pattern.

The Ecological Species Concept

"A species is a lineage (or a closely related set of lineages) which occupies an adaptive zone minimally different from any other lineage in its range and which evolves separately from all lineages outside its range". (van Valen, 1976).

The Ecological Species Concept sees the species as a set of organisms exploiting a single niche. The idea is that Nature is divided up into adaptive zones with non-adaptive gaps between them, and individual species are adapted to these zones in which gene combination can take place. The gaps in between the optimal ecological adaptive spaces cannot be adequately occupied by organisms, as natural selection will tend to eliminate such organisms. Hybrids in this concept will be selected against because they will tend to have characters which preadapt them to the gaps.

An important proponent of a precursor to this concept amongst ornithologists was David Lack who wrote much about the role of ecological isolation in species formation in birds (Lack, 1944, 1971). While Lack's ideas are clearly very similar to the more recently defined Ecological Species Concept his work predates the term. Lack (Lack, 1944) stated: "If the above mentioned species-formation [geographical isolation of Mayr (1942)] is the normal one, it seems puzzling that closely related species so often occupy separate but adjacent habitats, and with the significance of closely related but adjacent species which differ markedly in size". This interesting quote highlights the similarity between the ecological species concept and the concept of character displacement. Character displacement, while not usually associated with species concepts, has clear relevance to the ecological species concept. Character displacement was first discussed by Brown and Wilson (1956). It describes the ecological interaction between organisms, usually closely related, although not always (Schluter, 1986, 1988). This interaction may, however, lead either to character divergence, or convergence (Grant, 1972). The difference between the Ecological Species Concept and character divergence is that the former is more rigid and cannot accommodate change or variation. Therefore, the ecological species concept is similar to the phylogenetic species concept in being monotypic in emphasis, i.e. it does not allow for a great deal of variation within species. Character divergence also differs from the Ecological Species Concept in that it deals with two species in similar ecological niches exloiting the same resources, while Lack talks of adjacent habitats and therefore a degree of parapatry. Character divergence will be further discussed later in this chapter under the heading of intraspecific variation.

The implications of the Ecological Species Concept (particularly as expressed by Lack 1944, 1971) to the interpretation of Quaternary avian fossils is that size may be perceived as a reliable criterion for species identification. The concept implies further that adaptive characters in general are useful for species diagnosis. This would suggest that greater taxonomic significance could be given to temporal variation in osteological proportions than would be the case from the perspective of concepts such as the BSC.

It may be that this concept lay behind certain statements by Harrison (1988), who comes very close to expressing a belief in the ecological species concept when he states that "in birds, size is usually consistent among individuals of the same species" and further that because of this, bone size is a useful criterion for species identification. He acknowledges only a few instances where species might overlap considerably, such as the blackbird / ring ouzel *Turdus merula* / *T. torquatus* and crow / rook *Corvus corone* / *C. frugilegus* amongst others. This seems to imply that he believes that birds are generally divided up ecologically according to size. Such a belief certainly simplifies the identification of bird fossils, although the problem of overlap in size and shape undoubtedly exceeds Harrison's expectations (Stewart, 1992).

The Recognition Species Concept

"We can, therefore, regard as a species that most inclusive population of individual biparental organisms which share a common fertilization system". (Patterson, 1985).

The recognition concept of the species (RC) is a modification of the BSC which stresses the inclusivity of individuals of the same species rather than the exclusivity of individuals from other species. The concept also specifically identifies types of characters by which individuals of a species recognise each other as conspecific for the purpose of mating. The RC is similar to the BSC because it also relies on allopatry to produce changes in the Specific-Mate Recognition System (SMRS) of an organism leading to speciation.

Species-significant mate recognition characters are described by Paterson as making up an organism's Specific-Mate Recognition System (SMRS). A species' SMRS includes behavioural and physical traits, including plumage, song and courtship behaviour in birds, and in mammals such as cervids and bovids head ornamentation (antlers and horns). These characters promote recognition by individuals of the same species, leading them to select each other as potential mates. The SMRS's of birds are unlikely to enter the fossil record, while aspects of some mammal species will do. The recognition concept has therefore attracted much attention from palaeontologists studying mammalian fossils because the SMRS of certain groups may provide the potential for the unambiguous identification of discrete species, particularly extinct ones (e.g. Vbra, 1983; Turner and Chamberlain, 1989). This in turn may allow for more objective studies of speciation of these animals.

The implications of the recognition concept for work on birds, on the other hand, are more negative. Features of birds such as song and courtship display will never become part of the fossil record and plumage will only be of use in exceptional circumstances. Possible SMRS features in birds which may be preserved as fossils are Galliforme leg spurs, and wing spurs in the spur-winged goose *Plectropterus gambensis*, jacanas (Jacanidae), the spur-winged plover *Vanellus novaehollandiae* and sheathbills (Chionididae) (Davison, 1985). An example where spur shape has been used to diagnose a taxon is in *Alectoris graeca mediterranea* (Mourer-Chauviré, 1975a). In this case the spur ran into a prominent crest which led to the proximal end. This feature is, however, of debatable significance to the species' SMRS and illustrates the difficulties of identifying such species recognition features in bird fossils.

The concept may therefore suggest, conversely, that similar species need not or are unlikely to differ in aspects which do not relate to their SMRS. This in turn may explain the perception that certain closely related bird species, such as those within passerine families and genera, are difficult to distinguish osteologically (e.g. Lydekker, 1891b; Stewart, 1992). For example, it may be impossible to confidently identify Palaearctic members of the genus *Turdus* or their 'ancestors'. If members of these or other species do differ in their skeletons it would be because of adaptive differences between them which according to the recognition concept are of no actual significance to their species identity. Indeed, as pointed out by Vbra (1980) (reported by Masters and Rayner (1996)), and Mayr (1963) before her, signalling systems are often very similar amongst such closely related organisms. This does not mean that in all instances closely-related taxa cannot be identified because similar species do not readily exist in sympatry without being ecologically separated. This ecological separation often imparts differences to birds osteologies. Most commonly this is achieved by means of the species being differently sized (Lack, 1944) (see section on character displacement). Ratcliffe and Grant (1983) have shown, however, that adaptive characters may also be part of a bird's SMRS. In this instance they demonstrated experimentally that ground finches in the Galapagos use bill morphology and overall body size as recognition characters, although song was more important. This example, based on experimental field work, demonstrates that the components of an organism's SMRS should be tested rather than assumed to be known from a theoretical point of view.

A recent paper by Geist (1991) should be mentioned here. He was concerned with the identification of modern caribou subspecies for environmental legislative purposes. He favoured provable genetically determined characters such as pelage pattern, antler characteristics and glands to define subspecies. These types of characters, which he also considers useful in the delimiting of species (Geist 1992) would be considered SMRS's under the RC. Therefore, from the RC perspective, Geist is proposing a similar definition for species and subspecies. This would mean that decisions regarding specific and subspecific status would become arbitrary. While this may be realistic given the mutability of species, it is not particularly helpful. Finally, this belief is again not compatible with the Paterson model, for species should possess a stable mate recognition system, i.e. not varying sub-specifically (Masters and Rayner, 1996).

Pattern - related Concepts

The Typological, Morphological or Essentialist Species Concept

This concept is perhaps the oldest concepts of species and has its origins in the philosophies of Plato and Aristotle (Mayr, 1963). Early palaeontologists to a large extent followed this concept of the species which was formalised by Linnaeus and remained dominant until the advent of the biological species concept. The typological species concept defines species in relation to type specimens, and individuals which do not conform to the types potentially represent separate species. Therefore, morphological similarity is the criterion for allying individuals under one species. This concept was the forerunner to all the more complex pattern-based concepts.

The implications of this concept to the study of Quaternary birds is of importance as it helps explain the point of view of the palaeontologists who were naming fossil species before the biological species concept. During this time, any variation or difference in the fossil record was potentially a new species. It was predominantly a time of splitters, and many dubious species of fossil animal and plant were named (e.g. Quaternary rodent species by Hinton, 1926). Quaternary birds escaped this fate and conservatism was generally observed, particularly when compared to other animal groups such as mammals and beetles (Chaline, 1972; Sutcliffe and Kowalski, 1976; Lister, 1987 and Ellias, 1994). Table 1.1 gives the dates of authorship of some of the extinct European Quaternary birds and many can be seen to have been described since the 1970's. This may partially be explained by the increased use of measurements, but it may also signal the arrival of certain workers who appear to have a highly typological approach such as Jánossy who has named more Western Palaearctic species than anyone else. Table 1.1 lists only those named from France and Britain, most of the former being named by Mourer-Chauviré, who was greatly influenced by Jánossy (Mourer-Chauviré, 1975a).

The Phylogenetic Species Concept

"The smallest diagnosable cluster of self perpetuating organisms that have unique sets of characters" (Nelson and Platnick, 1981).

The phylogenetic species concept was proposed by cladists such as Eldredge and Cracraft (Eldredge and Cracraft, 1980; Cracraft, 1983) and has gradually increased in acceptance among neontologists since its first inception (Zink, 1997). The concept has its origins in the cladistic treatment of organisms in order to produce phylogenies. There are variants of this concept, but the one most commonly followed was first put forward by the pattern cladist Rosen (1979), and is completely process-free. Essentially, what is advocated is the treatment of species as groups of organisms sharing characteristics which are not found in other similar organisms. These characters are shared derived traits which are not present in their sister groups. Therefore, species formation occurs when a population aquires a new set of such shared characters. The latter has led to criticism because the consequence of this can be that any mutation which spreads through a population should be regarded as a speciation event (Masters and Rayner, 1996). The other criticism leveled at the PSC is that it is not process-free as it claims to be. This is because it still bases species delineations on interbreeding populations, otherwise males and females would be separate species.

There are two major implications of the PSC to the study of Quaternary fossil birds. First, elevating populations not hitherto so considered to full species will lead to more osteologically cryptic species, whose identification in the fossil record may be impossible. There are problems already in the identification of many closely related neospecies, so elevating subspecies to a species level will only compound the problem. Furthermore, if there are problems identifying neospecies from their skeletal remains, the recognition of the ancestral or sister-group species to a pair of such neospecies will prove equally difficult. Also, if speciations cannot be recognised, then any attempt to study them in the fossil record would be futile. These problems exist already, but adopting the PSC would only make things worse.

In certain instances, however, studies of evolutionary timing would be improved. An example is the case of the willow grouse / red grouse *Lagopus lagopus lagopus / L. l. scoticus*. If these were to be elevated to species, as they were once considered, we would have pinpointed a speciation in time and space (Chapter 9). This is because it is generally accepted that their separation took place at the end of the last Glaciation (*ca.* 10 000 yrs. BP) (Voous, 1960). The conditions required for the existence of *Lagopus lagopus*, formerly occuring all over Europe, receded in a northwards direction leading to the isolation of the species on a western European peninsula which was to become the British Isles when sea level rose. This in turn led to the differentiation of the population into the red grouse, which under the PSC could very well be considered a separate species from those which survived on the Continent. This phylogenetic event is distinct from attempts to gauge the timing of the separation of the red / willow grouse *Lagopus lagopus* from the ptarmigan *L. mutus*, where there are no zoogeographic circumstances which clarify events. Presumably something similar occurred during an earlier glacial/interglacial transition.

The Phenetic Species Concept

"We may regard as species (a) the smallest (most homogeneous) cluster that can be recognised upon some given criterion as being distinct from other such cluster, or (b) a phenetic group of a diversity somewhat below the subgenus category, whether or not it contains distinct subclusters." Sneath and Sokal (1973).

This concept sees a species as a group of organisms which share characters with each other and not with other groups. The concept therefore concentrates on an organism's phenotype in relation to other species, and defines a species statistically. In practical terms all variable characters among specimens under study are recorded and placed into a taxon-character matrix. These characters are then subjected to multivariate statistics to identify clusters of individuals which may be considered species. In many ways this concept is a more sophisticated version of the typological species concept (Smith, 1994). Sokal and Crovello (1970) proposed the phenetic species concept as an alternative to the BSC saying that the BSC was "non operational" because "it fails to provide the procedures for identifying species". In other words, before interfertility is demonstrated or inferred, groups of individuals need to be identified as potentially interfertile, and the BSC does not provide a means of doing so.

Opinions differ as to the usefulness and correctness of this method of identifying and defining species. It has been suggested that once clusters of organisms have been identified by methods such as principal component or canonical variate analysis there is no practical way of identifying new members of the species without performing the statistical procedure (Smith, 1994). However, Cooper and Ni (1986) state that once the species have been identified by this method it is easy to identify individuals which belong to the species clusters by the use of particular dominant characters. The phenetic species concept, however, suffers from lacking a philosophical basis. The concept ignores all ideas of process such as breeding or ancestry which dominate other concepts. The concept is therefore more of a practical method for distinguishing species which does not take into account problems such as taxonomic rank (species versus subspecies), or phylogenetic versus adaptive characters.

If Quaternary fossil birds are to be identified from the perspective of this concept there is the danger that subspecific populations would be given greater taxonomic rank than they merit. The concept therefore might promote splitting. There is also a danger that the techniques of multivariate statistics are considered an end in themselves and that the clusters resulting from such an analysis must be different species rather than groups of similar individuals which need to be further interpreted before taxonomic affinities are inferred. It is equally possible that the reverse could occur, species being considered subspecies because the resultant clusters are not sufficiently different from each other. This is dangerous because there is no quantitative morphological guide to the difference between species and subspecies. With the growth of morphometrics (e.g. Rolf and Marcus, 1993), there is a danger that the phenetic species concept, already effectively dominant in palaeontology, will be given spurious credibility.

In conclusion it may be stated that there is a range of different species concepts, and that a belief in any one of them will determine the taxonomic approach of an avian palaeontologist. Certain concepts may promote a "splitting" approach (particularly the pattern-based concepts and the ESC), others a "lumping" approach (particularly the BSC). The purpose of the above discussion of the various concepts is to illustrate the ways in which such concepts may influence workers. This volume shall endevour to consider the taxa chosen for study in the light of the different concepts, rather than to specifically adhere to any one of them.

2.3.2. Modes of Species Evolution

A central argument concerning the mode of species evolution is the degree to which it occurs in allopatry, or in sympatry. However, as Grant and Grant (1989) have stated, the allopatric speciation mode is the one accepted by most biologists, although the number of species within certain groups would appear to invoke an implausible number of range splittings and rejoinings. This has led to a variety of authors such as Bush (1994), Grant and Grant (1989) to consider other alternatives such as sympatric or parapatric speciation.

Studies on species formation started in the nineteenth century with Darwin and Wallace, who considered islands. The preoccupation with island bird species formation and the Galapagos Islands in particular has persisted until the present (e.g. Lack, 1947; Grant, 1986). This has also led to an interest in fossil island faunas due to the abundance of apparently extinct taxa (e.g. Olson and James, 1982; Steadman, 1981, 1985). Islands, however, provide unusual circumstances for speciation, particularly in the case of newly erupted volcanic islands, such as the Galapagos Islands and the Hawaiian Islands, where isolation is guaranteed and a great number of niches are likely to have arisen enabling speciation to take place. Allopatry is easily perceived in islands and this may have encouraged palaeontologists to name new species when differently-sized fossils have been found. Continents, on the other hand, are the power houses of bird diversification but it was not until such authors as Rand (1948) that a coherent model for speciation was developed for continental birds. Rand used the refined understanding of the complexity of the Quaternary, with its climatic cycles, and applied this to the problem.

The difficulties inherent in perceiving continental birds from the point of view of allopatry are that birds are very mobile organisms, and physical barriers are more difficult to conceive. This led Lack (1944, 1971) to suggest that birds had undergone ecological differentiation, this being a mechanism for sympatric speciation. However, this mechanism was not thought to be credible until Maynard-Smith (1966) provided a more detailed genetically-based explanantion of the mechanism.

The relevance of this argument to the understanding of Quaternary fossil birds is that, despite the existence of a theoretical alternative, the allopatric mode is almost ubiquitously assumed to have occurred (e.g. Tyrberg, 1991b and Randi *et al.*, 1992). Commensurate with this assumption, models have been devised to explain the speciation of northern latitude birds (a major subject of this volume). These have invoked the splitting of climatic and vegetational belts, giving rise to allopatric populations. For instance in Europe, colder climates are generally believed to have pushed temperate ecotones further south. The existence of peninsular land areas in the south of Europe such as Iberia and Italy signified that the birds became split longitudinally into subpopulations which may have subsequently diverged and speciated in allopatry. If, however, sympatric speciation is the dominant mode of speciation the convenient model of recurrent climate change may be unimportant in the formation of species, allowing for these events to have taken place during the more stable climes of the Pliocene.

A number of modern congeneric bird species have parapatric distributions (e.g. spotless and common starlings, *Sturnus unicolor* and *S. vulgaris*), which implies that speciation by geographic separation occurs at least some of the time. The parapatric distributions of the two species are presumably due to mutual competitive exclusion. However, other congeners live in sympatry which means that they either evolved while geographically isolated from one another and became reinforced ecologically after speciation, or they evolved in sympatry due to ecological diversification. The problem with trying to discern which occurred, is that the resultant patterns of distribution and ecological diversity are the same.

Recently, Klicka and Zink (1997) published some molecular data which they interpreted as revealing that 35 pairs of closely related song birds in North America which had always been considered to have speciated during the Late Pleistocene, were a great deal older. While they did not negate the role of cyclic glacial / interglacial refugial allopatric speciation, their view does imply that species are more resilient to such environmental events than previously believed. It also rekindles the possibility that sympatric speciation may have been prevalent (Rosenzweig, 1997).

Parapatric and / or peripatric speciation, the other alternatives to allopatry, involve the evolution of two or more populations of a species continuously distributed without a clear geographical break. They are in some ways intermediary, hypothetical situations between allopatric and sympatric speciation. Peripatric speciation is when a small population is not separated in the extreme sense as in allopatry but is in a peripheral part of a species' range. In parapatric speciation the two each have continuous ranges over relatively homogeneous environments which abut each other. The significance of these processes to the understanding of fossil birds is that it is not necessary to have disjunct populations for a speciation event to occur. Therefore, these ideas imply that two very similar congeneric species may exist in adjoining areas and that they evolved in those areas.

Another subject of concern is whether or not species form by anagenesis or solely by cladogenesis. This debate is centered on what a species is believed to be. Those using cladistics and following the phylogenetic species concept would regard only the results of phyletic branching events as true speciation. Chronospecies forming along a lineage through anagenesis would be termed pseudospeciations. In the study of Quaternary birds it is rare for lineages to be traced. Mourer-Chauviré (1993) is a notable exception, where ancestor-descendant lineages have been drawn up. In this instance various tetraonid (i.e. members of the genera *Tetrao*, *Lyrurus*, *Tetrastes* and *Lagopus*) lineages are pieced together from the fossil record and plotted against a temporal scale. A problem with such reconstructions is that heavy reliance is placed on the assumed completeness of the fossil record. Another possible reason for the lack of such lineage reconstructions, is the general difficulty in separating fossil bird species, particularly in species-rich clades such as the genera *Anas* and *Turdus*.

2.3.3. Rates of Species Evolution

Arguments concerning the rates of organic evolution have existed since the publication of the Origin of Species in 1859. In this instance Darwin explained the apparent jumps in the evolution of organisms, as witnessed by fossils, as gaps or discontinuities in the geological record. He firmly believed that evolution was gradual. This debate was given a new interpretative perspective when Eldredge and Gould (1972) published their ideas on punctuated equilibrium. This hypothesis suggested that the gaps in fossil morphology were real and the consequence of the variation of macroevolutionary rates. They explained the punctuations more fully by the suggestion that they were the product of evolution in allopatry, so that the missing rapid evolutionary events were occurring elsewhere (Gould and Eldredge, 1977). This linked punctuated equilibrium to the Mayrian ideas on speciation where such events require isolation (and displacement) from the original population to produced new species. However, Gould and Eldredge were keen to point out that punctuated equilibrium works equally well with sympatric speciation if the evolution takes place in a small section of the population. The latter would be less likely to become part of the fossil record thus allowing the event to appear punctuated (Gould and Eldredge, 1977). A great many arguments, both in defense of and attacking punctuated equilibrium, have been published since the theory was first proposed (e.g. Johnson, 1982; Brown, 1987; Williams, 1987). There also seems to be some confusion over terminology, because allopatry and peripatry have been at times used synonymously, as pointed out by Brown (1987) in discussing various works by Gould and Eldredge. It has, however, been pointed out that true punctuated equilibrium must signify the occurrence of punctuated evolution in situ, and not evolution at an unknown rate in a geographically separate area, followed by sudden immigration (Lister, 1984).

In the present context, the most important question concerns the determination of the age of our modern avifauna and particularly the constituent species. This may appear to be a separate subject, but as has been pointed out by Gould and Eldredge (1977), knowledge of stasis (the "equilibrium" part of their theory) is a genuine part of the study of evolutionary rates.

Quaternary palaeornithology has seen various views on the age of neospecies. The opposing ideas of Brodkorb and Wetmore have already been discussed in Chapter 1, and as has been illustrated, their views on what constitutes a species may have contributed to these. Brodkorb (1971) even produced a figure of approximately half a million years for the longevity of a [Late] Pleistocene species, while in the Tertiary it was 3 million years, the difference being attributed to differences in environmental stability. Recently, the subject has become largely buried and implicit within publications treating Quaternary fossil birds.

A recent exception to this rule is Emslie (1992), who has identified many neospecies in the Late Pliocene of Florida. This, he says, clearly shows that Brodkorb was wrong in his belief that all our modern species arose during the Pleistocene. However, the final comment by Emslie gives cause for concern. He states that "the fossil populations often differed in size and relative limb proportions from Recent populations of the same species" (Emslie, 1992). This implies that species can be reliably identified in the fossil record and that size and metric shape are not worthy identification criteria. Furthermore, it implies that there are characters over and above size and metric-shape which allow species to be identified, even in species-rich genera such as *Anas*, the dabbling ducks.

The difficulties inherent in the identification of closely related species have not been expressed in relation to Quaternary bird evolutionary rates, although the subject is of fundamental importance (Gould and Eldredge, 1993). There are considerable problems in separating congeneric avian neospecies on the basis of their osteology. Often, there can be as much, or even more, morphological variation within species as between them

(Stewart, 1992). In other words there is much overlap in the osteology of closely related bird species. This implies that the recognition of their phylogenetic ancestor will prove equally difficult. So a study of a certain bird taxon might produce a pattern of morphological evolution which is interpreted as a series of chronospecies, but which are merely adaptations equivalent to the intraspecific variation seen today. Equally, a speciation event might be completely invisible to the palaeontologist, particularly if it entailed new attributes of sexual recognition such as plumage or song. Of relevance here is the example, quoted by Emsly (1996), of the scrub jay species which was recognised in the fossil record on the basis of its bill, clearly an adaptive character and one which could change very quickly (Grant, 1986). Therefore, the similarity between the bills now and those in the late Pliocene is not evidence for stasis in a single lineage, because independent acquisition of such a feature is very possible.

2.3.4. Immigration versus speciation

The fossil record cannot always be taken at face value because apparent changes which are seen in the record at one place may be the result of immigration rather than evolution *in situ* (Lister, 1992 and 1993). It may be difficult to distinguish in the fossil record rapid punctuated evolutionary processes from immigrations of different forms. No examples of either are known to have been documented for Quaternary birds. A well known Quaternary mammal example is that of the water vole *Arvicola terrestris*, whose molar morphology in northern Europe appears to undergo a reversal between the Saalian and the Eemian (Röttger, 1987; van Kolfschoten, 1990). This has been interpreted as possibly being due to a replacement of populations by more southern forms as the climate warmed. While this example is one where a subspecific change occurs, it illustrates the possible process well and could complicate any interpretation of speciation in the fossil record.

2.3.5. Character displacement

Character displacement as proposed by Brown and Wilson (1956) describes situations where two species, with partly overlapping distributions, are more different from each other in the area of overlap. Examples usually involve closely related species, often in the same genera (Malmquist, 1985; Fenchel, 1975; Schluter, 1988). However, an example is known where the presence or absence of a very different organism determines the nature of another. The presence or absence of the carpenter bee *Xylocopa darwini* on Galapagos islands determines the beak shape of Darwin's finches because it can outcompete them when it comes to nectar feeding (Schluter, 1986). A further complication to the concept of character displacement is that it may cause either convergence or divergence (Grant, 1972).

The significance of the concept of character displacement to avian palaeontology is that it can be an explanation for differences in morphology. This point was made by Dayan *et al.* (1991), who pointed out that character displacement should first be eliminated before any interpretation of size change due to temperature regulation is deduced from the fossil record (see section 2.4.2.). The implications are even deeper than the interpretation of mere size, as other morphological traits can be affected, such as bill shape. The importance of this is that some consider bill morphology to be reliable in determining species (Emslie, 1996). However, it will not always be possible to identify character divergence, as this would at the very least require that the taxon causing the divergence be present in the fossil assemblage, and this cannot be relied upon.

2.3.6. Summary

It is clear from the above discussion of the various theoretical issues pertinent to the species-level, that such ideas will and have influenced workers beyond the scope of the data that they may be analysisng.

2.4. Intraspecific Variation

It is the intention here to discuss all sources of variation which occur below the level of the species. These are manifested across both time (temporal variation) and space (geographic variation). In certain respects time and geography are interchangeable with the important practical difference that temporal variation is less accessible to research, particularly if long durations are involved. Therefore temporal variation is perhaps best divided into macrotemporal and microtemporal durations. The former represents changes over periods which are outside the scope of human observation and require the use of fossils, while the second is usually within the limits of a human lifetime. The latter has resulted in a vast neontological literature treating a great many bird taxa, the most notable work being by Grant (1986). This details a long-term study of a whole population of Darwin's finches tracked through various environmental changes. The studies in palaeontology dealing with temporal change are more scarce although important exceptions are those by Northcote (1979a, 1980b, 1981a, 1981b, 1983) on mute swans *Cygnus olor* and Ericson (1986, 1987a) on eider duck *Somateria mollissima.*

Storer (1992) discussed this subject in some depth using the example of fossil grebes in North America. He suggested that size can only be used with confidence in the identification of Pleistocene and Pliocene grebes "when the amount of variation is understood within the same or related species". Storer, while he acknowledges that present-day variation must be known, does not mention any possibility that the modern variation may not include all variation which has existed within modern species. He says that cryptic species may have been present in the past but does not mention possible additional variation within neospecies in their history.

2.4.1. Subspecies

A knowledge of present-day bird subspecies is important to the interpretation of Quaternary birds. The importance lies less in the possible presence of extant subspecies in the past, than in gaining a view of the range of morphological variation in neospecies to enable an assessment of variation in the past. This uniformitarian perspective should not be applied too strictly, however, as similar variability need not be expected *a priori* in the past. I also feel that it is wrong to pre-judge, as others have done (e.g. Jehl, 1966) the possible occurrence of Recent subspecies in the Pleistocene. Jehl (1966) suggested that it is incorrect to assign Pleistocene fossils to modern subspecies because they are defined on charactersitics which are assumed to be adaptive to local conditions. Jehl stated that if these conditions had existed in the past the adaptions would be parallel ones and therefore if similar morphologies existed they would not necessarily relate to modern subspecies. Jehl also states that breeding range is an essential part of the definition of modern subspecies, rendering the use of modern subspecies for fossils unwarranted. This view is contradicted by Geist (1991), who dismisses the use of geographical range in the definition of subspecies and stresses the importance of "social insignia" at breeding time. This view rekindles the possible recognition of modern subspecies as fossils. (It has already been noted that Geist's use of "social insignia" in defining subspecies is completely at odds with the more commonly held belief that these characters distinguish species. That belief is encapsulated in the Recognition Concept of the species of Patterson (1981, 1985) and his followers (Masters and Rayner, 1996; Turner and Chamberlain, 1989; Turner, 1993). Nevertheless, whether temporal subspecies units are used in Quaternary palaeornithology is unaffected by this argument. Whichever view is taken, it will always be difficult to ascertain which are species-significant characters and which ones characterise subspecies.

2.4.2. Sexual dimorphism

Sexual dimorphism within species can result from either sexual or natural selection. Most of the more spectacular sexual dimorphisms, usually in breeding plumage of males, is a result of sexual selection. Natural selection can apparently produce sexual dimorphism in a species if competition for limited resources is high. Examples of the latter are hawks, where females are significantly larger than males (Storer, 1966); and woodpeckers, where a similar situation exists (Selander, 1966).

Northcote (1979a, 1980b), Ericson (1987a), as well as various of the Munich School dissertations (Woelfle, 1967; Bacher, 1967; Erbersdobler, 1968; Kraft, 1972; Fick, 1974; Langer, 1980; Otto, 1981; Schmidt-Burger, 1982; Kelner, 1986; Gruber, 1990; Lorch, 1992) have all examined the extent to which sexual dimorphism occurs in the osteology of various birds.

The importance of sexual dimorphism to Quaternary avian palaeontology is that the resulting bimodal variation within a single species could, especially if not known to exist in modern representatives, be mistaken for interspecific variation in the fossil record. It is tempting to assume that sexual dimorphism remains the same through time, although examples exist where fossils have been shown to have different degrees, or even types, of variation than present-day counterparts. Examples of this are the subspecies of black grouse *Lyrurus tertix* named by Mourer-Chauviré (1975a), where the females possessed longer legs; and the subspecies of snowy owl, *Nyctea scandiaca* whose females were larger while the males were the same size as modern counterparts (Mourer-Chauviré, 1975a) (Table 1.1 in Chapter 1). A more extreme example is the case of the sawbill species *Mergus connectens* (Jánossy, 1972), where no sexual dimorphism was found, in contrast to modern members of the genus whose males and females differ significantly in size (Cramp, 1977). This discrepancy between the present and past may, however, be explicable by theory. If natural selection is the dominant cause for the sexual dimorphism, this selection could vary through time. This might allow a theoretical explanation for the phenomenon, although it should merit comment when a new taxon is named. The longer legs of the female black grouse is perhaps the most unusual of the examples quoted, first because sexual dimorphism in tetraonids is usually thought to be due to sexual selection (Payne, 1984), and second because the dimorphism observed is usually in body size and not leg length.

In general, it is probably valid to assume that sexual dimorphism in species will remain through time, and only relatively small degrees of variation will take place. Furthermore, it is probably safe to expect that when an extinct species is discovered within a currently sexually dimorphic genus, sexual dimorphism will be present.

2.4.3. Pathological Variation

Very little work has been done on the palaeopathology of birds, and it is widely believed that a pathological wild bird would rarely survive long enough to leave a skeletal marker. A recent review by Brothwell (1993) looked at the significance of pathological conditions to the archaeological interpretation of bird remains. Of the six categories of pathology that he defines, those that are perhaps most significant to the present discussion are congenital abnormalities. All others are unlikely to be mistaken for non-pathological variation. The possibility for mistakes is important because certain conditions mentioned by Brothwell (1993), such as dwarfism, may give the false impression that a different, smaller species was present. A possible example of this in wild birds was seen in some skeletons of hooded crow *Corvus corone cornix* and a single carrion crow *C. c. corone* measured in Sweden and Britain respectively (Pers. Obs., see Chapter 10). In this instance, while the identities where unquestionable, the lengths of the birds' limb bones were very short in relation to their breadths compared to "normal" individuals.

Likewise, the validity of *Somateria gravipes*, the thick-legged eider, described by Harrison (1979a), was questioned by Mlikovsky (1982a), who concluded that the specimen was probably pathological. X-rays taken of the specimen by the present author to examine the possible degree of secondary bone thickening caused by the claimed pathology, failed to find any proof of a pathological condition. However, this is not to say that such potentially misleading pathological conditions never occur in the fossil record. The possibility should be borne in mind when naming species, particularly from single specimens.

2.4.4. Geographical Variation

Geographical variation within bird species has received much attention through the history of ornithology, partly because of its importance to the understanding of their evolution (Zink and Remsen, 1986). The variation can be in the form of morphological (size, shape and colouration), behavioural (migration, courtship behaviour or song) or biomolecular (proteins and DNA) differences between populations of a species over their geographic range. Two controls of such variation have been identified. The first is genetic (genophenotypic variation) while the second is non-genetic (ecophenotypic variation). It is important to make the distinction between the two as the second form of control has less significance to evolution. Few studies, however, have established the mechanism controlling geographical variation. This is largely due to the practical problems of demonstrating the mechanism itself. The most prominent exception is that by James (1983), where eggs of the North American red-winged blackbird *Agelaius phoeniceus* were transplanted between nests in different parts of their geographic range, where differently-sized and -shaped birds were present. There was a shift in the resulting adults toward the phenotypes of the foster population, indicating that an environmental component existed to the variation in this species. It is important to note that this experiment did not completely discount the role of a genetic control but showed that it was not alone in establishing the phenotype.

Another study was by Boag (1983), where heritability of morphological characters was shown to be strong among members of Darwin's ground finch genus *Geospiza*. However, the fact that, unlike James (1983), offspring were not displaced, prevented the study from answering the crucial problem to hand: whether genes or environment are most important in directly affecting morphology.

Most other studies on geographic variation are less concerned with the exact mechanism controlling the phenotypic state so much as the environmental conditions which are assumed to be directing the phenotype through natural selection. Many such studies have taken place and have led to various environmental factors being highlighted as causing the resultant morphologies. These studies are sometimes called ecomorphological studies, which should not be confused with the concept of ecophenotypic control of character states, as no mechanism is considered, although natural selection is usually implicit. Additionally, some of the studies have led to the definition of ecogeographic rules such as Bergmann's, Allen's and Gloger's Rules. These three rules are based on patterns observed among various animals, most prominently birds, across their geographic ranges in terms of size, shape and colouration respectively. In the present context Gloger's Rule, which predicts that populations of a species living further north will have darker plumage (or pelage in mammals) is not relevant as it will not affect the interpretation of fossils. Thus only Bergmann's and Allen's Rules will be discussed here.

There are, however, other parameters of natural selection which may affect a bird's size and shape, such as those which are involved in different degrees of intraspecific migratory behaviour, or those due to differing degrees of interspecific competition across a species' geographic range.

Table 2.1 lists some of the correlates of intraspecific body size variation. Clearly some of these covary (such as latitude and temperature) thus causing problems in interpreting observed patterns in the fossil record.

Size Variation

A number of different correlates of body size variation have been proposed by different authors (Table 2.1).

Bergmann's rule (Bergmann, 1847) is the best known of the ecogeographic rules proposed by biologists to explain the clinal variation in body size observed within species. As originally conceived the rule describes the relationship between warm-blooded vertebrate species body size and latitude (Bergmann, 1847), although it was redefined by Mayr (1956) as dealing with variation within single species in relation to warm/cold climates. The rule highlights the relationship between body volume and body surface area. If a body increases in size without changing shape its surface to volume ratio decreases and hence if temperature is to be conserved a larger body size is advantageous. However, there is no unanimous acceptance of the rule, and criticisms of both the empirical observations and the theoretical explanations have been made (e.g. McNab, 1971; Geist, 1987). The problem is that many selective forces, in addition to temperature regulation, may be acting on a set of organisms to influence their body size. These phenomena

Table 2.1. Correlates suggested for animal body size variation (not comprehensive).

Correlate	Taxon	Author
Altitude	Tits	Snow (1954)
Temperature	Various bird taxa	Mayr (1956, 1963)
Humidity	Birds	James (1970)
Intraspecific competition	Eider	Ericson (1987a)
Interspecific competition	Carnivorous mammals, Galapagos finches	Dayan *et al.* (1991), McNab (1971), Abbot *et al.* (1977)
Prey / food particle size	Galapagos finches and various bird taxa	Grant (1986), Wiens and Rottenburry (1980)
Home range	Various bird taxa	Wiens and Rottenburry (1980)
Migratory distance	*Sylvia* warblers	Winkler and Leisler (1992)
Sex - sexual selection	Galliformes	Payne (1984)
Sex - natural selection	Falcons	Storer (1966)
Plant productivity / productivity pulse	Large mammals	Geist (1987)
Seasonality	House sparrow, mammals	Boyce (1979), Linstedt and Boyce (1985)
Clutch size	Passerine birds	Ricklefs (1980)

are environmental conditions such as vegetation productivity, humidity, seasonality, prey size and competing species, in addition to variations of migratory behaviour and sexual dimorphism which themselves correlate with size (Table 2.1). Any of these might vary with latitude and thus produce "Bergmann's Rule", or vary in some other direction and thus might outweigh it.

Zink and Remsen (1986) recently reviewed the evidence for the "Rule" in North American bird species, and concluded that there were problems because many studies used wing length as an index of size. Wing length, although strongly correlated with body weight (James, 1970), also correlates with migratory distance (Salomonson, 1955; Hamilton, 1961; Leisler and Winkler, 1985) and with the degree of habitat openness (Temple, 1972; Aldrich, 1984). More northerly populations also tend to be more migratory; and altitude, often left out of studies of geographical variation, further complicates the problem. Another problem is that museum specimens rarely have weights recorded; on the other hand, body weight as an index of body size is questionable because it varies with reproductive condition and seasonal fat deposition.

The prediction that the "Rule" should be followed more often in sedentary than migratory birds because they are tied to a local climate was demonstrated by Rensch (1936). Of the 25 species he tested, 60% of the migratory birds conformed to the "Rule", while 76% of the sedentary birds did. Zink and Remsen's (1986) analysis on 92 North American species was more equivocal. They concluded that despite the fact that a statistically significant number of birds conformed to the "Rule", the degree to which they did was only weak. They were also more negative regarding the difference between migratory and sedentary birds.

Due to the climatic interpretation of Bergmann's Rule, it has been seized upon by vertebrate palaeontologists as an explanation for body size fluctuations in Quaternary mammals (Klein, 1986; Klein and Scott, 1989; Davis, 1981; Stuart,1982). Bochenski (1974), Northcote (1981a and 1983) and Ericson (1987a) have also given consideration of the possible application of the "Rule" to changes over time in ptarmigan /willow grouse, mute swan and eider duck respectively. Bochenski found that there may well be correspondence to the "Rule", while Northcote preferred wetland drainage and human disturbance of habitats, and Ericson favoured subsequent diminished competition for resources as explanations for size changes.

Dayan *et al.* (1991), working with carnivorous mammals, suggested that there are two major problems that should be overcome in applying Bergmann's Rule in palaeontology. First, the choice of proxy for body size, in their case tooth length which may be tenuous; and second, the effects of competing species (character displacement) should be taken into account.

The first point, when applied to birds, is similar to that made by Zink and Remsen (1986) for modern birds, where wing length is used as an index of size. In the case of fossil birds it is important that the index used be one which is not highly subject to other selective pressures. In birds, distal wing element measurements (e.g. carpometacarpi lengths) would not make a good index for body size, whereas humeral length would be better. This is because the carpometacarpus is subject to selection related to different migratory behaviour (Winkler and Leisler, 1992) (see below).

Metric-Shape Variation

There are two forms of metric-shape variation. That which is a direct consequence of size variation, called allometric variation, and that which is due to other forms of selection. The slection causing allometric variation is discussed above under size variation, while other forms of selection are described here.

Birds differ in the locomotor functions of their forelimbs and hindlimbs. The hindlimbs of birds can be treated in much the same way as the legs of mammals and are used for terrestrial locomotion, whereas the forelimbs, in most birds, are used for flight. Therefore, the forelimbs and hindlimbs will be discussed separately. However, there are some factors which could affect both wings and legs equally, and will be treated last. Intraspecific metric-shape differences of the skull elements also exist but will not be dealt with here as this volume only examines the post-cranial elements.

Winkler and Leisler's (1992) work on the ecomorphology of migratory birds has shown that certain bones such as the humerus, ulna and carpometacarpus vary according to migratory distance. The main finding is that the length of more proximal wing bones appears to correlate negatively with migratory behaviour, while more distal bones correlate positively. The phenomenon is no doubt linked to the observation that species within a genus which migrate have more pointed wings than less migratory species (Mönkkönen, 1995). Their study dealt with interspecific differences in birds, but these are undoubtedly the result of natural selection, and so should also work below the level of species. The prediction is therefore that the more migratory populations of a particular species will have longer carpometacarpi in relation to their humeri.

The lengths of bird leg bones, and particularly that of the tarsometatarsus, may be determined by locomotory factors, such as degree to which birds are terrestrial as opposed to spending time either in trees or in the air.

The length of the tarsus, the part of the leg incorporating the tarsometatarsus bone, is thought to be associated with perch characteristics (Grant, 1972). The observations are that species which forage in an upright position, on stable perches or larger branches, will have longer tarsi than those which forage on smaller twigs. Intraspecific relationships have also been found in this character, which suggests that natural selection is the source of the association (Fretwell, 1969). This is similar to the correlation found in *Anolis* lizards between hind limb lengths and vegetational perch diameters (Lossos *et al.*, 1997). The latter work was based on an experiment where lizards of known characteristics were released onto islands with different vegetations. It showed that the different populations evolved over the 10 -14 years between introduction and recapture.

Work on the locomotor adaptions of bovids illustrates the relationship between lengths of leg bones, particularly the distal elements, and adaptions for running (Scott, 1985) in other words longer distal elements signify fast running, shorter distal elements signify power locomotion (eg. climbing). Scott (1985) also demonstrated the relationship between leg bone diameter and body weight. This has also been shown for the femur and tibiotarsus of birds (Campbell and Marcus, 1992). This suggests that if bird populations were to change body weight, the leg bone breadths would be more helpful in elucidating such a change than their lengths.

Allen's Rule has been interpreted as a thermoregulatory rule like Bergmann's Rule. In cold climates, birds will tend to have shorter extremities, which will reduce their surface areas and hence reduce heat loss. There is a problem with testing this rule in neontology because, as mentioned, wing length is often used as a proxy for body size and so the reasoning becomes circular. The difficulty also lies in that an index of body size is required to standardise measurements so that metric-shape differences can be discerned (Zink and Remsen, 1986). Zink and Remsen (1986) surveyed the evidence in birds and found that, as for Bergmann's Rule, the results were conflicting.

Genetical control of the different limb proportions is also important in this context. Cock (1969) has shown experimentally that there may be incidental changes of characters. It was found that elongation of the nasal region of the skull, the bill, led to an associated effect on the tarsus in domestic fowl. Selection for long bills therefore might also produce long tarsi and *vice versa*. Such relationships, which may be numerous, have to be borne in mind, because they imply that metric-shape changes may purely be a consequence of selection for another unrelated trait.

The significance of the existence of metric-shape differences within species is that these differences should be used with great caution to indicate taxonomic status. By the same token, metric-shape differences which can occur between or within species can cause problems of identification. The degree of difference within species can even exceed that between species of the same genus. Clearly such phenomena will create problems for both the identification of neospecies in the fossil record and the recognition of extinct undescribed taxa. Metric-shape differences are often seen as more significant than mere size differences in this respect, but clearly are as potentially misleading as size.

Also important to the understanding of metric-shape variation within bird species is that such variation is not constant and may change according to differing environmental parameters. A good understanding of variation today is therefore paramount if that in the past is to be interpreted through the principle of uniformitarianism.

2.4.5. Temporal Variation

For the purposes of this discussion two categories of temporal variation can be defined. The first, macrotemporal variation, applies to changes over time which can only be directly detected by the use of fossils, while microtemporal variation is defined as that which has been directly observed within historical time or even the lifetime of a human observer. It is thought useful to distinguish between the two because the observations made during microtemporal studies are potentially far more accurate and allow an *a priori* predictive approach, compared to the *a posteriori* approach of studies with fossils.

Many factors influencing the variation in birds over time are the same as those operating across space. The advantage of adding the dimension of time is that it allows for the processes producing the variation to be observed.

Microtemporal variation

The best known study of morphological variation over a short time span in birds is that by the Grants on the Galapagos finches (Weiner, 1994). Many observations were made on the island of Daphne Major which was visited over more than a decade. Metric-shape changes were seen in the bills of *Geospiza fortis*, a species of ground finch, over a few seasons, due to strong selection for increased bill size because of the differential survival of large-seeded plants after a drought (Grant, 1986; Boag and Grant, 1981).

Another important study is on the introduced European house sparrow (*Passer domesticus*) in North America (Johnston and Selander, 1964). This work showed that significant adaptive differences in both plumage colouration and size had occured in a period possibly as short as 50 years. This demonstrated that differences large enough to warrant polytypic classification in native North American species can evolve faster than had previously been believed. The results indicated that after introduction in 1852 from England and Germany, the house sparrow has adapted to the various areas in which it now lives. A significant correlation between body size and climate was demonstrated, which appears to follow Bergmann's Rule. Similar agreement was found between colouration and geographical location of the populations, conforming to Gloger's Rule. A later study by one of the authors attempted to confirm Allen's Rule (Fleischer and Johnston, 1982). This yielded positive evidence for the relationship between core-to-limb proportions and climate, and supported Allen's Rule. Again these processes had occured since introduction of the species into North America.

Introduced birds provide ideal situations for the documentation of temporal change, because their time of introduction is often known and so timing of the changes can be approximated. A similar scenario is that of the rock dove (*Columba livia*), which was introduced into North America no earlier than 1600 AD as domestic doves. These birds escaped, as they had done earlier in Europe, forming colonies of feral pigeons. A study by Johnston (1992) showed that the feral birds in North America were more closely linked genetically to domesticates than are the feral birds in Europe. Despite this, the feral birds in North America were closer in skeletal size and shape to the feral birds of Europe than to domesticates. This showed that natural selection had produced similar morphologies in the feral birds of both Europe and North America, highlighting the importance of environment on such traits.

The importance of these studies through recent time is to show that natural selection may act far faster than previously believed. This implies that similar environmental tracking of species' morphologies could have occured in the more distant past and that the speed with which it took place may be effectively invisible in the fossil record. This should be borne in mind when there appear to be two sympatric, congeneric species in the fossil record. The time with which the deposit containing the fossils accumulated becomes enormously significant, as the two apparently contemporary "species" may in fact be chronologically separate parts of the same lineage.

Macrotemporal variation

Remakably few studies of taxa have been made over geological time in Quaternary avian palaeontology. The first to be done in Europe was by Bochenski (1974, 1985) on the genus *Lagopus*. As already mentioned in the historical section, widely geographically distributed samples of both species of *Lagopus* (*L. mutus* and *L. lagopus*) were compared osteometrically to fossil counterparts. He concluded that in the last glaciation both species had possessed longer carpometacarpi and shorter tarsometatarsi, as well as more massive articular ends to their humeri, coracoids and tarsometarsi. This was interpreted as evidence that the two species had maintained or slightly reduced their body size and that the birds had reduced their primary feathers over time. The cause of the changes was believed to be climate change. The mechanism was not, however, clarified, although it appeared that the metric-shape shape changes of the tarsometatarsi were due to Allen's Rule.

Ericson (1987a) studied the changes in eider (*Somateria gravipes*) between the present day (1960 - 1985 AD) and two past samples, one from 1895 - 1930 AD, the second an archaeological assembage of early to middle Viking Age (800 - 975 AD) from Birka in Sweden. The last hundred years have seen a moderate amount of change to the birds, although they clearly have decreased in body size during the period. The analysis of eider over the last thousand years shows that this size decrease has been taking place for some time. It is interesting that the change over the past hundred years has been relatively greater than during the previous thousand years. The cause of the size decrease was difficult to fathom, but climate change was thought to be an unlikely reason, because such changes were more dramatic over the longer time frame. The tentative conclusion is that the ecological balance may have been upset by recreational use of the sea and pollution, causing greater competition for resources and hence a decrease in eider size.

The two studies using fossils confirm that morphological change can be detected within taxa across time and that such changes are probably caused by environmental factors. They also suggest that the precise mechanism causing such changes may be difficult to elucidate.

2.5. Conclusion

It is clear that there are many theoretical perspectives from which to consider Quaternary fossil birds. Amongst these are alternative species concepts, modes of speciation, and sources of intraspecific variation. Most authors have described fossil birds, whether neospecies in the past, chronospecies, or truly extinct species, without any explicit attention to these issues. This volume attempts to help redress this situation by considering in detail the Quaternary fossil record of a limited range of taxa in the light of the theoretical considerations described above.

Chapter 3

Scope and Aims of the Book

3.1. Introduction

The review of past work on Quaternary birds (Chapter 1) and the theoretical issues (Chapter 2), have helped to highlight topics in need of further investigation, and taxa which would best suit their resolution. This chapter explains the geographical, temporal and taxonomic scope of the volume, and describes the specific aims of the investigation in terms of questions to be addressed.

3.2. Geographical scope

The geographic area covered by this volume, the Western Palaearctic, was partly dictated by the location of the institution where the analysis was carried out i.e. University College London in England. British material dominated the fossil samples analysed during the present study. The work was, however, not restricted to British material, either modern or fossil, because it was felt that this would give too a narrow a perspective of the range of possible taxa and morphologies which could be present today and in the past. In the light of the geographically based intraspecific variation discussed in Chapter 2, it was felt that widely spread samples of modern skeletons should be included in the analysis. To this end a north-south transect of European samples was analysed and institutions from Sweden to Spain were visited, via Poland, Denmark, Britain, Belgium and France.

3.3. Temporal scope

The temporal scope of the project was largely led by the availability of fossil material. The oldest material examined during the project was Middle Pleistocene in age (Cromerian *sensu stricto*) from West Runton in Norfolk. This is also the oldest deposit in Britain to yield a large number of bird fossils. The upper end of the temporal scale is the present day. Modern samples formed the baseline, of known taxonomic affinity, with which the fossils were compared. All available material was examined and the geological age range varied for each taxon. Most of the crane (*Grus* sp.) material was Holocene, most of the ptarmigan and grouse (*Lagopus* sp.) samples were Late Pleistocene, the bulk of the starling (*Sturnus* sp.) material was from oxygen isotope stage 5, while the raven (*Corvus corax* and ancestors) material was more evenly spread through time.

3.4. Taxonomic scope

The major criteria which influenced the eventual decision over which taxa would be studied were, first, that a taxon had to be relatively well represented in the fossil record; second, that sufficient widely spaced modern comparative specimens were available; and third, that there appeared to be some potential for meaningful results. The last criterion would be satisfied if previous work, in the literature, had indicated that fossil material did not conform in morphology to modern taxa and that the difficulties in identification of bones would not render such a study meaningless. For instance, while a variety of anatid (duck) species have been named in the Pleistocene of Europe (e.g. Jánossy, 1972), the relatively uniform osteology of that group was thought likely to create difficulties in interpretation. Similar problems existed in taxa such as the genus *Turdus* and the families Fringillidae and Charadriidae.

Grus: *Grus* was included in the present study because of the supposed former presence in western Europe of a large extinct species *Grus primigenia* Milne-Edwards (1867-71). Many opinions have existed regarding the identity of this species, although little consideration has been given to *Grus grus* (common crane), whose intraspecific variation has been poorly known. A better understanding of the variation within *G. grus* as well as *G. antigone* (sarus crane) (regularly proposed as the identity of *G. primigenia*) was needed before the identity of *G. primigenia* could be ascertained.

Lagopus: The genus *Lagopus*, with both its present day Western Palaearctic representatives *Lagopus lagopus* (red grouse / willow grouse) and *L. mutus* (ptarmigan), was included because various authors had recognised anomalously proportioned fossils in Quaternary deposits (Newton, 1924a; Mourer-Chauviré, 1975a; Jánossy, 1974a; Bochenski, 1974, 1985, 1991; Harrison, 1980b; Potapova, 1986; Bochenski and Tomek, 1994). In some cases these differently proportioned fossils were given separate taxonomic status, while in others they were not.

Corvus corax: The species *Corvus corax* (raven) was included because of the extinct late Middle Pleistocene ancestor *C. antecorax* named in France (Mourer-Chauviré, 1975a). *C. corax* has a wide distribution today and is polytypic, with seven subspecies in the Western Palaearctic alone. This implies that *Corvus antecorax* needed to be reconsidered and compared with a greater range of modern ravens than had previously been achieved.

Sturnus: Large starlings were recognised as present in the deposits of Tornewton Cave in Devon (Harrison, 1980b). However, no separate taxonomic status was attached to these fossils. First, it was felt that they should be confirmed as *Sturnus* (starlings) and not *Turdus* (thrushes), which are similar in osteological morphology (Stewart, 1992). Second, comparison of the fossils was necessary with a range of modern *Sturnus* samples, including the three extant European species.

3.5. Questions

A variety of specific questions regarding the taxa chosen for study arise, partly informed by the first two chapters:

***Grus*:**

- What is the pattern of size and shape in modern members of the genus *Grus*, including both *G. grus* and *G. antigone*?
- How sexually dimorphic are the species within *Grus* (and why)?
- What is the pattern of size and shape in fossil *Grus* in Europe?
- Is *Grus primigenia* a genuine extinct species or is it conspecific with one of the modern species?
- Are there any further conclusions to be made on the basis of the present study of *Grus* in the Quaternary of Europe?

***Lagopus*:**

- What is the pattern of morphology in *Lagopus lagopus* and *L. mutus* today throughout their geographical range and what is the explanation for that variation?
- What is the pattern of morphology in the fossil record of *Lagopus lagopus* and *L. mutus* throughout their geographical range and what does it signify?
- Is there any evidence for the timing of divergence of *Lagopus lagopus* and *L. mutus*?
- Are the subspecies of both *L. lagopus* and *L. mutus*, named by Mourer-Chauviré (1975a), valid when compared to modern subspecies?
- Are there any further conclusions to be made on *Lagopus* in the Quaternary of Europe?

***Corvus corax*:**

- What is the geographical variation in the osteology of modern Western Palaearctic ravens and what is the explanation for this variation?
- What variation can be seen in the fossil record and are there any temporal and geographical patterns in that record which may be explained by the modern patterns?
- Is *Corvus antecorax* a genuine species or should it be considered conspecific with *Corvus corax* as claimed by Pierce Brodkorb (1978)?
- Is there any evidence in the fossil record for the timing of the evolution of *Corvus corax*?
- Is there any evidence in the fossil record for the timing of the evolution of the modern subspecies of *Corvus corax*?
- Are there any further conclusions to be made on *Corvus corax* in the Quaternary of Europe?

***Sturnus*:**

- What is the pattern of size and shape in modern members of the genus *Sturnus*?
- What is the pattern of size and shape in fossil *Sturnus*?
- Are the large starlings at Tornewton Cave (Harrison, 1980b) sufficiently different to warrant a distinct taxon or are they representatives of a modern European species of *Sturnus*?
- Are the large *Sturnus* fossils present in any other sites and are there any further mophological patterns in the genus' history?
- Are there any further conclusions to be made on *Sturnus* in the Quaternary of Europe?

A number of general questions arise form the theoretical chapter (Chapter 2) with regard to Quaternary birds as a whole which shall be addressed using the study of the four chosen taxa. They are as follows:

- How do the different species concepts affect the interpretation of the four taxa?
- Can speciation modes be detected in the fossil record of the four taxa?
- Can evolutionary rates be detected in the fossil record of the four taxa?
- Does a better understanding of intraspecific variation today help interpret the fossils of the four taxa?
- Are there any theoretical issues which help distinguish subspecific from specific differences in the fossil record?

Section II

Materials And Methods

Chapter 4

Quaternary Chronological Framework

4.1. Introduction

This volume aims to examine the evolution of four bird taxa through the Quaternary in Europe. Due to this temporal perspective, it is necessary to outline the current views of the relative ages of sites where the taxa have been found, and to discuss the subdivisions of the Quaternary to which they belong. This chapter is included to provide the background to those views and to give a correlation across Europe, where different chronologies are, or have, been used. In Section 3, where the fossils are analysed, the statigraphic context and correlations of the sites are discussed. However, a broader setting for these sites is needed and follows here.

The Quaternary Era, including both the Pleistocene and Holocene epochs, started some 1.6 million years BP, and lasts until the present day. The boundary between the Pleistocene and Holocene is 10 000 radiocarbon years BP, and marks the end of the last glaciation. The period covered by this volume extended from the early Middle Pleistocene (*ca.* 600 K yr BP) up to and including the present day. Until recently, this period in Britain was thought to consist of four interglacials (Cromerian, Hoxnian, Ipswichian and Flandrian (or Holocene)), separated by glacials (Anglian, Wolstonian and Devensian) (Mitchell *et al*., 1973). The Mitchell *et al.* (1973) chronology was based on local and regional lithostratigraphic sequences which were then correlated by methods including pollen analysis. The method of correlation using pollen was based on the premise that the interglacials possessed distinctive vegetational sequences which would be reflected in their fossil pollen records. Glacials were identified by their lithostratigraphic relationship to interglacial deposits.

Interglacials, at their heights, were characterised by high sea-levels, temperate climates and climax deciduous woodlands. Glacials, on the other hand, had low sea-levels, cold climates and landscapes varying from ice sheet coverage to steppe-tundra or even boreal forests. In more detail, the interglacials are made up of vegetational sub-stages recognised from pollen zones representing the gradual warming and eventual cooling of the climate. The interglacial vegetational sequence has been thought to progress from a boreal type pine (*Pinus*) and birch (*Betula*) dominated chronozone (zone I) through mixed deciduous woodland chronozones (zones II and III), back to a boreal chronozone (zone IV) (Turner and West, 1968). Variations in these vegetational sequences were believed to distinguish the different interglacials from each other; for instance the Ipswichian stage III was dominated by hornbeam *Carpinus* (Phillips, 1973), a feature not thought to be found in other interglacial pollen sequences. However, it should be stated that rarely were all the pollen zones represented at individual sites. This signifies that many of the interglacial pollen sequences were composite sequences produced from a number of sites. The composite nature of these interglacials led to problems of correlation and emphasised the lack of continuous terrestrial sequences in the British Quaternary.

A publication which appeared the same year as Mitchell *et al.* (1973) was that by Shackleton and Opdyke (1973). They produced the oxygen isotope curve, derived from deep ocean sediments, which showed that the Mitchell *et al.* (1973) scheme is a simplification (Table 4.1). The oceanic record is thought to be more reliable in chronicling the passage of time because breaks in the sequence of sedimentation are less likely than in terrestrial sequences. The oxygen isotope curve is derived from cores through ocean sediments from which the shells of foraminifera (forams) are extracted. Forams are then analysed to gauge the chemical composition of their skeletons, proportions of the two isotopes of oxygen O^{16} and O^{18} are calculated, and ratios of these isotopes through the core are used to construct a curve. The curve is believed to represent climatic fluctuations because O^{16} evaporates from the oceans at a proportionately greater rate than O^{18}. As climates cool and more water is trapped in the ice caps the oceans become slightly enriched in O^{18}. The oxygen isotope curve confirmed what was already suspected, that more interglacials had existed than were allowed for in the Mitchell *et al.* (1973) chronology and hence that differently aged sites were being lumped together in the same interglacials. It had mainly been mammalian palaeontologists who had previously questioned the pollen based scheme.

The following sections describe the period covered by this volume starting with the oldest interglacial.

4.2. British Middle Pleistocene

4.2.1. The 'Cromerian'

The recent debate about the status of the Cromerian Interglacial was complex because it has involved correlation between Britain and continental Europe, and in particular the Netherlands (Gibbard *et al*., 1991 and Turner, 1996). The Dutch "Cromerian Complex", has a series of 4 cycles of warm and cold, while the British sequence initially had but one, represented by the Freshwater Bed at West Runton (West, 1980). Indications that sites representing more than one pre-Anglian interglacial had been referred to the British Cromerian, came from three main sites. At Westbury-sub-Mendip, Somerset, a cave was found to contain a fauna thought to post-date the Cromerian *sensu stricto* but pre-date the Anglian Glaciation (Bishop, 1975, 1982). This caused a rethink over the age of a site long known in Ostend, Norfolk which is now also believed to belong to this as

Table 4.1. Pleistocene Chronology: Early Middle Pleistocene to the Holocene (British terrestrial, European terrestrial, Alpine and Oceanic chronologies). Modified from Roberts *et al.*, 1995.

Conventional British stages	Modified scheme	British sites	Dutch/European sequence	Alpine sheme	OIS
Flandrian					1
Devensian	Devensian	**Four Ashes**	Weichselian	Würm	5d - 2
Ipswichian	Ipswichian *sensu stricto*	**Bobbitshole**, Trafalgar Square	Eemian	Riss - Würm	5e
These stages not recognised	Cold stage		Saalian complex	Riss	6
	Temperate / Interglacial stage	Marsworth (part), Ilford, Stanton Harcourt			7
	Cold stage				8
	Temperate / Interglacial stage	Purfleet, Grays Thurrock			9
Wolstonian	Cold stage				10
Hoxnian	Temperate / Interglacial stage	**Hoxne**, Swanscombe, Barnham, Clacton	Holsteinian	Mindel-Riss	11
Anglian	Anglian	Westbury-sub-Mendip (part)	Elsterian	Mindel	12
Cromerian	Temperate / Interglacial stage	Boxgrove, Westbury-sub-Mendip (part), Ostend	"Cromerian complex" (includes 4 interglacials and 3 glacials).		13 to 17
	Cold stage				
	Cromerian *sensu stricto*	**West Runton**			

Type-localities of conventional British stages (Mitchell *et al.*, 1973) in **bold**.

yet unnamed interglacial (Currant, 1989). Ostend had the additional advantage over Westbury that the deposits were stratified below the Anglian glacial till, further corroborating the age of this fauna (Stuart and West, 1976). Finally, the archaeological site of Boxgrove in Sussex yielded yet another similar fauna (Roberts, 1986; 1993). The biostratigraphic evidence for this interglacial is varied, but perhaps the most celebrated among them is the water vole *Arvicola terrestris cantiana,* which succeeds its presumed ancestor *Mimomys savini*, found at West Runton. *A. t. cantiana* sets these sites apart from Cromerian *sensu stricto* sites, but does not necessarily place the period to which they belong before the Anglian, because the vole survived into the succeding warm stage and beyond. The presence of the rhino *Stephanorhinus hundsheimensis*, Savin's shrew *Sorex savini* and, at Boxgrove, the giant deer *Megaloceros dawkinsi*, all of which are not known after the Anglian, provide the upper limit on the age this period (Currant, 1989; Roberts, 1993). Arguments over this biostratigraphic evidence have centred on whether *A. t. cantiana* could have evolved during a single interglacial and so sites with *M. savini* and *A. t. cantiana* represent periods earlier and later in the Cromerian *s. s.* (Stuart and West, 1976). Consensus is rapidly falling in favour of two separate interglacials which had been lumped together under the name Cromerian. A postscript to this is that Westbury-sub-Mendip may have evidence for more than one interglacial, or at least warm stage, post-dating the Cromerian *s. s.*, and predating the Anglian (Stringer *et al.*, 1996). This, together with work on molluscan assemblages (Meijer and Preece, 1996), could be the first evidence for an equivalent complex to that seen in the Netherlands, and require that Table 4.1 would need further amending.

4.2.2. The 'Hoxnian'

Most recently, the Hoxnian has come under scrutiny. The original recognition of this interglacial was also based on lithostratigraphy and pollen analysis (West, 1956). Indications that the sites placed in this interglacial might not be synchronous came from work on the Thames terrace sequence (Bridgland, 1988, 1994; 1995). There are four terraces which date from after the diversion of the Thames by Anglian ice. The river once ran north-east through the Valley of St. Albans, but was diverted to its present course by the Anglian Glaciation (Bridgland, 1988). If the number of terraces accurately records the number of interglacial / glacial cycles which have taken place, there is an extra interglacial, as yet unaccounted for, in the sequence (Table 4.1). River terraces are believed to form by the downcutting of a river during the middle of each glacial and so each terrace represents a whole interglacial / glacial cycle (Bridgland and Allen, 1996). The sites which appear to belong to this interglacial are Purfleet and Grays Thurrock. This interglacial would appear to correspond to OIS 9; while

Swanscombe, another Thames site which appears to belong to the preceding warm stage, is in OIS 11, the Hoxnian *sensu stricto* (Roberts *et al.*, 1995 and Bridgland 1988, 1994; 1995). This view, it must be stressed, has its dissenters. Gibbard (1995) believes that the correlations between Thames terraces and the oxygen isotope curve are premature. Indeed there is great scope for revision of the present scheme as a large degree of circularity is likely to be present in the new predominant litho-biostratigraphic model of Britain largely based on Bridgland's Thames terrace sequence.

4.2.3. The 'Ipswichian'

The first controversy in the correlation between the oceanic and terrestrial records concerned various sites thought to be Ipswichian in age, in particular Trafalgar Square and Ilford in the Thames Basin (West, 1976; Sutcliffe, 1976; Sutcliffe and Kowalski, 1976). These sites are very different in terms of their mammalian faunas. Trafalgar Square has a fauna with *Hippoppotamus* and no horse while Ilford includes a mixture of apparently temperate (*Palaeoloxodon antiquus* and *Stephanorhinus hemitoechus*) and boreal (*Mammuthus primigenius*) animals and includes horse. The reason for this faunal difference, suggested by palynologists and others, was that the faunas derived from different parts of the Ipswichian (Stuart, 1976). Conversely, there were arguments from stratigraphy to support the case for distinct interglacials, the sites of Trafalgar Square and Ilford being interpreted as within two different Thames terraces. The debate concerning status of sites said to be Ipswichian lasted for some time and the consensus is now that two interglacials are represented by these two sites, which correspond to oxygen isotope stages 7 and 5e for Ilford and Trafalgar Square respectively. An important site to give further weight to this argument has been Marsworth in Buckinghamshire where the characteristic mammalian faunas of both interglacials were found in superposition, separated by a cold climate deposit (Green *et al.*, 1984). This sequence of climatic events is almost universally accepted today, although in a recent text the two alternative interpretations are described (Jones and Keen, 1993).

4.2.4. The 'Wolstonian'

To further complicate matters, there has been a reinterpretation of the deposits thought to be of Anglian and Wolstonian age. In particular, it was demonstrated that sediments believed to be lateral equivalents of the Baginton - Lillington Gravels, which are at the base of the Wolstonian type section, are found beneath Anglian glacial till in East Anglia (Rose, 1987). The Wolstonian had previously been thought to represent a glaciation between the Hoxnian and Ipswichian interglacials. There is now, therefore, no good evidence for a severe glaciation between the Anglian and Devensian in Britain.

4.3. Continental chronologies

Table 4.1 shows the relationship between the different chronological schemes for the early Middle Pleistocene until the present. It includes those discussed above as well as ones in use in mainland Europe: both the Alpine and northern European chronologies. The Alpine system, while out of date, is still used by some authors (e.g. Mourer-Chauviré, 1993) and so has been included. Correlation of the Alpine scheme with those of northern Europe, Britain and the oxygen isotope curve is difficult because the Alpine and north European schemes recognise fewer warm and cold stages.

4.4. The Late Pleistocene

Late Pleistocene sites (OIS 5 to 2) require further description, as more detail is known. As already explained, the last interglacial (OIS 5e), the Ipswichian / Eemian, has now been distinguished from previous interglacials (OIS 7). However, the oxygen isotope curve reveals two further warm stages, OIS 5c and 5a, subsequent to stage 5e. The existence of these additional stages in the deep sea core may be causing problems of correlation between the OIS curve and the terrestrial sequence, although certain sites have been implicated as recording parts of the detail of OIS 5 (Currant, 1996) (Table 4.2). Of greatest interest here are the Gower Peninsula sites of Bacon Hole and Minchin Hole. Bacon Hole possesses a sequence which probably runs from OIS 6 through to 5c or 5a (Chapter 12). The deposit considered to be of OIS 6 age has a mammalian fauna which includes large northern vole *Microtus oeconomus* and small horse *Equus* sp. Many sites are now known with this characteristic fauna and are believed to securely date from OIS 6 or the transition between 7 and 6 (Parfitt and Owen in Bates *et al.*, 2000). Next in the sequence of Bacon Hole there is a series of deposits with a temperate mammalian fauna including wood mouse *Apodemus sylvaticus* and bank vole *Clethrionomys glareolus*; as well as Cory's shearwater *Calonectris diomeda*, which today does not breed as far north as the Gower coast (Sutcliffe, *et al.*, 1987; Harrison, 1977a). These deposits are correlated with the "Patella Beach" of nearby Minchin Hole which is believed to represent a high sea level. This is thought to be the climatic optimum of OIS 5e, dated by Uranium Series, in Minchin Hole, at between 107 and 127 Kyrs. This correlates well with the date derived from the *Hippoppotamus* fauna at Victoria Cave in Yorkshire (Gascoyne *et al.*, 1981). The Gower sediments, correlated with 5e, lack hippo but are thought to belong to the suite of Stage 5e '*Hippopotamus*' sites including Joint Mitnor in Devon, Barrington in Cambridgeshire and Trafalgar Square in London. However, there are higher temperate deposits in Bacon Hole which are separated from the OIS 5e deposits by a unit which represents a cooler climate. These deposits are believed to represent one of the other warm stages of OIS 5 and are no younger than 81 Kyrs. old (Stringer *et al.*, 1986). In the strict sense, being post-Ipswichian, these

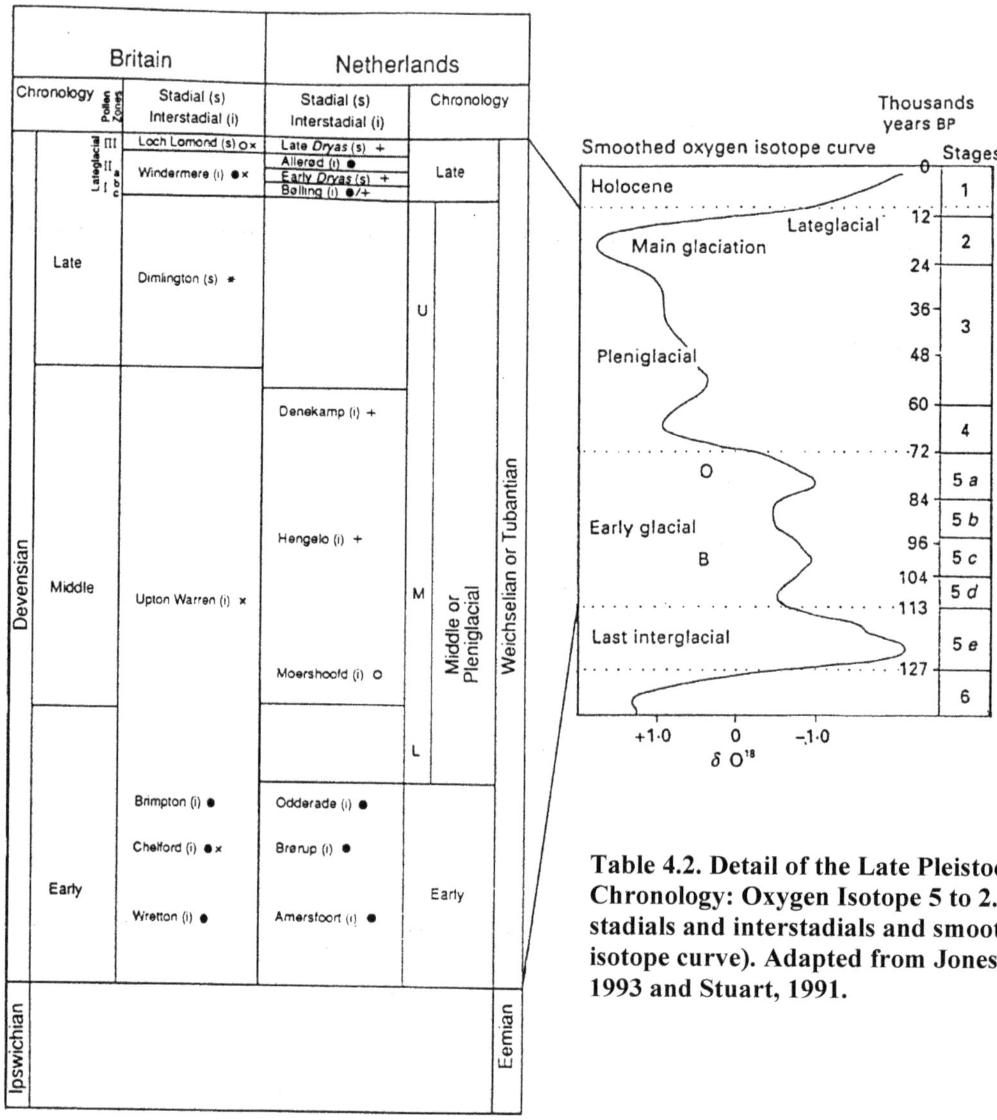

Table 4.2. Detail of the Late Pleistocene Chronology: Oxygen Isotope 5 to 2. (Includes stadials and interstadials and smoothed oxygen isotope curve). Adapted from Jones and Keen, 1993 and Stuart, 1991.

deposits are early Devensian, although Currant and Jacobi (1997) have recently confused matters by suggesting that the Devensian should start after OIS 5. An alternative view is given by Stuart (1991), based on Guiot *et al.* (1989) and Behre (1989), who correlate OIS 5c and 5a with the Brørup and Ooderade interstadials respectively of the continental sequence. The latter is, however, completely at odds with the possible ages, in years, of those interstadials given in Jones and Keen (1993). Interestingly, Behre's paper indicates that while apparently relatively warm, the Brørup and Ooderade interstadials do not possess the vegetations of interglacials, i.e. mixed deciduous woodlands. This may be difficult to reconcile with mammals thought to be living in such a warm stage on the Gower coast, such as the rhino *Stephanorhinus hemitoechus* although of the two *Stephanorhinus rhinos* it is the browser. (Sutcliffe, *et al.*, 1987).

The problem with correlating the climatic oscillations of the oxygen isotope curve with the terrestrial sequence, and pollen based terrestrial sites with cave sites, is that there are no completely reliable means of doing so. No absolute dates are available for the pollen sites at this period, and lithostratigraphy is not applicable to correlating between caves and open air sites. Finally, there are no open sites in OIS 5c or 5a with mammals, and no caves with meaningful pollen at this time.

According to the oxygen isotope curve (Table 4.2) the next stages of the Devensian (OIS 4 and 3) are considerably colder albeit variable. Again there are problems in the correlation of sites with the various oscillations in climate, and a number of interstadials and stadials have been named (Behre, 1989). The chief problem is that the much of the period is beyond the reasonable application of C^{14} dating. An attempt has been made by Currant and Jacobi (1997) to establish a chronology based on mammalian assemblages from British cave sites. However, without any independent absolute or relative dating this scheme remains completely untested. In essence these sites, although Devensian, have not been dated more accurately.

Table 4.3. Detail of chronology of the Late Glacial through to early Flandrian. From Barton et al. (1991).

Radio-carbon years BP	Pollen zones (after Iversen 1954)	NW Europe chrono-zones (based on Mangerud *et al* 1974)	NW Europe climato-stratigraphic units (after Lowe & Gray 1980)	British biozones (based on Gray & Lowe 1977; Coope & Pennington 1977)
	IV	PREBOREAL	FLANDRIAN INTERGLACIAL	FLANDRIAN
10,000			TRANSITION	
10,500	III	YOUNGER DRYAS		LOCH LOMOND STADIAL
			YOUNGER DRYAS STADIAL	
11,000				
11,800	II	ALLERØD	TRANSITION	WINDERMERE INTERSTADIAL
	Ic	OLDER DRYAS		
12,000	Ib	BØLLING	LATE GLACIAL INTERSTADIAL	
13,000	Ia			
			TRANSITION	
14,000		WEICHSELIAN STAGE	LATE DEVENSIAN/MAIN STADIAL	LATE DEVENSIAN

Immediately after this was the Glacial Maximum, from 23 – 16 k BP, from which time there are no dated bird fossils in Britain. This is either because conditions were too harsh for birds to live in significant numbers, or because of the lack of suitable deposits which would preserve avian material. The last major phase of the Pleistocene, the Late Glacial, includes some of the best dated samples studied here. This is largely because C^{14} dating is appropriate and has been widely used, so the Late Glacial is the best understood part of the Pleistocene. The dates which follow are in uncalibrated radiocarbon years BP.

The Late Glacial itself begins at approximately 16000 BP and ends at 10000 BP. Overall the period is one of deglaciation and warming in Britain and Europe; however, there was a period of climatic reversal from *ca.* 11000 to 10000 BP. The period of reversal is known as the Younger Dryas, and climatic deterioration most strongly affected areas nearest the Atlantic seaboard, being less marked further east and south. The Younger Dryas means that the preceding period of warming can be regarded as a 'false start' to the Holocene, and is known as the Late Glacial Interstadial. This interstadial comprises the Bølling *ca.* 13700 - 12200 BP, the Older Dryas *ca.* 12200 - 12000 BP and the Allerød *ca.* 12000 - 11000 BP (Table 4.3). The Older Dryas was probably drier rather than colder than either the Bølling or the Allerød (Kolstrup, 1991).

Evidence from radiocarbon-dated beetles faunas in Britain has enabled reconstructions of temperatures during the Late Glacial (Atkinson *et al.*, 1987). This work has shown that the initial warming trend took place quickly, while the cooling was slower. Temperatures went from about –8°C at *ca.* 14000 BP to 8°C at *ca.* 13000 BP, then around 12500 BP temperatures declined, steepening at *ca.* 11000 BP, to –7°C at *ca.* 10400 in the Younger Dryas. This model, while essentially correct, has however been shown to be oversimplified, as minor climatic fluctuations are superimposed on these overall trends as seen in the Greenland GRIP ice core record (Lowe *et al.*, 1995).

The vegetation during the Late Glacial has been studied using pollen diagrams which must, however, be interpreted with caution. This is because they over-represent the pollen from wet areas and hence will under-represent dry steppic grassland habitats (Kolstrup, 1991). Nevertheless, the general view of vegetation during this period, over much of Europe, has been summarised by Huntley and Birks (1983). At 13000 BP Fennoscandia was ice covered, while the rest of Europe was vegetated by tundra in the north and steppe in the south. Trees were confined to southern Iberia (deciduous woodland) and western Russia (boreal woodland). The Fennoscandian ice retreated from 13000 BP onwards, and a progressive invasion by forests from the east took place, wedging between the shrinking northern tundra and southern steppe. The tundra expanded again briefly during the Younger Dryas at *ca.* 10500 BP. At 10000 BP only the extreme north-west and south-east of Europe remained unforested and deciduous trees moved north into southern England, displacing the boreal type forests completely by about 9000 BP.

^{14}C *years BP*	STAGE	*England and Wales* Blytt – Sernander periods	*England and Wales* Godwin (1940a) Pollen zones	*England and Wales* West (1980b) Chronozones	*Scotland* West (1977a) Pollen zones
0	FLANDRIAN				
1000		Sub-Atlantic	VIII		
2000					
3000		Sub-Boreal	VIIb	Fl III	VIII–VIIb
4000					
5000		Atlantic	VIIa		
6000					VIIa
7000				Fl II	
8000		Boreal	VI c b a		V–VI
9000			V	Fl I	
		Pre-Boreal	IV		IV
10 000					

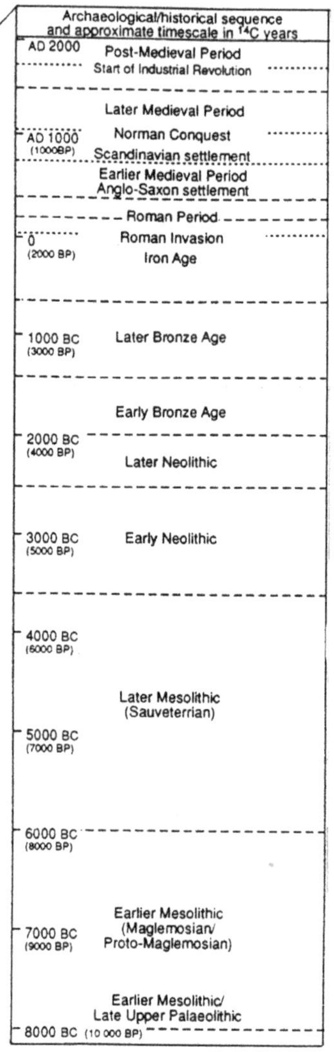

Table 4.4. Detail of the chronology from the early Flandrian through to present day. (Inclusdes archaeological technologies and radiocarbon dates). Adapted from Jones and Keen (1993).

4.5. The Flandrian or Holocene

The Flandrian, or Holocene, started some 10000 radiocarbon years ago and is unlike previous interglacials of the Pleistocene in not being based on a type site, which may not matter as correlation of sites can easily be achieved by means of radiocarbon dates (West, 1979). Pollen zones for the Flandrian temperate stage have been established by Godwin (1940) and West (1980) for England and Wales, and West (1977) for Scotland, and these differ significantly from one another (Table 4.4). As pointed out by Turner and West (1968), it is the detailed knowledge of regional and local diversity that has produced these contrasting zonation schemes. The most widely used scheme for the Holocene in continental Europe is the Blytt - Sernander scheme (Blytt, 1876 and Sernander, 1908). This derived from studies of plant macrofossils in Scandinavia (Table 4.4).

At the beginning of the Holocene, temperatures rose by 1.7°C per century, and maximum summer temperatures were attained between 9000 and 8000 BP (Kutzbach and Guetter, 1986). This is not universally accepted and some consider the Climatic Optimum to have occurred later, at *ca.* 7000 BP (Lamb, 1977). Osborne (1976) postulated, on the basis of fossil insects, that after the Climatic Optimum major climatic trends towards warmer temperatures occurred at *ca.* 5000 BP and 3000 BP. Between 3000 and 2300 BP there was cooling of about 2°C average temperature in Britain, accompanied by an increase in precipitation. Then from 2350 BP to 700 BP there was a warmer, drier phase which was eventually succeeded by the period known as the Little Ice Age between *ca.* 500 BP and 300 BP. The Little Ice Age saw temperatures colder than at any time since the end of the Devensian. After that time there has been an overall amelioration in climate until the present day.

4.6. Conclusion

The above discussion emphasises the wealth of current work that is being carried out to further elucidate the chronological sequence in Britain and Europe during the Quaternary. Further details of the climatic and vegetational changes are also being constantly filled in, providing a rich and varied backdrop to the study of avifaunal change studied here.

Chapter 5

Materials

5.1. Introduction

This chapter describes the modern and fossil samples used in the present study. The crane *Grus* sp., ptarmigan / grouse *Lagopus* spp., raven *Corvus corax* and starling *Sturnus* spp. sample locations are illustrated in Figures 5.1 to 5.4. Meanwhile Tables 5.1 and 5.2 show the relative ages of fossil samples. The samples shall be described in greater detail in the chapters dealing specifically with the respective taxa (Chapters 8 to 11). This section illustrates the general distribution of the samples in relation to the geographic and temporal scopes described in Chapter 3.

Wherever possible skeletal material used, whether modern or fossil, was adult. This was because the additional variation seen in juvenile or immature bird skeletons, both in terms of morphology and measurement would complicate the analysis. Juvenile or immature individuals were identified by the state of ossification seen in the bones. Incompletely ossified bones have a tendency to be grainy in texture and the articular ends may in the youngest individuals appear to be incompletely formed and ill defined.

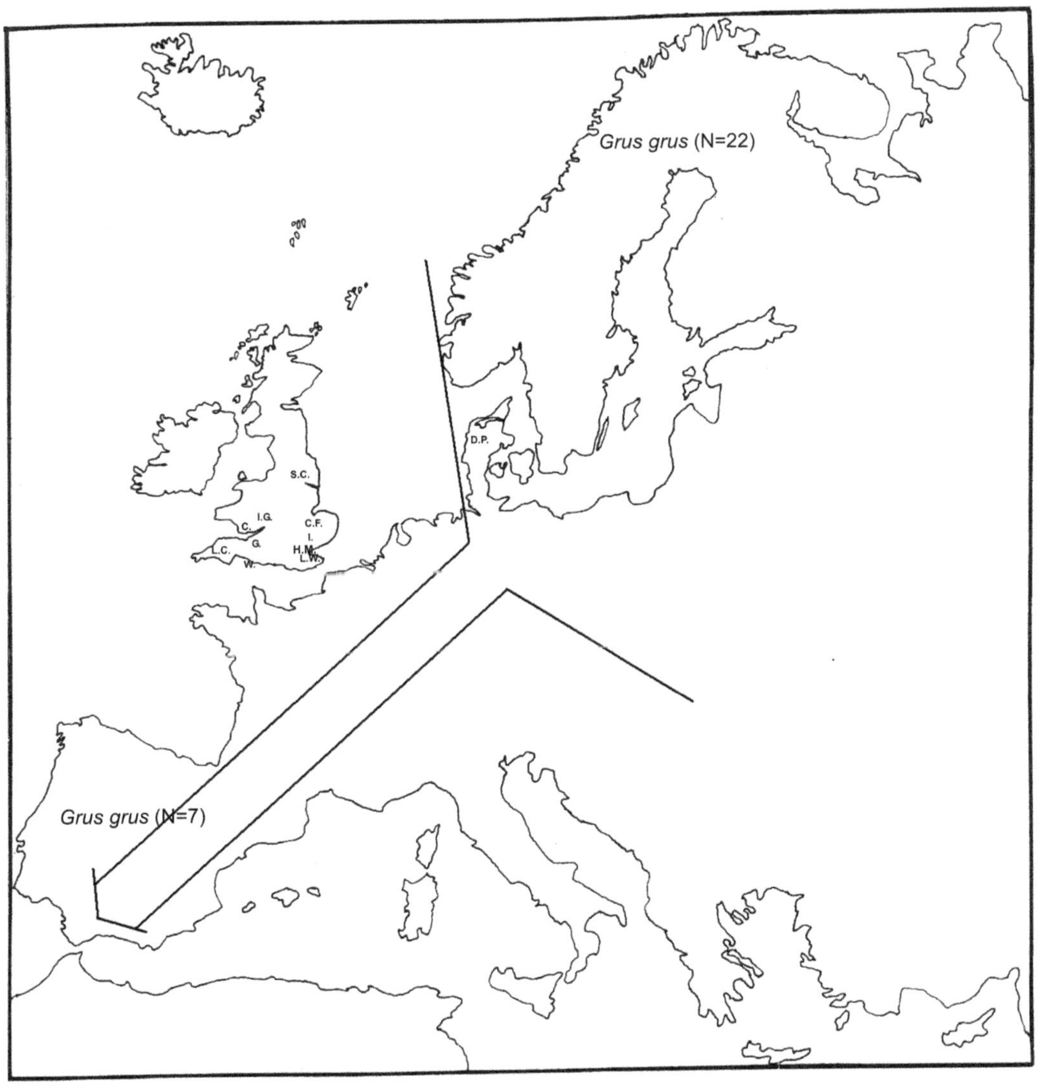

Figure 5.1. Distribution of modern and fossil European *Grus* sp. samples. Also shown are the dominant migratory corridors of *Grus grus* today. Note – Sarus crane *G. antigone* samples are not shown on the map because they came from zoos.

Key: S.C. –Star Carr, D.P. –Danish Peat, C.F. –Cambridgeshire Fens, L.W. –London Wall, I. – Illford, H.M. –Hackney Marshes, W. – Woodbury, G. – Glastonbury, L.C. – Launceston Castle, C. – Caerleaon, I.G. – Isca Grange.

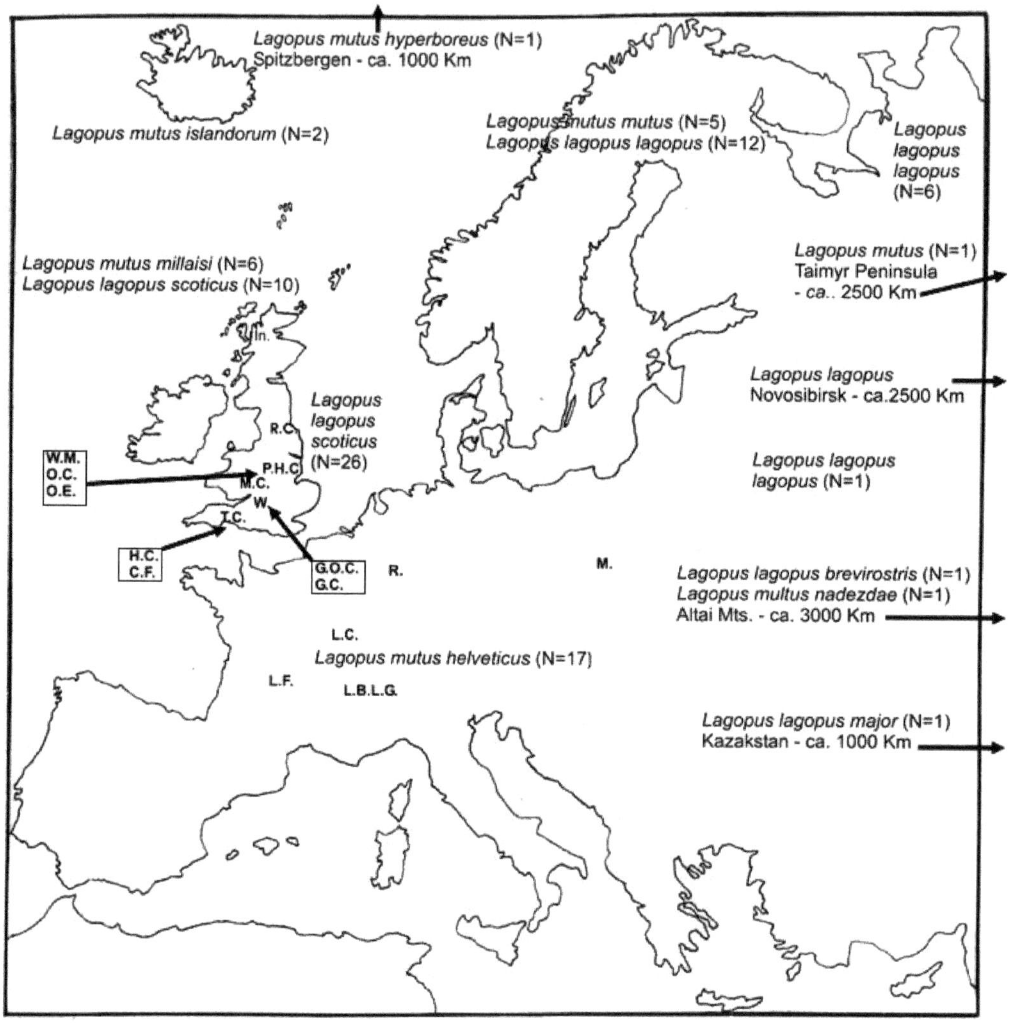

Figure 5.2. Distribution of modern and fossil European *Lagopus* sp. samples.
Key: In. – Inchnadamff, R.C. – Ravencliffe Cave, W.M. – Wetton Mill Rockshelter, O.C. – Ossum's Cave, O.E. – Ossum's Eyrie, P.H.C. – Pin Hole Cave, M.C. – Merlin's Cave, W. – Westbury-sub-Mendip, T.C. – Tornewton Cave, H.C. – Hoppaway Cave, C.F. – Chudleigh Fissure, G.O.C. – Gough's Old Cave, G.C. – Gough's Cave, R. – Remouchamps, L.C. – La Colombière, L.F. – La Fage, L.B.L.G. – La Balme des Grottes, M. - Mamutowa

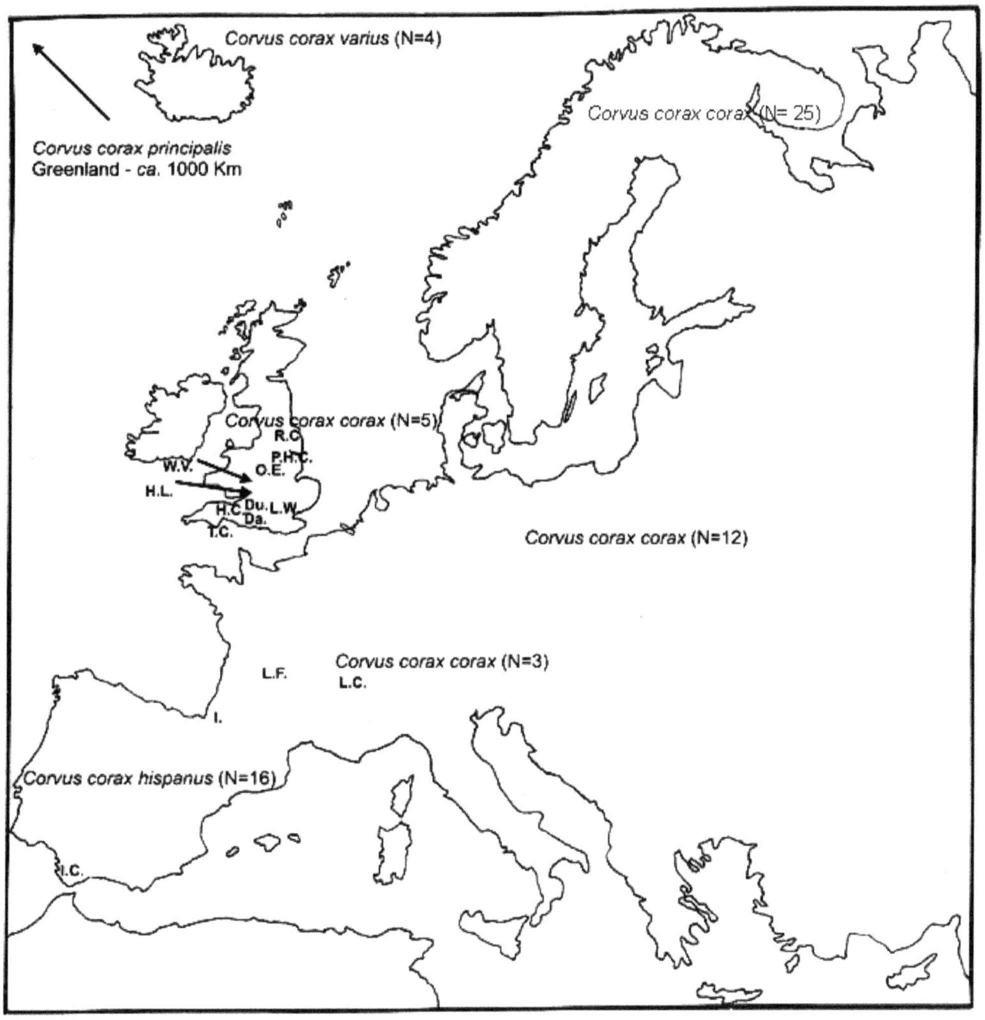

Figure 5.3. Distribution of modern and fossil European *Corvus corax* samples.
Key: R.C. – Ravencliffe Cave, O.E. – Ossum's Eyrie, P.H.C. – Pin Hole Cave, T.C. – Tornewton Cave, H.C. – Happaway Cave, L.C. – La Colombière, L.F. – La Fage, Da. – Danebury, Du. – Dunstable, H.L. – Hangong Langford, L.W. – London Wall, W.V. – Witcombe villa, I.C. – Ibex Cave, I – Grotte d'Isturitz.

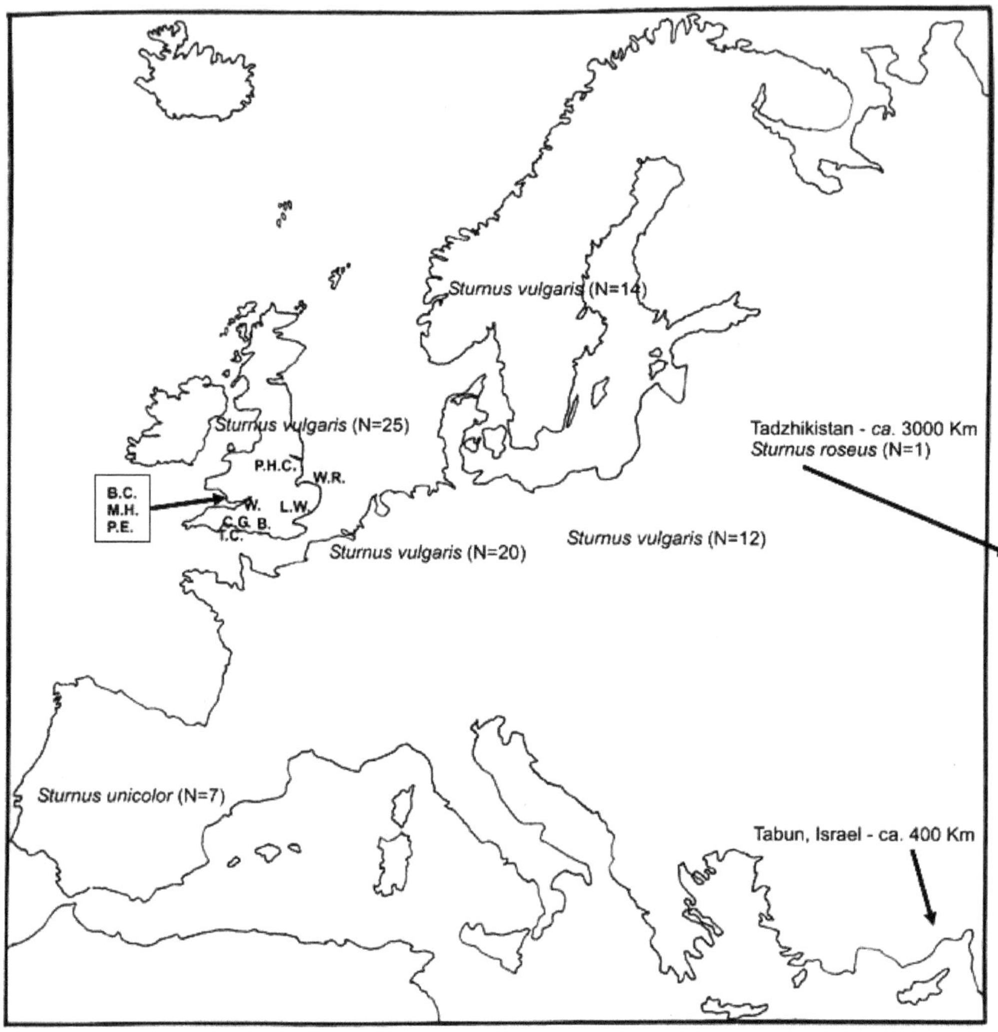

Figure 5.4. Distribution of modern and fossil European *Sturnus* sp. samples. Also shown are the dominant migratory corridors of *Sturnus vulgaris* today.

Key: L.W. – London Wall, P.H.C. – Pin Hole Cave, W. –Westbury-sub-Mendip, T.C. – Tornewton Cave, C.G. – Chudleigh Gorge, W.R. – West Runton, B. – Boxgrove, P.E. – Port Enyon, B.H. – Bacon Hole, M.C. – Minchin Hole.

5.2. Modern Samples

The geographical provenance of the modern samples is illustrated in Figures 5.1 to 5.4. Also noted on the maps are the sample number (N) for each geographical area. A few specimens were not included on these maps because they consisted of single individuals and it was felt that to expand the area covered by the maps would decrease their clarity. Other specimens were not included because they originated as captive birds in zoos. This included all the rose-coloured starlings *Sturnus roseus* and the sarus cranes *Grus antigone*.

Specimens from zoos were included in the analysis by necessity. Ideally these would be replaced by wild individuals. This is because, theoretically, additional or different physical and physiological stresses may be imposed on captive animals. Such stresses are known to cause a variety of pathological conditions which can affect bird bones (Brothwell, 1993). However, this was not possible due to a general lack of comparative specimens in museum collections of certain rarer species.

5.3. Fossil Samples

The fossil material which receives attention in this volume comprises samples of bird fossils from different parts of the Quaternary, and the ages of these samples are known with varying degrees of accuracy. Therefore, with the benefit of the chronology for the Quaternary as delineated in Chapter 4, the samples of the four taxa are described in terms of their relative ages. Table 5.1 lists those which can be most easily placed in the chronology while Table 5.2 gives the probable ages of those which remain problematic. Also detailed in the two tables are the institutions in which the fossil samples are housed.

Table 5.1. Ages of Samples from British Sites and mainland European sites.

Oxygen Isotope Stage	Conventional British Stages	British Sites	European Sites
1	Holocene / Flandrian	Ossum's Eyrie (L, C, 2), Cambridge Fens (G, 7), Glastonbury Lake Dwelling (G, 1), London Wall (G, S, C, 1), Woodbury (G, 1), Whitcombe (C, 1), Danebury (C, 15), Dunstable (C, 11), Hanging Langford (C, 1), Star Carr (G, 14), Launceston Castle (G), Isca Grange (G), Caerleon (G)	Danish Peat Deposits (G, 6)
2		Merlin's Cave (L, 4), Gough's Cave (L, 1), Gough's Old Cave (L, 4), Inchnadamff (L, 5), Ossum's Cave (L, 3), Tornewton Cave (Reindeer Stratum) (L, C, S, 1)	La Balme-les-Grottes (L, 11), La Colombière (C, L, 11), Grotte d'Isturitz (C, 11), Remouchamps (L, 8), Mamutowa Cave (L, 13)
3			
4	Devensian		
5a			
5b			
5c		Tornewton Cave (Hyaena Stratum) (S, 1), Bacon Hole (S, 1), Minchin Hole (S, 1) *	
5d			
5e	Ipswichian		Tabun (S, 1)
6			
7		Ilford (G, 1)	
8			
9			
10			
11	Hoxnian		
12			
13		Westbury-sub-Mendip (L, 1), Boxgrove (1)	
14			
15	Cromerian (*sensu stricto*)	West Runton (1)	

Note: The last three sites in the British Holocene box were not seen but came from literature sources. Also note that the age of the Tabun *Sturnus* fossils is uncertain as different dating techniques have indicated very different ages (see Chapter 11).

Table 5.2. Probable ages of sites whose dates remains obscure.

Sites	Probable ages
Pin Hole Cave (L, C, S, 2)	Holocene or Devensian
Chudleigh Fissure (L, S, 1)	Holocene or Devensian
Happaway Cave (C, L, 1)	Holocene or Devensian
Ravenscliffe Cave (C, L, 1)	Holocene or Devensian
Torbryan Caves (S, L, 1)	Holocene or Devensian
Port Eynon (S, 1)	Holocene or Devensian
Hackney Marshes (G, 1)	Holocene or Devensian
Wetton Mill Rockshelter (L, 3),	Holocene or Devensian
Devil's Tower, (C, 1)	Weichselian
La Fage (L, C, 12)	Rissian (equivalent to O.I.S. 10 – 6)

Keys to Tables 5.1 – 5.2 :-

Taxa:	G - *Grus* L - *Lagopus* C - *Corvus corax* S - *Sturnus*
Museums:	1. Department of Palaeontology, Natural History Museum, London. 2. Manchester University Museum. 3. Stoke-on-Trent Museum. 4. University of Bristol Speleological Society Museum. 5. Geological Survey Museum, 6. Zoologisk Museum Copenhagen. 7. Sedgwick Museum, Cambridge University. 8. Musee Royal d'Art d'Histoire, Bruxelle. 9. Biology Department, U.C.L. 10. Natural History Museum, Tring, Herts. 11. Université Claude-Bernard, Lyon. 12. Muséum d'histoire naturelle de Lyon. 13. Institute of Systematics and Evolution of Animals, Cracow. 14. Zoology Department, Natural History Museum, London. 15. Faunal Remains Unit, Southampton University.

Chapter 6

Mensural Techniques and Statistical Tools

6.1. Introduction

This chapter's role is to describe in detail the measurements taken on the postcranium of the four taxa in this study and the methods by which these measurements have been analysed. The measurements have been described for each individual taxon, which was considered preferable to general descriptions. This is because the most important aspect of measurements is that they can be accurately reproduced, and each taxon requires slightly different techniques.

The first publications to make use of measurements where Quaternary birds are concerned were those by Richard Lydekker (1891) and E.T. Newton (1883). However, these authors did not describe the exact means by which they took the measurements and only a limited number were taken on elements. More recently, von den Driesch (1976) illustrated some general measurements to be taken on bird bones for use in zooarchaeology. The most useful works are those detailing measurements for specific taxa such as Woelfle (1967), Bacher (1967), Erbersdobler (1968), Kraft (1972), Fick (1974), Langer (1980), Otto (1981), Schmidt-Burger (1982), Kelner (1986), Gruber (1990) and Lorch (1992). Of the taxa dealt with in this volume only the genus *Lagopus* has received previous attention (Kraft, 1972).

6.2. Measurements taken

When deciding which measurements to take during this study the foremost subject of concern was whether they are reproducible. Without reproducibility comparative metrical studies are meaningless. Measurements are either defined as "point to point", or are a maximum or minimum in the plane illustrated.

Some measurements require that the bone be twisted between the jaws of the calliper. "Twisting is necessary" is stated after those measurements.

6.2.1. *Grus* species (Figures 6.1 - 6.3)

Humerus (Figure 6.1 A)

Greatest length - GL - Taken from the proximal edge of the caput humeri to the distal edge of the condylus ventralis. Twisting is necessary.

Proximal breadth - Bp - Taken from the dorsal edge of the crista pectoralis to the ventral edge of the crista bicipitalis. Twisting is necessary.

Distal breadth - Bd - Taken from the ventral edge of the epicondylus ventralis to the dorsal edge of the epicondylus dorsalis. Twisting is necessary.

Shaft breadth - SC - Taken across the narrowest part of the midshaft in the plane illustrated in Figure 6.1 A. No twisting.

Ulna (Figure 6.1 B, Bi, Bii)

Greatest length - GL - Taken from the proximal edge of the olecranon to the condylus ventrale. Twisting is necessary

Proximal breadth - Bp - Taken with one jaw of the caliper resting against the ventral edge of the cotyla ventralis and the olecranon and the other jaw against the dorsal edge of the cotyla dorsalis. No twisting.

Proximal diagonal - DP - Taken from the olecranon to the cranial-most edge of the cotyla dorsalis. Twisting is necessary.

Distal diagonal depth - Did - Taken from the labrum condyli to the tuberculum carpale. Twisting is necessary.

Shaft breadth - SC - Taken across the midshaft in the plane illustrated in Figure 6.1 B. No twisting.

Radius (Figure 6.2 C)

Greatest length - GL - Taken from the extreme proximal edge to the extreme distal parts of the radius. Twisting is necessary.

Proximal breadth - Bp - Taken across the proximal end in the manner illustrated in Figure 6.2 C. Twisting is necessary.

Distal breadth - Bd - Taken across the distal end in the plane illustrated in Figure 6.2 C. Twisting is necessary.

Carpometacarpus (Figure 6.2 D, Di)

Greatest length - GL - Taken from the proximal edge of the trochlea carpalis to the distal edge of the facies articularis minor. Twisting is necessary.

Proximal breadth - Bp - Taken from the tip of the process extensorius to the edge of the facies articularis ulnocarpalis in manner detailed in Figure 6.2 D. Twisting is necessary.

Distal diagonal no. 1 - Did1 - Taken across the distal end in the plane illustrated in Figure 6.2 Di. Twisting required.

Distal diagonal no. 2 - Did2 - Taken across the distal end in the plane illustrated in Figure 6.2 Di. No twisting.

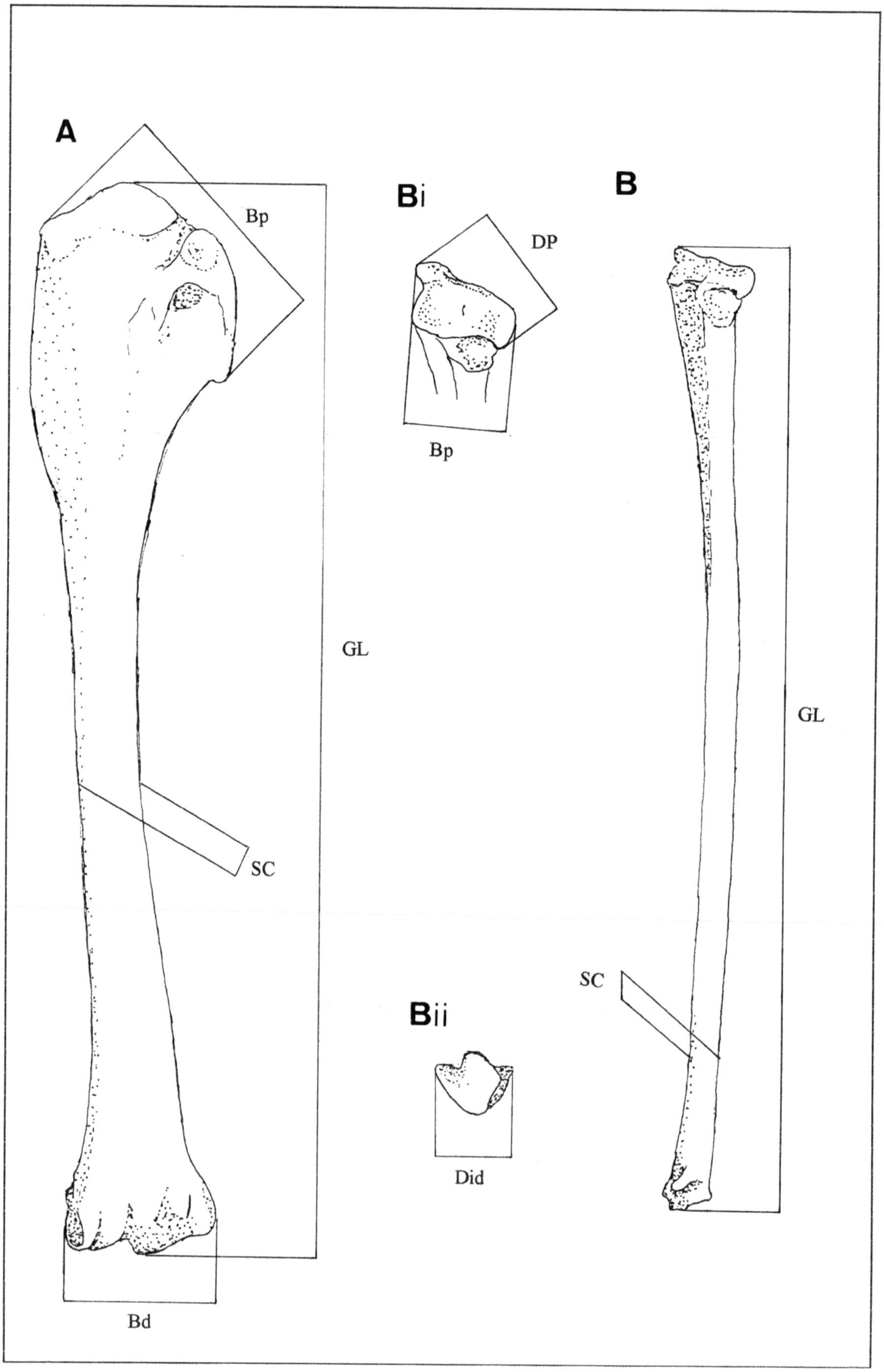

Figure 6.1. Measurements taken on the humerus (A) and ulna (B, Bi, Bii) of *Grus*.

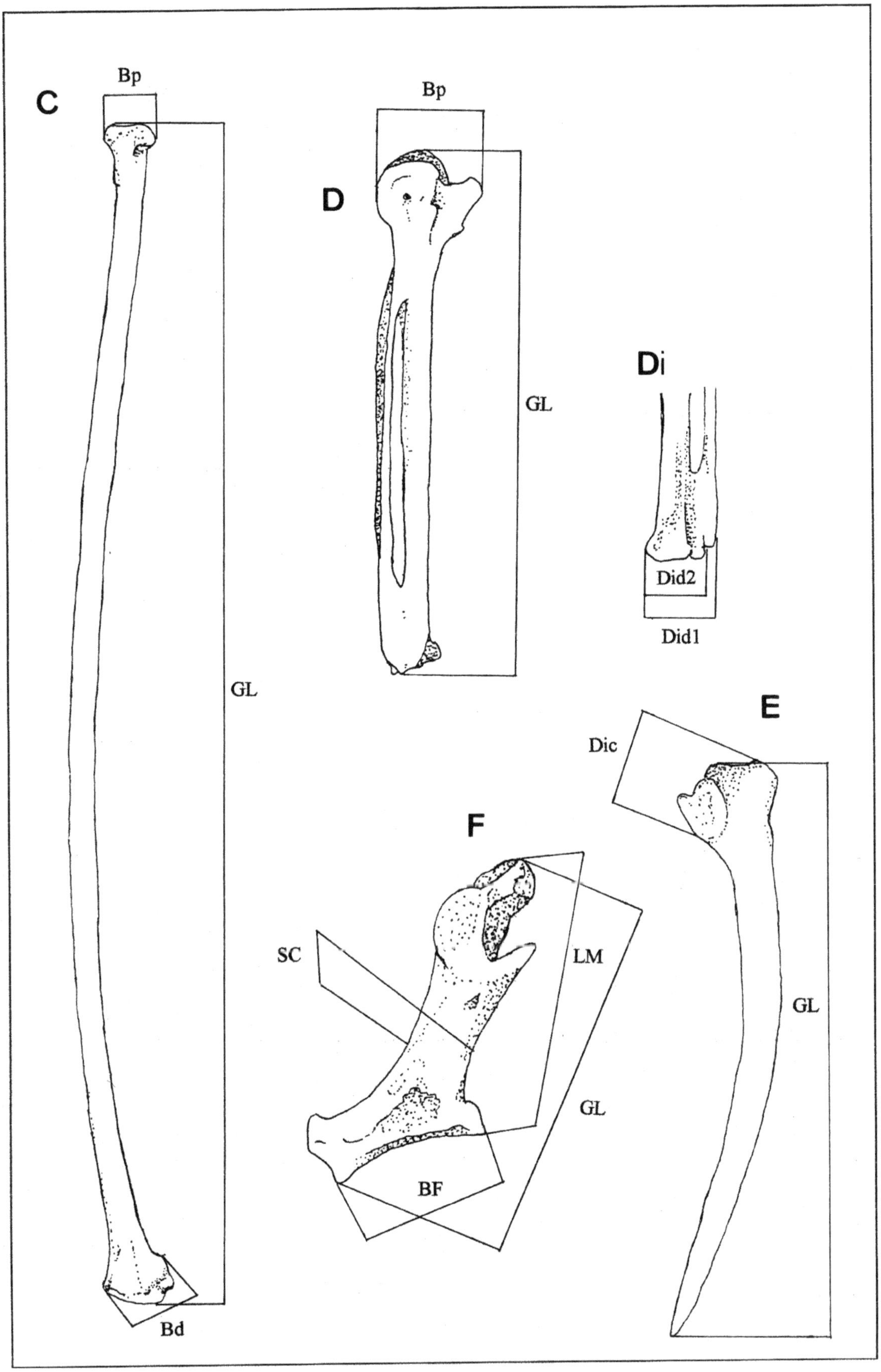

Figure 6.2. Measurements taken on the radius (C), carpometacarpus (D, Di), scapula (E) and coracoid (F) of *Grus*.

Scapula (Figure 6.2 E)

Greatest length - GL - Taken from the from the tuberculum coracoideum to the extremitas caudalis. Twisting required.

Greatest cranial breadth - Dic - Taken from the acromiom to the caudal edge of the facies articularis humeralis as illustrated in Figure 6.2 E. Twisting required.

Coracoid (Figure 6.2 F)

Greatest length - GL - Taken from the process acrocoracoideum to the lateral edge of the facies articularis sternalis. Twisting necessary.

Medial length - LM - Taken from the process acrocoracoideum to the medial edge of the facies articularis sternalis. Twisting necessary.

Breadth of the facies articularis sternalis - BF - Taken from the medial edge to the lateral edge of the facies. Care must be taken here so as not to damage the process lateralis. Twisting necessary.

Shaft breadth - SC - Taken across the midshaft in the plane illustrated in Figure 6.2 F. Twisting necessary.

Femur (Figure 6.3 G)

Greatest length - GL - Taken from the proximal edge of the crista trochanteris to the distal condylus lateralis. Twisting necessary.

Proximal breadth - Bp - Taken from the medial edge of the facies articularis acetabularis to the lateral edge of the trochanter femoris. Twisting necessary.

Proximal depth - Dp - Taken with one jaw against the cranial edge of the crista trochanteris and the facies articularis acetabularis and one jaw against the opposing caudal side of the femur. No twisting.

Distal breadth - Bd - Taken from the epicondylus medialis with one caliper jaw square against that side to the opposite epicondylus lateralis. This measurement does not involve twisting as the flattened medial edge of the distal femur provides a secure rest for the jaw.

Shaft breadth - SC - Taken across the midshaft in the plane illustrated in Figure 6.3 G. Twisting necessary.

Tibiotarsus (Figure 6.3 H, Hi, Hii)

Greatest length - GL - Taken from the proximal edge of the crista patellaris to the distal edge of the condylus medialis. Twisting necessary.

Proximal diagonal depth - Dip - Taken from the medial edge of the facies articularis medialis to the crista cnemialis lateralis. Twisting necessary.

Distal breadth - Bd - Taken from the lateral edge of the condylus lateralis to the cranial edge of the condylus medialis. Jaw flat against the lateral edge and no twisting.

Distal depth - Dd - Taken with one caliper jaw flat against both the condylus medialis and the condylus lateralis and the other jaw against the opposing side. No twisting.

Shaft breadth - SC - Taken across the midshaft in the plane illustrated in Figure 6.3 H. Twisting necessary.

Tarsometatarsus (Figure 6.3 I, Ii)

Greatest length - GL - Taken from the eminentia intercondylaris to the trochlea metatarsi tertii. Twisting required.

Proximal breadth - Bp - Taken from the lateral to medial edges of the proximal end. No twisting. Make sure that the prominence indicated in Figure 6.3 Ii is not included in the measurement.

Distal breadth - Bd - Taken with one jaw flat against the medial edge of the trochlea metatarsi II and the other against the distal edge of the trochlea metatarsi IV. No twisting.

Shaft breadth - SC - Taken across the midshaft in the plane illustrated in Figure 6.3 I. No twisting.

6.2.2. *Lagopus* species (After Kraft, 1972) (Figures 6.4 and 6.5)

Humerus (Figure 6.4 A and Ai)

Greatest length - GL - Taken from the proximal edge of the caput humeri to the distal edge of the condylus ventralis. Twisting is necessary.

Greatest proximal breadth - TP - Taken from the dorsal edge of the crista pectoralis to the ventral edge of the crista bicipitalis. Twisting is necessary.

Greatest distal breadth - TD - Taken from the ventral edge of the epicondylus ventralis to the dorsal edge of the epicondylus dorsalis. Twisting is necessary.

Smallest shaft depth - KB - Taken across the shaft immediately above the condylus dorsalis in the plane illustrated in Figure 6.4 Ai. No twisting.

Smallest shaft breadth - KT - Taken across the narrowest part of the midshaft in the plane illustrated in Figure 6.4 A. No twisting.

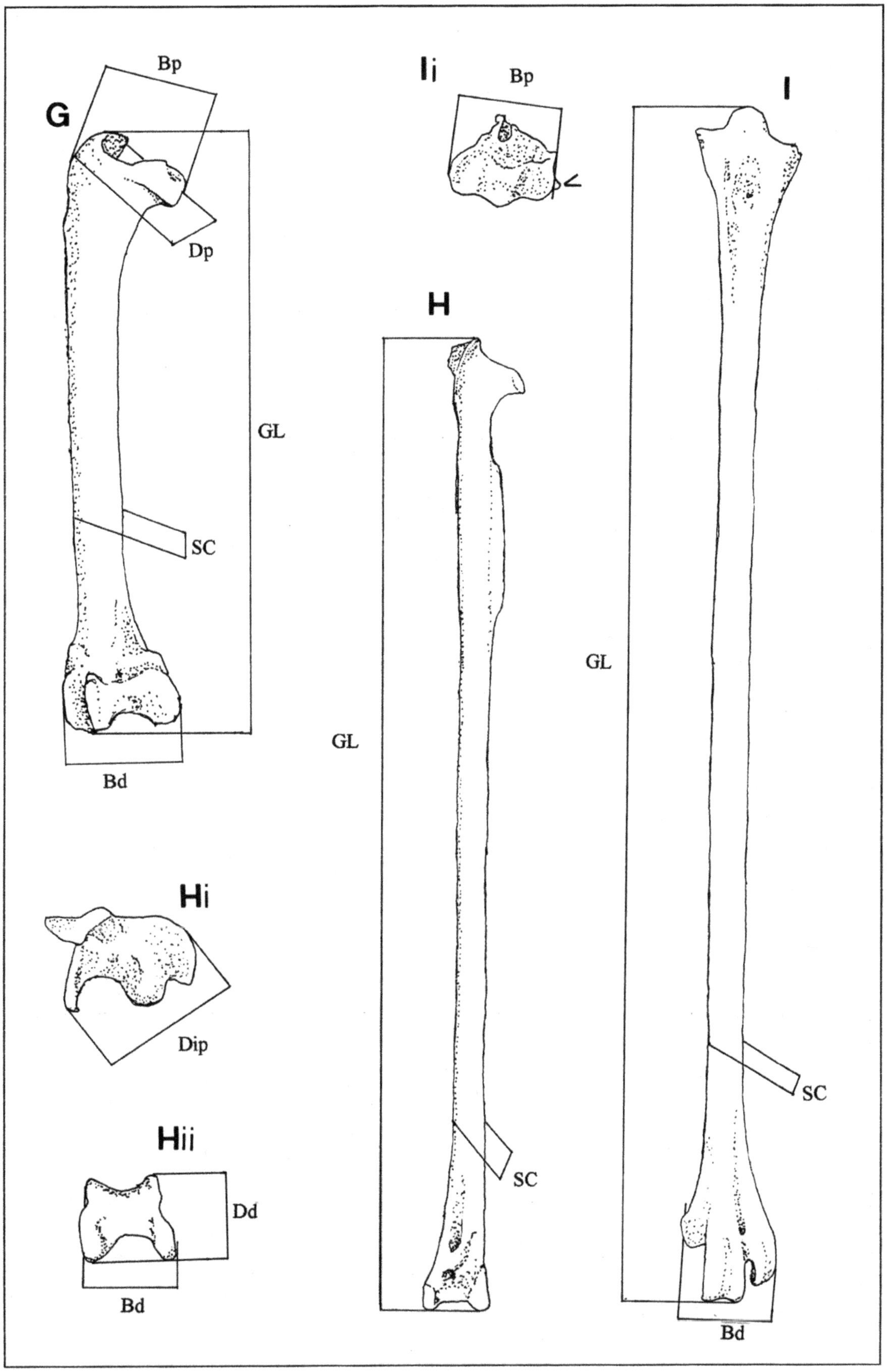

Figure 6.3. Measurements taken on the femur (G), tibiotarsus (H, Hi, Hii) and tarsometatarsus (I, Ii) of *Grus*.

Radius (Figure 6.4 B and Bi)

Greatest length - GL - Taken from the extreme proximal edge to the extreme distal parts of the radius. Twisting is necessary

Greatest proximal depth - TP - Taken across the proximal end in the manner illustrated in Figure 6.4 Bi. This measurement is taken in a plane perpendicular to that of BP. No twisting.

Greatest proximal breadth - BP - Taken across the proximal end in the manner illustrated in Figure 6.4 Bi. This measurement is taken in a plane perpendicular to that TP. No twisting.

Smallest breadth of the shaft - KT - Taken at the narrowest part of the midshaft in the plane illustrated in Figure 6.4 B. No twisting.

Greatest distal breadth - TD - Taken across the distal end in the plane illustrated in Figure 6.4 B. Twisting is necessary.

Ulna (Figure 6.4 C, Ci and Cii)

Greatest length - GL - Taken from the proximal edge of the olecranon to the condylus ventrale. Twisting is necessary.

Greatest proximal diagonal - DP - Taken from the olecranon to the cranial most edge of the cotyla dorsalis. Twisting is necessary.

Greatest proximal breadth - BP - Taken with one jaw resting against the ventral edge of the cotyla ventralis and the olecranon and the other jaw against the dorsal edge of the cotyla dorsalis. No twisting.

Greatest proximal depth - TP - Taken with one jaw resting against the dorsal edge of the cotyla dorsalis and the olecranon and the other jaw against the ventral edge of the cotyla ventralis. No twisting.

Distal diagonal distal depth - DD - Taken from the labrum condyli to the tuberculum carpale. Twisting is necessary.

Smallest shaft depth - KT - Taken across the midshaft in the plane illustrated in Figure 6.4 Ci. No twisting.

Carpometacarpus (Figure 6.4 D)

Greatest length - GL - Taken from the proximal edge of the trochlea carpalis to the distal edge of the facies articularis minor. Twisting is necessary.

Greatest proximal breadth - TP - Taken from the tip of the process extensorius to the edge of the facies articularis ulnocarpalis in manner detailed in Figure 6.4 D. Twisting is necessary.

Breadth accross process intermetacapalis* - GT - Taken from the tip of the process intermetacaparpalis to the opposing edge of the os metacarpale majus. No twisting.

Greatest distal diagonal - DD - Taken across the distal articulation in the plane and angle illustrated in Figure 6.4 D. No twisting.

Height of the symphysis - HS - Taken from the proximal edge of the symphysis to the flattened surface of the facies articularis digit minor. No twisting.

* This is a variant on the measurement taken by Kraft (1972). The original GT was across both metacarpal bones.

Coracoid (Figure 6.4 E)

Greatest length - GL - Taken from the process acrocoracoideum to the lateral edge of the facies articularis sternalis. Twisting is necessary.

Medial length - LM - Taken from the process acrocoracoideum to the medial edge of the facies articularis sternalis. Twisting is necessary.

Breadth of the facies articularis sternalis - BB - Taken from the medial edge to the lateral edge of the facies articularis sternalis. No twisting.

Basal breadth - BD - Taken from the medial edge of the process lateralis to the medial edge of the facies articularis sternalis. Twisting is necessary.

Scapula (Figure 6.4 F)

Greatest length - GL - Taken from the tip of the acromion to the extremitas caudalis. Twisting is necessary.

Greatest cranial breadth - BC - Taken from the acromion to the opposing edge of the facies articularis humeralis. Twisting is necessary.

Smallest breadth of the collum scapulae - KC - Taken across the collum scapulae in the plane illustrated in Figure 6.4 F. Twisting is necessary.

Femur (Figure 6.5 G, Gi, Gii)

Greatest length - GL - Taken from the proximal edge of the trochanter femoris to the distal edge of the condylus lateralis. Twisting is necessary.

Medial length - LM - Taken from the proximal edge of the facies articularis acetabularis to the distal edge of the condylus medialis. Twisting is necessary.

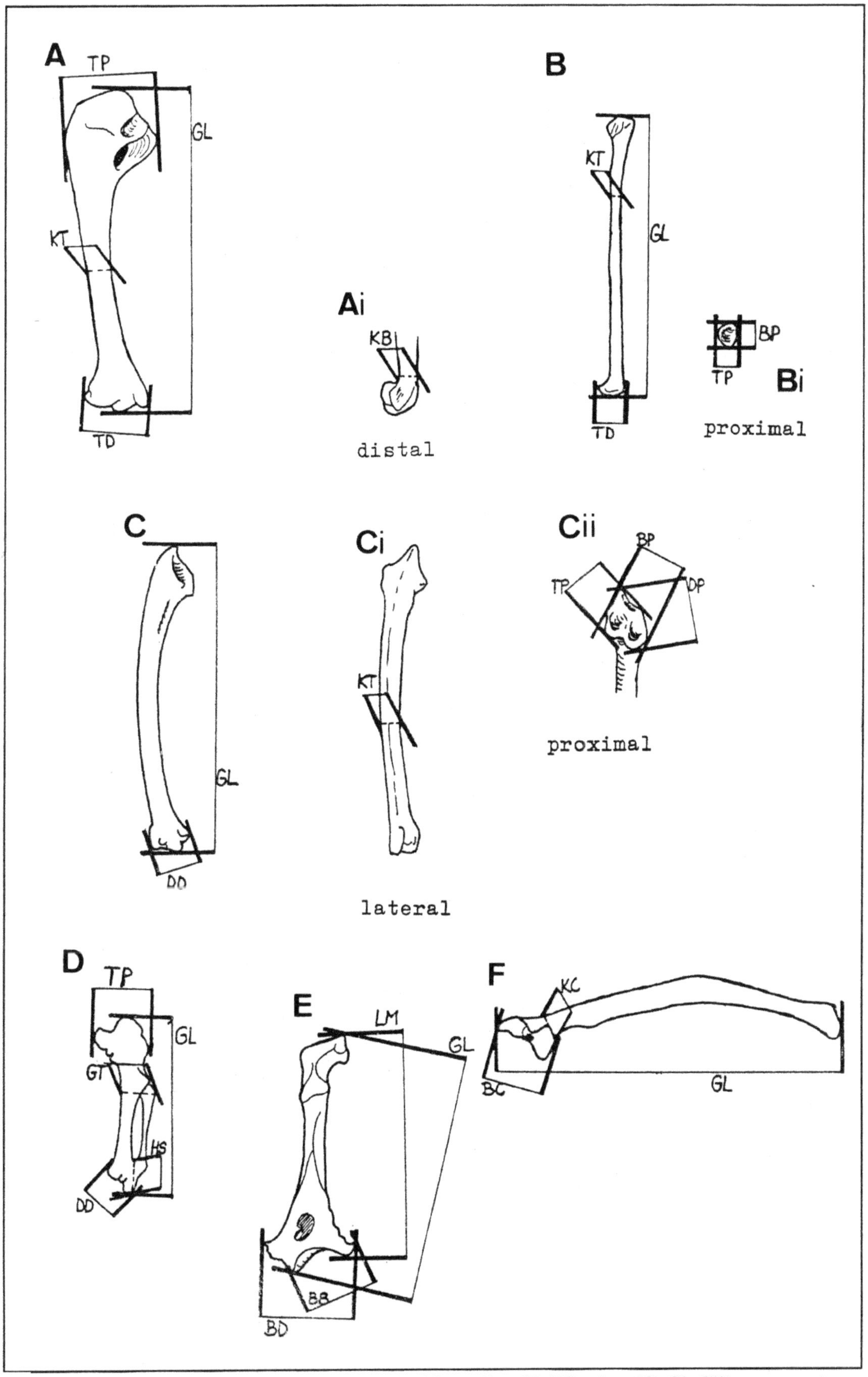

Figure 6.4. Measurements taken on the humerus (A, Ai), radius (B, Bi), ulna (C, Ci, Cii), carpometacarpus (D), coracoid (E) and scapula (F) of *Lagopus* after Kraft 1972.

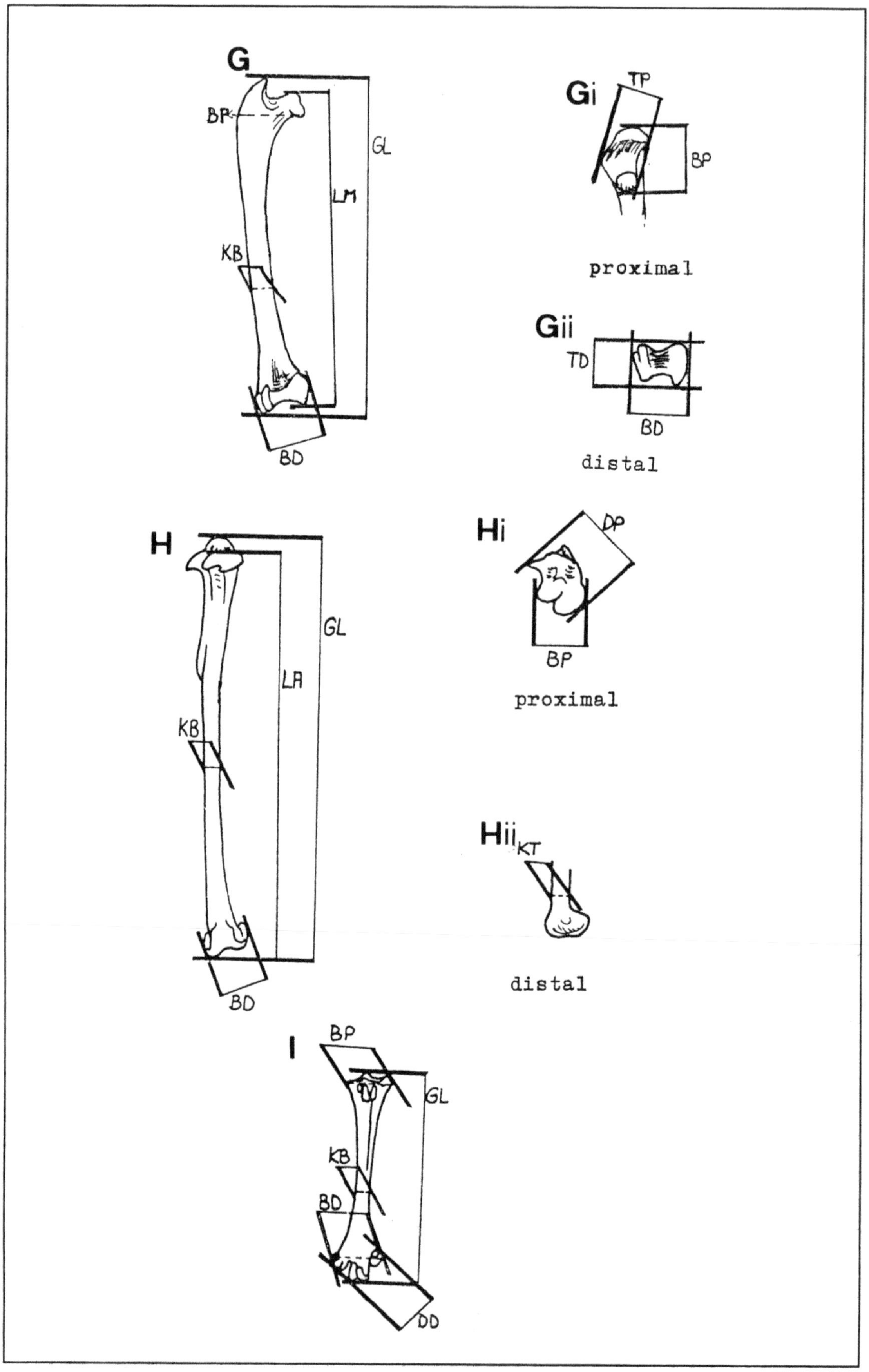

Figure 6.5. Measurements taken on the femur (G, Gi, Gii), tibiotarsus (H, Hi, Hii) and tarsometatarsus (I) of *Lagopus* after Kraft 1972.

Greatest proximal depth -TP - Taken with one jaw against the cranial edge of the crista trochanteris and the facies articularis acetabularis and one jaw against the opposing caudal side of the femur. No twisting.

Greatest proximal breadth -BP- Taken from the lateral edge of the trochanter femoris to the medial edge of the facies articularis acetabularis. Twisting is necessary.

Smallest breadth of the shaft - KB - Taken across the narrowest part of the shaft in the plane illustrated in Figure 6.5 G. No twisting.

Greatest distal breadth - BD - Taken with one jaw flat against the medial edge of the condylus medialis and the other against the lateral edge of the condylus lateralis. No twisting.

Greatest distal depth -TD - Taken with one jaw against the caudal edges of both the condylus medialis and lateralis and the other jaw against the cranial edge of the condylus medialis. No twisting.

Tibiotarsus (Figure 6.5 H, Hi, Hii)

Greatest length - GL - Taken from the proximal edge of the crista cnemialis cranialis to the distal edge of the epicondylus lateralis. Twisting is necessary.

Axial length - LA - Taken from the proximal edge of the area interarticularis to the distal edge of the epicondylus lateralis. Twisting is necessary.

Greatest proximal breadth - BP - Taken from the medial edge of the facies articularis medialis to the lateral edge of the facies articularis lateralis near to the attachment for the fibula. No twisting.

Greatest proximal diagonal - DP - Taken from the medial edge of the facies articularis medialis to the edge of the crista cnemialis lateralis. Twisting is necessary.

Smallest breadth of the shaft - KB - Taken at the narrowest part of the shaft in the plane illustrated in Figure 6.5 H. No twisting.

Smallest depth of the shaft - KT - Taken immediately above the distal articulation in the plane illustrated in Figure 6.5 Hii. No twisting.

Greatest distal breadth - BD - Taken with one jaw flat against the lateral edge of the epicondylus lateralis and the other jaw against the cranialmost part of the epicondylus medialis as illustrated in Figure 6.5 H. No twisting.

Tarsometatarsus (Figure 6.5 I)

Greatest length - GL - Taken from the proximal edge of the eminentia intercondylaris to the distal edge of the cotyla metatarsi III. Twisting is necessary.

Greatest proximal breadth - BP - Taken with one jaw flat against lateral side of the proximal articulation and the other jaw on the opposing side. No twisting.

Smallest shaft breadth - KB - Taken across the narrowest part of the mid shaft in the plane illustrated in Figure 6.5 I. No twisting.

Greatest distal breadth - BD - Taken with one jaw flat against the lateral side of the trochlea metatarsi IV and the other against the medial edge of the trochlea metatarsi II. No twisting.

Greatest distal diagonal - DD - Taken with one jaw against both the trochlea metatarsi IV and III and the other jaw against the trochlea metatarsi II. No twisting.

6.2.3. *Corvus corax* (Figures 6.6 – 6.8)

Humerus (Figure 6.6 A)

Greatest length - GL - Taken from the caput humeri to the process flexorius. Twisting necessary.

Proximal breadth - Bp - Taken accross the proximal articulation from the dorsal edge of the crista bicipitalis to the proximal part of the tuberculum dorsale as seen in Figure 6.6 A Twisting necessary.

Distal breadth - Bd - Taken from the process supracondylaris dorsalis to the process flexorius. Twisting necessary.

Shaft breadth - SC - Taken across the midshaft in the plane illustrated in Figure 6.6 A. No twisting.

Ulna (Figure 6.6 B, Bi)

Greatest length - GL - Taken from the olecranon to the condylus ventrale. Twisting necessary.

Proximal breadth - Bp - Taken from the edge of the cotyla ventralis to the dorsal edge of the cotyla dorsalis. Care should be taken during twisting of the bone between the caliper jaws that the tuberculum ligamentum collateralis ventrale is not included in the measurement.

Proximal diagonal depth - Dip - Taken from the olecranon to the cranialmost edge of the cotyla dorsalis. Twisting is necessary.

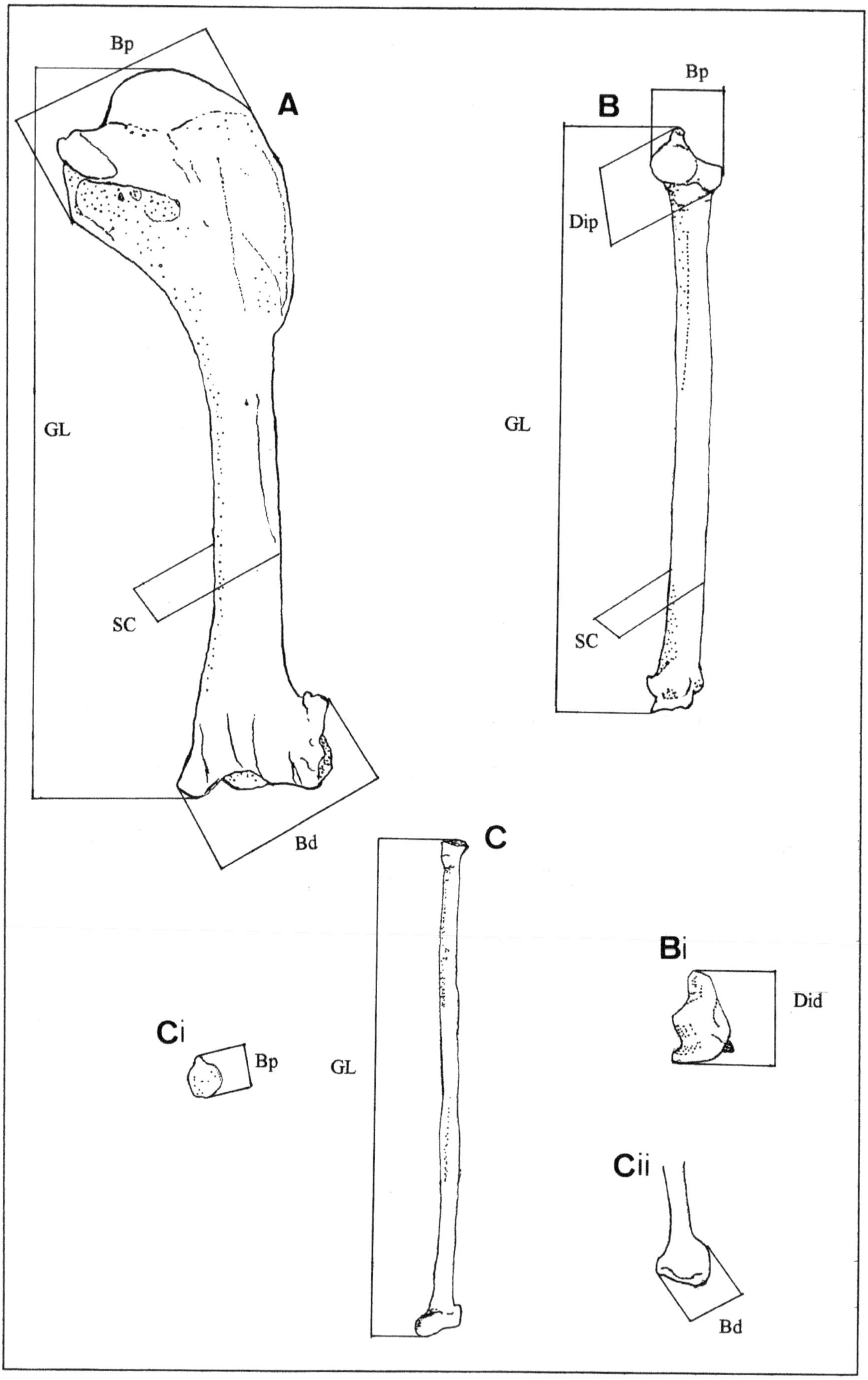

Figure 6.6. Measurements taken on the humerus (A), ulna (B, Bi) and radius (C, Ci, Cii) of *Corvus*.

Distal diagonal depth - Did - Taken from the labrum condylii to the tuberculum carpale. Twisting necessary.

Shaft breadth - SC - Taken across the midshaft in the plane illustrated in Figure 6.6 B. No twisting.

Radius (Figure 6.6 C, Ci, Cii)

Greatest length - GL - Taken from the extreme proximal edge to the extreme distal parts of the radius. Twisting is necessary.

Proximal breadth - Bp - Taken across the proximal ariculation from the tuberculum bicipitale to the opposing edge. Twisting necessary.

Distal breadth - Bd - Taken across the distal end as illustrated in Figure 6.6 Cii. Twisting necessary.

Carpometacarpus (Figure 6.7 D)

Greatest length - GL - Taken from the trochlea carpalis to the facies articularis minor. Twisting necessary.

Proximal breadth - Bp - Taken from the process extensorius to the distal edge of the facies articularis ulnocarpalis in manner detailed in Figure 6.7 D. Twisting necessary.

Distal breadth - Did - Taken across the distal end in the manner illustrated in Figure 6.7 D. Twisting necessary.

Scapula (Figure 6.7 E)

Greatest length - GL - Taken from the cranial process indicated in Figure 6.7 E to the extremitas caudalis. Twisting necessary.

Cranial diagonal breadth - Dic - Taken from the cranial process indicated in Figure 6.7 E to the opposing edge of the facies articularis humeralis. Twisting necessary.

Coracoid (Figure 6.7 F)

Greatest length - GL - Taken from the process acrocoracoideum to the lateral edge of the facies articularis sternalis. Twisting necessary.

Medial length - LM - Taken from the process acrocoracoideum to the medial edge of the facies articularis sternalis. Twisting necessary.

Breadth of the facies articularis sternalis - BF - Taken from the medial edge to the lateral edge of the facies. Care must be taken here so as not to damage the process lateralis. Twisting necessary.

Shaft breadth - SC - Taken across the midshaft in the plane illustrated in Figure 6.7 F. Twisting necessary.

Femur (Figure 6.7 G)

Greatest length - GL - Taken from the proximal edge of the crista trochanteris to the distal condylus lateralis. Twisting necessary

Proximal breadth - Bp - Taken from the medial edge of the facies articularis acetabularis to the lateral edge of the trochanter femoris. Twisting necessary

Proximal depth - Dp - Taken with the caliper jaw against the cranial edge of the crista trochanteris and facies articularis acetabularis and the other on the opposing caudal edge. No twisting.

Distal breadth - Bd - Taken from the epicondylus medialis with one caliper jaw square against that side to the opposite epicondylus lateralis. This measurement does not involve twisting as the flattened medial edge of the distal femur provides a secure rest for the jaw.

Shaft breadth - SC - Taken across the midshaft in the plane illustrated in Figure 6.7 G. No twisting.

Tibiotarsus (Figure 6.8 H, Hi, Hii)

Greatest length - GL - Taken from the proximal edge of the crista patellaris to the distal edge of the condylus medialis. Twisting necessary

Proximal diagonal depth - Dip - Taken from the medial edge of the facies articularis medialis to the crista cnemialis lateralis. Twisting necessary.

Distal breadth - Bd - Taken from the lateral edge of the condylus lateralis to the cranial edge of the condylus medialis. Jaw flat against the lateral edge and no twisting.

Distal depth - Dd - Taken with one caliper jaw flat against both the condylus medialis and the condylus lateralis and the other jaw against the opposing side. Twisting necessary

Tarsometatarsus (Figure 6.8 I, Ii)

Greatest length - GL - Taken from the eminentia intercondylaris to the trochlea metatarsi tertii. Twisting required.

Proximal breadth - Bp - Taken from the lateral to medial edges of the proximal end. No twisting.

Distal breadth - Bd - Taken from the lateral edge of the trochanter metatarsi IV to the medial edge of the trochlea II. The caliper jaw should rest against the lateral edge. No twisting.

Shaft breadth - SC - Taken across the midshaft in the plane illustrated in Figure 6.8 I. No twisting.

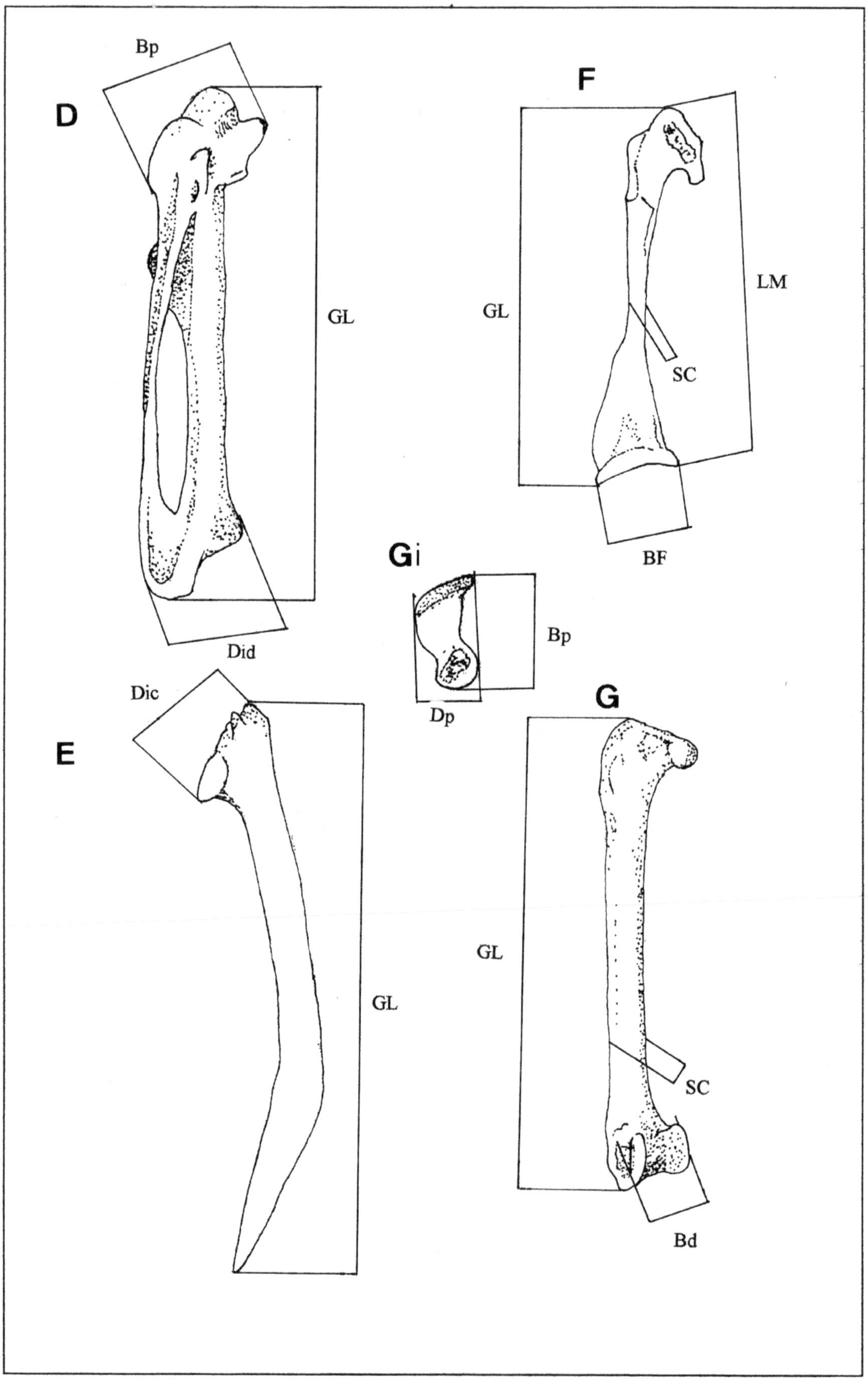

Figure 6.7. Measurements taken on the carpometacarpus (D), coracoid (F), scapula (E) and femur (G, Gi) of *Corvus*.

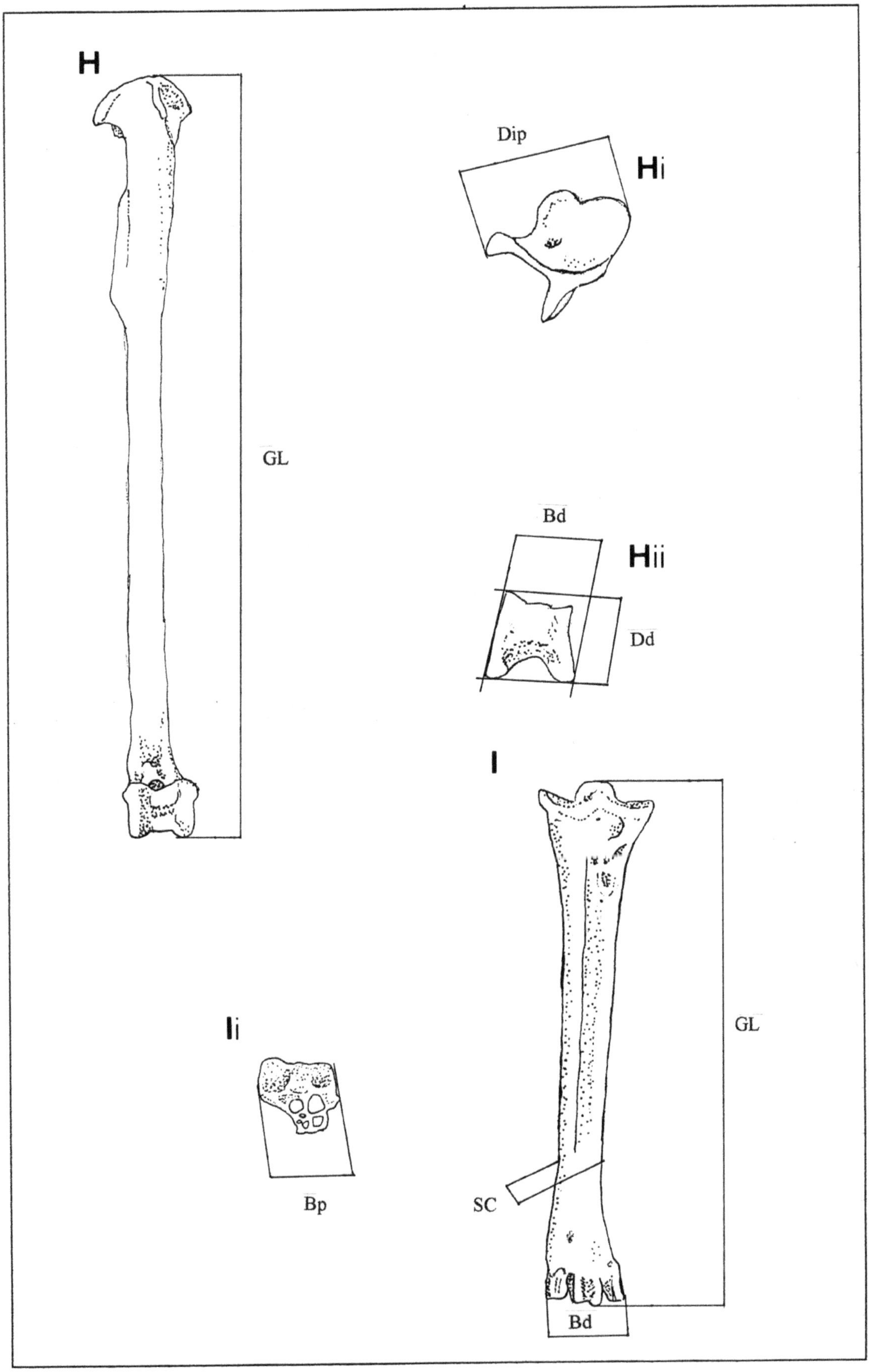

Figure 6.8. Measurements taken on the tibiotarsus (H, Hi, Hii) and tarsometatarsus of *Corvus*.

6.2.4. *Sturnus* species (Figures 6.9 – 6.11)

Humerus (Figure 6.9 A)

Greatest length - GL - Taken from the caput humeri to the process flexorius. Twisting necessary.

Proximal breadth - Bp - Taken from the dorsal edge of the crista pectoralis to the ventral edge of the crista bicipitalis. When taking this measurement the twisting of the bone between the caliper jaws should not be done to the extent of reaching the tuberculum ventrale.

Distal breadth - Bd - Taken from the process supracondylaris dorsalis to the process flexorius. Twisting necessary.

Shaft breadth - SC - Taken across the midshaft in the plane illustrated in Figure 6.9 A. No twisting.

Radius (Figure 6.9 B and Bi)

Greatest length - GL - Taken from the extreme proximal edge to the extreme distal parts of the radius. Twisting is necessary

Proximal breadth - Bp - Taken across the proximal articulation from the tuberculum bicipitale to the opposing edge. Twisting necessary.

Distal breadth - Bd - Taken across the distal end as illustrated in Figure 6.9 B. Twisting necessary.

Ulna (Figure 6.9 C and Ci)

Greatest length - GL - Taken from the olecranon to the condylus ventrale. Twisting necessary.

Proximal breadth - Bp - Taken from the edge of the cotyla ventralis to the dorsal edge of the cotyla dorsalis. Care should be taken during twisting of the bone between the caliper jaws that the tuberculum ligamentum collateralis ventrale is not included in the measurement.

Distal diagonal distal depth - Did - Taken from the labrum condylii to the tuberculum carpale. Twisting necessary. (Figure 6.9 Ci).

Shaft breadth - SC - Taken across the midshaft in the plane illustrated in Figure 6.9 C. No twisting.

Carpometacarpus (Figure 6.9 D)

Greatest length - GL - Taken from the trochlea carpalis to the facies articularis minor. Twisting necessary.

Proximal breadth - Bp - Taken from the process extensorius to the distal edge of the facies articularis ulnocarpalis in manner detailed in Figure 6.9 D. Twisting necessary.

Distal breadth - Did - Taken across the distal end in the manner illustrated in Figure 6.9 D. Twisting necessary.

Scapula (Figure 6.10 E)

Greatest length - GL - Taken from the cranial process indicated in Figure 6.10 E to the extremitas caudalis. Twisting necessary.

Cranial diagonal breadth - Dic - Taken from the cranial process indicated in Figure 6.10 E to the opposing edge of the facies articularis humeralis. Twisting necessary.

Coracoid (Figure 6.10 F)

Greatest length - GL - Taken from the process acrocoracoideum to the lateral edge of the facies articularis sternalis. Twisting necessary.

Medial length - Lm - Taken from the process acrocoracoideum to the medial edge of the facies articularis sternalis. Twisting necessary.

Breadth of the facies articularis sternalis - BF - Taken from the medial edge to the lateral edge of the facies. Care must be taken here so as not to damage the process lateralis. Twisting necessary.

Shaft breadth - SC - Taken across the midshaft in the plane illustrated in Figure 6.10 F. Twisting necessary.

Femur (Figure 6.11G)

Greatest length - GL - Taken from the proximal edge of the crista trochanteris to the distal condylus lateralis. Twisting necessary.

Proximal breadth - Bp - Taken from the medial edge of the facies articularis acetabularis to the lateral edge of the trochanter femoris. Twisting necessary.

Distal breadth - Bd - Taken from the epicondylus medialis with one caliper jaw square against that side to the opposite epicondylus lateralis. This measurement does not involve twisting as the flattened medial edge of the distal femur provides a secure rest for the jaw.

Shaft breadth - SC - Taken across the midshaft in the plane illustrated in Figure 6.11 G. Twisting necessary.

Tibiotarsus (Figure 6.11 H, Hi and Hii)

Greatest length - GL - Taken from the proximal edge of the crista patellaris to the distal edge of the condylus medialis. Twisting necessary.

Proximal diagonal depth no.1 - Dip1 - Taken from the medial edge of the facies articularis medialis to the crista cnemialis lateralis. Twisting necessary.

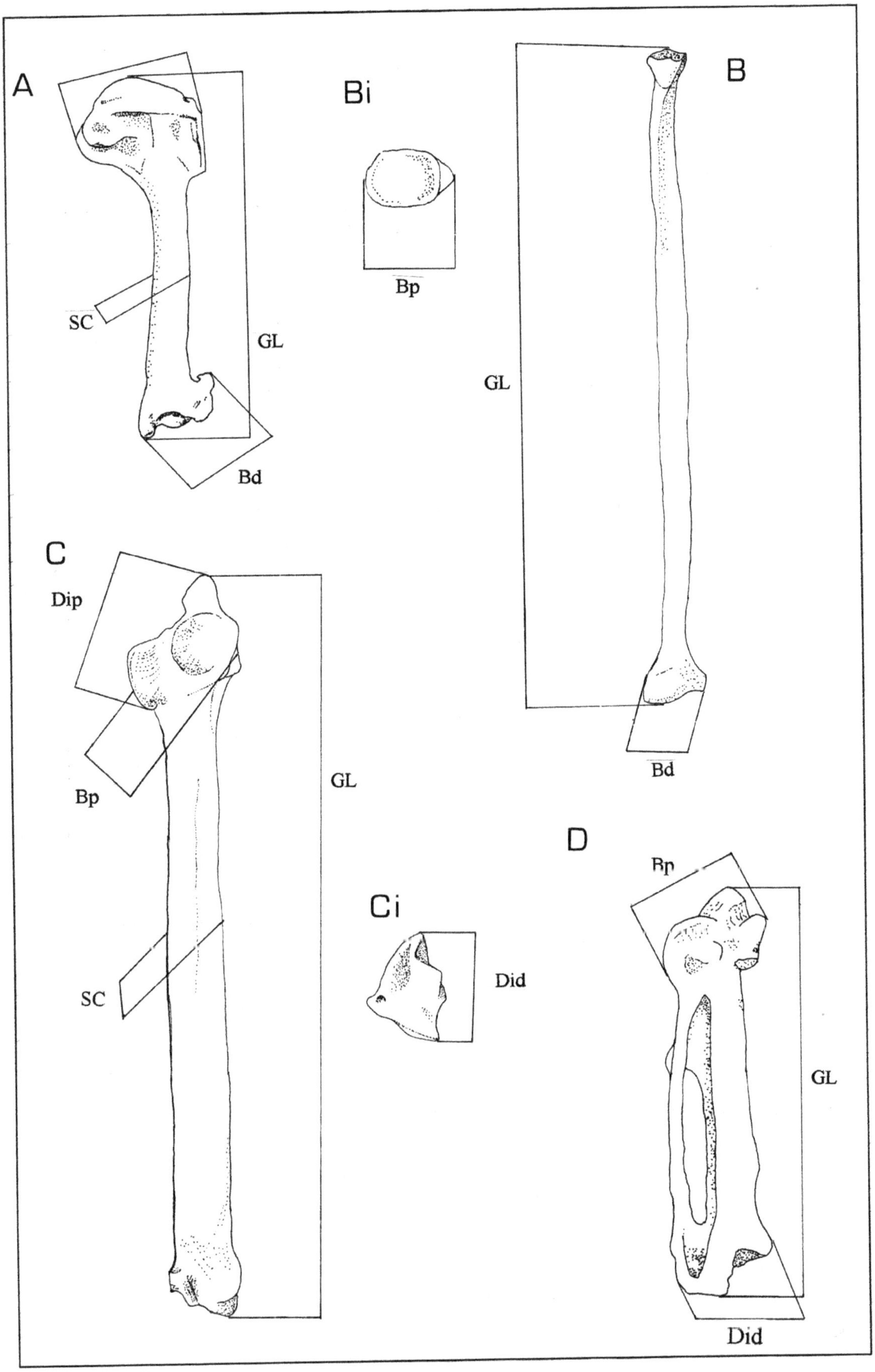

Figure 6.9. Measurements taken on the humerus (A), radius (B, Bi), ulna (C, Ci) and carpometacarpus (D) of *Sturnus*.

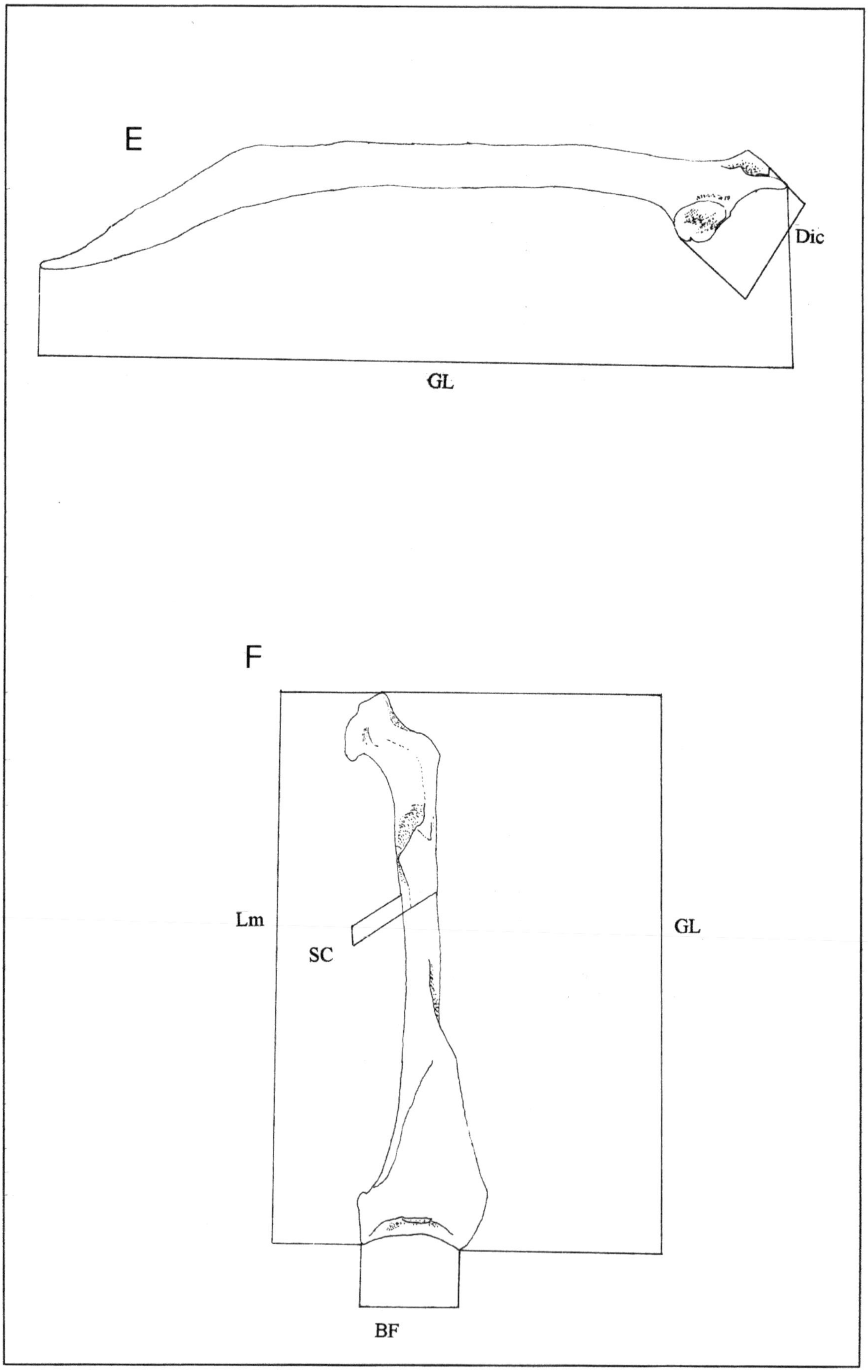

Figure 6.10. Measurements taken on the scapula (E) and coracoid (F) of *Sturnus*.

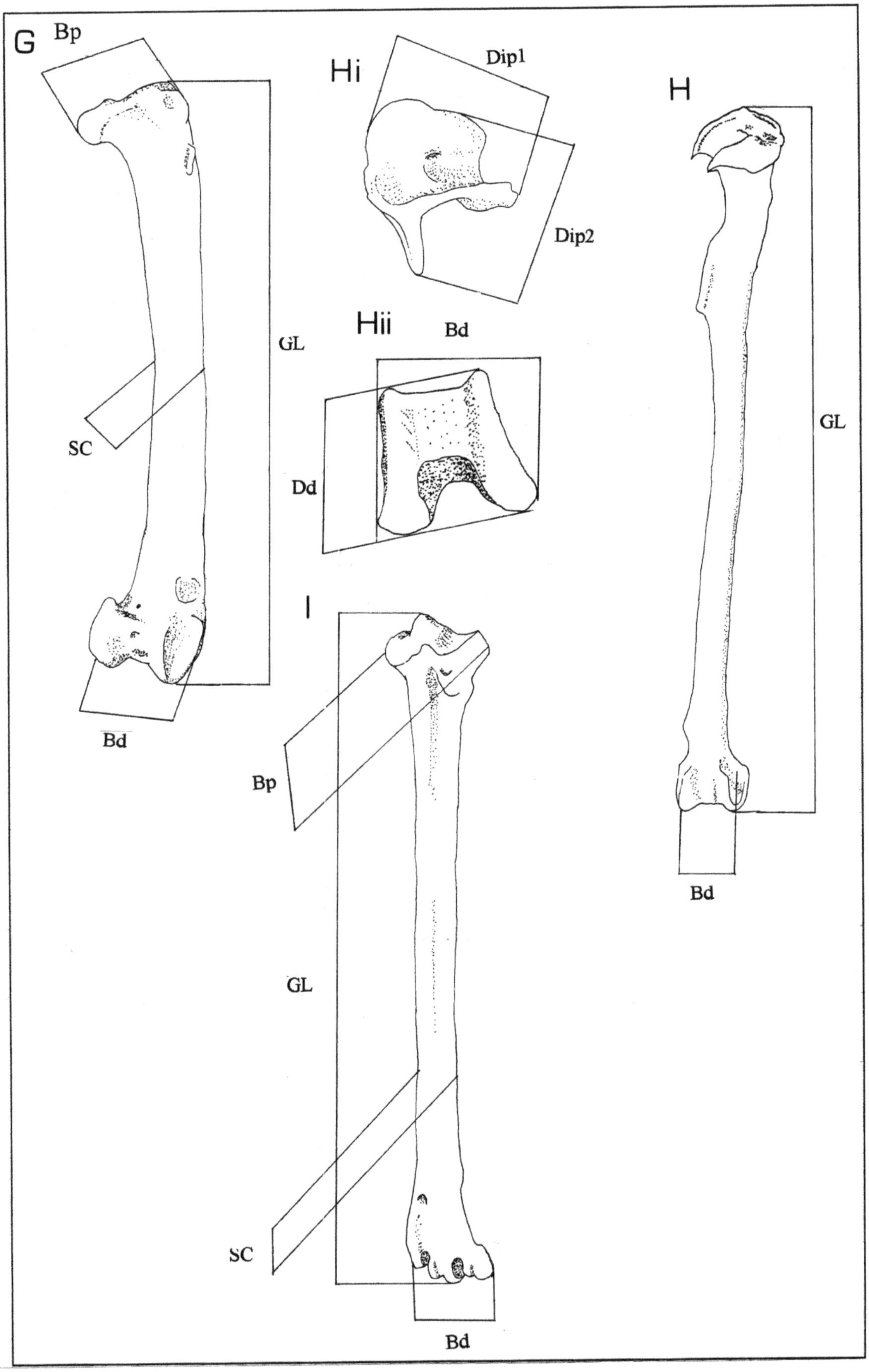

Figure 6.11. Measurements taken on the femur (G), tibiotarsus (H, Hi, Hii) and tarsometatarsus (I) of *Sturnus*.

Proximal diagonal depth no.2 – Dip2 - Taken from the caudal edge of the facies articularis lateralis to the crista cnemialis cranialis. Twisting necessary.

Distal breadth - Bd - Taken from the lateral edge of the condylus lateralis to the cranial edge of the condylus medialis. Jaw flat against the lateral edge and no twisting.

Distal depth - Dd - Taken with one caliper jaw flat against both the condylus medialis and the condylus lateralis and the other jaw against the opposing side. No twisting.

Tarsometatarsus (Figure 6.11 I)

Greatest length - GL - Taken from the eminentia intercondylaris to the trochlea metatarsi tertii. Twisting required.

Proximal breadth - Bp - Taken from the lateral to medial edges of the proximal end. No twisting.

Distal breadth - Bd - Taken from the lateral edge of the trochanter metatarsi IV to the medial edge of the trochlea II. The caliper jaw should rest against the lateral edge. No twisting.

Shaft breadth - SC - Taken across the midshaft in the plane illustrated in Figure 6.11 I. No twisting.

6.3. Data Presentation and Methods of Analysis

All graph plotting and calculations were done using Microsoft Excel™.

6.3.1. Data tables

Data collected during the course of the present study is tabulated throughout the relevant parts of Section 3. This includes both the modern data which are presented as tables of minima, maxima, means and standard deviations, as well as of the fossils which is usually presented raw, i.e. the measurements of each individual specimen are given.

6.3.2. Graphs

In order to compare the mensural data collected from the various skeletal samples of the four taxonomic groups studied, bivariate scattergrams were plotted. This gave a means of demonstrating visually the similarities and differences between the samples. Bivariate scattergrams are useful to show both shape differences as well as size differences. For example length / breadth or breadth / depth differences of a particular bone; or the relationship between different bones (in the case of complete skeletons) can be seen visually from the graph. Metric-shape differences are distinguished from simple size differences by looking at the regression lines drawn. Isometric differences will all fall on the same line, while metrical-shape differences in a sample will fall on a line which is offset on one axis. To further support the analysis, regression lines were drawn for the larger samples. Accompanying the regression lines R^2 values and the line fomulae are given. This enables the relationship between the two variables plotted of the samples to be compared.
In addition to the scattergrams, on one occasion, when only one fossil measurement was available a univariate comparison was done by means of a bar chart.

6.3.3. *T*-testing

Samples which appeared to differ in the graph were tested using Student's *t*-test. This technique is a univariate statistical test of the difference between the means. The results of the *t*-testing are presented as tables with indications of the degree of significance between the means expressed as $p< 0.001$, $p < 0.01$, $p < 0.5$ or not significant at the 5% level. It was not always possible to *t*-test the difference between samples because they included too few individuals.

6.3.4. Tables of limb bone ratios expressed as percentages

In a number of instances it was felt that differences between samples were best expressed by comparing measurement ratios. Therefore, tables of the ratios between certain limb bone dimensions are given where relevant.

6.3.5. Why no multivariate analysis

Two main reasons influenced the decision not to use multivariate analysis. Firstly, and most importantly, the fossils were in the main individual, fragmentary bones. This signified that the number of variables which were available for these fossils was limited often to two or three. The project is primarily a palaeontological one and so it seemed illogical to perform a different form of analytical statistics on the modern material which possessed more variables than the fossils. The second reason is that the aim of the work was to identify specific differences, whether in size or metric shape, with a view to explaining them. Multivariate analyses, such as principal component analysis maximises the difference between samples but does not allow immediate interpretation of differences. It is easier to consider a scattergram with length versus breadth than PC1 and PC2.

Chapter 7

Morphological Comparative Methodology

When bones are identified there is often the temptation to do so in a non-formalised way. In some cases, where there are no other possible candidates for specific identity, this may be justified. However, in many instances this is not the situation, as similarly sized species in the same genus, family or even order can present serious difficulties to identification.

Various authors have designed methodologies to overcome this problem (Bochenski and Tomek, 1995; Lister, 1981 and 1996; Gee, 1993; Clutton-Brock *et al.*, 1990). Most of these papers deal with examples where similar mammals are distinguished, but it is the methods which are important. Essentially, the methodology utilised by the last three sets of authors is the same, where distinguishing characters between two taxa are established by testing them for consistency among several individuals. Clutton-Brock *et al.* (1990) is an improvement on Boessneck's (1969) first attempt at differentiating sheep *Ovis* and goat *Capra* bones, by scoring the consistency of the diagnostic traits. However, the study only scored the diagnostic traits on a population of soay sheep, presenting two problems. First, we are left not knowing how often these traits appear in goat bones, therefore leaving them incompletely tested, and second soay sheep are but one small breed and the traits could very well differ in occurrence in other, particularly in larger, breeds. The paper by Gee dealing with *Bos / Bison* illustrates yet another problem, which is that the specimens used for comparative purposes were fossils, and thus not from animals of strictly known identity. This was unavoidable because modern reference material is unavailable for *Bos primigenius* or *Bison priscus*. The paper dealing with red deer *Cervus elaphus* / fallow deer *Dama dama* distinctions (Lister, 1996) was clear of the above flaws, and further added a test of the characters against a population of fallow deer from the Late Pleistocene of Tabun in Israel. The investigation by Bochenski and Tomek (1995) differed significantly from the others in focusing on the question "How many comparative skeletons do we need to identify a bird bone?" Here three pairs of congeneric species were considered in terms of how easily correct identifications could be achieved. Like the methods above, variability in the occurrence of diagnostic traits was acknowledged, although traits were not scored as such. The method instead appears to assess what is the minimum number of skeletons which would be required to eliminate the possibility that variation among individuals was giving a false appearance of distinctiveness. This method does not provide a systematic means of establishing the reliability with which two bones, belonging to two closely related species, can be distinguished, as it does not test the consistency of particular morphological character states.

The means used in this volume to overcome problems of variation in the occurrence of discriminating characters is that devised by Lister (1981). In the present analysis, where 4 genera have been studied, only *Sturnus* presented a situation where the method was required: in the distinction between *Sturnus* and *Turdus*. The method shall therefore be described in detail using the example of *Sturnus* versus *Turdus*.

The first stage of the process to establish discriminating features between two taxa (*Sturnus* and *Turdus*) was to compare a relatively small number of each skeletal element of each taxon with the other. Potentially useful traits, confined to one or other taxon, were then illustrated and described. Individuals of the two taxa were then scored on a scale of morphology as follows:

S	*Sturnus* condition
T : S	Intermediate
T	*Turdus* condition

The frequency of each condition in the test samples was counted and expressed as a percentage. This was achieved by first turning each specimen's observed condition into a percentage:

S	100%
T:S	50%
T	0%

These individual percentages were then converted into a sample mean by totalling the scores for each specimen and dividing by the total number of specimens. For each character, this was done for each species of the genus *Sturnus*, and finally for all specimens of the genus regardless of species. The equivalent was then done, on each *Turdus* species, and subsequently for the whole *Turdus* genus.

The first trait to be tested was the presence of an angle in the crista pectoralis of the proximal humeri of 24 specimens of *Sturnus vulgaris* (Figure 7.1) This produced:

S	23
S:T	1
T	0

$$\frac{(23 \times 100) + (1 \times 50) + (0 \times 0)}{24} = 97.92\,\%$$

This signifies that the average expression of the trait among *Sturnus vulgaris* is 97.92%. The trait was then tested among *S. unicolor* and *S. roseus* specimens, and

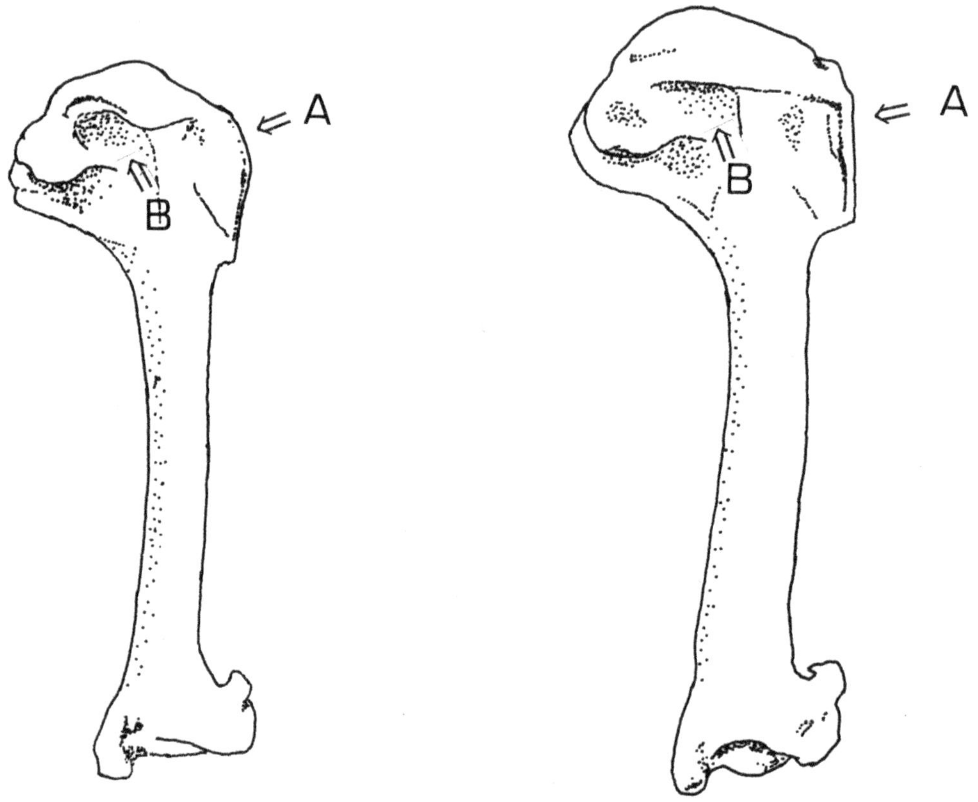

Key : A - The crista pectoralis which in *Sturnus* has a pronounced angle while in *Turdus* there is a curve. B - The size of the second, more proximal, pneumatic fossa in *Sturnus* is small when compared to the first fossa, in *Turdus* it is as large as the first.

Figure 7.1. Morphological differences between *Sturnus* and *Turdus* humeri in caudal view.

finally all *Sturnus* specimens were amalgamated (Table 7.1). Next, the *Sturnus* condition was tested against *Turdus* specimens and produced the results seen in Table 7.1. For a character to be reliable, the percentage presence of *Sturnus* characters should be high in *Sturnus* and the percentage absence in *Turdus* should also be high. If it is high in one taxon and low in the other it would be an ambiguous or potentially misleading character and of little use.

This procedure was carried out for all major postcranial limbones of *Sturnus* and *Turdus* (humeri, ulnae, radii, carpometacarpi, coracoidea, scapulae, femora, tibiotarsi and tarsometatarsi). However, only the humerus, carpometacarpus, tibiotarsus and coracoid provided testable characters. The characters can be seen in Figures 7.1 - 7.4 while the results of the scoring can be seen in Tables 7.1 - 7.8.

The humeri of *Sturnus* and *Turdus* have the potential for being distinguished solely at the proximal end. The best known character in need of testing is that first described by Bock (1962) and subsequently by Ballman (1978) and Janossy (1983), where the relative size of the two pneumatic fossae are used. In *Sturnus* the second, more proximal, fossa is small in relation to the first fossa. In *Turdus* the second fossa is as large as the first (Figure 7.1 B). The other character is the shape of the crista pectoralis described above (Ballman, 1978; Stewart, 1992) (Figure 7.1 A).

The carpometacarpus has potential characters at both the proximal and distal ends. The proximal end in *Turdus* may exhibit a notch or discontinuity in the curve of the facies articularis ulnocarpalis, while in *Sturnus* the notch may be absent and the curve of the facies is rounded and continuous (Figure 7.2 A). This was first described in Stewart (1992). The distal end appears to differ in the

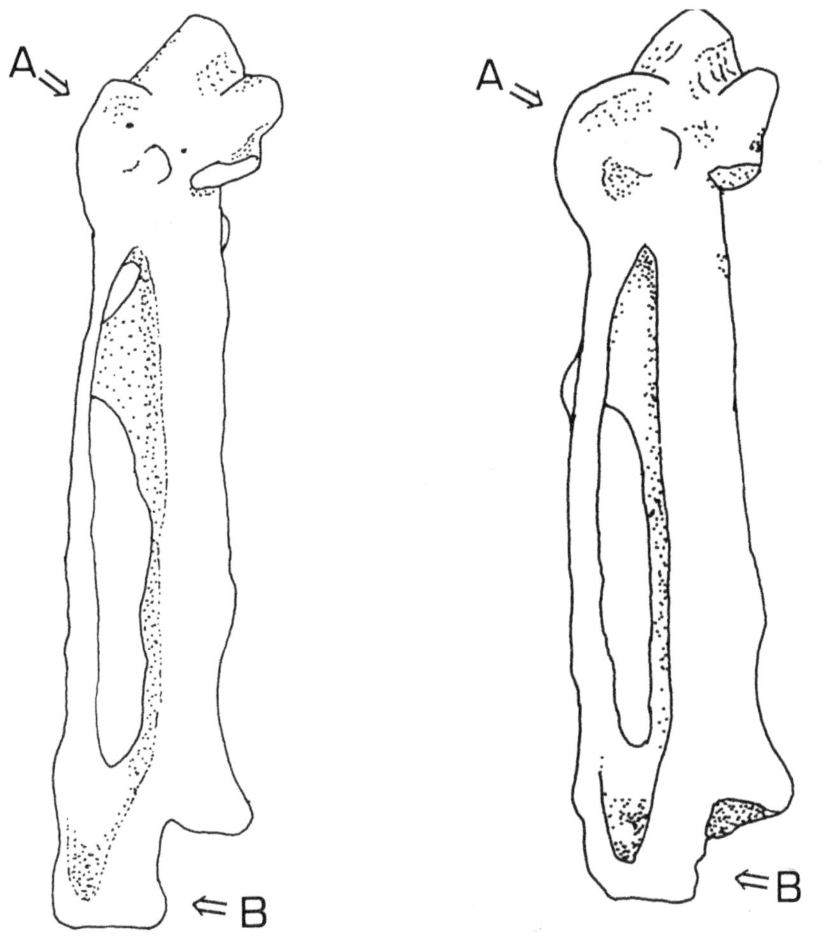

Turdus *Sturnus*

Key: A - The facies articularis ulnocarpalis has a notch in *Turdus*, while in *Sturnus* the facies is smooth and continuous. B - The edge of the facies for the phalanx digit minor overhangs the facies for the phalanx digit major in *Sturnus* and not in *Turdus*.

Figure 7.2. Morphological differences between *Sturnus* and *Turdus* carpometacarpi in ventral view.

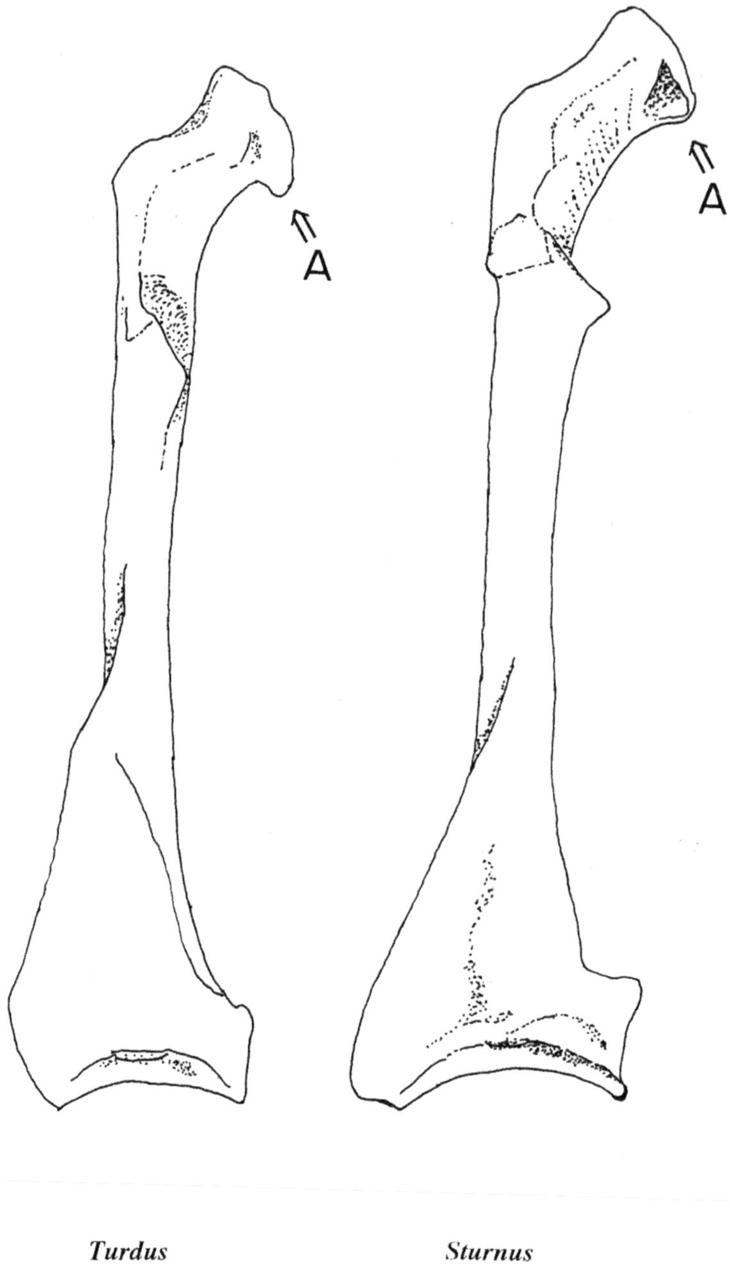

Key: A - The facies articularis clavicularis projects sternally in *Turdus* and is blunt in *Sturnus*.

Figure 7.3. Morphological differences between *Sturnus* and *Turdus* coracoids in dorsal view.

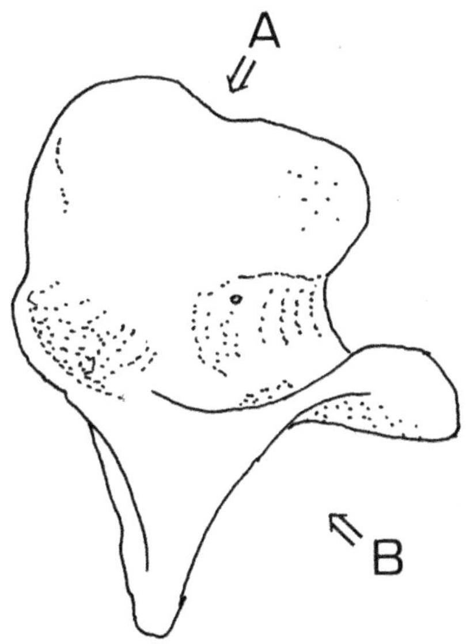

Turdus

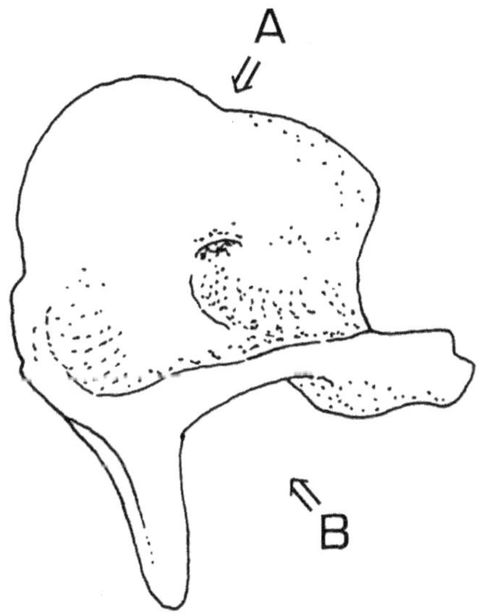

Sturnus

Key: A - The edge of the area interarticularis, which is concave, may be more V-shaped in *Turdus* than in *Sturnus*. B - The sulcus intercristalis is wide with straight edges in *Sturnus* while in *Turdus* is narrow and curved.

Figure 7.4. Morphological differences between *Sturnus* and *Turdus* tibiotarsi in proximal view.

degree to which there is an overhang on the edge of the facies for the phalanx digit minor in *Sturnus* and not in *Turdus*. This overhang projects over the facies for the phalanx digit major (Figure 7.2 B).

In the coracoid there may be but one potential difference. The facies articularis clavicularis projects sternally in *Turdus* and is blunt without projection in *Sturnus* (Figure 7.3 A).

The tibiotarsus is probably confidently separated only on the proximal end. The edge of the area interarticularis, which forms a concave surface when seen from above, may have a V-shaped outline in *Turdus* and is less pronounced and not V-shaped in *Sturnus* (Figure 7.4 A). In addition to this, *Sturnus*, when seen in proximal view, has a wide sulcus intercristalis with straight edges, while in *Turdus* it is narrower and more curved (Figure 7.4 B).

Results

The results of the testing of the potential discriminatory morphological characters are shown in Tables 7.1 - 7.7:

Humerus

Table 7.1. Angle in the crista pectoralis.

% present in *Sturnus*.	
S. vulgaris (Britain)	97.92% (N=24)
S. unicolor	100% (N=11)
S. roseus	100% (N=3)
All *Sturnus*	98.68% (N=38)

% absent in *Turdus*.	
Turdus merula	96.43% (N=14)
T. philomelos	93.18% (N=22)
T. iliacus	91.67% (N=24)
T. pilaris	93.3% (N=15)
T. torquatus	100% (N=3)
T. viscivorus	100% (N=9)
All *Turdus*	94.25% (N=87)

Table 7.2. Second pneumatic fossa is smaller than first.

% present in *Sturnus*.	
S. vulgaris (Britain)	100% (N=24)
S. unicolor	90.91% (N=11)
S. roseus	100% (N=3)
All *Sturnus*	97.37% (N=38)

% absent in *Turdus*.	
Turdus merula	82.14% (N=14)
T. philomelos	63.16% (N=19)
T. iliacus	72.92% (N=24)
T. pilaris	80% (N=15)
T. torquatus	100% (N=3)
T. viscivorus	100% (N=7)
All *Turdus*	76.83% (N=82)

Carpometacarpus

Table 7.3. No discontinuity in facies ulnare carpalis.

% present in *Sturnus*.	
S. vulgaris (Britain)	88% (N=25)
S. unicolor	86.33% (N=86)
S. roseus	83.33% (N=3)
All *Sturnus*	87.18% (N=39)

% absent in *Turdus*.	
Turdus merula	82.14% (N=14)
T. philomelos	93.18% (N=22)
T. iliacus	100% (N=23)
T. pilaris	66.67% (N=15)
T. torquatus	66.67% (N=3)
T. viscivorus	75% (N=8)
All *Turdus*	87.06% (N=85)

Table 7.4. Overhang on edge of facies for phalanx digit minoris.

% present in *Sturnus*.	
S. vulgaris (Britain)	98% (N=25)
S. unicolor	Not done
S. roseus	100% (N=2)
All *Sturnus*	98.15% (N=27)

% absent in *Turdus*.	
Turdus merula	89.29% (N=14)
T. philomelos	50% (N=22)
T. iliacus	65.22% (N=23)
T. pilaris	93.33% (N=15)
T. torquatus	66.67% (N=3)
T. viscivorus	50% (N=8)
All *Turdus*	68.82% (N=85)

Tibiotarsus

Table 7.5. Wide sulcus intercristalis with a straight edge.

% present in *Sturnus*.	
S. vulgaris (Britain)	97.83% (N=23)
S. unicolor	100% (N=11)
S. roseus	50% (N=2)
All *Sturnus*	98.61% (N=36)

% absent in *Turdus*.	
Turdus merula	86.67% (N=15)
T. philomelos	100% (N=21)
T. iliacus	97.92% (N=24)
T. pilaris	96.67% (N=15)
T. torquatus	100% (N=3)
T. viscivorus	100% (N=9)
All *Turdus*	96.55% (N=87)

Table 7.6. No V-shaped caudal part of edge of interarticular area.

% present in *Sturnus*.	
S. vulgaris (Britain)	80.95% (N=21)
S. unicolor	75% (N=10)
S. roseus	50% (N=3)
All *Sturnus*	76.47% (N=34)

% absent in *Turdus*.	
Turdus merula	83.33% (N=15)
T. philomelos	61.91% (N=21)
T. iliacus	77.08% (N=24)
T. pilaris	70% (N=15)
T. torquatus	100% (N=3)
T. viscivorus	94.49% (N=9)
All *Turdus*	75.86% (N=87)

Coracoid

Table 7.7. No sternal projection of facies articularis clavicularis.

% present in *Sturnus*.	
S. vulgaris (Britain)	64.58% (N=24)
S. unicolor	33.33% (N=9)
S. roseus	83.33% (N=3)
All *Sturnus*	58.33% (N=36)

% absent in *Turdus*.	
Turdus merula	100% (N=15)
T. philomelos	90.9% (N=22)
T. iliacus	95.83% (N=24)
T. pilaris	93.33% (N=15)
T. torquatus	66.67% (N=3)
T. viscivorus	100% (N=9)
All *Turdus*	94.32% (N=88)

The results signify that the proximal humeri of *Sturnus* and *Turdus* can be distinguished with some confidence on two characters, both the angle of the crista pectoralis and the size of the second pneumatic fossa. The proximal carpometacarpi are also reasonably consistently different, although the distal ends are less so. The proximal tibiotarsi are diagnosable to a good level of confidence (>95% confidence) on one character, although less so on the presence or absence of a V-shape in the caudal part of the interarticular area. The coracoid character is probably not consistently different enough to distinguish *Sturnus* and *Turdus*.

The method used for distinguishing between taxa in their morphology provides a realistic recognition that variation exists in the characters used. Nevertheless, the method was successful in pinpointing reliable characters in three post-cranial bones which could then be used in fossil identifications. An interesting aspect of the results has been the fact that the characters varied in their reliability between species. This further highlights the importance of testing putative species specific or genus-specific characters.

Section III

Results

Chapter 8

The Genus *Grus* (Cranes)

8.1. Introduction

Cranes of the genus *Grus*, known as the white cranes, include ten species world-wide and are large terrestrial / wading birds. Today the genus *Grus* in the Western Palaearctic includes a single regular nesting species, the common crane, *Grus grus*. The common crane was formerly more widespread in Europe, particularly as a breeder, and once bred regularly in Britain (Southwell, 1901), but is now mostly restricted to northern and eastern Europe, although it occurs elsewhere, as a migrant on its way to Iberia and north Africa for the winter, although rarely in Britain (Boisseau and Yalden, 1998). It has recently started breeding in Britain again although only one or two pairs a year since 1981. Other species which occur in the Western Palaearctic from time to time include the Siberian white crane (*Grus leucogeranus*), which occurs rarely on migration in the east of the Western Palaearctic, and the sandhill crane (*G. canadensis*), which has occurred on at least one occasion as a vagrant (Cramp and Simmons, 1980).

Cranes of the genus *Grus* are not uncommon in the Quaternary of Europe, although they are most often found in Holocene archaeological and peat deposits. Most of these fossils have been attributed to the common crane although a large extinct Palaearctic crane, *Grus primigenia*, was described by Milne-Edwards (1867 - 71) on the basis of its morphological differences when compared to the Sarus crane, *Grus antigone*. These and other differences have been variously rejected and accepted by different authors since then. Size has, however, mostly been thought to rule out the possibility that this crane is merely a large common crane, *Grus grus*. In addition, there have been two other extinct taxa named from fossils in the Western Palaearctic, *Grus melitensis* Lydekker, 1890 from Malta and *Grus bohatshevi* Serebrovsilj, 1940 from Azerbaijan.

The questions which shall be addressed here are as follows:

- What is the pattern of size and shape in modern members of the genus *Grus*?
- How sexually dimorphic are the species within *Grus*, and why?
- What is the pattern of size and shape in fossil *Grus*?
- Is *Grus primigenia* a genuine extinct species or is it conspecific with one of the modern species?

8.2. Systematics

This section describes all the species of crane which are pertinent to the questions outlined above as well as the higher order taxa to which they belong.

Family Gruidae

The cranes (Aves: Gruidae) are a widely distributed group of 15 (or arguably 14) species in four genera. The species and genera within the family have been studied by a number of authors with a view to understanding their systematic relationships. Techniques used have been the examination of unison calls as taxonomic characters (Archibald, 1976a, 1976b), morphological characters (Wood, 1979) and proteins and DNA (Ingold *et al.* 1987a, 1987b, 1989; Krajewski, 1989). The different techniques ally the crane species in different ways. Biomolecular methods agree, however, that the crowned cranes (*Balearica*) are the most ancient lineage and that all the remaining species form a cluster within which the Siberian white crane *Grus leucogeranus* forms a sister group to the others. The latter implies that genera such as *Anthropoides* and *Bugeranus* should be synonymised with *Grus* (Krajewski, 1989) although this has not been widely adopted.

Genus *Grus*

The genus *Grus* contains 9 species according to Peters (1934), although Krajewski (1989) suggested adding the species within *Anthropoides* and *Bugeranus* to the genus. The nine species traditionally placed within *Grus* are the common crane *Grus grus*, the hooded crane *G. monacha*, the sandhill crane *G. canadensis*, the red-crowned crane *G. japonensis*, the whooping crane *G. americana*, the white naped crane *G. vipio*, the sarus crane *G. antigone*, the broiga *G. rubicunda* and the Siberian white crane *G. leucogeranus*.

Grus grus (Linnaeus, 1758) - Common crane

The common crane is a terrestrial bird about 20% larger than the grey heron *Ardea cinerea*. Their plumage is grey, getting darker around the head and neck, with a white band behind the eye and black flight feathers. Their bills are sharply pointed and their necks and legs are long (Cramp and Simmons, 1980).

The common crane is polytypic with two subspecies, the nominate *G. g. grus* Linnaeus, 1758 which occurs in Europe east to the Urals and *G. g. lilfordi* Sharpe, 1894 which is found in Turkey, Transcaucasia, Armenia and Siberia (Cramp and Simmons, 1980).

The common crane is now concentrated in upper and middle latitudes: in the arctic tundra and steppe zones, in temperate and boreal taiga, and in deciduous forests (Cramp and Simmons, 1980). In northern parts of their range they will breed where there are no trees, such as on moorlands and *Sphagnum* bogs. In Sweden they typically breed in swampy clearings in coniferous forests, and in eastern Germany in reedy wetlands. In other areas they make use of alder carr and quaking bogs. They are prone to human disturbance and their breeding range has been particularly reduced further south, the birds having at one time bred in areas such as Andalucia in Spain and Norfolk in Britain. During migration, part of the Scandinavian population of *G. g. grus* flies south to Spain and North Africa while the rest fly to south-eastern Europe, parts of the Middle East and Sudan. The Siberian race *G. g. lilfordi* moves into southern Asia. They move in family groups during September.

Their food includes both animal and plant material, although plants are their primary food and include roots, rhizomes, tubers, stems leaves, fruits and seeds. Animal foods, mostly insects but also small vertebrates at times, are taken more often during the summer.

The species has been recorded regularly in the fossil record of Europe, although commonly only on later Holocene archaeological sites (Boisseau and Yalden, 1998).

Grus antigone (Linnaeus, 1758) - Sarus crane

The adults of both sexes have bare heads, throats and necks. This bare skin is mostly orange-red and covered in coarse granulations except for the crown which is smooth and pale ashy green. In the nominate subspecies there is a ring of white feathers which is replaced by grey plumage further down the neck and the rest of the body. The other subspecies lack the white ring.

The sarus crane is a resident of India, south-east Asia and Australia and is the world's tallest flying bird. Three subspecies are currently recognised, the Indian sarus crane (*G. antigone antigone*), the eastern sarus crane *(G. a. sharpii*) and the Australian sarus crane (*G. a. gilli*). The nominate subspecies lives today in northern India although its range has been reduced. The eastern sarus crane has also had its range much reduced over historical times and is now known only from Cambodia, Vietnam and Laos. The Australian sarus crane, which was not recognised until the 1960s is limited to north-eastern Australia. There may have been a fourth subspecies in the Philippines which, however, was never named (Johnsgard, 1983). The subspecies differ in size, *G. a. antigone* being taller than the other two (Mein and Archibold, 1996).

Their habitat seems to vary slightly according to the area they occur in. In India they nest in forestry areas with flooded marshes which are pastured by cattle and water buffalo. Nesting is tied to the monsoon period, and when the rains fail they may not nest at all and simply remain in the area until the following year (Johnsgard, 1983). In Burma the species is more likely to be found on isolated swamps and marshes than in India. In Australia the sarus crane is found near freshwater and cultivated areas. Non-breeding habitats are the same as those while nesting because the species is mostly sedentary, only dispersing in times of extreme drought (Walkingshaw, 1973).

Their food consists of tubers and corms of aquatic vegetation, green shoots of grasses, as well as grasshoppers and other insects and even on occasion aquatic snails and vertebrates such as snakes and fish (Johnsgard, 1983).

The fossil record of this species is subject of debate due to questions of the identity of large crane fossils found in the Western Palaearctic. Some authorities believe the fossils to be sarus cranes while others have preferred to assign them to an extinct species *Grus primigenia* (see below).

Grus leucogeranus (Pallas, 1773) - Siberian white crane

Grus leucogeranus is a large white crane with a raw looking face and black outer wings. The species is 10% larger than the common crane (Cramp and Simmons, 1980).

The Siberian white crane exists in the Western Palaearctic only in small numbers on migration. The species' breeding range has become much reduced over the last 100 years and is now disjunct across the Eastern Palaearctic. They breed from middle to high latitudes on dry inaccessible parts of marshland, on reeds by low lying shores of lakes, or on brushlands with occasional trees. When not breeding they occur by lakes, river mouths or on steppes near water. The species is more likely than other cranes to be wading in shallow water. They reach very high altitudes in flight when on migration. They are omnivorous although most of their diet consists of aquatic vegetation which they take by probing in the mud with their heads submerged (Cramp and Simmons, 1980).

A single record is known from the Western Palaearctic fossil record, described by Boessneck (1987) from the early Bronze Age of Hassek Höyuük in south-east Turkey. In this instance three bones were found including a scapula, a coracoid and a posterior phalanx I of digit II.

Grus primigenia (Milne-Edwards, 1867 - 71)

In the nineteenth century Milne-Edwards (1867-71) recognised that a large crane had existed in the Quaternary of the Western Palaearctic which he named *Grus primigenia.* The species was based on a distal end of a tibiotarsus from a cave site, Les Eyzies in the Dordogne in France. His diagnosis stated that the distal tibiotarsus was wider mediolaterally than in *G. antigone.* Later he referred to this species a humerus from the Grotte de la Madeleine and a mandible from the Grotte Gourdan (Milne-Edwards, 1875). This crane has been the subject of controversy ever since, with suggestions that the remains represent the modern Asiatic sarus crane

Grus antigone (Lydekker, 1891b; Lambrecht, 1933; Soergel, 1955; Brodkorb, 1967; Northcote, 1979b and 1982c; Mourer-Chauviré *et al.*, 1975) or simply a large common crane *G. grus* (Driesch and Boessneck, 1969). Harrison and Cowles (1977) and two of the above authors have subsequently accepted the validity of *Grus primigenia* (Northcote and Mourer-Chauviré, 1985, 1988). Harrison and Cowles (1977) concluded that the large crane bones were unlikely to belong to *Grus antigone* because of their geographical distribution, the species being an isolated one, found only south of the Himalayas. Northcote and Mourer-Chauviré (1985, 1988) based their deduction on a comparison of the morphology of certain skeletal elements of *Grus primigenia* with *Grus antigone.*

Tyrberg has produced a useful review of the literature and listed all the purported occurences of *Grus primigenia* (Tyrberg, 1985). Hamilton-Dyer (1993) studied a large collection, 62 bones, of Roman crane remains of which some were as large as the sarus crane. She considered that too few remains of the common crane were available to her to resolve the problem, but stated that it was possible that the common crane had decreased in size over time.

The status of this crane is therefore still unresolved, and bones have been ascribed to *Grus primigenia* from the Holocene of Germany and England (Soergel, 1955 and 1961; Harrison and Cowles, 1977) as well as the Pleistocene of France, England and Spain (Saint-Périer, 1920, 1927; Delpech, 1967, 1975; Harrison and Cowles, 1977; Northcote and Mourer-Chauviré, 1988).

***Grus melitensis* (Lydekker, 1890)**

Lydekker named this species on the basis of crane fossils collected in Zebbug Cave on Malta and presented to the Natural History Museum in London by Admiral Spratt in 1878. The diagnosis of the species was that it was of a similar size to *G. antigone* but with a smaller and narrower coracoid head and a shorter bar over the extensor groove of the tibiotarsus (Lydekker, 1890). Subsequent to being named, the species has remained unchallenged except in details of the interpretation of its lifestyle and what material is referable to it rather than to other species such as *Grus grus* (Brodkorb, 1967; Harrison and Cowles, 1977; Northcote, 1980; Northcote, 1982a; Northcote and Mourer-Chauviré, 1988). For instance, the original suggestion that the species had poor powers of flight was challenged when the proportionately small coracoid and humerus of the species were shown to belong to the common crane (Northcote, 1984). This crane has also been found in two further Maltese caves since the original discovery, Mnaidra Cave and Tal Gnien Cave.

***Grus bohatshevi* (Serebrovskilj, 1940)**

A fragmentary crane skull was found in the Pleistocene Binagady Kir layers near Baku in Azerbaijan. The specimen was found to resemble that of a Siberian white crane *G. leucogeranus*, but was considered distinct enough to warrant designation as a separate species in the same genus (Serebrovskilj, 1940).

8.3. *Grus* Osteology

8.3.1. Identification of the genus *Grus*

There are no significant difficulties in distinguishing the major long bones of *Grus* from other avian groups. Their extreme size eliminates most birds in the Western Palaearctic and birds of comparative size (e.g. swans and eagles) are very distinct morphologically.

8.3.2. Modern samples

Grus grus

The best collection of common crane skeletons in Europe is housed in the Swedish Museum of Natural History in Stockholm. This collection of cranes is both well provenanced and sexed, including ten of each sex. An additional two males and a single female were available from collections in Denmark (Zoological Museum, Copenhagen), Belgium (Musée d'Histoire Naturelle in Brussels) and Spain (Universidad Autonoma de Madrid) respectively. Finally, a further seven complete individuals and a trunk skeleton were measured in Spain (Universidad Autonoma de Madrid and Museo de Cienca Naturale de Madrid), but were of unknown sex. The Scandinavian birds were breeding birds while those from Spain were on migration in their wintering quarters. The single skeleton measured in Brussels was from Antwerp Zoo and originally came from Africa, although exactly where is not known. All except for perhaps the specimen from Africa are likely to be of the nominate subspecies *Grus grus grus*. The common crane specimens were mostly collected between 1989 and 1994 except for the specimen from Brussels which was dated as 1897.

Grus antigone

The sarus cranes came from the collections of both the Musée d'Histoire Naturelle in Brussels (N=12) and the sub-department of ornithology of the Natural History Museum in Tring in Hertfordshire (N=6). All the specimens from Belgium and three of the specimens from Britain were recorded as having died in captivity, at Antwerp and London Zoos respectively. Some of the birds from Antwerp Zoo had subspecific data. There were six specimens of *Grus antigone antigone* and four of *G. a. sharpii*, but a further two did not possess this information. Only four of the sarus cranes were sexed, and were all males of the *G. a. antigone* subspecies. The sarus crane skeletons were acquired by the two museums between 1853 and 1967.

All the specimens measured were adults to the extent that none had unfused or visible fusion lines on their tibiotarsi or tarsometatarsi (see Chapter 6).

8.3.3. Morphometric comparison of *Grus grus* and *Grus antigone*

A metrical comparison between the major limb bones of *Grus grus* and *Grus antigone* was necessary due to the problems of the identity of European Quaternary crane fossils. These are the two modern species that have been considered in relation to the identity of *Grus primigenia*. Therefore, before the fossils were considered, a study of the morphometric difference between the two was necessary. No previous study had attempted to compare these two species in terms of size and metric shape. Previous analyses had also tended to be based on relatively small numbers of individuals. The present analysis attempts to redress both these flaws. The measurements taken for this analysis are described in detail in Chapter 6.

The humeri of *Grus grus* and *G. antigone* are on the whole very distinct in size, *Grus grus* being almost consistently smaller than *G. antigone* (Figure 8.1, Table 8.3). The only exception was a single small specimen of *G. antigone* which fell within the *G. grus* range in every measurement.

The ulnae showed good separation in length between the two species, although breadth and depth measurements overlapped to a greater extent (Figure 8.2). The radius was similar in its proximal breadth although a more satisfactory separation between *G. grus* and *G. antigone* exists in the distal measurement (Figure 8.3).

The carpometacarpi, once more, differ greatest in their length and overlap in breadth measurements, although all dimensions are statistically significantly different (Figure 8.4 and Table 8.3).

The coracoidea of *G. antigone* overlap in the breadth of the facies articularis basalis (BF) measurement, indicating that they are not simply scaled up versions of *G. grus* (Figure 8.5, Table 8.3).

The two scapulae of *G. antigone* are distinctly larger in both greatest length and the greatest cranial diagonal (Dic) than all those of *G. grus* (Figure 8.6).

The femora of *Grus antigone* are consistently larger than those of *G. grus* (Figure 8.7, Table 8.3).

The tibiotarsi and tarsometatarsi exhibit the greatest difference in all measurements taken (means are 25 – 32 % larger cf. Tables 8.1 and 8.2). They overlap slightly only on one measurement, that of the tarsometatarsal shaft breadth (Figure 8. 8 and 8.9).

Figure 8.10 shows the relationship between the two species' humeral lengths and carpometacarpal lengths. This reveals that *G. antigone* has a very similar carpometacarpus to humerus ratio to *G. grus*. The regression lines in Figure 8.10 are not very different.

In summary, it is clear that *G. antigone* is a significantly larger bird than *G. grus* (Table 8.3: BF). Means for limb bone length dimensions of *G. antigone* range from 10.7 % to 32.3 % greater than for *G. grus*. The breadth dimensions of *G. antigone* vary from 9.6 % to 32.7 % greater than for *G. grus*. In addition, there maybe some evidence for differences in shape between the two species. This is particularly apparent in the breadth measurements of the facies articularis basalis and the shaft of the coracoid (Table 8.3) although this may be due to low sample size of the *G. antigone* coracoid measurements (i.e. N=4). Interestingly there is no difference in the humerus to carpometacarpus greatest length ratios of the two species. This is contrary to expectations because *G. grus* is migratory while *G. antigone* is sedentary. These proportions would normally be expected to differ, with migratory taxa having longer relative carpometacarpi to humeri than sedentary species cf. Chapter 2.

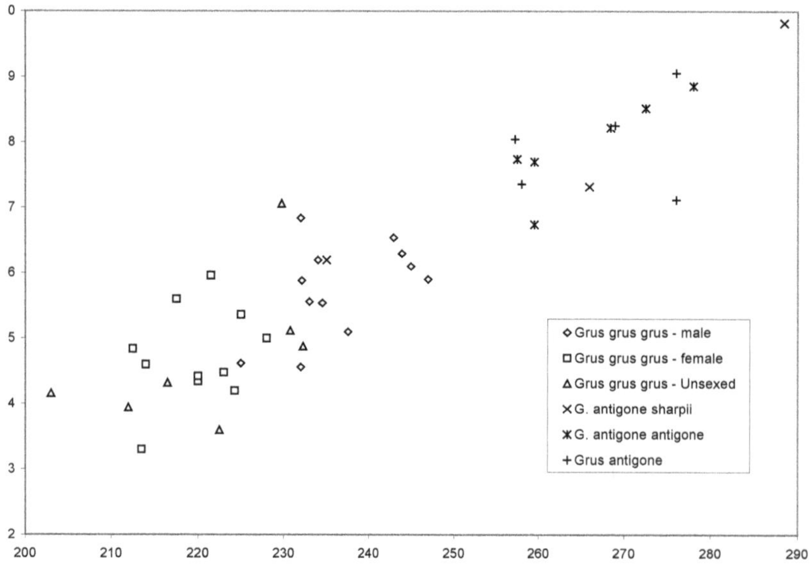

Figure 8. 1. *Grus grus* and *Grus antigone* humeri. Greatest length (GL) plotted against shaft breadth (SC).

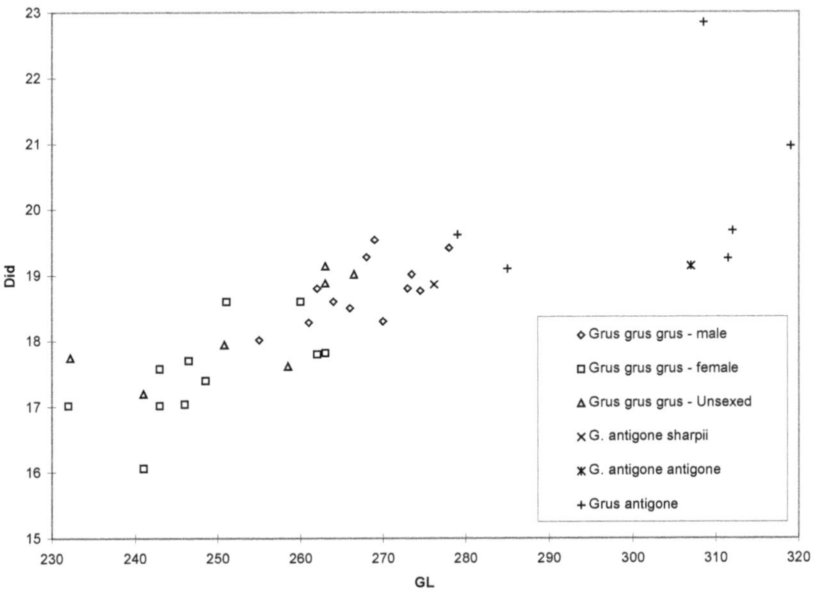

Figure 8. 2. *Grus grus* and *Grus antigone* ulnae. Greatest length (GL) plotted against distal diagonal (Did).

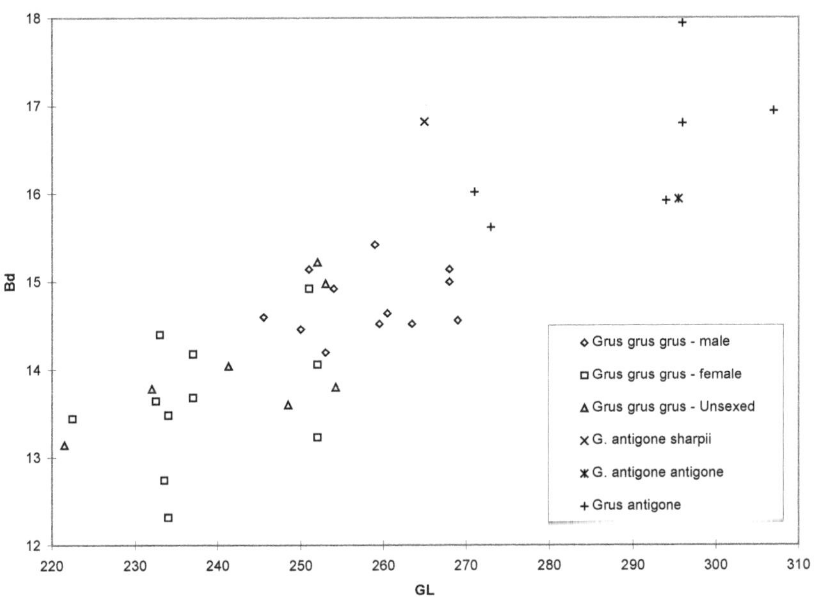

Figure 8. 3. *Grus grus* and *Grus antigone* radii. Greatest length (GL) plotted against distal breadth (Bd).

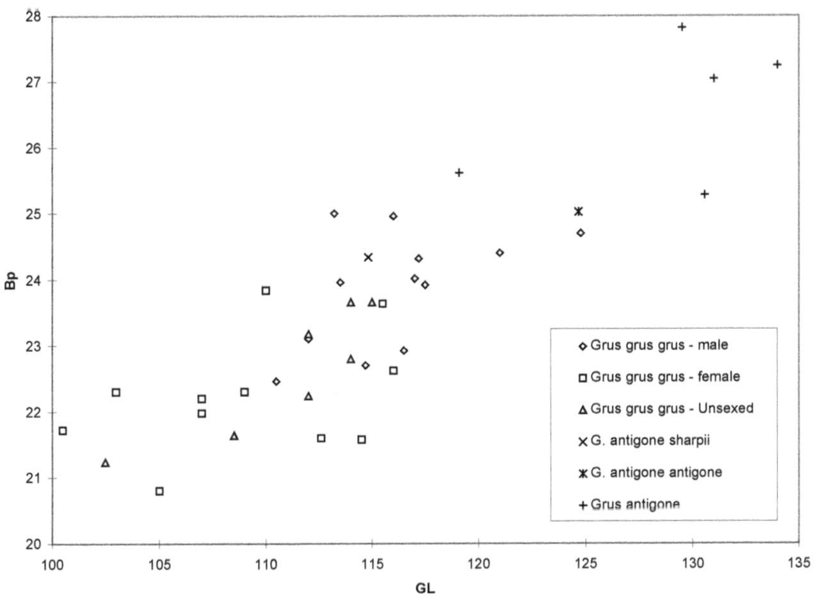

Figure 8. 4. *Grus grus* and *Grus antigone* carpometacarpi. Greatest length (GL) plotted against proximal breadth (Bp).

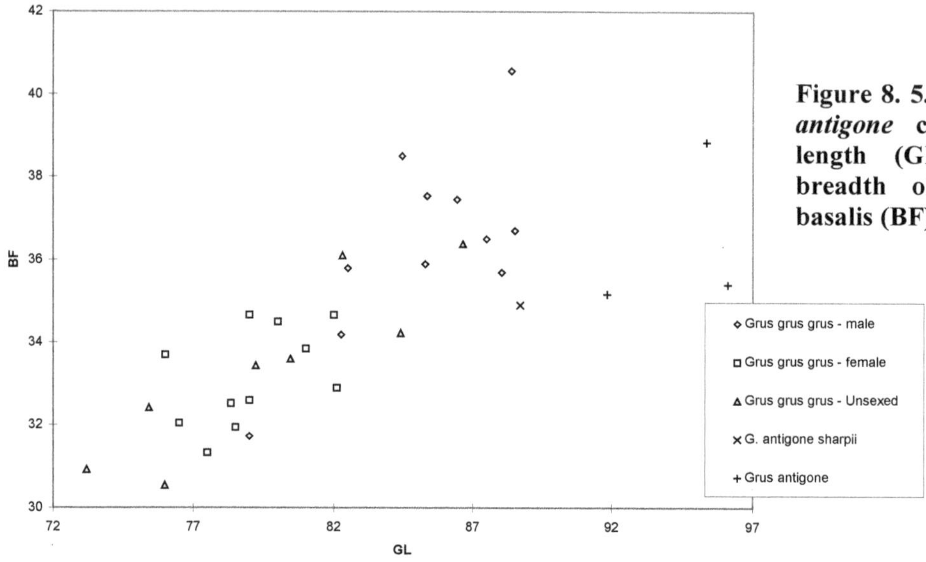

Figure 8. 5. *Grus grus* and *Grus antigone* coracoidea. Greatest length (GL) plotted against breadth of facies articularis basalis (BF).

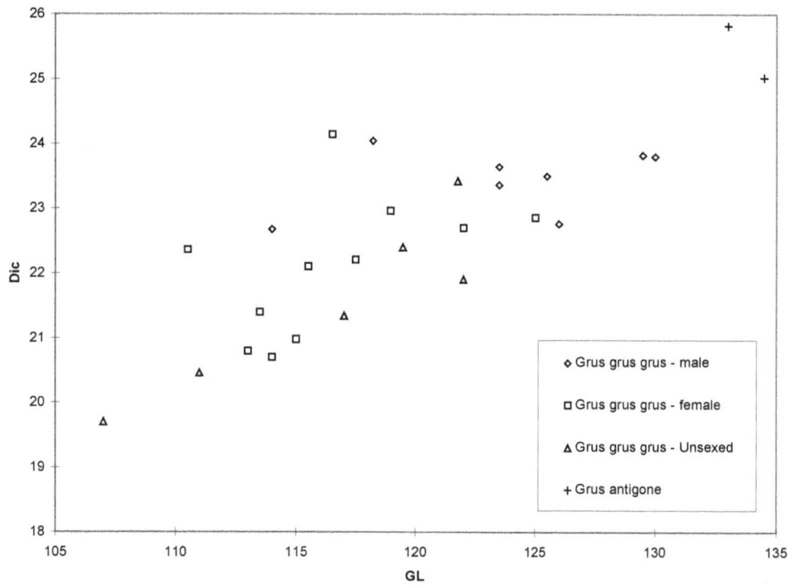

Figure 8. 6. *Grus grus* and *Grus antigone* scapulae. Greatest length (GL) plotted against greatest cranial diagonal (Dic).

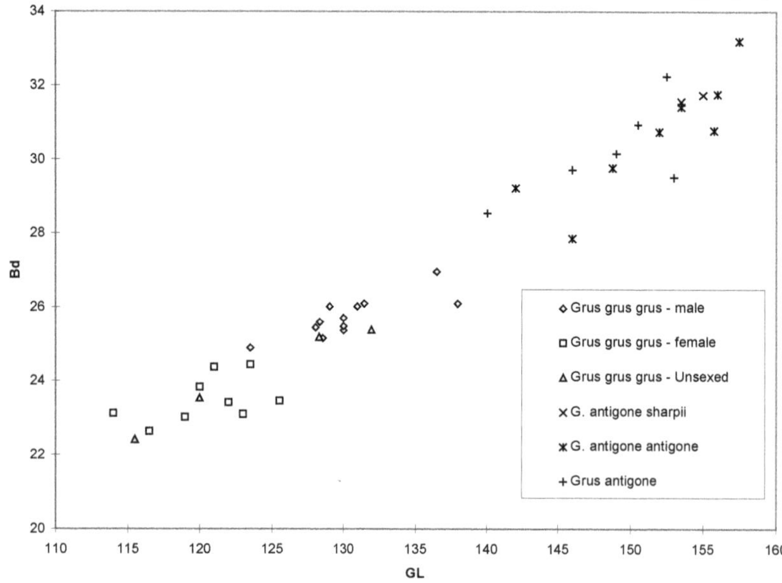

Figure 8. 7. *Grus grus* and *Grus antigone* femora. Greatest length (GL) plotted against distal breadth (Bd).

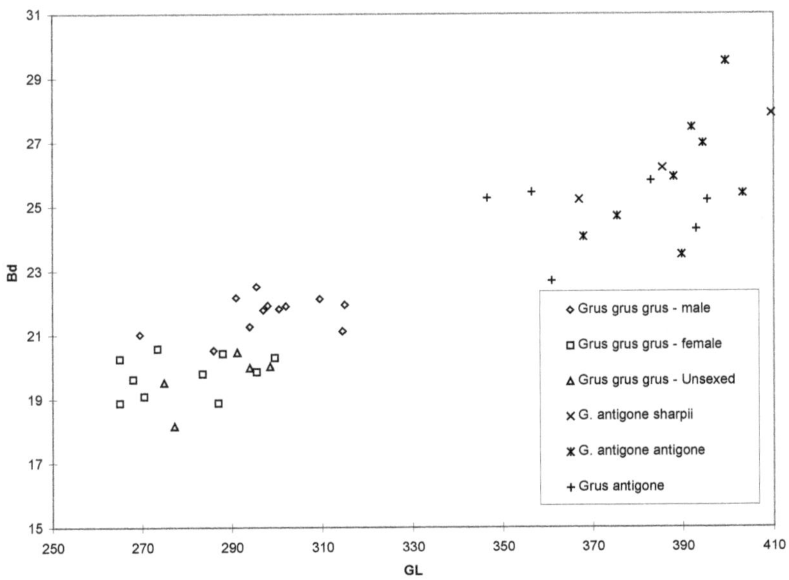

Figure 8.8. *Grus grus* and *Grus antigone* tibiotarsi. Greatest length (GL) plotted against distal breadth (Bd).

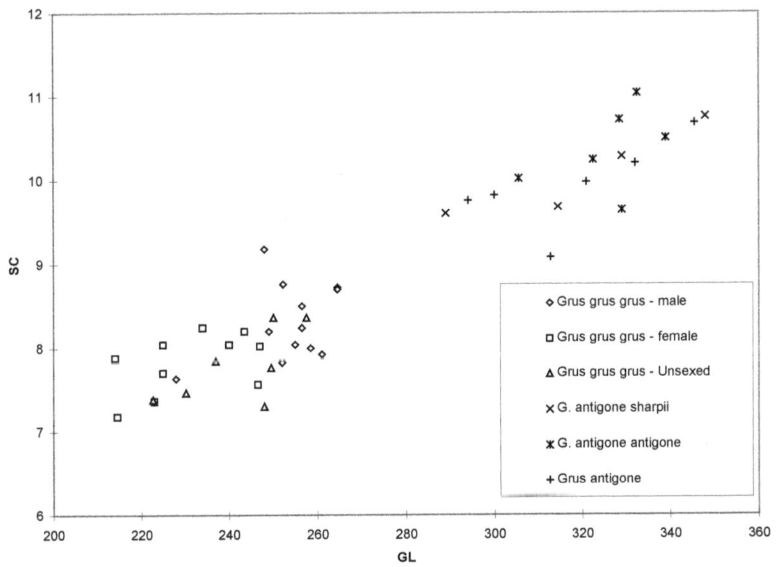

Figure 8. 9. *Grus grus* and *Grus antigone* tarsometatarsi. Greatest length (GL) plotted against shaft breadth (SC).

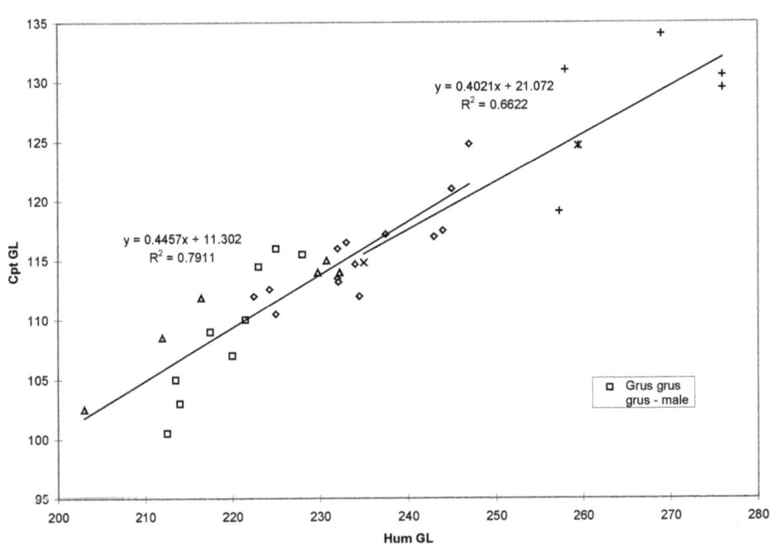

Figure 8.10. *Grus grus* and *Grus antigone* humerus to carpometacarpus greatest length ratios with regression lines.

Table 8. 1. Mensural data for *Grus grus* (mixed sexes).

Measurement	N	Min Max	Mean	Stand. dev.
Hum GL	30	203 – 247	226.84	10.81
Hum Bp	30	45.22 – 56.22	49.84	3.16
Hum Bd	30	30.72 – 36.04	33.31	1.55
Hum SC	30	13.3 – 17.06	15.14	0.96
Uln GL	30	232 – 278	257.5	12.64
Uln Bp	30	20.8 – 24.7	22.36	1.04
Uln Dip	23	25.2 – 29.34	27.28	1.23
Uln Did	30	16.06 – 19.54	18.18	0.84
Uln SC	29	7.08 – 9.34	8.56	0.51
Rad GL	30	221.5 – 269	246.9	12.83
Rad Bp	29	9.72 – 11.94	10.8	0.55
Rad Bd	30	12.32 – 15.42	14.19	0.78
Cpt GL	30	100.5 – 124.78	112.4	5.42
Cpt Bp	30	20.8 – 25	22.32	1.15
Cpt Bd1	30	14.42 – 16.7	15.44	0.64
Cpt Bd2	30	11.32 – 13.78	12.72	0.63
Fem GL	30	112.25 – 138	125.26	6.35
Fem Bp	30	24.68 – 30.14	27.31	1.31
Fem Dp	30	15.3 – 19.9	17.43	1.23
Fem Bd	30	22.42 – 26.96	24.55	1.29
Fem SC	28	9.8 – 11.92	10.97	0.69
Tib GL	30	256 – 315	287.97	15.02
Tib Dip	30	27.66 – 36.04	31.73	1.91
Tib Bd	30	18.16 – 22.5	20.48	1.16
Tib Dd	30	18.02 – 22.42	19.78	1.2
Tib SC	28	8.16 – 10.26	9.34	0.55
Tmt GL	30	214 – 264.5	243.51	14.71
Tmt Bp	28	20.6 – 25.7	22.77	1.22
Tmt Bd	29	19.74 – 26.54	23.41	1.51
Tmt SC	30	7.18 – 9.18	8	0.48
Cor GL	30	73.2 – 88.52	81.52	4.28
Cor Lm	30	57.84 – 70.74	64.22	3.27
Cor BF	30	30.54 – 40.56	34.43	2.4
Cor SC	31	12.08 – 16.18	13.75	0.99
Scap GL	26	107 – 130	118.97	5.93
Scap Dic	28	19.7 – 24.4	22.35	1.3

Table 8. 2. Mensural data for *Grus antigone* (mixed sexes and subspecies).

Measurement	N	Min Max	Mean	Stand. dev.
Hum GL	14	235 - 288.5	265.8	12.91
Hum Bp	14	51.36 - 62.16	57.42	3.31
Hum Bd	14	35.58 - 41.28	38.67	1.75
Hum SC	14	16.2 - 19.82	17.93	0.97
Uln GL	8	276 - 319	299.78	16.86
Uln Bp	8	23.8 - 27.44	25.19	1.26
Uln Dip	8	28.52 - 33.98	30.51	1.76
Uln Did	8	18.86 - 22.84	19.93	1.34
Uln SC	8	8.9 - 10.42	9.68	0.59
Rad GL	8	265 - 307	287.19	15.21
Rad Bp	8	11.48 - 13.56	12.08	0.71
Rad Bd	8	15.62 - 17.94	16.5	0.77
Cpt GL	8	111.44 - 134	124.39	8.35
Cpt Bp	8	24.34 - 27.82	26.05	1.31
Cpt Bd1	8	16.4 - 19.14	17.64	0.91
Cpt Bd2	8	13.06 - 14.92	14.09	0.62
Fem GL	18	140.05 - 157.5	149.81	5.44
Fem Bp	18	29.56 - 35.92	32.79	1.65
Fem Dp	18	19.02 - 22.98	20.98	1.18
Fem Bd	18	27.86 - 33.2	30.44	1.41
Fem SC	18	12.26 - 15.68	13.61	0.92
Tib GL	18	346.5 - 409.5	380.94	18.89
Tib Dip	18	37.08 - 43.1	40.26	1.76
Tib Bd	18	22.68 - 29.52	25.58	1.63
Tib Dd	18	22.72 - 27.92	25.52	1.57
Tib SC	18	10.62 - 13.9	12.39	0.75
Tmt GL	16	289 - 348	321.42	17.58
Tmt Bp	15	25.38 - 33.5	28.59	1.98
Tmt Bd	15	27.22 - 31.62	29.76	1.31
Tmt SC	16	9.08 - 11.04	10.13	0.53
Cor GL	4	88.7 - 96.1	93	3.42
Cor Lm	4	72.7 - 79	75.14	2.76
Cor BF	4	34.9 - 38.84	36.08	1.85
Cor SC	4	14.66 - 15.2	14.79	0.42
Scap GL	3	127.2 - 134	131.67	3.86
Scap Dic	2	25.02 - 25.82	25.42	0.57

Table 8.3. *t*- tests between *Grus grus* and *G. antigone*.

Measurements	
Hum GL	***
Hum Bp	***
Hum Bd	***
Hum SC	***
Uln GL	***
Uln Bp	***
Uln Dip	***
Uln Did	***
Uln SC	***
Rad GL	***
Rad Bp	**
Rad Bd	***
Cpt GL	***
Cpt Bp	***
Cpt Bd1	***
Cpt Bd2	***
Fem GL	***
Fem Bp	***
Fem Dp	***
Fem Bd	***
Fem SC	***
Tib GL	***
Tib Dip	***
Tib Bd	***
Tib Dd	***
Tib SC	***
Tmt GL	***
Tmt Bp	***
Tmt Bd	***
Tmt SC	***
Cor GL	***
Cor Lm	***
Cor BF	ns
Cor SC	ns
Scap GL	**
Scap Dic	**

Key: *** $p< 0.001$, ** $p< 0.01$, * $p< 0.05$, ns Not significant at the 5% level.

8.3.4. Sexual Dimorphism

Sexual dimorphism within *Grus grus* has been described as small in magnitude by Northcote (1979b) from her analysis of coefficients of variation for her sample of modern common cranes. Figures 8.1 - 8.10, however, clearly show that the difference between males and females in terms of size is impressive. Table 8.4 gives the percentage differences between the males and females of *G. grus* for all the limb bones measured in this study. All are between 5.72 % and 8.14 %.

Table 8. 4. Percentage differences between the means of male and female *Grus grus*.

Measurement	n Males	n Female	% difference
Humerus GL	12	11	7.04
Ulna GL	12	11	7.13
Radius GL	12	11	7.88
Carpometacarpus GL	12	11	5.95
Coracoid GL	11	11	7.04
Scapula GL	9	11	5.82
Femur GL	12	11	6.94
Tibiotarsus GL	12	11	5.72
Tarsometatarsus GL	12	11	8.14

Table 8. 5. Mensural data for male *Grus grus*.

Measurement	N	Min Max	Mean	Stand. dev.
Hum GL	12	225 - 247	236.59	6.72
Hum Bp	12	48.8 - 56.88	52.73	2.34
Hum Bd	12	32.68 - 36.04	34.64	1.03
Hum SC	12	14.56 - 16.84	15.76	0.72
Uln GL	12	255 - 278	267.83	6.58
Uln Bp	12	22.2 - 24.7	23.25	0.71
Uln Dip	12	26.8 - 29.34	28.1	0.82
Uln Did	12	18.02 - 19.54	18.78	0.47
Uln SC	12	8.52 - 9.34	8.95	0.27
Rad GL	12	245.5 - 269	258.42	7.8
Rad Bp	12	10.66 - 11.82	11.16	0.35
Rad Bd	12	14.2 - 15.42	14.78	0.39
Cpt GL	12	110.5 - 124.78	116.16	3.92
Cpt Bp	12	22.46 - 25	23.87	0.88
Cpt Bd1	12	15.24 - 16.7	15.97	0.44
Cpt Bd2	12	12.26 - 13.78	13.18	0.51
Fem GL	12	123.5 - 138	130.36	3.82
Fem Bp	12	27.26 - 30.14	28.33	0.87
Fem Dp	12	17.58 - 19.9	18.47	0.73
Fem Bd	12	24.9 - 26.96	25.74	0.55
Fem SC	12	10.4 - 11.9	11.17	0.51
Tib GL	12	269.5 - 315	297.7	12.59
Tib Dip	12	31.48 - 36.04	33.42	1.39
Tib Bd	12	20.52 - 22.5	21.67	0.57
Tib Dd	12	19.62 - 22.42	20.94	0.77
Tib SC	12	8.72 - 10.24	9.61	0.43
Tmt GL	12	220-264	253.81	9.76
Tmt Bp	12	21.98 - 25.7	23.69	0.98
Tmt Bd	12	23.1 - 26.54	24.54	0.96
Tmt SC	12	7.63 - 9.18	8.31	0.46
Cor GL	12	79 - 88.52	85.26	3.02
Cor Lm	12	62.4 - 70.74	66.97	2.57
Cor BF	12	31.72 - 40.56	36.41	2.28
Cor SC	12	13.06 - 16.18	14.46	0.91
Scap GL	12	114 - 130	123.7	5.07
Scap Dic	12	22.68 - 24.4	23.56	0.56

Table 8. 6. Mensural data for female *Grus grus*.

Measurement	N	Min Max	Mean	Stand. dev.
Hum GL	12	212.5 - 228	219.93	5.09
Hum Bp	12	45.22 - 50.68	47.55	1.58
Hum Bd	12	30.8 - 34.02	32.29	1.02
Hum SC	12	13.3 - 15.96	14.74	0.74
Uln GL	12	232 - 263	248.73	9.64
Uln Bp	12	21 - 22.64	21.65	0.55
Uln Dip	12	25.2 - 27.94	26.39	0.96
Uln Did	12	16.06 - 18.6	17.51	0.74
Uln SC	12	7.08 - 9.04	8.27	0.56
Rad GL	12	222.5 - 252	238.05	9.54
Rad Bp	12	9.96 - 11.94	10.54	0.54
Rad Bd	12	12.32 - 14.92	13.64	0.74
Cpt GL	12	100.5 - 116	109.1	5.17
Cpt Bp	12	13.84 - 23.64	21.33	2.58
Cpt Bd1	12	14.58 - 16.04	15.12	0.4
Cpt Bd2	12	11.32 - 13.3	12.39	0.6
Fem GL	12	114 - 127	121.31	3.8
Fem Bp	12	25.64 - 28.02	26.8	0.8
Fem Dp	12	15.58 - 17.84	16.53	0.73
Fem Bd	12	22.64 - 24.44	23.57	0.61
Fem SC	12	9.9 - 11.58	10.45	0.56
Tib GL	12	265 - 299.5	280.68	12.69
Tib Dip	12	27.66 - 31.94	30.34	1.1
Tib Bd	12	18.88 - 20.58	19.72	0.63
Tib Dd	12	18.02 - 20.06	18.94	0.62
Tib SC	12	8.16 - 9.9	9.07	0.47
Tmt GL	12	214 - 252	233.14	13.53
Tmt Bp	12	20.6 - 23.32	21.98	0.8
Tmt Bd	12	20.88 - 25.5	23.03	1.24
Tmt SC	12	7.18 - 8.24	7.8	0.35
Cor GL	12	76 - 82.1	79.09	2.04
Cor Lm	12	60.52 - 66.44	62.67	1.74
Cor BF	12	31.32 - 34.66	33.15	1.18
Cor SC	12	12.3 - 15.34	13.34	0.87
Scap GL	12	110.5 - 125	116.5	4.2
Scap Dic	12	20.7 - 24.14	22.11	1.06

8.3.5. Geographical and subspecific differences within *Grus grus* and *Grus antigone*

Grus grus

The specimens of *Grus grus* measured in this study all belong to the nominate subspecies *Grus grus grus*. No comparative material of the other subspecies *G. g. lilfordi*, from the east of the common crane's geographic range, were seen during the study. The specimens from Spain and Scandinavia cannot be treated as separate populations because they consist of birds collected in their wintering and breeding ranges respectively (Cramp and Simmons, 1980). The Spanish specimens appear to be smaller on average than the Scandinavian examples, although because they all fall inside the range of the females from Scandinavia, it may be that they are predominately females.

Grus antigone

Two subspecies of *Grus antigone* were available in the Brussels Museum collection, *Grus antigone antigone* and *Grus antigone sharpii*. These two subspecies' ranges are India and south-east Asia, respectively. However, complete overlap exists between the two subspecies as can be seen in Figures 8.1 to 8.10.

8.3.6. Morphometric comparison of fossil samples

The fossil samples of *Grus* analysed here come in the main from Britain and Denmark. One of the fossils is from the penultimate interglacial of the Pleistocene (*ca.* 200 Kyr BP); all others are dated to between the Mesolithic and Roman periods. The fossils have been treated the same regardless of previous assignments to neospecies such as *G. grus* or *G. antigone* or to the extinct species *G. primigenia*.

Roman (Britain)

London Wall, London

The bird remains from London Wall are Roman in age (Harrison, 1980a) although very little other information is available about the site. It appears that the bird remains found at the site formed part of a refuse pit deposit.

Table 8. 7. *Grus* fossils and their measurements from London Wall.

Specimen No.	Skeletal element	Measurements		
		GL	Bp	Bd
A1033	Radius Dist. radius	254.0 -	10.92 -	14.18 14.14

The two radii from London Wall conform in dimensions to present day small males and/or large females of *Grus grus* (Figure 8.11; Table 8.5).

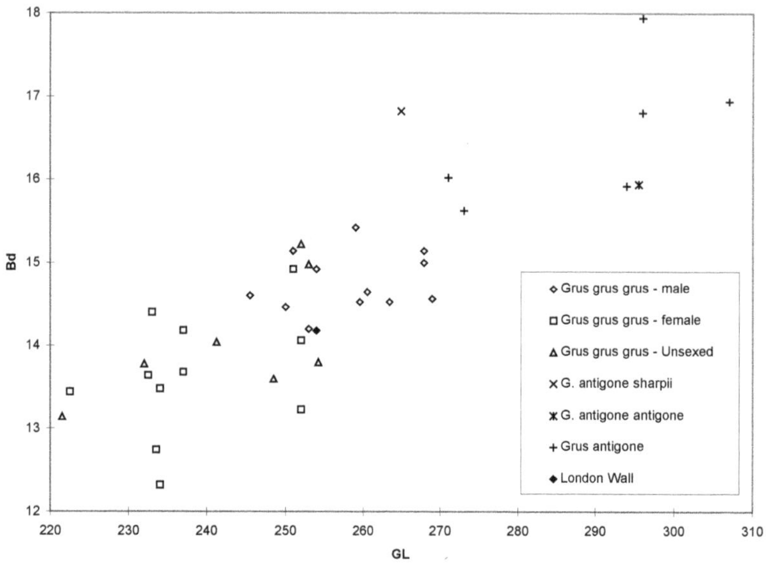

Figure 8.11. *Grus* radius from London Wall plotted with modern samples of *Grus grus* and *Grus antigone*. Greatest length (GL) plotted against distal breadth (Bd).

Iron Age (Britain)

Glastonbury Lake Dwelling, Somerset

The lake dwelling at Glastonbury was excavated during the last century and described by Bulleid and Gray (1917). A number of bird bones were found during the excavation and identified at the time by C.W. Andrews (Bulleid and Gray, 1917). The age of the site is thought to be Iron Age based on artifactual evidence. Most of the other bird taxa were water birds including Dalmatian pelican (*Pelecanus crispus*), bittern (*Botaurus stellaris*), mute swan (*Cygnus olor*), mallard (*Anas platyrhynchos*), wigeon (*A. penelope*), pintail (*A. acuta*), pochard (*Aythya ferina*), tufted duck (*A. fuligula*), smew (*Mergus albellus*) and possible moorhen (*Gallinula chloropus*?) in addition to white-tailed sea eagle (*Haliaeetus albicilla*), marsh harrier (*Circus aeruginosus*), barn owl (*Tyto alba*) and crow / rook (*Corvus corone / C. frugilegus*) (Harrison, 1980a).

The right ulna (A1248) from Glastonbury Lake Dwelling has been referred to the species *G. primigenia* by Harrison and Cowles (1977) on the basis of its extreme size. Figure 8.12 and Table 8.5 confirm that the specimen is larger than all modern specimens of *Grus grus* measured in the present study. The specimen falls close to those of modern *Grus antigone* in Figure 8.12. The distal right tibiotarsus is the size of a small male *Grus grus* while the tarsometatarsus appears to be the size of a modern female common crane (see Figures 8.13 and 8.14). There is, therefore, a broad spectrum of size within the three *Grus* fossils from Glastonbury.

Table 8. 8. *Grus* fossils and their measurements from Glastonbury Lake Dwelling

Specimen No.	Skeletal element	Measurements				
		GL	Bp	Dip	Did	SC
A1248	R. ulna	303.50	25.0	31.74	20.52	9.54
		GL	Dip		Bd	Dd
A1250	Dist. R. tibiotarsus	-	-		20.84	20.34
		GL	Bp		Bd	SC
A1251	L. tarsometatarsus	245 +	22.4		22.1	8.32

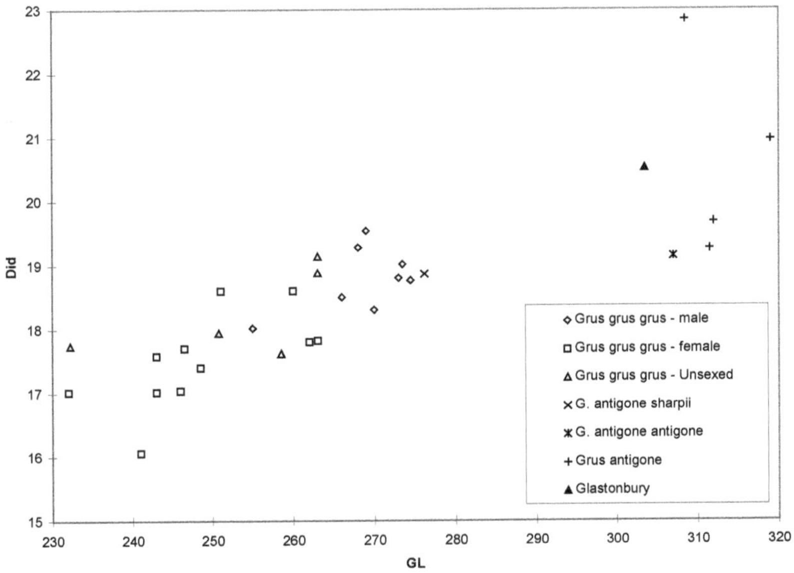

Figure 8. 12. *Grus* ulna from Glastonbury plotted with modern samples of *Grus grus* and *Grus antigone*. Greatest length (GL) plotted against diagonal breadth (Did).

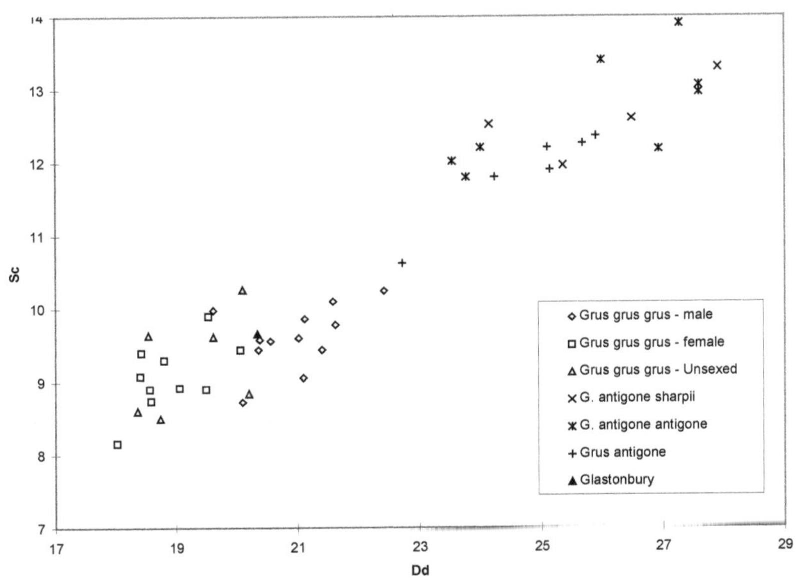

Figure 8. 13. *Grus* tibiotarsus from Glastonbury plotted with modern samples of *Grus grus* and *Grus antigone*. Distal depth (Dd) plotted against shaft breadth (SC).

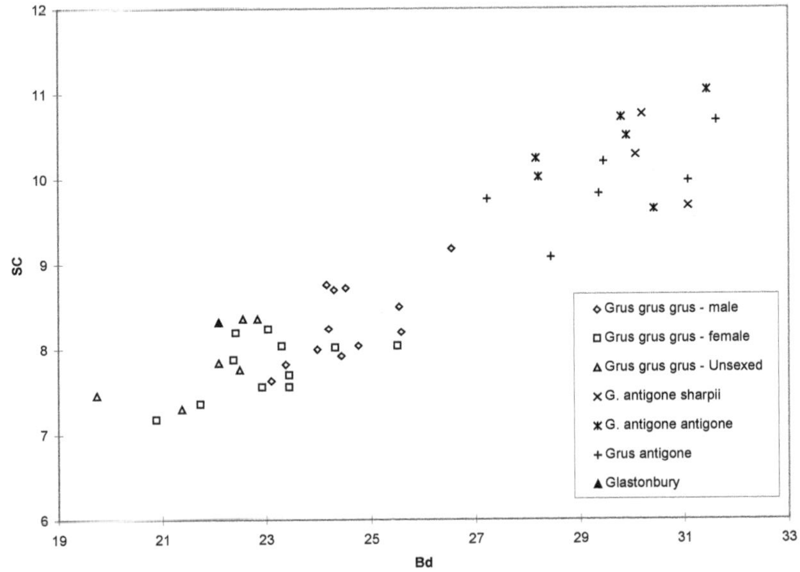

Figure 8. 14. *Grus* tarsometatarsus from Glastonbury plotted with modern samples of *Grus grus* and *Grus antigone*. Distal breadth (Bd) plotted against shaft breadth (SC).

Table 8.9. *Grus* fossils and their measurements from Woodbury.

Specimen No.	Skeletal element	Measurements				
		GL	Bp	Dp	Bd	SC
A3304	Dist R. femur	-	-	-	26.18	-
		GL	Dip		Bd	Dd
A3303	Prox. R. tibiotarsus	-	32.58		-	-

Woodbury, Devon

The two crane bones from the settlement site at Woodbury in Devon were donated to the Natural History Museum in 1945 by Dr. J. W. Jackson. The site is Iron Age and produced mallard *A. platyrhynchos* in addition to the remains of crane (Harrison, 1980a).

The distal femur (Table 8.9) was the size of a male *Grus grus* (Table 8.5), as was the fragment of proximal tibiotarsus. Both were too incomplete to allow more than one measurement on each specimen (Table 8.9) and so they have not been plotted on a scatterplot. However, comparison of their dimensions with the values in Tables 8.5 and 8.6 confirm that both are in the size range of modern male *G. grus* and exceed that of females.

Mesolithic to Bronze Age (Britain)

Cambridgeshire Fens

The crane fossils from the Cambridgeshire Fens examined are housed in both the Sedgwick Museum in Cambridge and the Natural History Museum in South Kensington. The material in the Sedgwick Museum was presented by a number of individuals around the turn of the century (1890 - 1907). The bones were derived from excavations undertaken to construct water courses and to retrieve peat for fuel, as well as during ploughing. The localities from which individual specimens or associated skeletons came from are Reach Fen and Burwell Fen. In addition, there was some material which was said to have come from the Cambridgeshire Fens without further details. The fossil cranes from the Cambridgeshire Fens have been treated as a single sample because they apparently came from similar levels and in a relatively confined area (Northcote, 1979b).

It appears that the crane fossils from the Cambridgeshire Fens date from zone VII of the Holocene, 7000 - 3000 years BP, equivalent to the Mesolithic to Bronze Age (Northcote, 1979b, 1980a). The evidence for this comes from both radiocarbon dating and pollen analysis. The pollen evidence (Godwin, 1978) has shown that East Anglia changed from a forested area during zone VI to being covered in sedge fen from the beginning of zone VII (7000 BP). This sedge fen led to the build up of alkaline peat over which a marine transgression deposited a layer of clay towards the middle of zone VII (4500 BP). Subsequent to the transgression an acidic peat was deposited. In the area of Reach Fen the alkaline peat is continuous with the earlier peat, as the intermediary transgression did not extend that far inland. Finally, just before the beginning of zone VIII during the Iron Age, a second transgression occurred and covered the whole area in a silt up to 5 metres in thickness. Since Roman times, due to drainage and cultivation the upper levels have been lost signifying that all fossils found are likely to come from zone VII (Northcote, 1979b). Pollen analysis on some peat attached to a mute swan (*Cygnus olor*) bone gave confirmation for this date and beaver bones have been dated to 2677± 123 BP (Mayhew, 1975) and 3079 ± 99 BP (Northcote, 1979b) giving further credence to the proposed age of the crane bones.

These crane remains were identified by Northcote (1979b) as those of common cranes because they were significantly shorter than those of the sarus crane. Southwell (1901) noted that cranes were known to have bred in East Anglia within more recent times.

Other taxa found in the Fenland peats include Dalmatian pelican (*P. crispus*), bittern (*B. stellaris*), mute swan (*C. olor*), whooper swan (*C. cygnus cygnus*), greylag goose (*Anser anser*), smew (*Mergus albellus*), red-breasted merganser (*M. serrator*), mallard (*A. platyrhynchos*), white-tailed eagle (*H. albicilla*), moorhen (*G. chloropus*), lapwing (*Vanellus vanellus*), woodcock (*Scolopax rusticola*) and razorbill (*Alca torda*). All these avian taxa have an association with water. Mammals found in the Fenland peats include beaver (*Castor fiber*), aurochs (*Bos primigenius*), and red deer (*Cervus elaphus*).

Table 8. 10. *Grus* fossils and their measurements from the Cambridgeshire Fens.

Specimen No.	Skeletal elements and measurements
Burwell Fen, 1907	L. humerus GL: 243; Bp: 54.86; Bd: 36.74; SC: 16.08 L. ulna GL: 271; Bp: 24.38; Dip: 29.06; Did: 20.14; SC: 9.5 L. radius GL: 258.25; Bp: 11.26; Bd: ±13.82 L. carpometacarpus GL: 118.24; Bp: 24; Bd1: 16.4; Bd2: 13.62 L. coracoid GL: 89.1; LM: 69.3; BF: 38.64; SC: 14.52 L. scapula GL: - ; Dic: 24.92 L. femur GL: 135.5; Bp: 29.36; Dp: 18.7; Bd: 27.32; SC: 12.74 L. tibiotarsus GL: 305; Dip: 33.98; Bd: 21.5; Dd: 21.76; SC: 10.74 L. tarsometatarsus GL: 248; Bp: 24.32; Bd: 25.78; SC: 9.04
Burwell Fen, 1912	L. humerus GL: 240; Bp: 51.02; Bd: 34.44; SC: 15.92 R. ulna GL: 271; Bp: 22.78; Dip: 27.06; Did: 18.82; SC: 9.34 Radius GL: 258.5; Bp: 10.58; Bd: 13.08 R. carpometacarpus GL: 121.92; Bp: 23.62; Bd1: 15.7; Bd2: 13.22 R. coracoid GL: - ; LM: 65.74; BF: - ; SC: 12.98 L. scapula GL: - ; Dic: 22.72 L. femur GL: 130.6; Bp: 27.02; Dp: 16.96; Bd: 25.34; SC: 11.72 L. tibiotarsus GL: 301; Dip: 30.76; Bd: 20.64; Dd: 20.2; SC: 10.32 L. tarsometatarsus GL: 269; Bp: 22.88; Bd: 25.02; SC: 8.76
Burwell Fen, 1900a	L. humerus GL: 233.5; Bp: 50.82; Bd: 33.58; SC: 15.12 L. ulna GL: 265.5; Bp: 22.82; Dip: 26.68; Did: 17.88; SC: 8.88 Radius GL: 256.5; Bp: 9.9; Bd: 13.54 R. femur GL: 130; Bp: 27.96; Dp: 19.06; Bd: 25.42; SC: 11.52 R. tibiotarsus GL: 300.5; Dip: 31.96; Bd: 21.18; Dd: 21.2; SC: 10.62 R. tarsometatarsus GL: 256; Bp: 23.4; Bd: 25.16; SC: 8.92
Burwell Fen, 1890	L. humerus GL: 241; Bp: - ; Bd: 36.3; SC: 17.02 Prox. L. ulna Bp: 22.84 Dist. L. carpometacarpus Bd1: 15.96; Bd2: 13.32 L. coracoid SC: ±13.94 L. scapula GL: - ; Dic: 22.84 R. femur GL: 132; Bp: 27.96; Dp: 15.98; Bd: 24; SC: 11.38
Cambridge Fens	Radius GL: 244.5; Bp: 9.36; Bd: 13.42 R. femur GL: 118; Bp: 25.88; Dp: 13.94; Bd: - ; SC: 10.46 R. tarsometatarsus GL: 233.5; Bp: 21.6; Bd: 21.2; SC: 8.3
Cambridge Fens A844	Dist. R. humerus Bd: 35.92; SC: 16.56
Cambridge Fens A845	Dist R. tibiotarsus Bd: 20.86; Dd: 22.32; SC: 10.5 L. tibiotarsus GL: 318.5; Dip: 34.58; Bd: 20.8; Dd: 22.2; SC: 10.8
Cambridge Fens A846	R. femur GL: 137.25; Bp: 29.46; Dp: 19.1; Bd: 26.48; SC: 11.94

Table 8. 10. (Cont.). *Grus* fossils and their measurements from the Cambridgeshire Fens.

Burwell Fen, 1909a	R. humerus GL: - ; Bp: - ; Bd: 35.08; SC: 15.94 L. & R. ulna (combined) GL: - ; Bp: 23.1; Dip: - ; Did: 19.18; SC: 9.26 R. scapula GL: - ; Dic: 23.8 R. tibiotarsus GL: 302; Dip: - ; Bd: 22.86; Dd: 21.8; SC: 10.24 L. tarsometatarsus GL: 261; Bp: 25.98; Bd: 25.88; SC: 8.48
Reach Fen, 1894	R. tibiotarsus GL: 288.5; Dip: 28.02; Bd: 20.18; Dd: 20.34; SC: 10.1 R. tarsometatarsus GL: 256; Bp: 22.8; Bd: 25.04; SC: 8.38
Burwell Fen, 1900b	R. tibiotarsus GL: - ; Dip: - ; Bd: 21.68; Dd: 20.84; SC: 11.5 L. tarsometatarsus GL: 249; Bp: 24.52; Bd: 25.36; SC: 9.48
Cambridge Fens	L. humerus GL: 223; Bp: 48.64; Bd: 33.5; SC: 15.82 L. ulna GL: - ; Bp: 21.84; Dip: - ; Did: 18.02; SC: 8.64 L. tibiotarsus GL: 286; Dip: 31.08; Bd: 18.98; Dd: 19.56; SC: 9.9 L. tibiotarsus GL: 295; Dip: - ; Bd: 18.44; Dd: 19.3; SC: 9.22

All but one of the five *Grus* humeri from the Cambridgeshire Fens fall in the size range of the modern males of *Grus grus* (Figure 8.15). The four ulnae were also of the size of modern male *G. grus* (Figure 8.16). The radii are all the length of modern males of the nominate species, although they include one which has a low value for its proximal breadth (Figure 8.17). The three carpometacarpi are once again the size of modern male common cranes as are the femora (Figures 8.18 - 8.19). The tibiotarsi mostly fall in the range of male *Grus grus*, while two are the size of larger females (Figure 8.20). The tarsometatarsi again are mostly modern common crane sized and distinguish themselves additionally by possessing relatively robust shaft (Figure 8.21) (*t*-test on SC p = 0.00043, Table 8.11).

The Cambridgeshire Fen cranes in most instances conform in size to those of modern males of *Grus grus* and therefore few of their bone dimensions are stastistically different from modern *G. grus* (Table 8.11). They never approach the size of the significantly larger *G. antigone*. However, there are metric-shape difference seen in various breadth measurements, which are wider in the fossils than the modern population. This is particularly true of the shaft dimensions of the tibiotarsus and tarsometatarsus which may reflect a trend towards greater body weight in the Fen cranes compared to the modern birds from Scandinavia.

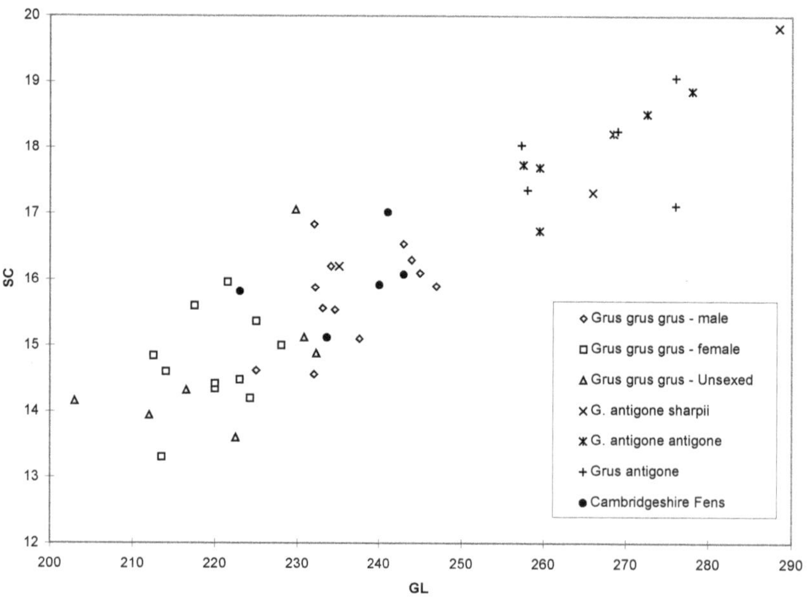

Figure 8. 15. *Grus* humeri from the Cambridgeshire Fens plotted with modern samples of *Grus grus* and *Grus antigone*. Greatest length (GL) plotted against shaft breadth (SC).

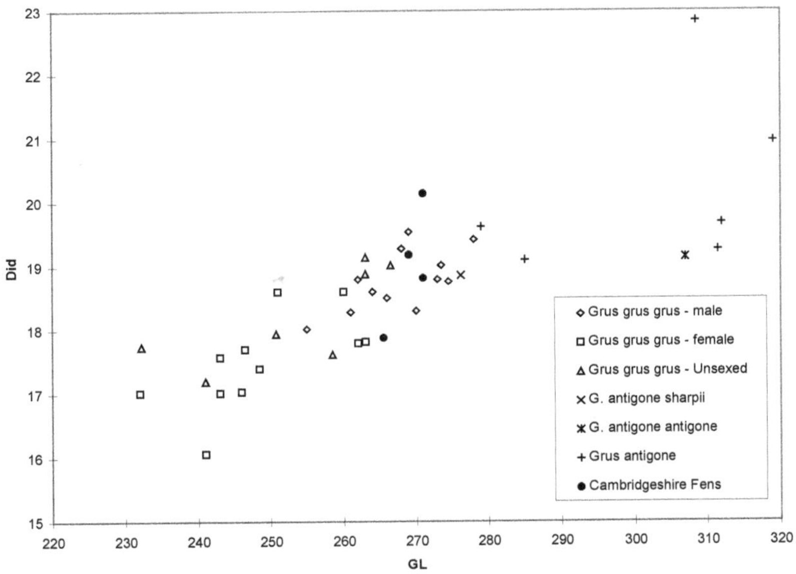

Figure 8. 16. *Grus* ulnae from the Cambridgeshire Fens plotted with modern samples of *Grus grus* and *Grus antigone*. Greatest length (GL) plotted against distal diagonal (Did).

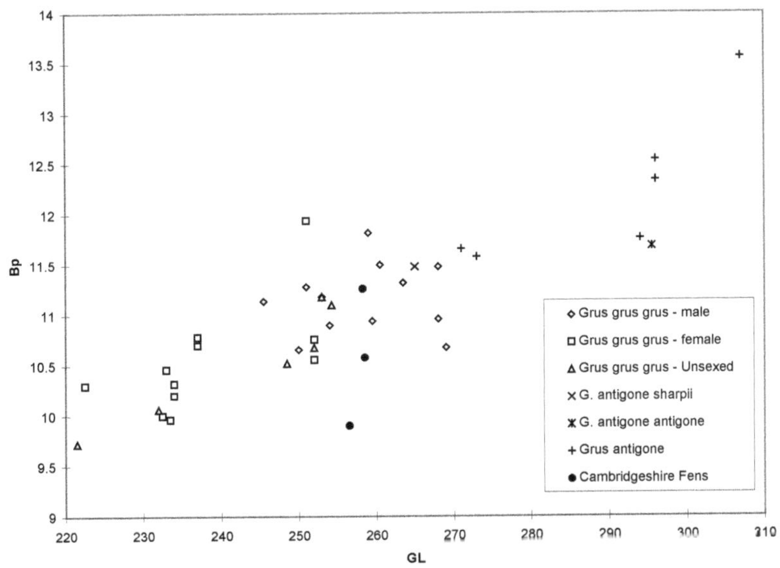

Figure 8. 17. *Grus* radii from the Cambridgeshire Fens plotted with modern samples of *Grus grus* and *Grus antigone*. Greatest length (GL) plotted against proximal breadth (Bp).

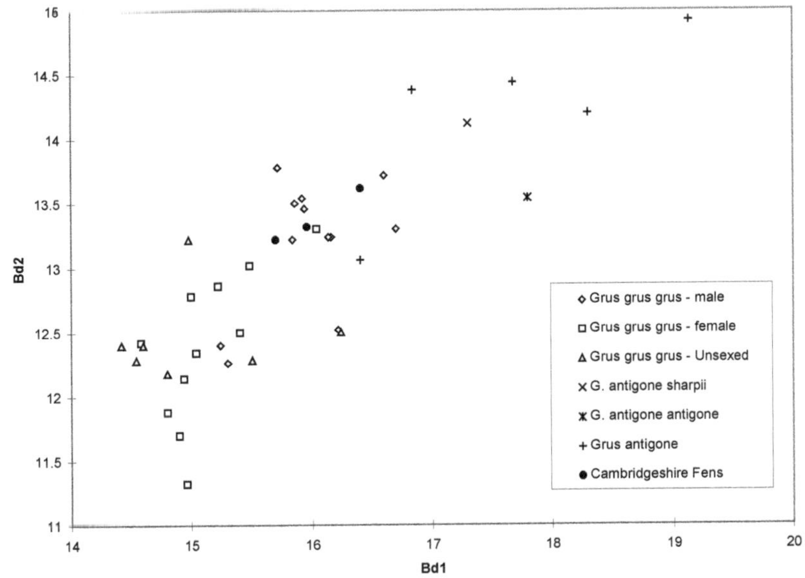

Figure 8.18. *Grus* carpometacarpi from the Cambridgeshire Fens plotted with modern samples of *Grus grus* and *Grus antigone*. Distal breadth No. 1 (Bd1) plotted against Distal breadth No. 2 (Bd2).

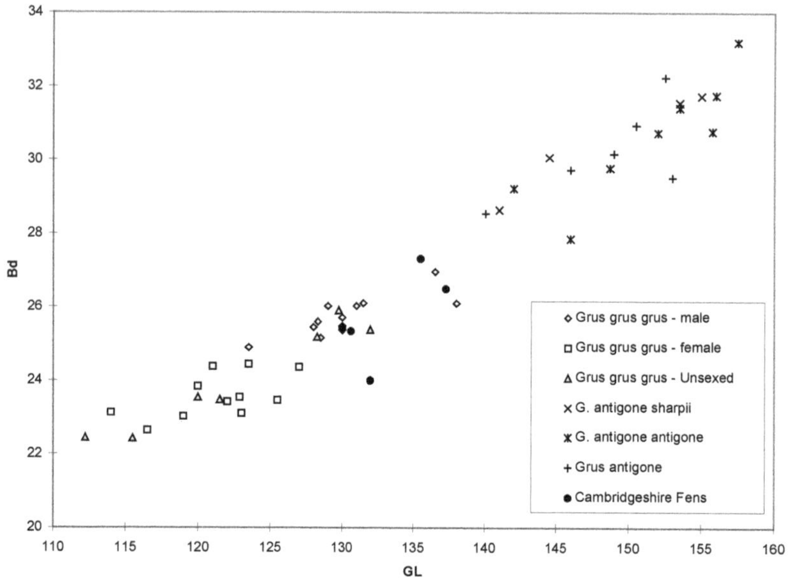

Figure 8.19. ***Grus*** **femora from the Cambridgeshire Fens plotted with modern samples of** ***Grus grus*** **and** ***Grus antigone.*** **Greatest length (GL) plotted against distal breadth (Bd).**

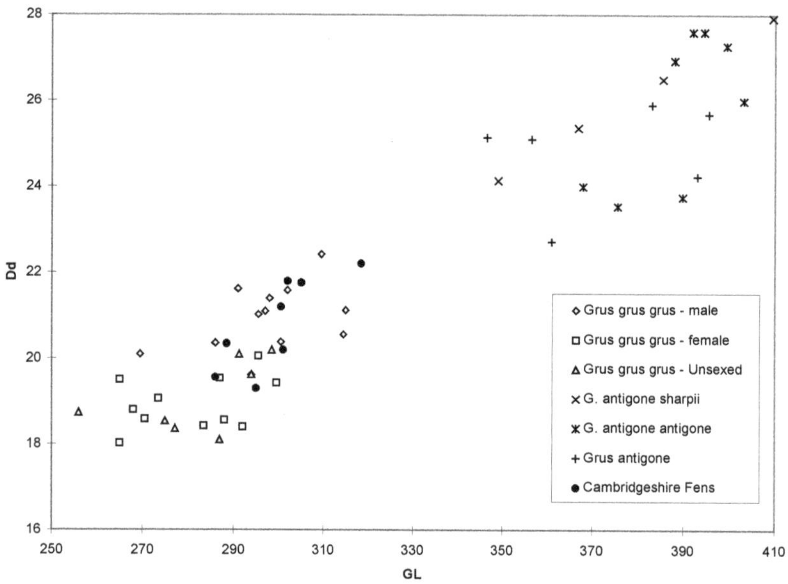

Figure 8. 20. ***Grus*** **tibiotarsi from the Cambridgeshire Fens plotted with modern samples of** ***Grus grus*** **and** ***Grus antigone.*** **Greatest length (GL) plotted against distal depth (Dd).**

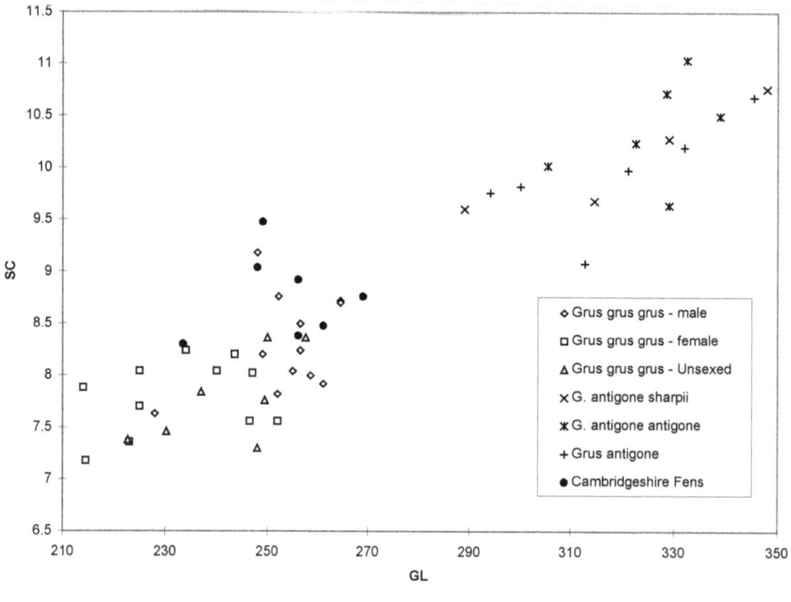

Figure 8. 21. ***Grus*** **tarsometatarsi from the Cambridgeshire Fens plotted with modern samples of** ***Grus grus*** **and** ***Grus antigone.*** **Greatest length (GL) plotted against shaft breadth (SC).**

Table 8.11. *t*- tests between Cambridgeshire Fen *Grus* sp. and *Grus grus*.

Measurements	
Hum GL	ns
Hum Bp	ns
Hum Bd	**
Hum SC	*
Uln GL	ns
Uln Bp	ns
Uln Dip	ns
Uln Did	ns
Uln SC	*
Rad GL	ns
Rad Bp	ns
Rad Bd	ns
Cpt GL	ns
Cpt Bp	ns
Cpt Bd1	ns
Cpt Bd2	ns
Fem GL	ns
Fem Bp	ns
Fem Dp	ns
Fem Bd	ns
Fem SC	**
Tib GL	*
Tib Dip	ns
Tib Bd	ns
Tib Dd	**
Tib SC	***
Tmt GL	ns
Tmt Bp	ns
Tmt Bd	*
Tmt SC	***
Cor GL	ns
Cor Lm	ns
Cor BF	ns
Cor SC	ns
Scap GL	ns
Scap Dic	ns

Key: *** $p < 0.001$, ** $p < 0.01$, * $p < 0.05$, ns Not significant at the 5% level.

Star Carr, North Yorkshire

The crane specimens from the excavations at Star Carr have been described by both Fraser and King (in Clark, 1954) and Harrison (1987c). Star Carr is an archaeological site which was formerly on the edge of a lake in the Vale of Pickering in North Yorkshire. A radiocarbon date of approximately 9500 BP (Middle Mesolithic) has been obtained from the mid-silt deposit at the site. This can be taken as the approximate age of the crane remains.

All other bird species found at the site are water birds such as red-throated diver (*Gavia stellata*), great crested grebe (*Podiceps cristus*), little grebe (*Tachybaptus ruficollis*), brent goose (*Branta bernicla*), red-breasted merganser (*Mergus serrator*) and common scotter (*Mellanita nigra*) (Harrison, 1987c). These species, together with the common crane, as identified by Fraser and King (1954) and confirmed by Harrison (1980a), are not unexpected for a fresh water site. Other specimens, identified by Fraser and King (1954) as species such as buzzard (*Buteo buteo*), pintail (*Anas acuta*) and lapwing (*Vanellus vanellus*), were reassigned by Harrison. The only cause for concern regarding the taxa present are the red-throated diver and red-breasted merganser whose diets consist solely of fish. This is because Wheeler (1978) has suggested that fresh water fish, which were not found at the site, were absent because they had not yet arrived in Britain from the Continent by this time. The resolution of this paradox might be that other fish such as salmonids, eels and sticklebacks, which are euryhaline and can travel across sea water, may have been present. This does not affect the present study but is never-the-less of interest and might call into question the hypothesis that the absence of fresh water fish could restrict the distribution of piscivorous birds such as Dalmatian pelicans (*Pelecanus crispus*) (Northcote, 1980a).

The two humeri are different in size, although only one was complete enough for a confident comparison to be made with modern *Grus* humeri (see Figure 8.22). The complete left humerus is the size of a modern male *Grus grus*, while the right humerus appears to be smaller, although still within the range of modern male common cranes. The two ulnae again differ in measurements from each other; one specimen has a slightly larger value for its shaft dimension than the available sample of modern male *G. grus*, while the other is slightly smaller in this respect. None of the Star Carr specimens was the size of modern *Grus antigone*.

Table 8. 12. *Grus* fossils and their measurements from Star Carr.

Specimen No.	Skeletal element	Measurements				
		GL	Bp	Bd		SC
87.5011	L. humerus	229.5	46.32 +	32.74		17.02
87.5010	Prox. R. humerus	-	45.7 +	-		14.96
		GL	Bp	Dip	Did	SC
87.5013	L. ulna shaft	-	-	-	-	9.46
87.5012	R. ulna shaft	-	-	-	-	8.42

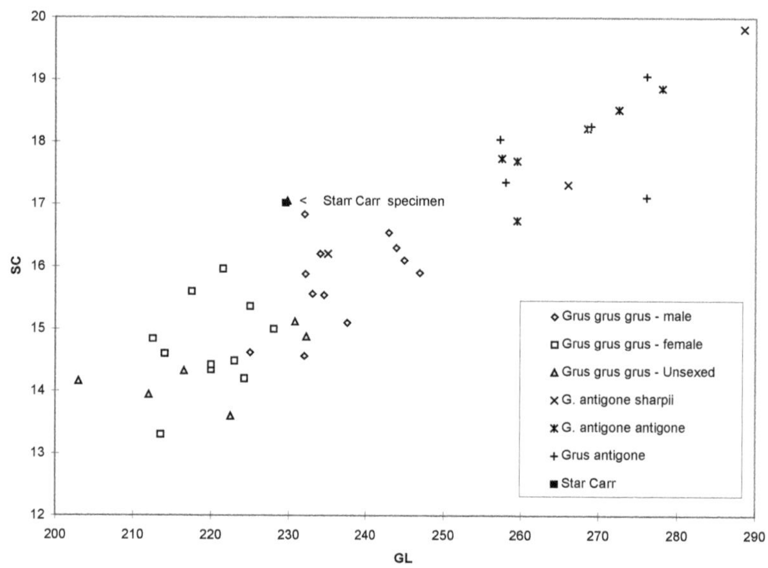

Figure 8. 22. *Grus* humerus from Star Carr plotted with modern samples of *Grus grus* and *Grus antigone*. Greatest length (GL) plotted against shaft breadth (SC).

Pleistocene (Britain)

Ilford, Essex

The age of the fossils from the site or sites at Ilford in Essex has been a source of controversy (Chapter 4). At one time they were believed by most Quaternary palaeontologists to correspond to the Ipswichian or last interglacial (e.g. Harrison and Walker, 1977; Stuart, 1982). This view has been challenged and the present belief is that material from the brick pits which were being excavated during the last century is from the penultimate interglacial (Currant, 1989). This interglacial has yet to be formally named.

The birds from Ilford were described and identified by Harrison and Walker (1977). Included among these was a distal end of a left radius (A4961) of a crane which appeared to be too big for the modern common crane and was therefore assigned to the extinct crane *Grus primigenia* (Harrison and Cowles, 1977).

Table 8. 13. *Grus* fossil and measurement from Ilford.

Specimen No.	Skeletal element	Measurements		
		GL	Bp	Bd
A4961	Dist. Radius	-	-	16.24

The distal radius considered by Harrison and Cowles (1977) to be an example of the extinct species *G. primigenia* falls above the modern range of *Grus grus* (Figure 8.23). It was not, however, very much larger, although its distal breadth falls in the range of the values for *G. antigone*.

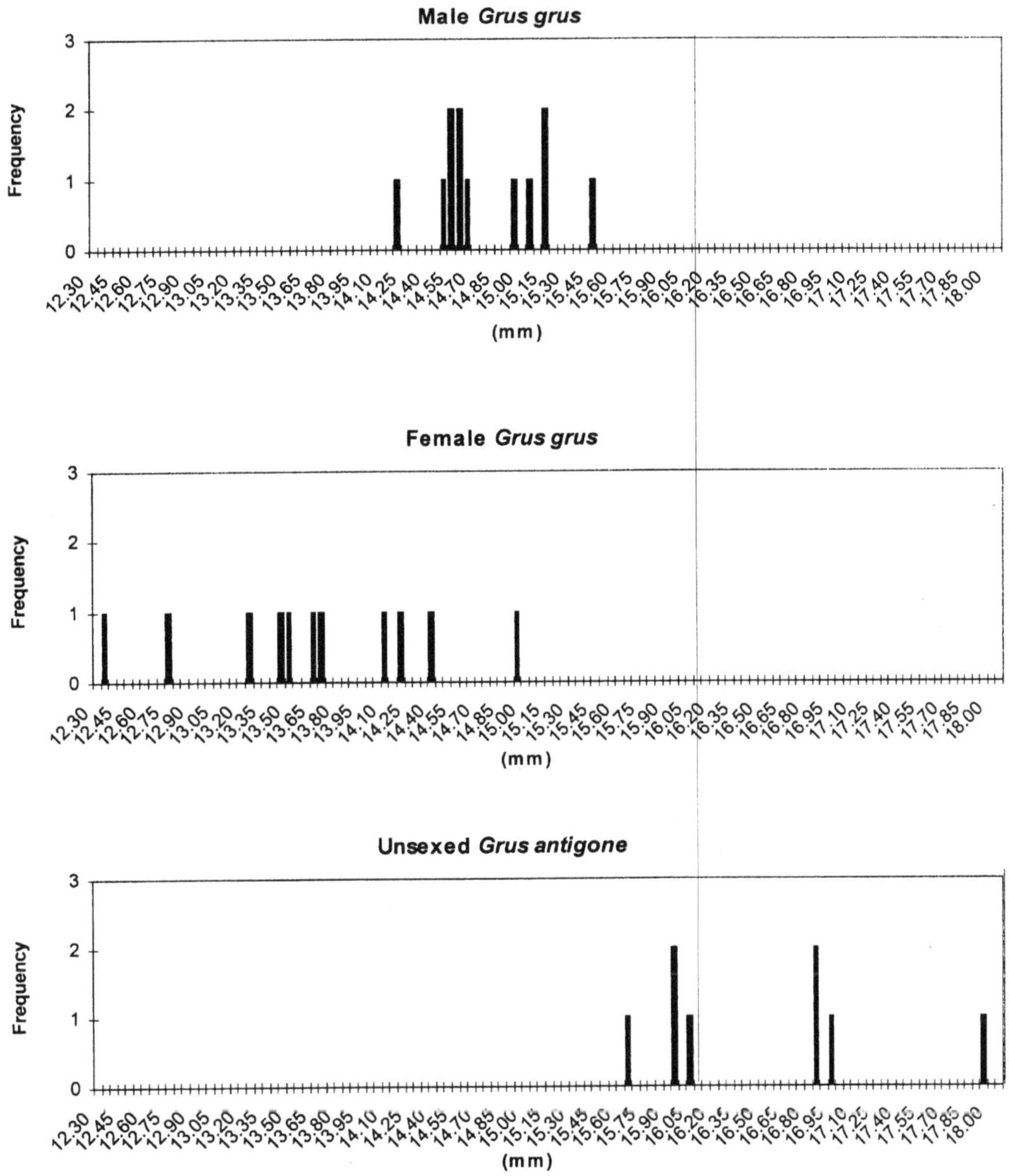

Figure 8. 23. Barcharts of male and female *Grus grus* distal radii dimensions together with *G. antigone* and indication of the size of the Ilford radius (vertical line).

Undated (Britain)

Windsor Road, Hackney Marshes, London

The site at Windsor Road, Hackney Marshes, in the Lea Valley, is poorly documented. Harrison (1980) mentions the site and the presence of crane there and states that the remains could be either Pleistocene or Holocene, since deposits covering both ages occur in the Lea Valley (Gibbard, 1994)

The left tibiotarsus from Hackney Marshes is the size of a modern male specimen of *Grus grus*. It is considerably smaller than modern *G. antigone* measured (Figure 8.24 and Table 8.13).

Table 8. 14. *Grus* fossils and their measurements from Hackney Marshes.

Specimen No.	Skeletal element	Measurements			
		GL	Dip	Bd	Dd
A718	L. tibiotarsus	311.0	33.36	22.22	22.04

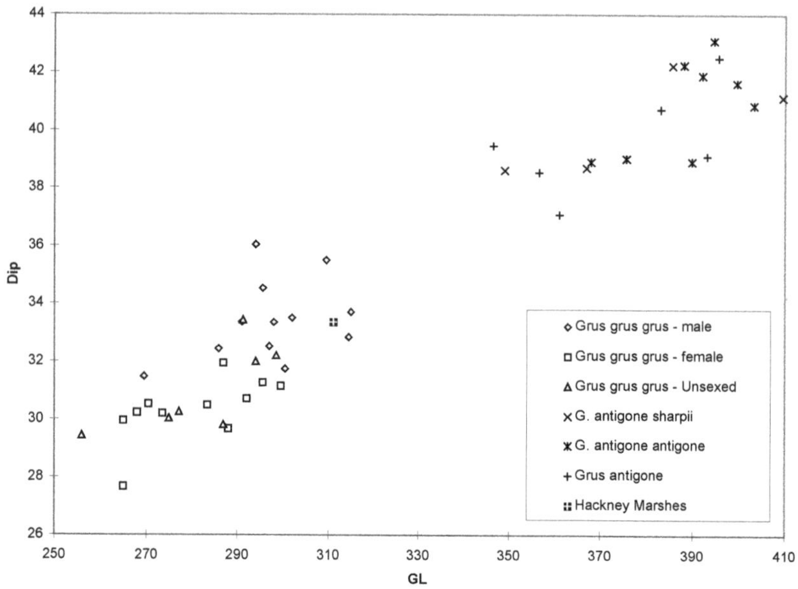

Figure 8. 24. *Grus* tibiotarsus from Hackney Marshes plotted with modern samples of *Grus grus* and *Grus antigone*. Greatest length (GL) plotted against proximal diagonal (Dip).

Holocene (Denmark)

Danish Peat Deposits

The crane fossils from the Danish peat deposits are all housed in the Zoological Museum in Copenhagen. These fossils came from a variety of sites, most of which are poorly dated. They have been dealt with as a single sample of Holocene finds of mixed age. Two of the sites were dated, Nyrup Højby, by pollen analysis to the late Sub-Boreal or Bronze Age (*ca.* 3000 BP) and the other, Sejerø Borreblerg, to the mid Sub-Atlantic or Early Roman Iron Age (*ca.* 2000 BP).

Table 8. 15. *Grus* fossils and their measurements from the Danish peat deposits.

Specimen No.	Skeletal elements and measurements.
Nørre Brøby	
ZMK 7/1927 (a)	L. humerus GL: 241; Bp: 52.22; Bd: 34.88; SC: 16.4 R. humerus GL: 242.5; Bp: 52.52; Bd: 35.1; SC: 16.38 L. ulna GL: 278; Bp: ±23.52, Dip: - ; Did: 18.84; SC: 8.66 R. ulna GL: 278.5; Bp: 23.76, Dip: - ; Did: 19.14; SC: 8.46 Radius GL: 265.5; Bp: 10.9; Bd: 14.5 Radius GL: 266; Bp: 11.1; Bd: 14.34 L. carpometacarpus GL: 120.84; Bp: 23.5; Bd1: 16.1; Bd2: 13.38 L. coracoid GL: 88.32; LM: 67.48; BF: 37.86; SC: 14.36 R. coracoid GL: 88.7; LM: 68.2; BF: 38.1; SC: 15.02 Scapula GL: - ; Dic: 22.7 Dist. L. Femur Bd: 25; SC: 12.2. Proximal L. tibiotarsus Dip: 32.02
ZMK 7/1927 (b)	Dist. R. humerus Bd: 31.5; SC 14.2
Bramdrup Mose	
ZMK 3/1924	R. tibiotarsus GL: 300.5; Dip: 31.92; Bd: 20.6; Dd: 20.04; SC: 10.62
Over Dråby Strand	
ZMK 11/1974	L. humerus GL: 232.5; Bp: 49.68; Bd: 33.94; SC: 16.94 Dist. R. humerus Bd: 34.2; SC 16.9
Skavngaards Mose	
No no.	Dist. L. ulna Did: 18.12; SC: 8.16 R. coracoid GL: ±82.32; LM: 63.36; BF: 37.2; SC: 14.88 L. femur GL: 123.2; Bp: 26.98; Dp: 17.22; Bd: 14.54; SC: 11.06 R. femur GL: - ; Bp: ±26.44; Dp: 16.96; Bd: - ; SC: 11.8 R. tibiotarsus GL: 300.5; Dip: 31.92; Bd: 20.6; Dd: 20.04; SC: 10.62

Table 8. 15 (contd.) *Grus* fossils and their measurements from the Danish peat deposits.

Specimen No.	Skeletal elements and measurements.
Faaborg ZMK 16/1953	L. humerus GL: 232.5; Bp: 47.24; Bd: 32.2; SC: 13.9 Dist. R. humerus Bd: 32.8; SC: 14.1 Dist. L. ulna Did: 16.7; SC: 8.08 Dist. R. ulna Did: 16.02; SC: 8.12 Radius GL: 250.5; Bp: 10.2; Bd: 13.02 Radius GL: - ; Bp: - ; Bd: 12.74 R. carpometacarpus GL: 116; Bp: 22.1; Bd1: 14.52; Bd2: 12.16 L. femur GL: 124.5; Bp: 26.18; Dp: 17.04; Bd: 22.4; SC: 9.72 R. femur GL: 126; Bp: 25.48; Dp: - ; Bd: 22.28; SC: 9.86 L. tibiotarsus GL: - ; Dip: - ; Bd: 18.8; Dd: 17.9; SC: 9.04 R. tibiotarsus GL: - ; Dip: 28.64; Bd: 18.66; Dd: - ; SC: 8.86 L. tarsometatarsus GL: - ; Bp: 20.52; Bd: - ; SC: 7.7
Pinstrup Mose ZMK 9/1927	R. femur GL: 128; Bp: 27.5; Dp: 17.66; Bd: 24.54; SC: 10.42 L. tibiotarsus GL: 283; Dip: 32.24; Bd: 19.74; Dd: 19.64; SC: 9.52 L. tarsometatarsus GL: 247.5; Bp: 21.42; Bd: 22.74; SC: 7.8
Hesselballe ZMK 5/1915	Dist. R. humerus Bd: 32.7; SC: 14.54 Dist. L. ulna Did: 12.12; SC: 8.76 Prox. L. ulna Bp: ±20.96; Dp: ±25.28 Radius GL: 243; Bp: 9.9; Bd: ±13.72 Dist. R. femur Bd: 23.84; SC: 10.6 R. tibiotarsus GL: - ; Dip: 31.06; Bd: 20.42; Dd: 19.12; SC: 9.36 L. tarsometatarsus GL: 251; Bp: 22.44; Bd: 24.32; SC: 7.6
Allers Tovermose ZMK 25/1962	L. humerus GL: 216; Bp: 46.8; Bd: 31.52; SC: 14.98 R. ulna GL: 239.5; Dip: 26.06 ; Bp: 21.44; Did: 17.28; SC: 8.38 Prox. radius Bp: 9.94 Dist. L. tibiotarsus Bd: 20.98; Dd: 19.04; SC: 9.52 L. tarsometatarsus GL: 249.5; Bp: 22.14; Bd: 22.56; SC: 7.98
Kulmosen, Pinstrup ZMK 85/1941	Dist. R. humerus SC: 15.06 R. ulna GL: 269; Bp: ±22.76; Dip: 27.42; Did: 18.42; SC: 8.76 R. tibiotarsus GL: 289.5; Dip: 31.98; Bd: 21; Dd: 19.34; SC: 10.2 L. tarsometatarsus GL: 249; Bp: 23.44; Bd: 23.4, SC: 8.44
Orting Mose 27/1909	Dist. L. tibiotarsus Bd: 21.2; Dd: 20.92; SC: 10.36 L. tarsometatarsus GL: 263; Bp: 23.6; Bd: 25.4; SC: 8.8
Norre Mose 55/1943	Dist. L. humerus Bd: 31.84; SC: 15.84 R. ulna GL: 251; Bp: ±22.3; Dip: 27.42; Did: 18.72; SC: 8.7 Dist. L. tibiotarsus Bd: 20.38; Dd: 19.42; SC: 9.82 Dist. L. tarsometatarsus Bd: 23.38; SC: 8.16
Tovermose No no.	Dist. R. humerus SC: 14.86 Dist. L. ulna Did: 12.18; SC: 8.44 Prox. radius Bd: ±9.54 L. femur GL: 125.5; Bp: 26.02; Dp: 16.72; Bd: - ; SC: 11.12 Dist. L. tibiotarsus Bd: 18.92; Dd: 19.1; SC: 10.22 L. tarsometatarsus GL: 251; Bp: 21.18; Bd: 23.28; SC: 8.7
Nyrup Højby ZMK 45/1940	R. tarsometatarsus GL: 283; Bp: 25.36; Bd: 26.7; SC: 9.3

Table 8. 15 (contd.) *Grus* fossils and their measurements from the Danish peat deposits.

Specimen No.	Skeletal elements and measurements.
Aspendam No no.	R. ulna GL: 266.5; Bp: 21.92; Dip: 26.28; Bd: 17.82; SC: 8.92 R. carpometacarpus GL: 116; Bp: 22.8; Bd1: 15.9; Bd2: 12.4
Sejerø Borreblerg ZMK 16/1899	L. tarsometatarsus GL: 241; Bp: 22.4; Bd: 23.7; SC: 8.5 Prox. R. tarsometatarsus Bp: 25.48 Dist. R. tibiotarsus Bd: 19.4; Dd: 18.68
Ordrup Mose No no.	Dist. L. humerus Bd: 30.1; SC: 10.58
Glumso No no.	L. tarsometatarsus GL: 253; Bp: 20.82; Bd: 23.08; SC: 8.02

The humeri of the cranes from the Danish peat deposits include specimens whose sizes span the range of male and female common cranes (Figure 8.25). They are not within the range of size of the sarus crane. The ulnae similarly are within the range of size and shape of modern *G. grus* and distinctly smaller than *G. antigone* (Figure 8.26). The two radii plotted in Figure 8.27 are the size of modern female *Grus grus*, while the carpometacarpi, femora and tibiotarsi are in the zone of overlap between the male and female *G. grus* (Figures 8.28 - 31). There is a single tarsometatarsus which is larger than all the modern *G. grus* measured. This tarsometatarsus is roughly equidistant in size between modern *G. grus* and *G. antigone* (Figure 8.32). The other tarsometatarsi are the size of male *G. grus*.

In summary, it appears that the fossil cranes from the Danish peat deposits are mostly the size of modern cranes breeding in Scandinavia with the exception of a single tarsometarsus. The only statistical difference found between the Danish fossils and the modern sample are the breadth dimensions of the radius which is difficult to explain (Table 8.15). The single large tarsometatarsus was an isolated find rather than part of an associated skeleton therefore unfortunately the relationship between the individual birds different limb bone dimensions could not be ascertained.

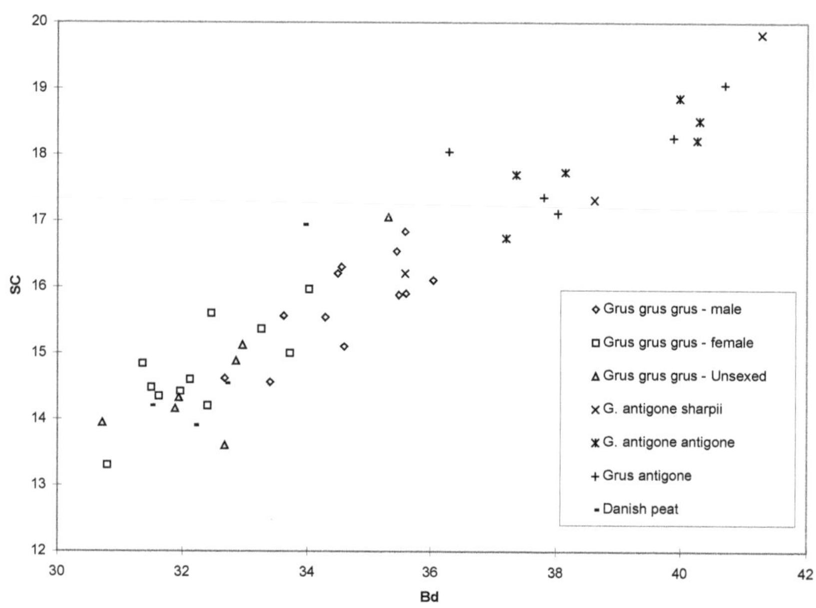

Figure 8. 25. *Grus* humeri from Danish Peat sites plotted with modern samples of *Grus grus* and *Grus antigone*. Distal breadth (Bd) plotted against shaft breadth (SC).

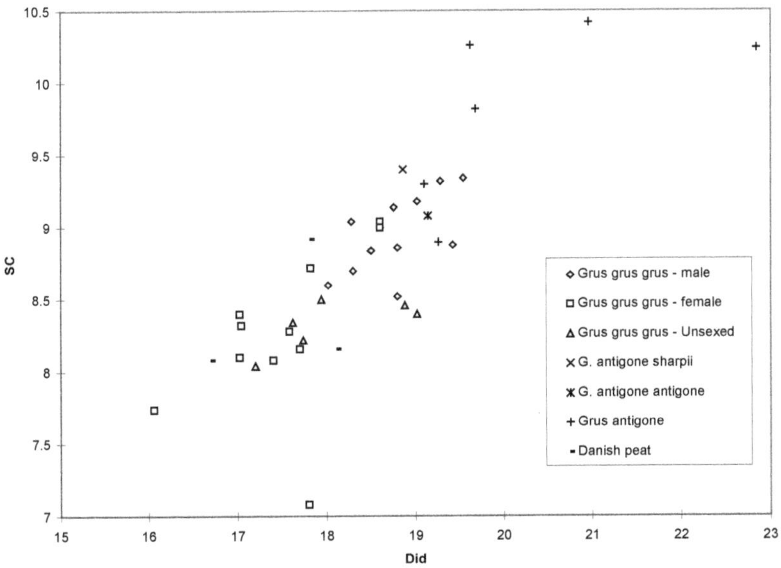

Figure 8. 26. *Grus* ulnae from Danish Peat sites plotted with modern samples of *Grus grus* and *Grus antigone*. Distal diagonal depth (Did) plotted against shaft breadth (SC).

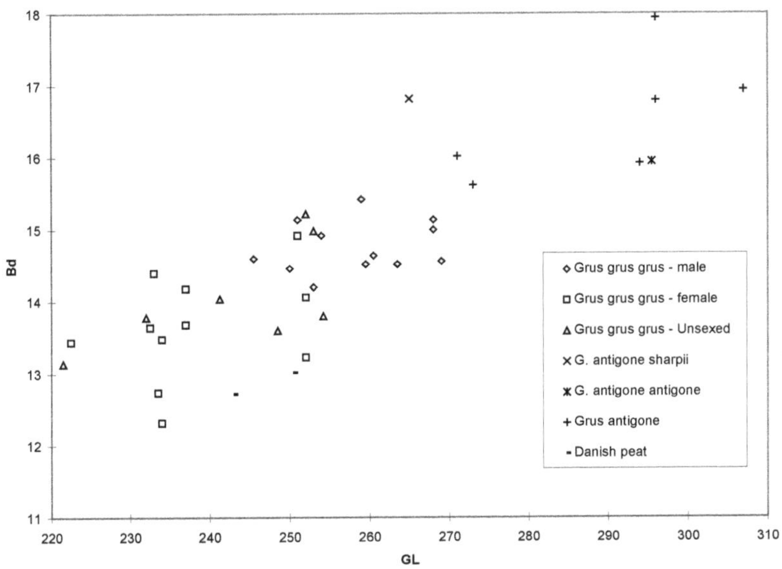

Figure 8. 27. *Grus* radii from Danish Peat sites plotted with modern samples of *Grus grus* and *Grus antigone*. Greatest length (GL) plotted against distal breadth (Bd).

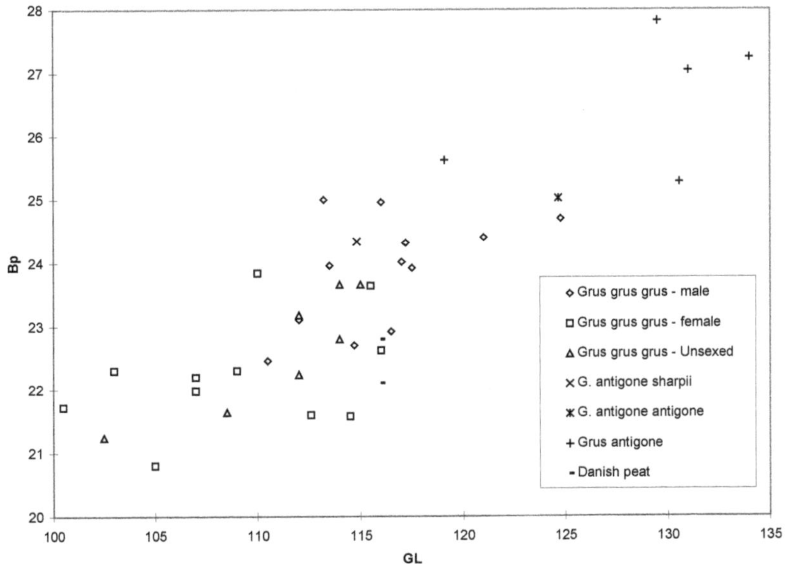

Figure 8. 28. *Grus* carpometacarpi from Danish Peat sites plotted with modern samples of *Grus grus* and *Grus antigone*. Greatest length (GL) plotted against proximal breadth (Bp).

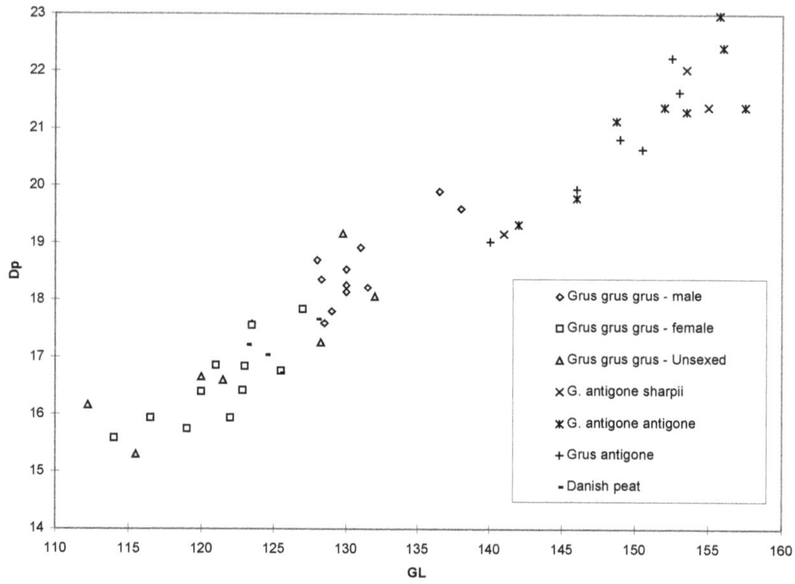

Figure 8. 29. *Grus* femora from Danish Peat sites plotted with modern samples of *Grus grus* and *Grus antigone*. Greatest length (GL) plotted against proximal depth (Dp).

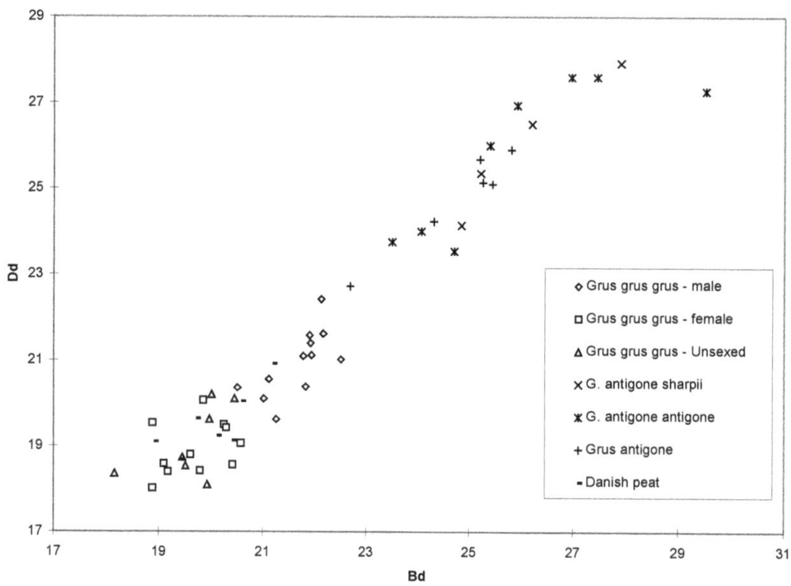

Figure 8. 30. *Grus* tibiotarsi from Danish Peat sites plotted with modern samples of *Grus grus* and *Grus antigone*. Distal breadth (Bd) plotted against distal depth (Dd).

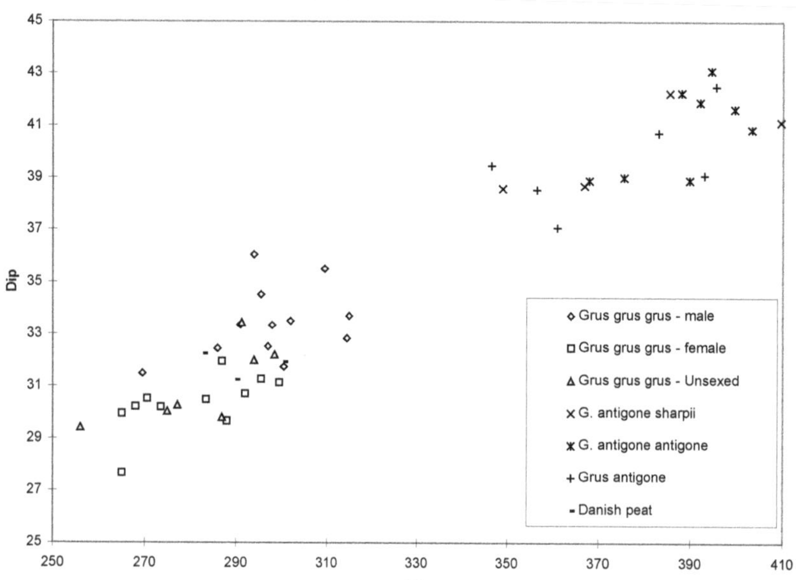

Figure 8. 31. *Grus* tibiotarsi from Danish Peat sites plotted with modern samples of *Grus grus* and *Grus antigone*. Greatest length (GL) plotted against proximal diagonal depth (Dip).

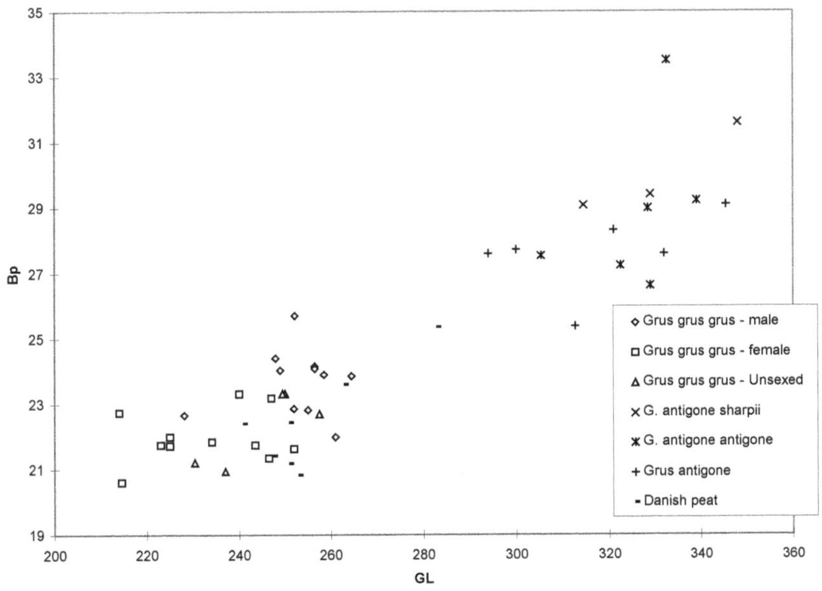

Figure 8. 32. *Grus* tarsometatarsi from Danish Peat sites plotted with modern samples of *Grus grus* and *Grus antigone*. Greatest length (GL) plotted against proximal breadth (Bp).

Table 8.16. *t*- tests between Danish Peat *Grus* sp. and *Grus grus*.

Measurements	
Hum GL	ns
Hum Bp	ns
Hum Bd	ns
Hum SC	ns
Uln GL	ns
Uln Bp	ns
Uln Dip	ns
Uln Did	ns
Uln SC	ns
Rad GL	ns
Rad Bp	**
Rad Bd	*
Cpt GL	ns
Cpt Bp	ns
Cpt Bd1	ns
Cpt Bd2	ns
Fem GL	ns
Fem Bp	ns
Fem Dp	ns
Fem Bd	ns
Fem SC	ns
Tib GL	ns
Tib Dip	ns
Tib Bd	ns
Tib Dd	ns
Tib SC	ns
Tmt GL	ns
Tmt Bp	ns
Tmt Bd	ns
Tmt SC	ns
Cor GL	ns
Cor Lm	ns
Cor BF	ns
Cor SC	ns
Scap GL	ns
Scap Dic	ns

Key: *** $p < 0.001$, ** $p < 0.01$, * $p < 0.05$, ns Not significant at the 5% level.

8.3.7. Measurements from the literature

In addition to the measurements taken during this investigation it seemed sensible to consider those quoted in the literature. Such an enterprise presents one very significant potential stumbling block, that of intercomparability. If the measurements were not taken in exactly the same manner by two workers then their measurements cannot be meaningfully compared. Most workers quote von den Driesch (1976) as a source of their measurements, but the crane was not specifically dealt with in that publication. Variations in the interpretation of how to apply the von den Driesch (1976) measurements to other taxa will inevitably arise. However, the problem is not necessarily universal, as certain measurements are less prone to different interpretations than others. A comparison between the measurements taken during the present study and the ones depicted in von den Driesch (1976) was made to assess which would be the least likely to cause problems. The measurements thought to be most useful are listed below.

Humerus: GL, Bd, SC. Ulna: GL, Bp, Dip Radius: GL
Carpometacarpus: GL Coracoid: GL, Lm, BF. Scapula: GL.
Femur: GL, Bp, Dp, Bd, SC. Tibiotarsus: GL, Dd. Tarsometatarsus: GL, SC
(Of these the greatest lengths are the most reliable).

Table 8. 17. Measurements from other workers.

Site	Age	Skeletal elements and measurements.
Launceston Castle, Cornwall. (Albarella and Davis, 1996).	Mediaeval, Late 13th to 15th century.	Dist. tibiotarsus Dd: 19.3 Tarsometatarsus SC: 9.5 Tarsometatarsus SC: 8.6
Isca Grange, Wales (Hamilton-Dyer, pers comm.)	Roman, 2nd. century AD	Dist. humerus Bd: 34.6
Caerleon, Wales (Hamilton-Dyer, pers comm.)	Roman, ca. 75 - 100 AD	Humerus GL: 235; Bd: 33.7; SC: 16.1 Humerus GL: 236; Bd: 34.5; SC: 16.5 Humerus SC: 17.6 Humerus SC: 16.1 Humerus SC: 18.6 Humerus SC: 17 Humerus SC: 16 Radius GL: 282 Ulna Bp: 24.8 **Carpometacarpus GL: 130.6** Tarsometatarsus GL: 265; SC: 8.5 **Tarsometatarsus GL: 285; SC: 9** Tarsometatarsus SC: 9.1 Tarsometatarsus SC: 8.9

The measurements when compared to the modern samples in the present study reveal that the Launceston Castle specimens comprise two specimens which are the size of the comparative sample of modern common cranes and one which is slightly larger. These measurements are not greatest lengths and thus should be treated with caution. The Isca Grange specimen, which is Roman, is the size of a modern male *G. grus*. Finally, the Caerleon material comprises specimens which are the size of modern male common cranes and others which are larger. Some of the skeletal elements are complete bones and are much larger than modern *G. grus*. Of particular note are a complete tarsometatarsus and a carpometacarpus (highlighted in Table 8.16). The carpometacarpus is 130.6 mm long while the tarsometatarsus is 285 mm compared to a modern range of 100.5 - 124.78 mm for the carpometacarpi and 214 - 264 mm for the tarsometatarsi. This, therefore, represents evidence for cranes larger than modern *G. grus* in the 1st. century AD which is possibly the latest date for such large cranes in the European fossil record. The significance of this will be discussed in section 8.6.

8.4. The possible occurrence of *Grus leucogeranus* (Siberian white crane)

At a relatively late stage in the analysis it became apparent that no mention had been made in the literature of the possible occurrence of *Grus leucogeranus* in the fossil record of western Europe, particularly with respect to the identity of the anomalously large cranes found. A single record of the species was known to the author, that in the Early Bronze Ages of south east Turkey (Boessneck, 1987). There is little modern comparative skeletal material available of this rare species. Therefore, rather than ignore it completely, it was decided that estimates of their bone lengths would be made using a method similar to that described by Bochenski and Bochenski (1992). While it is not envisaged that these calculated estimates are completely accurate they will at least allow for gross comparisons to be made. The estimates were then compared to the measurements of three specimens of the species available.

Standard ornithological measurements of the wing, the tail, the tarsus, the bill and the toes were available for both the common crane and the Siberian white crane (Cramp and Simmons, 1980). Therefore, the known mean lengths of the common crane limb bones were expressed as a fraction of the most appropriate ornithological mean measurements (wing bones and pectoral girdle to wing measurement, and leg bones to tarsus). The resulting figures were then used to estimate the limb bone lengths of the Siberian white crane. This exercise was done separately for both the males and females which, it was hoped, would give us a better idea of the size range of the Siberian white crane.

Formula for the estimation of limb bone lengths:

(GL G.g. ÷ OM G.g.) x OM G.l. ≈ GL G.l.

Key:- GL G.g. = Mean greatest length of limb bone of *Grus grus*. OM G.g. = Mean ornithological measurement of *Grus grus*. OM G.l. = Mean ornithological measurement of *Grus leucogeranus*. GL G.l = Mean limb bone greatest length estimate for *Grus leucogeranus*.

Table 8. 18. Mean ornithological measurements (mm) of *Grus grus* and *Grus leucogeranus* from Cramp and Simmons (1980).

	Male	Female
Grus leucogeranus		
Wing	626	594
Tarsus	284	268
Grus Grus		
Wing	593	557
Tarsus	258	238

Table 8. 19. Estimated mean limb bone lengths (mm) of *Grus leucogeranus*.

	Male	Female
Humerus	250.4	234.54
Ulna	282.78	264.47
Radius	272.8	253.81
Carpometacarpus	122.62	116.35
Femur	143.5	136.6
Tibiotarsus	327.7	316.06
Tarsometatarsus	278.32	262.53
Coracoid	90	84.34
Scapula	130.58	124.24

These estimates were then compared to the greatest length measurements of three available specimens of *G. leucogeranus* (Table 8.19). This comparison could only be made in a very general way as sample numbers are too low for statistical testing. However, the dimensions of the three specimens do not differ markedly from the estimates in Table 8.18.

Table 8.20. Limb bone lengths (mm) of *Grus leucogeranus*.

	S/ 1952.2.150	344 R	344D
	Unsexed	Unsexed	Female
Humerus	227.3	243	229
Ulna	252.1	273.5	256
Radius	242.3	268	248
Carpometacarpus	111.4	121.96	113.56
Femur	129.6	139	128
Tibiotarsus	303.3	317.5	289
Tarsometatarsus	258.3	282	241
Coracoid	81.5	87.9	86.14
Scapula	120.4	121.84	114.72

When these estimated means (Table 8.18) were compared with the graphs plotted in order to compare *Grus grus* with *G. antigone* (Figures 8.1 to 8.10), it was apparent that the Siberian white crane bones are likely to be longer on average than those of the common crane, although with a considerable overlap with the males of the nominate species. It is likely that the leg bone measurements would overlap less than those of the wing and pectoral girdle bones. This is because the tarsus measurements of *G. leucogeranus* differ from those of *G. grus* more than their wing measurements, possibly because of their greater tendency towards wading. They are unlikely to overlap greatly in length with the bones of the sarus crane *G. antigone*.

On the basis of the bone size estimates above *G. leucogeranus* could possibly represent the identity of the large cranes in the Quaternary fossil record in Western Europe.

8.5. Non-metrical morphological comparative study

A dominant aspect of previous investigations into the identity of large crane fossils in the Quaternary of Europe has been the use of non-metric morphological traits (Milne-Edwards, 1867 - 71; Northcote and Mourer-Chauviré, 1985 and 1988). These traits or osteological characters were derived from a comparison between modern sarus crane *Grus antigone* and fossils attributed to *G. primigenia*. They belong solely to two skeletal elements: the tibiotarsus and the humerus. They were described as follows:

The tibiotarsus

1/ The canalis extensorius and the incisura intercondylaris are wider in *Grus primigenia* than in *G. antigone* (Milne-Edwards, 1867 - 71).

2/ "In *G. primigenia* both the medial and the lateral epicondyles are small and discrete whereas in *G. antigone* both are large and each forms a proximodistal ridge; the epicondylus medialis is especially well developed and prominent in the latter species." (Northcote and Mourer-Chauviré, 1985).

The humerus

1/ "[I]n *G. primigenia* the shaft slopes gradually to epicondyles (tubercules for elbow ligaments) that are not pronounced, whereas, in *G. antigone* the shaft curves sharply towards epicondyles that are very pronounced." (Northcote and Mourer-Chauviré, 1985).
2/ "[I]n *G. antigone* the epicondylus ventralis is distally placed and the tuberculum supracondyl are ventrale is discrete and prominent; in *G. primigenia*, on the other hand, this epicondyle is more proximally placed so that it merges with the tuberculum which is, therefore, neither discrete nor prominent." (Northcote and Mourer-Chauviré, 1985).

The frequency with which these features occur has been scored among a maximum of 11 examples of these elements belonging to individuals of *G. antigone*. This number is greater than that examined by Northcote and Mourer-Chauviré (1985). Note that the tibiotarsus and humerus of one specimen were such that the characters could not be scored .

The tibiotarsus

1/ Northcote and Mourer-Chauviré (1985) were correct in their cautious dismissal of these characters due to difficulties in their clear definition.

2/ The two epicondyles were tested separately.

Epicondylus medialis.	
Prominent proximodistal ridge:	9/11
Small and discrete epicondyles:	2/11
	11/11
Epicondylus lateralis.	
Prominent proximodistal ridge:	2/10
Shallow proximodistal ridge:	7/10
Small and discrete epicondyles:	1/10
	10/10

The humerus

1/ This was found to be difficult to judge due to parallax factors and the subjective nature of this character difference.
2/ These characters were not tested

Of the non-metric traits described by Northcote and Mourer-Chauviré (1985 and 1988) only one of the tibiotarsal characters (the prominence of the epicondylus medialis forming a proximodistal ridge) is potentially useful for distinguishing *G. primigenia* and *G. antigone* and that character is not 100% consistent. A prominent epicondylus medialis, forming a proximodistal ridge in the distal tibiotarsus, was present in 9 out of 11 specimens, which signifies that there is an 18% chance that the character is misleading. The lateral epicondyle was confirmed to be generally less prominent, with most specimens (7/10) having a shallow proximodistal ridge for an epicondyle and only 2/10 being prominent and even 1/10 being small and discrete. It should be borne in mind that the original definition of these characters was based on a small series of specimens of *Grus primigenia*, which further calls into question their validity. The humeral character was found to be difficult to judge and adequately compare. Parallax, or the angle from which the distal humerus is viewed, affects the apparent shape of the slope towards the epicondyles.

The character which when tested produced a relatively high descrimination, the distal tibiotarsus, may be explained by loccomotory differences (Northcote and Mourer-Chauviré, 1985) which, however, need not be species-defining, especially in view of the observed variation.

A flaw in the present analysis, which cannot be amended here, is that these non-metric traits were not tested against a large series of common cranes *G. grus*. This cannot be achieved because the large series of specimens used in the morphometric analysis came from institutions in Spain and particularly Sweden. However, those few specimens of *G. grus* in Britain which were checked for the *Grus primigenia* characters appeared to suggest again that the features are ambiguous. Indeed there are general logistical difficulties with non-metric analyses because they are difficult to conduct without comparing the specimens directly. This is not easily achieved because the modern comparative specimens and the fossils are dispersed throughout Europe. Ideally all the modern crane skeletons (*G. antigone*, *G. grus* and *G. leucogeranus*) should be brought together for a comparison with each other and as many fossils as possible, including specimens attributed to both *Grus grus* and *Grus primigenia*. It would therefore seem to be unwise to accept at this stage the validity of the species *Grus primigenia* on the basis of the morphological characters described by Northcote and Mourer-Chauviré (1985 and 1988).

8.6. Conclusion

- **What is the pattern of size and shape in modern members of the genus *Grus*?**

The members of the genus *Grus* considered in this study are those deemed relevant to Quaternary fossil cranes in the Western Palaearctic. The common crane *Grus grus* held centre stage in the analysis as it is the only species currently breeding in the area. The sarus crane, *Grus antigone*, from Asia was also included because of the debate as to the identity of the large western European *Grus* fossils, which many authors have attributed to that species (Lydekker, 1891b; Lambrecht, 1933; Soergel, 1955; Brodkorb, 1967; Northcote, 1979 and 1982; Mourer-Chauviré *et al.*, 1975). A further possibility, the Siberian white crane *G. leucogeranus*, had not formerly been considered in the context of the anomalously large cranes. Unfortunately, the possible occurrence of that species was not realised until late in the study after the European survey had been completed. This meant that few bone measurements were available. Therefore, estimates of the male and female bone lengths were based on ornithological measurements in the literature (section 8.4.4.).

The order of increasing mean size for these three species is *Grus grus*, *G. leucogeranus* and *G. antigone.* There is, however, overlap in certain measurements between the species, including some between the smallest, *G. grus*, and largest, *G. antigone*. The overlap between *G. grus* and *G. antigone* occurs particularly among the skeletal elements of the wing. *Grus leucogeranus*, based on the estimates of mean limb bone length, appears to be closer in size to *G. grus* than to *G. antigone* and overlaps with *G. grus* most in measurements of the wing and the pectoral girdle.

There are also differences in metric shape between *Grus grus* and *Grus antigone*. *Grus grus* has a wider facies articularis basalis of the coracoid. This difference may be due to differences in locomotion between the two species. The sedentary nature of the sarus crane may account for the smaller facies arcticularis basalis breadths. In the light of the difference between *G. grus* and *G. antigone*, which appears to reflect flight capabilities, it is interesting that there appears to be no difference in the humerus to carpometacarpus greatest length proportion (Table 8.17). The lack of difference is perhaps because *G. antigone*, while sedentary, is still a very accomplished flyer (Walkinshaw, 1973).

Table 8.21. Humerus to carpometacarpus proportions.

Taxon	Carpometacarpus % of humerus
Grus grus	49.31%
Grus antigone	46.8%

Note: These proportions are based on data in Tables 8.1, 8.2.

- **How sexually dimorphic are the species within *Grus* and why?**

In her thesis Northcote (1979) showed that her *Grus grus* long bone sample had a relatively low coefficient of variation which she took to signify that the species was not very sexually dimorphic. She also stated that limb lengths for the two sexes of *Grus grus* given by Walkinshaw (1973) overlapped, which again demonstrated a lack of sexual dimorphism. The data from the present study would appear to contradict this conclusion. Figures 8.1 - 8.10 all show that in mean measurements, males are larger than females in *Grus grus*. There is overlap and this varies according to the specific measurement. The likelihood is therefore that Northcote's unsexed sample was not large enough to include the whole variation within the species. The large overlap present in the Walkinshaw (1973) feathered limb lengths is not unexpected as such measurements are prone to much variation due to changes in seasonal plumage condition.

In addition to the size difference observed between the males and females of *Grus grus* there appear to be differences in metric shape. Figure 8.10 shows that the females have a lower humerus to carpometacarpus ratio than males. This could indicate that the sexes differ in their migratory pattern, but there is no evidence of this.

Unfortunately the lack of sexed specimens of *G. antigone* has precluded a similar analysis of that species.

- **What is the pattern of size and shape in fossil *Grus*?**

The measurements supplied by Albarella and Davis (1996) from Mediaeval Cornwall (see 8.3.8) are based on fragmentary material and so should be treated with caution. They include specimens which are the size of modern *G. grus* and one which may be slightly larger.

The youngest fossil material in Britain examined during this study was that from London Wall, which is Roman in age. These two Roman specimens fall in the middle of the size distribution of modern common cranes. Another Roman site in Britain (Caerleon) has yielded some specimens which are the size of modern male *G. grus* and some which are larger than the modern common cranes (Table 8.16) (Hamilton-Dyer, 1993 and pers. comm.). These measurements are important because they include greatest lengths for certain elements which are likely to be reliable. This appears to be the latest period (1st century AD) at which cranes are present which exceed the size of modern *Grus grus*.

The next youngest cranes which were examined are those from Glastonbury and Woodbury settlements, in Somerset and Devon respectively, which are Iron Age and have a wide size range and include both specimens which are within and above the modern common crane size distribution. Most significantly is a right ulna A1248 from Glastonbury which is the size of modern sarus crane ulnae (Figure 8.12). The other specimens are either the size of modern male *Grus grus* or in the overlap range between males and females.

The next age category of crane fossils examined are Mesolithic to Bronze Age (7000 - 3000 BP). They are from the Cambridgeshire Fens and Star Carr in Yorkshire. All these specimens were the size of modern male *Grus grus* and none fell above this range.

A single left tibiotarsus of *Grus* from the Hackney Marshes which is undated may well be Early Holocene or Late Devensian in age. This specimen is the size of a modern male *Grus grus*.

The oldest specimen from Britain, which appears to be the oldest fossil of the genus *Grus* in Europe, is relatively large. The specimen probably dates from the penultimate interglacial or Oxygen Isotope Stage 7 (*ca.* 200 k BP). It is larger than all the modern *Grus grus* and within the range of *Grus antigone*.

The fossils attributable to *Grus* from Denmark, unlike the material from Britain, included material which was comfortably within the range of female *G. grus*. However, there was also a single tarsometatarsus which was considerably larger than all the modern specimens of that species. This specimen was, however, smaller than all the *Grus antigone* tarsometatarsi measured.

The general pattern seen among the British *Grus* fossils, regardless of age, is that they are either the size of male *Grus grus* or, in a few cases, they are larger. Certainly there are no indisputable female-sized fossils. In addition there is an indication that the Fen sample possess broader leg breadths particularly those of the shafts of the tibiotarsus and tarsometatarsus possibly reflecting greater body size of these birds. The Danish fossils, however, include specimens which are relatively small and well within the range of modern female *G. grus* as well as ones which are male sized, and a single bone which is considerably larger than all the *G. grus* measured.

- **Is *Grus primigenia* a genuine extinct species or is it conspecific with one of the modern species?**

The original diagnosis of *Grus primigenia* by Milne-Edwards (1864 - 71) was based on a distal tibiotarsus from Les Eyzies in the Dordogne which was said to be larger than those of the extant *Grus antigone* as well as possessing differences in morphology.

As was briefly mentioned earlier, a number of authors have used a variety of methodologies to address the question of the validity of *Grus primigenia*. These arguments have in the main been based on analyses of non-metric osteological traits and on the zoogeographic distributions of the fossils in relation to modern taxa, together with perceptions about the significance of these types of information. The present study has looked at both of those techniques as well as an in-depth morphometric analysis of both fossil and modern cranes in order to find a solution to this question.

The main hypotheses that have been proposed regarding the taxonomic identity of the anomalously large cranes in western Europe have been as follows:

1 - They belong to an extinct species named *Grus primigenia*.

Milne-Edwards (1867 - 71) believed that both non-metric traits and size distinguished the fossil tibiotarsus from Les Eyzies from that of the sarus crane (see 8.2). Other material was subsequently assigned to that species by him and others (Milne-Edwards, 1875; Harrison and Cowles, 1977 and Northcote and Mourer-Chauviré, 1985 and 1988). A review of all *Grus primigenia* material described until 1985 was published by Tyrberg (1985).

Harrison and Cowles (1977) believed that the large fossil cranes in Europe were best referred to the extinct *Grus primigenia* because they were unlikely to belong to *G. antigone* because the latter is distributed further to the East. Furthermore, Harrison (1977) went on to speculate that *G. primigenia* would have possessed white plumage. He claimed that in the northern hemisphere there are usually two breeding cranes, one smaller and grey in coloration while the other is large and white. Certainly, this is true in North America with the large white whooping crane *G. americana* and the small grey sandhill crane *G. canadensis*. In Europe the common crane would form the small grey counterpart to the extinct large white *G. primigenia*. In the Central Palaearctic it is the common crane *G. grus* which forms the small grey partner to the Siberian white crane *G. leucogeranus*, while further east it is the red-crowned crane *G. japonica*. Harrison's interesting suggestion does not, however, automatically imply that there must be an extinct white species in Europe. It is equally plausible that the Siberian white crane may have had a greater western extent, in the same way that *Grus grus* currently extends accross Eurasia accompanying different large species in different areas. Indeed there may never have been two species in the western Palaearctic. In any case it is hard to see how any suggestion as to plumage coloration in extinct species could be tested.

Northcote and Mourer-Chauviré (1985; 1988) claimed to have resolved the identity of the large cranes, concluding that they were distinguishable from both the sarus crane *G. antigone* and *G. melitensis*, and hence belonged to *G. primigenia*. However, the distinguishing characters used were not checked against a very large series of *G. antigone*, and *G. grus* was not compared at all.

2 - They belong to the neospecies *Grus antigone* whose geographical range has become much reduced.

The first author to suggest that the large crane bones in the European Quaternary represented the neospecies *Grus antigone* was Lydekker (1891b), although Milne-Edwards had eliminated the possibility on morphological grounds. Lydekker stated that there were no characters by which *Grus primigenia* could be distinguished from *G. antigone*. Since then others have agreed with this belief, whether they had examined the fossils or not (Lambrecht, 1933; Soergel, 1955; Brodkorb, 1967; Mourer-Chauviré *et al.*, 1975; Northcote, 1979, 1982). The publications by both Lambrecht and Brodkorb were catalogues and so their judgements were not based on the fossils themselves.

Soergel (1955 and 1961) described some remains of *Grus* from the Mesolithic and Neolithic of Germany. The presence of two differently-sized cranes led to his conclusion that two sympatric congeneric species were present. The smaller of the two he took to be *G. grus* while the larger was identified as *G. antigone*.

Mourer-Chauviré, in Mourer-Chauviré *et al.* (1975), initially believed the large cranes to be conspecific with the sarus crane although later, together with Northcote, came to challenge this belief (Northcote and Mourer-Chauviré, 1985 and 1988). Morphological characters were also dismissed as too variable to reliably distinguish the two species. Mourer-Chauviré *et al.* (1975) therefore agreed with others who had synonymised the species *Grus primigenia* with *Grus antigone*.

Northcote (1979) concentrated her remarks on the argument put forward by Harrison and Cowles (1977). She countered the biogeographic argument of the latter authors, which said that the sarus crane was unlikely to have existed beyond its present eastern sub-Himalayan distribution, stating that many other birds (dabchick *Tachybaptus ruficollis*, grey heron *Ardea cinerea*, moorhen *Gallinula chloropus*, coot *Fulica atra* and stone curlew *Burhinus oedicnemus*) are found both in western Europe and in India. She said that future extinctions of these in western Europe would no doubt lead to a similar problem. Also used in support of this argument was the fact that the sarus crane need not have had a continuous distribution between the two areas in the past because present day species, such as the twite *Carduelis flavirostris*, have two completely isolated populations in central Asia on the one hand and Scandinavia and Britain on the other. These facts were believed to eliminate the barriers to the possible conspecific identity of the large fossils with the sarus crane.

3 - They belong to the neospecies *Grus grus* whose size may have changed or whose modern variation was not known.

Newton (1901) discussed the large difference between two crane tibiotarsae from the Alexandra Dock in Kings Lynn, although he did not consider the possibility that *G. primigenia* was present. He stated that there was no possibility that another species other than *G. grus* was present and that sexual dimorphism was undoubtedly the reason for the difference although he did not have access to comparative material to confirm this supposition.

Von den Driesch and Boessneck (1969) found crane bones in Iberia which were larger than those of the modern common crane. However, unlike all who preceded them they considered that the fossils probably belonged to the common crane because the variation within this species was too little known. This view had not been given much attention until Hamilton-Dyer (1993) published an account of a large assemblage of Roman cranes from Caerleon in Wales. Among this assemblage were bones which were as large as those of the sarus crane, in particular a carpometacarpus which measured 130.6 mm. in length (Hamilton-Dyer, 1993) (see above). Hamilton-Dyer also stated that too little was known of the common crane to be sure that the large cranes were simply not large individuals of that species. She also said, rather oddly, that there was a possibility that the species *Grus grus* had "included more large individuals in the past than now, similar to the red deer *Cervus elaphus*". The latter comment appears to suggest that the coefficient of variation of *G. grus* may have changes since the Roman period rather than simply that the species had changed size. (The red deer is in fact generally considered to have changed size rather than to have changed in the proportions of differently sized individuals (Lister, pers. comm.)). However, the possibility that the large cranes represent a distinct subspecies from those today should be considered.

Milne-Edwards (1867-71) commented on the difference between the Scandinavian and temperate European nutcrackers (*Nucifraga caryocatactes*) and realised the implications this would have for the study of their bones. This important realisation was very enlitened for its time although he did not apply the knowledge to his extinct crane species, despite the fact that all the material he named was from the last glaciation in France i.e. possibly living in a different climate from the modern examples. The finding of large *Grus* material in Holocene deposits might counter any suggestion that a Bergmannian effect had taken place since size change by Bergmann's Rule would not be predict larger animals in a temperate context. However, the possibility that the two forms represent two subspecies should be considered. In migratory birds it is possible for two subspecies to occur sympatrically during the non-breeding season. This is seen for example in the cormorant *Phalacrocorax carbo* in Europe today. In the winter months both the nominate subspecies *P. c. carbo* and *P. c. sinensis* regularly occur together.

In North America, the sandhill crane *Grus canadensis* has currently four subspecies: the larger *G. c. pratensis* and *G. c. tabida*, and the smaller *G. c. canadensis* and *G. c. nesiotes*. These two sets of subspecies (large and small) occur with each other today on migration, so were the larger subspecies to become extinct, this would lead to a similar situation to that observed in the European fossil record.

It could be argued that this is not much of a contribution to our understanding of *Grus* in the Quaternary, as the word subspecies has merely been substituted for species. However, if a new species is to be upheld, the possibility that intraspecific variation is involved should be eliminated.

4 - They belong to the neospecies *Grus leucogeranus*.

To date no mention has been made of the Siberian white crane *G. leucogeranus* with respect to the identity of the large cranes in Europe. There is alway the possibility that this species does represent the identity of some of the large cranes, however, this would not explain the apparent absence of female-sized *G. grus* in the British Holocene record. *G. leucogeranus* is therefore not thought to be very significant in the argument. Having said this the large cranes in the last glacial, including the holotype not examined in this study, could possibly represent this species.

5 - The answer lies in a combination of two or more of the above scenarios.

It is possible that *Grus grus* was polytypic in the past in western Europe, with different sized subspecies (or simply populations), while at the same time, the whole species eventually changed size, and / or another sympatric congener occurred from time to time or in lower numbers. The migratory nature of cranes makes the interpretation of the apparent sympatrics more difficult. In situations where there are two differently sized birds in the fossil record which can be assumed to be sedentary, then it may be reasonable to assume that they represent different species, but in migratory taxa this is not possible.

Given the possible explanations above, the data collected during the present study can now be re-considered. It was hoped that the metrical survey undertaken on fossil crane bones, regardless of whether they were previously assigned to *Grus grus*, *G. antigone* or *G. primigenia*, might help clarify the identity of the larger specimens. In the initial stages of work on the British material, it seemed possible that the lack of female-sized specimens in the fossil record could be part of the answer. If the species *Grus grus* had diminished in size since about the Roman times, so that ancient females were approximately the size of modern males, then the large specimens, which have been called *Grus primigenia*, may be large ancient males. However, the female-sized Danish material would seem to counter this argument, unless the Danish specimens belong to a differently sized population or subspecies from those found in Britain. That again may be contradicted by the large tarsometatarsus found among the Danish crane fossils (Figure 8. 31). In addition the record of a large crane in Germany from Ehrenstein bei Ulm is considerably further south (Soergel, 1955). The age of the Danish material may enlighten the problem because if, as likely, they are of different dates, then time-averaging will be taking place when treating them as a single sample. In other words, the Danish fossils may fall before and after the time when *Grus grus* became smaller therefore presenting a confusing diachronuous view of Danish crane sizes. An alternative explanation is that the Danish sample includes birds from two different populations which are wintering together, which may be unlikely because the area is too far north to represent a wintering ground for two separate populations of cranes.

The likelihood that the large crane bones represent *G. antigone* is probably low because this crane is a sedentary Asian species whose breeding is tied to monsoon climates. Sarus cranes do not breed if the monsoons fail (Johnsgard, 1983). If climate is a determinant of breeding, then *G. antigone* could not represent large crane in Europe. The bird's sedentary nature also makes it unlikely that this species was once present in Europe, since becoming a tropical monsoon-adapted sedentary species was probably intrinsically linked to how the species arose in the first place. To suggest that both these aspects of the sarus crane's nature had changed, while possible, would be special pleading. *Grus antigone* is therefore unlikely to be the identity of the large crane fossils.

The British fossils, however, do appear to show that the size distribution of cranes is different to that of the modern common cranes measured. The modern cranes include Scandinavian breeding birds and possibly some more north-easterly breeding birds (the Spanish sample). The British Holocene fossils are probably breeding birds since they are certainly known to have bred in Britain (Southwell, 1901 and Gurney, 1921) and Britain is outside the modern migratory corridor (Cramp and Simmons, 1980). It is therefore interesting to speculate that the British breeding population may have been made up of larger birds. The alternative explanation, that another species, extinct or extant, cannot be wholly eliminated. However, to suggest that the large cranes are conspecific with the extant common crane is a more parsimonious explanation given the lack of fossils the size of modern females. It would seem more likely that the overall size of the common crane in Britain was larger than the modern northern population, than to suggest that there were consistently no small females preserved in the fossil record.

The difference seen in the fossil cranes may have had similar causes to the size reduction documented for the mute swan (*Cygnus olor*) by Northcote (1981 and 1983) which was attributed to the effects of habitat destruction by wetland drainage. The swans, however, persisted in Britain whereas the cranes were lost. Indeed the wetlands drainage, which has been extensive in Britain (Gurney, 1921), would almost certainly have caused the extirpation of the British crane population. The apparent size reduction in the common crane throughout Europe, if correctly deduced from the present data, may have occured when the species became extinct in its more southern and western breeding areas, such as Britain, due to habitat destruction. The large cranes in the last glaciation of France, where *G. primigenia* was named, have not been seen in the present study. However, analysis of other authors' metrical data, in the light of the extensive modern data in the present study make it is probable that they too were common cranes of a larger size than today. They were perhaps adapted to a different environment to that of the large Holocene birds because conditions were considerably different in the last cold period. A parallel to this phenomenon may be the Devensian and Holocene red deer in Britain, which were larger than those at the present day. The Holocene and Devensian red deer differ from those of today for different specific reasons, although both are connected with environmental conditions. The red deer of the Devensian were probably large due to the high carrying capacity of the vegetation of the steppe-tundra, which dominated at the time, while the red deer in the earlier Holocene was subsisting on climax forest because humans had not as yet depleted their prefered habitats. Neither of the large forms of red deer warrant distinct specific names, although one was proposed (*Stongyloceros spelaeus* Owen) for the Devensian red deer of Kent's Cavern but has subsequently found little favour (Lister, 1987). The difference between the large red deer and large cranes, in terms of the willingness for palaeontologists to accept the taxonomic separation of the latter and not the former, is that red deer fossils are more common than those of the crane, contributing to a better understanding of their size distribution at different times. This distribution could be seen to shift statistically through time (Lister, 1987). The smaller sample size of fossil cranes, some of which are the size of modern common cranes and others which are larger, has led to a belief in two distinct sympatric congeners, one of which is extinct. However, when crane fossils are viewed as a whole the large specimens attributed to *G. primigenia* can alternatively be seen as outliers of *G. grus*.

The possibility that some of the crane fossils represent evidence for *G. leucogeranus* in Europe should be borne in mind. This is perhaps the next most likely crane to be found in the area, particularly during the last glaciation, although no evidence exists for it as yet. The large Maltese cranes (*G. melitensis*) have not received attention here: although the lack of cranes of a similar age (Oxygen Isotope Stage 5e?) in Europe outside Malta will seriously hinder their interpretation. It is possible to speculate that they were even larger common cranes which for some reason were floundering in relatively large numbers on migration through Malta.

In summary, it appears that the differently-sized cranes in Britain in the past were probably of the same species as *G. grus*. The common crane in Britain was larger than modern Scandinavian birds, certainly until Roman times, and may have remained so until it became extirpated during the sixteenth century due to the effects of extensive wetland drainage. Whether all cranes became smaller across Europe or, as perhaps more likely, the more southern breeding populations which dissapeared were larger than the present Scandinavian birds, is uncertain. However, there is evidence that the extensive wetland environment present in Britain once supported a rich and varied bird fauna, elements of which have since disappeared or have become rare (Northcote, 1980). It is probable that the common crane thrived in such environments and was able to attain sizes which exceed that possible in the more northern and eastern European areas where they have persisted until today.

Chapter 9

The Genus *Lagopus* (Ptarmigan / Grouse)

9.1. Introduction

In the western Palaearctic there are two species within the genus *Lagopus*, the red / willow grouse *Lagopus lagopus* and the ptarmigan *L. mutus*. These species are small to medium sized tetraonid Galliformes and are distributed around the northern fringes of the both the Palaearctic and Nearctic with more southerly-distributed populations of *L. mutus* on higher ground such as the Alps and the Pyrenees. They are essentially cold-adapted birds with a preference for high latitude open tundra and steppe environments.

The Quaternary fossil record demonstrates that both the willow grouse and ptarmigan were at one time more widely distributed throughout the Western Palaearctic. These records include low lying areas in southern France (Mourer-Chauviré, 1975a). The earliest record of the genus is Upper Pliocene from Poland (Jánossy, 1974; Bochenski, 1991) where an ancestor of both modern taxa has been identified and named, *Lagopus atavus*. The next records are from the early Middle Pleistocene at Westbury-sub-Mendip, where both species have been recognised (Harrison, unpublished notes). Following this there are late Middle Pleistocene records in France, where extinct subspecies were named of both taxa *L. lagopus noaillensis* and *L. m. correzensis* (Mourer-Chauviré, 1975a, b). In addition to the morphologically different fossils which have been named, various authors such as Newton (1924), Mourer-Chauviré (1975a), Jánossy (1974, 1976), Bochenski (1974, 1985, 1991), Harrison (1980), Potapova (1986) and Bochenski and Tomek (1994) have described differently sized and proportioned post-cranial bones of both *Lagopus mutus* and *Lagopus lagopus* from Europe. Specifically, samples of both species from the Pleistocene were seen to differ in size and metric-shape from skeletons of their recent counterparts. These metric-shape differences were seen in the relative proportions of their tarsometatarsi, carpometacarpi, humeri, coracoidea, phalanx 1 digit III pedis and phalanx 1 digit II alae, as well as cranial elements.

The foremost questions regarding *Lagopus* skeletal morphology now and in the past are as follows:

- What is the pattern of morphology in *Lagopus lagopus* and *L. mutus* today throughout their geographical range and what is the explanation for that variation?
- What is the pattern of morphology in the fossil record of *Lagopus lagopus* and *L. mutus* throughout their geographical range and what does it signify?
- Is there any evidence for the timing of divergence of *Lagopus lagopus* and *L. mutus*?
- Are the subspecies of both *L. lagopus* and *L. mutus* named by Mourer-Chauviré valid when compared to modern subspecies?
- Are there any further conclusions to be made on the basis of the present study of *Lagopus* in the Quaternary of Europe?

9.2. Systematics and Ecology

Family Tetraonidae
Genus *Lagopus* (Brisson, 1760)

***Lagopus lagopus* (Linnaeus, 1758) - Red Grouse and Willow Grouse.**

Lagopus lagopus is a small headed, rotund game bird about 10% larger than the ptarmigan *L. mutus* with triangular wings best seen in flight. Their plumage varies in colour according to race but is basically brown with black tail feathers in the breeding season and, except for *L. l. scoticus*, turns mostly white in winter.

The willow grouse and red grouse together form the species *Lagopus lagopus* which is polytypic and has a northern circumpolar distribution. The red grouse *L. l. scoticus* was formerly considered to be a distinct species because of its isolated distribution in the British Isles and its lack of white plumage in winter (Witherby *et al.*, 1941). However, the intermediate subspecies in West Norway *L. l. variegatus*, whose population has only partial white winter plumage, has caused the two to be regarded as conspecific (Cramp, 1980).

In addition to *L. l. lagopus, L. l. scoticus* and *L. l. variegatus* there are more than 10 further subspecies. The Western Palaearctic subspecies and their distributions are as follows:

L. l. lagopus: Scandinavia, Finland and N. Russia south to c. 60° N.
L. l. scoticus: Britain and Ireland.
L. l. variegatus: Islands off Trondheim Fjord.
L. l. rossicus: Russia south of ca. 60° N.
L. l. koreni: N. Siberia.
L. l. maior: South west Siberia.
L. l. brevirostris: South east Siberia.
(+ ca. 9 other subspecies in Eastern Palaearctic and North America).

These subspecies differ markedly in their habitat requirements although *L. lagopus* is "essentially a bird of tundra and interrupted taiga, tolerant of scrub but not woodland " (Cramp, 1980). *L. l. scoticus* is now primarily made up of populations under human mangement and is more associated with heather moors than other subspecies. *L. l. maior*, a large form, lives in wooded steppes, much as does the black grouse *Tetrao tetrix* today (Dement'ev and Gladkov, 1967). *L. l. lagopus*, however, in Russia prefers stands of willow and dwarf birch in tundra vegetation, but will avoid thick tall vegetation in the breeding season (Dement'ev and Gladkov, 1967).

The species' food is almost entirely made up of plant matter although the components of it are determined by seasonal and geographic variations in availability as well as selection for most nutritious food sources and competition with the congener *L. mutus*. In the case of *L. l. scoticus* the diet is mostly heather *Calluna vulgaris* throughout the year. In most other subspecies snow covers the ground vegetation in winter and willow *Salix* and birch *Betula* catkins, buds and twigs dominate the diet. Willow is first choice followed by birch. In summer, spring and autumn the diet is more varied and berries, new shoots and leaves are taken from a variety of other plant species. These plants include bilberry *Vaccinium myrtillus*, alpine bistort *Polygonum viviparum*, crowberry *Empetrum nigrum*, bog bilberry *V. uliginosum* and the sedges *Carex* and *Scirpus*.

The species is a ground bird but will undertake movements when forced to do so by the weather.

Together with *Lagopus mutus*, *L. lagopus* fossils are the most common Quaternary bird fossils in Europe (Tyrberg, 1995). They are at their most common in Late Pleistocene sites. The *L. lagopus* lineage first makes its appearance in the Upper Pliocene as *L. atavus*. An extinct subspecies of willow grouse *Lagopus lagopus noiaillensis* was named by Mourer-Chauviré (1975a) on the basis of "Rissian" material from La Fage in France. The diagnosis of this subspecies was that it possessed more robust tarsometatarsi than modern members of the species.

***Lagopus mutus* (Montin, 1776) - Ptarmigan.**

The ptarmigan is very similar to the willow and red grouse but smaller and slighter in build with narrower wings.

The species is polytypic with many subspecies distributed around the northern fringes of both the Palaearctic and Nearctic as well as isolated subspecies on islands and more southern mountain ranges. The subspecies and their distribution are as follows:

L. m. mutus: Scandinavia and the Kola peninsula.
L. m. millaisi: Scotland.
L. m. helveticus: Alps.
L. m. pyrenaicus: Pyrenees.
L. m. nelsoni: Urals, east through northern Siberia into North America to Mackenzie, where it merges with *L. m. rupestris*.
L. m. hyperboreus: Bear Island, Spitsbergen and Franz Josef Island.
L. m. islandorum: Iceland.
L. m. rupestris: South Greenland and arctic N. America to Mackenzie.
L. m. saturatus: Central West coast of Greenland.
L. m. captus: North Greenland.

(+ ca. 12 other subspecies in the Eastern Palaearctic, Commander and Aleutian islands and North America.)

The habitat of *L. mutus* ranges from sea-level at the highest latitudes to 1700 to 2400 m in the Alps. The species breeds at higher altitudes the lower the latitude. They prefer mosaic habitats of rock boulders, detritus, bare pavements with carpets of moss and lichen and dwarf heath plants such as bilberry *Vaccinium*, crowberry *Empetrum*, heather *Calluna*, mountain azalea *Loiseleuria* and dwarf willow *Salix herbacea*. They may occur in the same areas as *L. lagopus,* but are ecologically separated from them, prefering desolate tundra, mountain tops and other barren areas, and do not tolerate woodland or scrubland.

The species is resident but may irrupt in arctic latitudes. They often move uphill as temperatures rise in summer, tracking the melting snow, and some races make greater movements in winter. *L. m. hyperboreus* moves between islands in Spitzbergen and *L. m. islandorum* can make journeys of up to 275 km. Other subspecies such as *L. m. saturatus* and *L. m. captus* in Greenland may migrate south in flocks similar to the movements of other irruptive species such as the snowy owl *Nyctea scandiaca* and lemming *Lemmus*.

The food of the ptarmigan is plant matter and their diet is determined, as with *L. lagopus*, by seasonal and geographic availability, selection for most nutritious foods and competition with its congener. The diets reflect their generally higher, more exposed habitats than *L. lagopus*. In Scotland they eat shoots of heather *Calluna vulgaris*, bilberry *Vaccinium myrtillus* and *Empetrum*, which form ca. 90% of their dry weight food intake and ca. 60% in summer. In summer this is supplemented by berries of *Vaccinium* and *Empetrum*. In Iceland these food elements are more rarely taken and shoots of dwarf willow *S. herbacea* and catkins and buds of downy birch *Betula pubescens* and dwarf birch *B. nana* form the dominant foods despite an abundance of alternatives. Other plant species taken at different times by different populations include *Dryas octopetala*, *Polygonum viviparum*, *Vaccinium vitis-idaea*, *V. uliginosum*, *Loiseleuria procumbens*, *V. myrtillus*, *B. tortuosa*, *Arctous alpina*, *Carex* and *Geum*.

L. mutus is a very common bird in the Quaternary fossil record of Europe and particularly in the later Pleistocene (Tyrberg, 1995). *Lagopus mutus correzensis*, an extinct subspecies, was named by Mourer-Chauviré (1975a) on the basis of "Rissian" material from La Fage in France. The diagnosis of this subspecies was that it possessed more robust tarsometatarsi than modern members of the species.

***Lagopus atavus* (Jánossy, 1974).**

Lagopus atavus was originally described as a new subspecies of willow grouse *Lagopus lagopus atavus* by Jánossy in 1974 but was subsequently elevated to species (Jánossy, 1976) without explanation. The specimen from which the species was diagnosed came from Rebielice Królewskie 1, a site now considered to be Upper Pliocene in age. *Lagopus atavus* is believed to be ancestral to both *Lagopus lagopus* and *L. mutus* (Bochenski, 1991).

The holotype is a proximal humerus (AF/26 - RK1:22) and differences between it and modern *L. lagopus* and *L. mutus* were given in Jánossy (1974). The other 31 bone fragments ascribed to the species by Jánossy, together with the holotype, have recently been reassessed by Bochenski (1991). Bochenski also examined other Upper Pliocene *Lagopus* fossils in the light of *L. atavus* and ascribed a fossil from Kielniki 3B to the species.

9.3. *Lagopus* Osteology.

The means by which the post cranial bones of the two European *Lagopus* species can be distinguished from those of similar sized Galliformes are detailed in Kraft (1972), although there are two problems. First, it was found to be difficult in the present study to confirm the criteria defined for the ulna, radius, carpometacarpus, femur, tibiotarsus, coracoid and scapula. This means that only the humerus and tarsometatarsus could be confidently identified in the fossil record to the genus *Lagopus*. Futhermore, only the proximal humerus possessed a reliable character, distinguished from *Tetrastes*, *Alectoris* and *Perdix* on the basis of the shape of the fossa tricipitalis. In *Lagopus* it is moderately shallow with a proximally attenuation, while in *Alectoris* the depression is deep and rounded. *Perdix* could be distinguished from *Lagopus* by having only a very shallow fossa tricipitalis. *Tetrastes* is similar to *Perdix* in having a shallower fossa, which however is similarly pointed proximal to *Lagopus*. The tarsometatarsus of *Lagopus* was distinguished from that of *Alectoris* because in *Alectoris* the tarsometatarsi are much longer and flattened dorsoventrally. In *Perdix* the tarsometatarsi are also very noticeably longer while in *Tetrastes* they are very long and thin. All the differences were present in 100% of the specimens of all genera examined by the present author.

The second problem is the distinction between the two *Lagopus* species, which is even less satisfactorily achieved. There appear to be no details of morphology that can aid determination, the two species being identified simply on the basis of size: *L. lagopus* is consistently larger than *L. mutus* (Kraft, 1972). However, considerable overlap is present, which varies according to the skeletal element in question. The tarsometatarsus is the most easily identified due to greater than usual separation in size between the two species (Kraft, 1972; Mourer-Chauviré, 1975a; Bochenski, 1985). However, this is in need of verification with a larger sample, more widely distributed geographically.

9.4. *Lagopus* Samples

***Lagopus lagopus lagopus* - Scandinavia**

The Scandinavian sample of willow grouse consists of 12 individual skeletons, 10 of which are housed in the Swedish Museum of Natural History in Stockholm while two others come from the Institute of Systematics and Evolution of Animals in Krakow and the Zoological Museum in Copenhagen respectively. Of these 7 were male, 4 female, and one unsexed. They were collected between 1832 and 1985. They originated from Kasepakt (5), Vassijaure (1) and Dalrisa (1), all in Sweden; Melatus in Finland (1); and unspecified locations in Norway (1) and Sweden (3).

***Lagopus lagopus lagopus* - Russia**

The Russian sample is housed in the Institute of Systematics and Evolution of Animals in Krakow. Of the 8 specimens two are male and one female, the others are unsexed. Six of the specimens came from northern Russia in Arkhangelsk while one came from Sanichord near Moscow and another from Novosibirsk in Western Siberia. The birds were collected between 1965 and 1991.

***Lagopus lagopus scoticus* - Derbyshire Peak District**

The sample of red grouse from the Derbyshire Peak District in England forms part of the comparative collection of Derek Yalden of Manchester. They total 26 specimens and were collected as dead carcases found during walks between 1969 and 1983. Of the 26 very few were sexed, but include four males and three females.

***Lagopus lagopus scoticus* - Scotland**

The Scottish sample totals 10 individuals, of which 2 are male and 1 female. They were collected between 1882 and 1984 and are in the collections of the Ancient Monuments Laboratory of English Heritage, the Sub-department of Ornithology, Natural History Museum in Tring, and D. Yalden of Manchester University. They include one female and two males while the rest are unsexed. They came from Banchory (2), Beauly (2), Berwickshire, Sterling (1), Glen Shee (1) and one unspecified.

***Lagopus lagopus brevirostris* - Altai Mountains, Siberia**

The single skeleton of *L. l. brevirostris* was measured in the collections of the Institute of Systematics and Evolution of Animals in Krakow. This specimen was collected in the Altai Mountains in Siberia during 1960 and is a female.

Lagopus lagopus maior **- Kazakhstan**

The individual skeleton of *L. l. maior* is in the the Institute of Systematics and Evolution of Animals in Krakow and was collected in Kazakhstan in 1956. It is a male.

Lagopus mutus mutus **- Scandinavia**

The four Scandinavian ptarmigan skeletons are housed in the Institute of Systematics and Evolution of Animals in Krakow (1), the Swedish Museum of Natural History in Stockholm (1) and the Zoological Museum in Copenhagen (2). They include two females, one male and one unsexed individual, and were collected between 1904 and 1990. The localities from which they were collected were Brodo, Tellmarken and a non-localised place in Norway as well as Kiruna, Sweden.

Lagopus mutus millaisi **- Scotland**

The sample of the Scottish subspecies of ptarmigan comprised six specimens. Of these three were sexed, two males and one female. They are in the collection of the Natural History Museum at Tring. They were collected between 1922 and 1952 in localities in Banffshire (1), Rosshire (2), Beauly (1) and two unspecified locations.

Lagopus mutus islandorum **- Iceland**

Three skeletons from Iceland comprise one female and two unsexed specimens. They are housed in the Zoological Museum Copenhagen (1) and the Natural History Museum in Tring (2). Their collection dates are between 1924 and 1925. One came from Frederickshaal, while the other two had no precise locality data.

Lagopus mutus helveticus **- Alps**

The Alpine ptarmigan sample came from the Zoological Museum in Copenhagen (1), the Institute of Systematics and Evolution of Animals in Krakow (1) and the Université Claude Bernard in Lyon (16). The specimens from Lyon included all the post-cranial elements usually measured, although due to time constraints only the tarsometatarsi were measured on 12 of them and a further two had their coracoids and humeri measured. Six of the specimens had been sexed: two were males and four females. The material came from the Laurichard (1); Vallée de Servière, Haute Alpes (8); Val d'Isere, Savoie (1); Arsine, Oisans, Haute Alpes (1); Roche Brune d'Archeboc, Savoie (1); Villar d'Arene, Oisans, Haute Alpes (1) and three which were not further provenanced. They were collected between 1895 and 1974.

Lagopus mutus hyperboreus **- Spitzbergen**

A single specimen from Spitzbergen was measured in the Institute of Systematics and Evolution of Animals in Krakow. The specimen was collected in 1985 and was unsexed.

Lagopus mutus nadezdae **- Altai Mts., Siberia**

A single specimen of this subspecies was measured in the collections of the Institute of Systematics and Evolution of Animals in Krakow. The specimen is a female and was collected in 1960 in the Altai mountains in Siberia.

Lagopus mutus kellogae **- Taimyr Peninsula, Siberia**

One specimen in the collections of the Institute of Systematics and Evolution of Animals in Krakow came from the Taimyr Peninsula, north Siberia. The specimen was unsexed.

Lagopus mutus **- Russia**

One specimen came from Moscow (presumably the market?); the specimen was unsexed.

9.5. Osteometric comparison of species and subspecies within *Lagopus* in Europe

It can be seen from Figures 9.1 - 9.9 as well as Tables 9.1 and 9.2 that in Europe today *L. lagopus* is larger on average than *L. mutus*. Table 9.3 demonstrates that most dimensions of the bones measured of these two species yield highly significant differences. However, despite the significant difference in many skeletal elements there is considerable overlap. This overlap in measurements seems to be most marked amongst the wing and pectoral girdle bones and less so in the leg elements. There also appears to be progressively less overlap distally in the leg (Figures 9.7 - 9.9) and proximally in the wing (Figure 9.1 - 9.4). Therefore, the tarsometatarsus in the leg and the humerus in the wing present the greatest interspecific difference in those limbs.

There are also differences in mean between subspecies and populations within the two taxa (Tables 9.1 – 9.2). However, only the samples of the subspecies *L. lagopus lagopus* and *L. l. scoticus* were found to have many statistically significant differences (Tables 9.4 – 9.6). Sample sizes for *L. mutus* are relatively small, so intraspecific comparisons are provisional.

In the larger species *L. l. lagopus* from Russia and Scandinavia possess humeri that are smaller on average than in *L. l. scoticus* (Table 9.1); however, these differences are not statistically significant (Table 9.4). The single *L. l. brevirostris* is relatively small (Figure 9.1). In *L. mutus*, *L. m. millaisi*, the Scottish subspecies, have larger humeri than *L. m. mutus* from Scandinavia. The small sample sizes of *L. m. helveticus* and *L. m. islandorum* make it difficult to make reliable comparisons. However, *L. m. helveticus* include the smallest humeri, and the two *L. m. islandorum* specimens have very different humeral lengths (Figure 9.1).

In *L. lagopus*, the single ulna of *L. l. maior* is the largest and the single *L. l. brevirostris* ulna appears to be the smallest (Figure 9.2). In *L. l. scoticus* the ulna is significantly larger than in *L. l. lagopus* (Table 9.4). While the lengths differ significantly, the breadths are not appreciably different (Table 9.4). The pattern is the same in the radius with no appreciable difference. The two ulnae of *L. m. millaisi* are long and narrow. The single specimens of *L. m. hyperboreas* and *L. m. kellogae* are the largest, while *L. m. mutus* from Scandinavia and *L. m. islandorum* include the shortest and longest ulnae and those from the Alps, *L. m. helveticus*, are intermediate. All these samples are, however, very small and can not be *t*-tested.

The *Lagopus lagopus* carpometacarpi overlap greatly, although *L. l. maior* is the largest and the single *L. l. brevirostris* is relatively small. The only significant difference found between the carpometacarpi was that of the breadth across the process intermetacarpalis (GT) where the Scottish subspecies was larger than that from Scandinavia. Of the carpometacarpi in *L. mutus*, the single specimen of *L. m. hyperboreas* is the largest, followed by *L. m. kellogae*. *L. m. millaisi* from Scotland may be relatively robust (Figure 9.4), the Icelandic sample *L. m. islandorum* is again variable, while the Alpine, *L. m. helveticus*, and Scandinavian, *L. m. mutus*, specimens are on average smallest (Table 9.2).

Among the *L. lagopus* coracoidea there is much overlap with *L. l. maior* forming a large outlier. *L. l. scoticus* while larger on average than *L. l. lagopus* is not statistically different (Table 9.4). The coracoidea of *L. mutus* fall broadly into two clusters. The two specimens of *L. m. millaisi*, the *L. m. hyperboreas* and *L. m. kellogae* individuals forming the large-sized group. Meanwhile the three specimens of *L. m. helveticus*, the two of *L. m. mutus* and the individual coracoidea of *L. m. nadezdae*, *L. m. islandorum* and the bird from Russia form the smaller-sized cluster.

Figure 9.6 appears to gives the greatest overlap between the two species of all the graphs, largely because it shows two breadth measurements of the scapula. The only measurement to produce significant differences is the smallest breadth of the scapular collum. Comparisons of the two samples of *L. l. scoticus* and the whole samples of *L. l. lagopus* and *L. l. scoticus* produced differences (Table 9. 4 and 9.4). In *L. mutus*, *L. m. hyperboreas*, *L. m. millaisi* and , *L. m. kellogae* are larger; *L. m. mutus*, *L. m. islandorum* and the Russian specimens are smaller, while *L. m. helveticus* spans both.

The largest femur in *L. lagopus* is that of *L. l. maior*. The other samples overlap although the British subspecies, *L. l. scoticus*, is significantly larger than the nominate from both Russia and Scandinavia, *L. l. lagopus* (Table 9.4). In *L. mutus* the femur is largest in the two *L. m. millaisi* specimens from Scotland; all the other samples overlap considerably. There are no data for *L. m. hyperboreas*, although the other subspecies with larger fore-limbs, *L. m. kellogae* has small a relatively small femur. This is probably meaningful because the measurements derive from the skeleton of a single individual.

The tibiotarsus of the single individual of *L. l. maior* is again the largest of that species and the British subspecies, *L. l. scoticus*, is significantly larger than *L. l. lagopus* (Table 9.4). The tibiotarsi of *L. mutus* form an almost identical pattern of size as do the femora for that species. The two *L. m. millaisi* are large while the others show no clear pattern of size difference.

In the tarsometatarsi of *L. l. lagopus* a great deal of overlap is apparent in Figure 9.9, although Table 9.1 shows that *L. l. scoticus* is on average longer than *L. l. lagopus*. The two dimensions which produced statistically significant differences between *L.l.scoticus* and *L.l.lagopus* were the greatest length and the proximal breadth (Table 9.4). The single specimen of *L. l. maior* again forms a large outier. As with more proximal limb bones of the leg, the largest individual ptarmigan in terms of tarsometatarsus is the specimen of *L. m. millaisi*. There is some suggestion that *L. m. helveticus* has relatively long and thin tarsometatarsi while those of *L. m. mutus* are on average short and relatively robust. *L. m. kellogae* has a comparatively small tarsometatarsus in relation to its other fore-limb bones and pectoral girdle.

Very few significant differences exist between the populations measured of either *L. l. scoticus* or *L. l. lagopus* (Table 9.5 – 9.6). None were found between the two populations of *L. l. lagopus* from Scandinavia and Russia. However, the shaft breadth and the smallest breadth of the scapular collum each differed significantly between the populations of *L. l. scoticus* from Scotland and Derbyshire. The meaning of these differences is difficult to explain.

It can be concluded that in both *Lagopus mutus* and *L. lagopus* there is variation in size and shape. The analysis of the data given in Tables 9.1 and 9.2 is hampered by the sample sizes being relatively small for certain populations. However, it is clear certain differences do exist. For intance, in *L. mutus*, the single individual of *L. m. hyperboreas* is the largest, followed by *L. m. millaisi*. *L. m. helveticus* and *L. m. mutus*, meanwhile, overlap greatly in many skeletal element dimensions, and are smaller. In *L. lagopus*, the single available individual of *L. l. maior* hints that as its name suggests, it is large and forms an outlier on all the scatters on which it is plotted. Equally, the smallest appears to be *L. l. brevirostris* although only one specimen was examined. Finally, there is clear evidence that *L. l. scoticus* is larger on average than the nominate subspecies *L. l. lagopus*. The dimensions which proved to be statistically different were mostly bone lengths and particularly those of the leg. This means that the red grouse has longer legs than the willow grouse. The significance of this presumably lies in some aspect of the birds' locomotion. The relative winter snow cover in the geographical ranges of the two subspecies may explain the longer legs of red grouse. As with the white plumage of willow grouse shorter legs may be an advantage to living in a snow covered landscape. The white plumage acts as camouflage in winter while the shorter legs may make terrestrial locomotion easier, although it is not clear how.

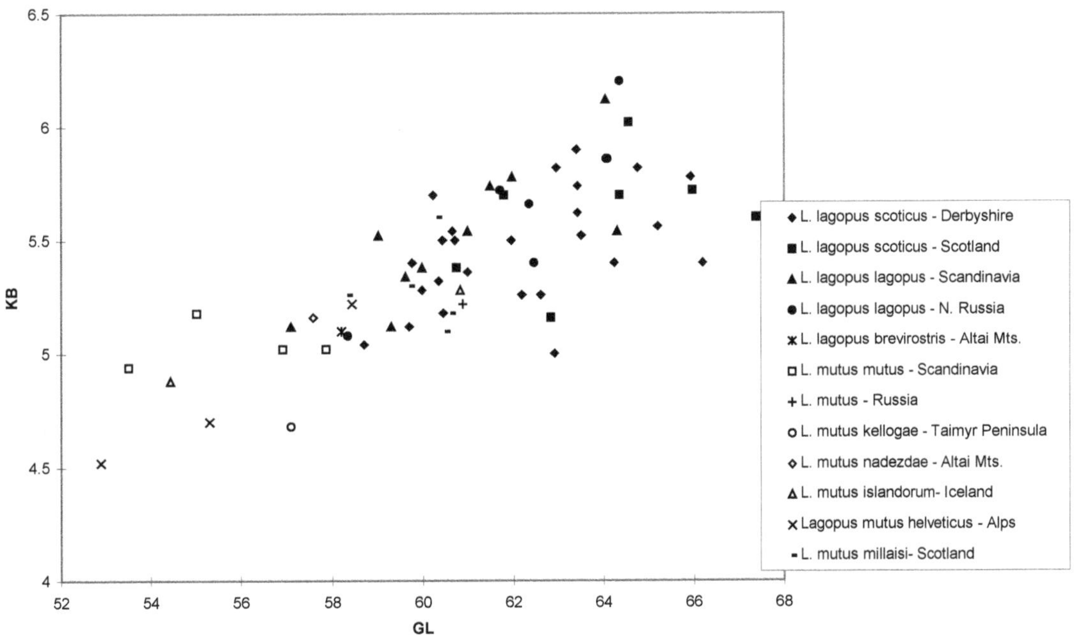

Figure 9.1. Scattergram of European *Lagopus* species and subspecies humeri. Humerus greatest length (GL) plotted against shaft breadth (KB).

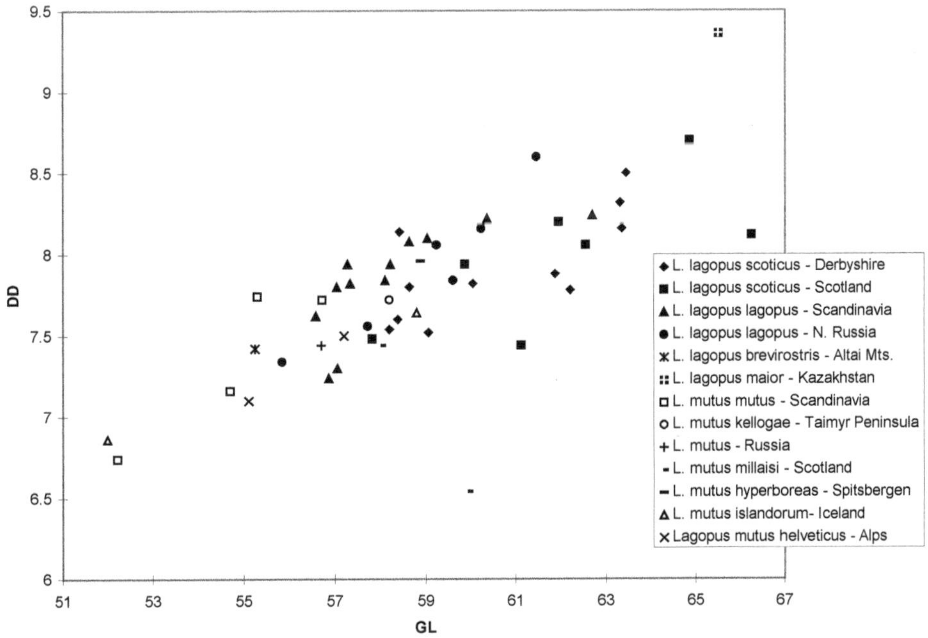

Figure 9.2. Scattergram of European *Lagopus* species and subspecies ulnae. Ulnar greatest length (GL) plotted against distal diagonal depth (DD).

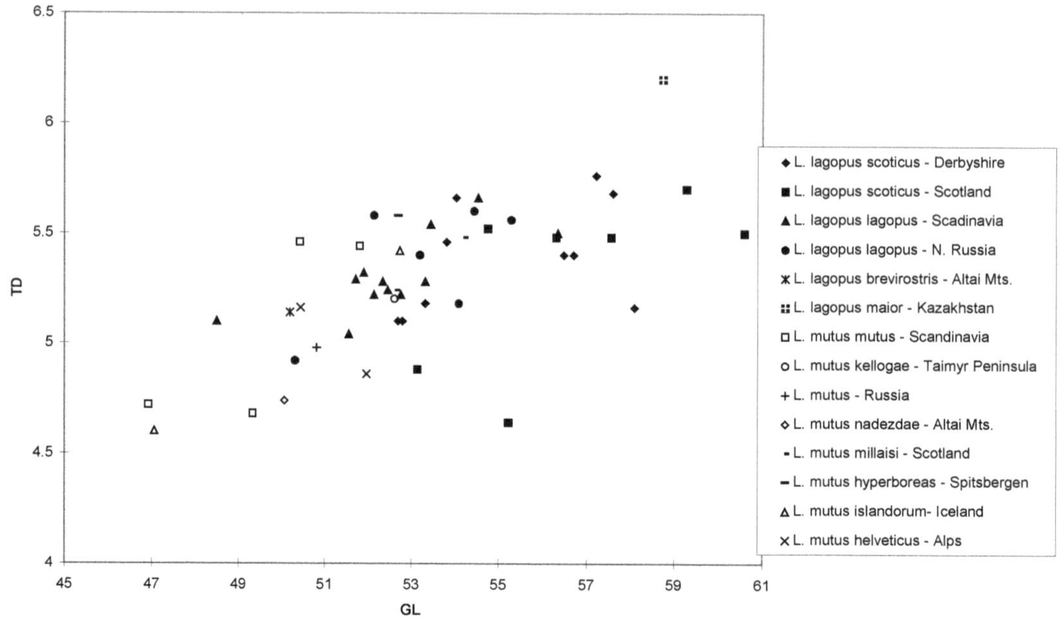

Figure 9.3. Scattergram of European *Lagopus* species and subspecies radii. Radial greatest length (GL) plotted against distal breadth (TD).

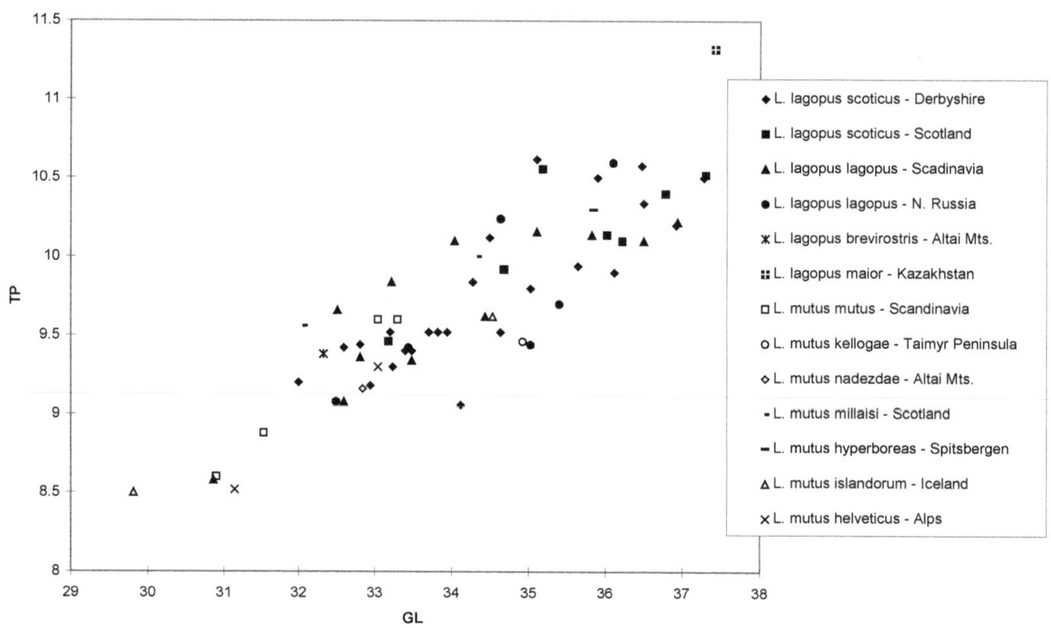

Figure 9.4. Scattergram of European *Lagopus* species and subspecies carpometacarpi. Carpometacarpal greatest length (GL) plotted against proximal breadth (TP).

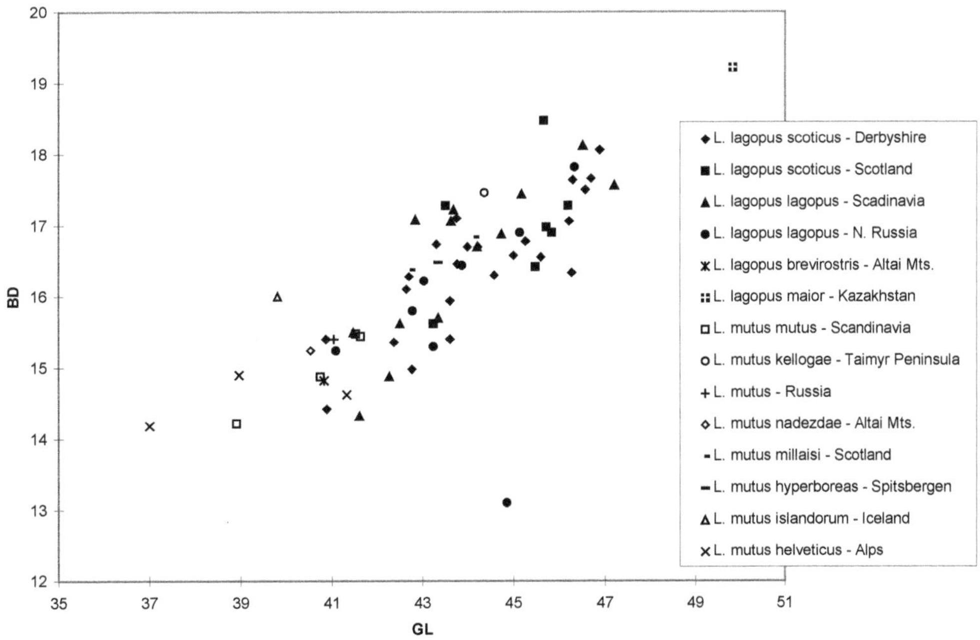

Figure 9.5. Scattergram of European *Lagopus* species and subspecies coracoidea. Coracoideal greatest length (GL) plotted against distal diagonal depth (BD).

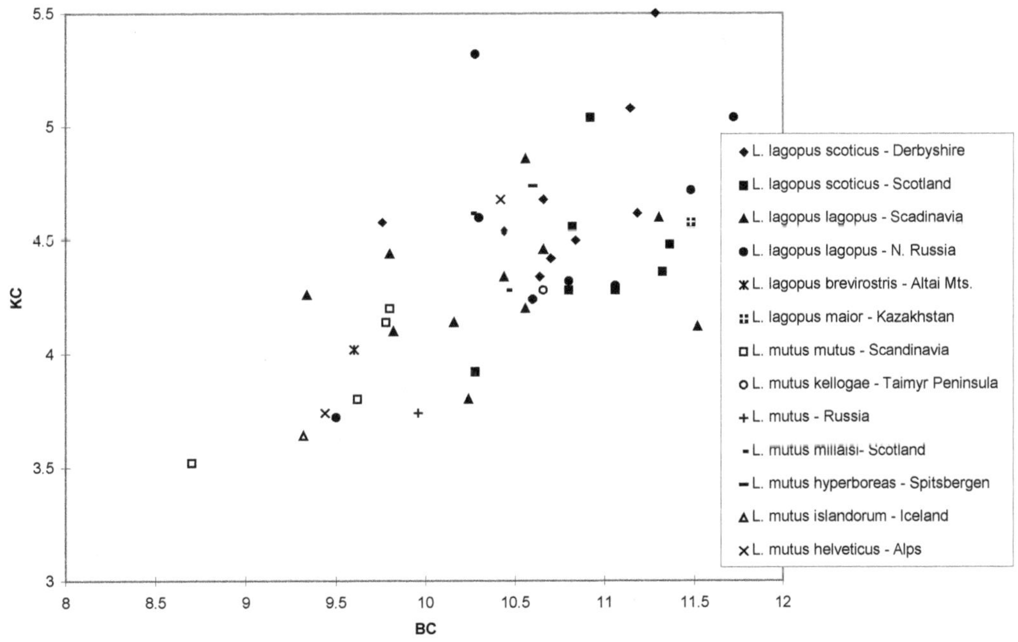

Figure 9.6. Scattergram of European *Lagopus* species and subspecies scapulae. Scapular greatest cranial breadth (BC) plotted against the smallest breadth of the collum scapulae (KC).

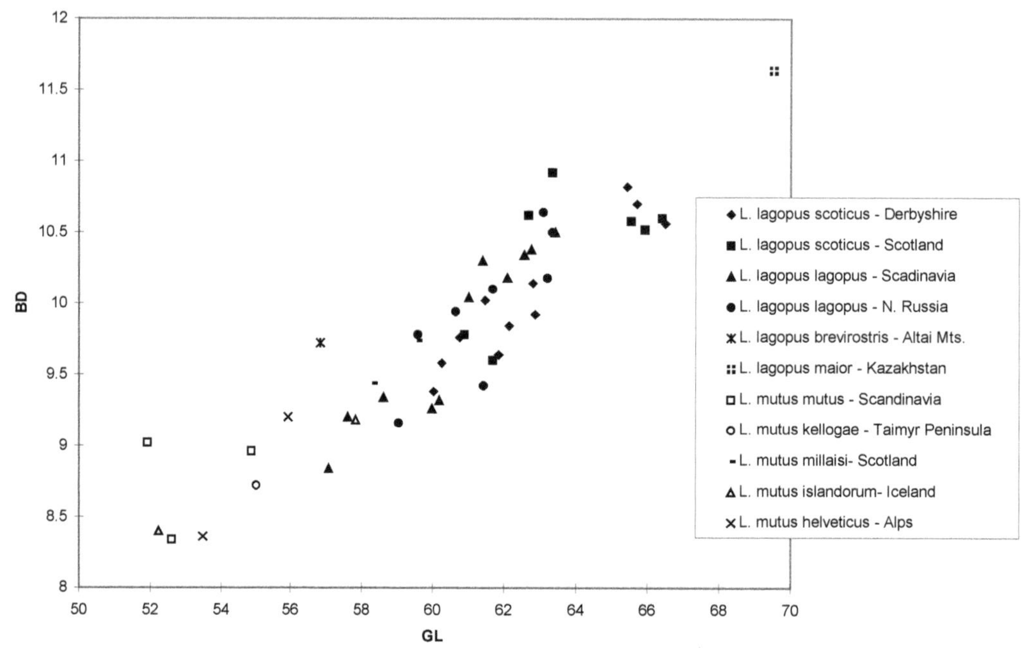

Figure 9.7. Scattergram of European *Lagopus* species and subspecies femora. Femoral greatest length (GL) plotted against distal breadth (BD).

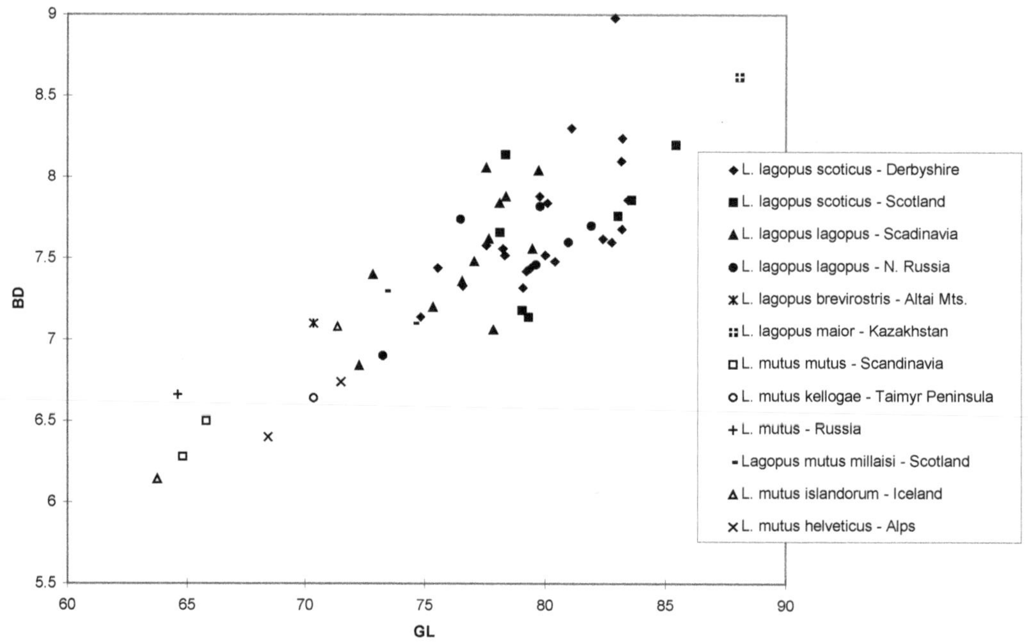

Figure 9.8. Scattergram of European *Lagopus* species and subspecies tibiotarsi. Tibiotarsal greatest length (GL) plotted against distal breadth (BD).

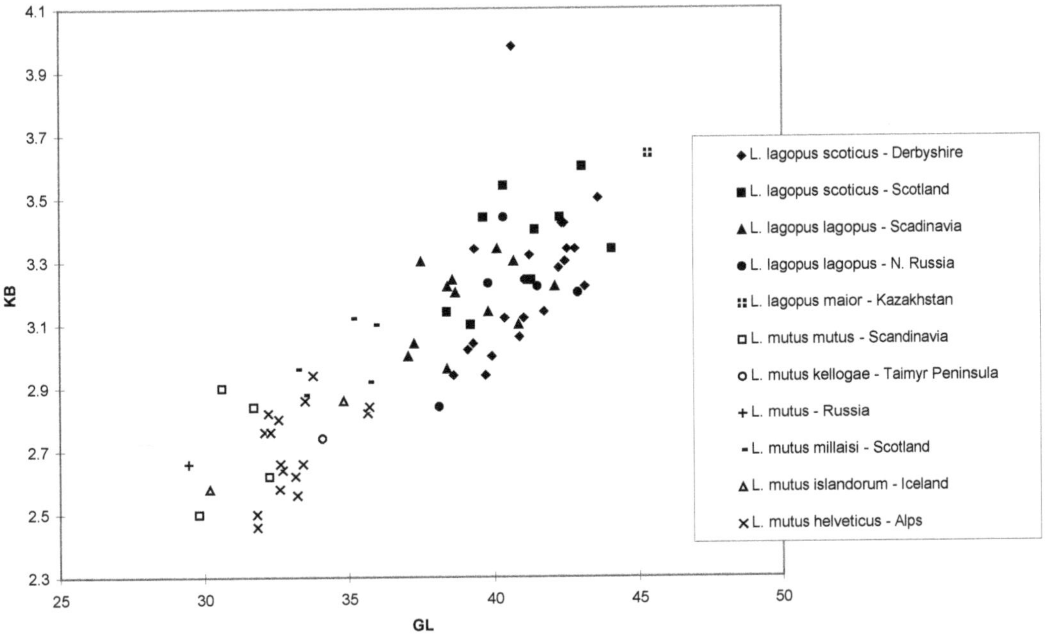

Figure 9.9. Scattergram of European *Lagopus* species and subspecies tarsometatarsi. Tarsometatarsus greatest length (GL) plotted against shaft breadth (KB).

Table 9.1. Measurement data of the modern *Lagopus lagopus* samples.

Lagopus lagopus samples	*L. l. scoticus* - Derbyshire					*L. l. scoticus* - Scotland					*L. l. lagopus* - Scandinavia				
Measurements	N	Min	Max	Mean	St. dev.	N	Min	Max	Mean	St. dev.	N	Min	Max	Mean	St. dev.
Hum GL	25	58.7	66.18	62.18	2.1	7	60.74	67.38	63.94	2.33	12	57.1	64.3	60.87	2.05
Hum TP	25	15.12	17.16	16.11	0.57	7	15.96	17.16	16.57	0.42	12	14.52	17.42	16.27	0.96
Hum KT	25	5.0	5.9	5.46	0.25	7	5.16	6.02	5.61	0.27	11	5.12	6.12	5.52	0.29
Hum KB	25	3.64	5.28	4.09	0.32	7	3.8	4.46	4.24	0.23	12	3.82	4.46	4.1	0.24
Hum TD	25	10.26	11.9	10.98	0.46	7	10.28	11.56	11.26	0.45	12	10.2	11.74	11.03	0.5
Uln GL	11	58.2	63.44	60.63	2.22	7	57.82	66.26	62.06	2.87	12	56.58	62.7	58.27	1.77
Uln DP	11	10.06	11.08	10.50	0.37	7	9.92	11.54	10.84	0.51	12	9.58	11.06	10.41	0.58
Uln BP	11	7.7	8.58	8.15	0.3	7	7.6	8.6	8.15	0.37	12	7.24	8.72	7.86	0.5
Uln TP	11	6.1	6.96	6.57	0.27	7	6.1	6.76	6.61	0.23	12	6.04	6.84	6.39	0.25
Uln KT	11	3.12	3.8	3.44	0.23	7	3.4	3.9	3.61	0.18	12	2.92	3.68	3.4	0.21
Uln DD	11	7.52	8.5	7.91	0.32	7	7.44	8.7	7.94	0.43	12	7.24	8.24	7.84	0.32
Rad GL	10	52.7	58.1	55.27	2.13	7	53.14	60.6	56.69	2.63	12	48.5	56.34	52.58	1.87
Rad TP	10	3.54	4.7	3.84	0.34	7	3.48	4.7	4.02	0.46	12	3.54	3.94	3.71	0.15
Rad BP	10	4.18	4.74	4.43	0.2	7	3.9	4.58	4.23	0.31	12	4.22	4.8	4.48	0.21
Rad KT	10	1.88	2.4	2.07	0.18	7	2.02	2.42	2.17	0.13	12	1.86	2.32	2.11	0.16
Rad TD	10	5.1	5.76	5.39	0.25	7	4.64	5.7	5.3	0.39	12	5.04	5.66	5.31	0.18
Cpt GL	24	32.0	37.28	34.48	1.48	7	33.18	37.3	35.62	1.4	12	30.86	36.94	34.02	1.81
Cpt TP	24	9.06	10.62	9.76	0.48	7	9.46	10.56	10.16	0.39	12	8.58	10.22	9.68	0.52
Cpt GT	23	5.04	6.1	5.56	0.26	7	5.06	5.9	5.54	0.29	12	4.7	5.82	5.31	0.33
Cpt DD	24	6.26	7.48	6.94	0.33	7	6.8	7.74	7.2	0.36	12	6.3	7.38	6.89	0.33
Cpt HS	24	3.26	5.02	4.41	0.43	7	4.02	5.54	4.61	0.46	12	4.06	5.16	4.62	0.26
Fem GL	21	60.02	66.5	62.71	2.24	7	60.9	66.4	63.78	2.18	10	57.06	63.42	60.91	1.99
Fem LM	21	56.46	62.74	59.46	2.02	7	57.82	62.82	60.18	1.88	11	54.1	59.62	57.52	1.69
Fem TP	20	6.7	7.6	7.04	0.3	7	6.62	7.84	7.3	0.45	11	6.08	8.4	7.17	0.49
Fem BP	20	11.1	12.7	11.73	0.51	7	10.9	12.54	11.82	0.62	11	10.94	13.12	11.69	0.65
Fem KB	20	4.34	5.14	4.64	0.26	7	4.52	5.14	4.79	0.25	11	4.14	5.28	4.65	0.34
Fem TD	20	7.8	8.9	8.31	0.38	7	7.5	9.02	8.29	0.62	11	7.76	9.1	8.32	0.46
Fem BD	20	9.8	10.82	10.03	0.48	7	9.6	10.92	10.37	0.49	11	8.84	10.5	9.79	0.6
Tib GL	11	74.84	83.44	80.07	2.61	7	78.16	85.42	80.99	2.94	12	72.26	79.74	76.93	2.35
Tib LA	11	73.58	82.42	78.78	2.65	7	76.54	84.32	79.61	3.02	11	70.82	78.3	75.31	2.39
Tib BP	11	7.72	11.96	8.89	0.88	7	8.14	9.42	8.99	0.49	11	7.74	9.24	8.59	0.52
Tib DP	11	7.84	13.86	12.6	1.19	7	11.62	14.1	12.9	0.89	12	12.02	13.94	12.88	0.65
Tib KB	11	3.12	3.82	3.49	0.19	7	3.28	3.96	3.66	0.3	12	3.36	4.02	3.64	0.19
Tib KT	11	3.14	3.86	3.47	0.17	7	3.2	3.76	3.46	0.19	12	3.2	3.7	3.45	0.17
Tib BD	11	7.14	8.98	7.71	0.42	7	7.14	8.2	7.71	0.42	12	6.84	8.06	7.53	0.38
Tmt GL	21	38.6	43.58	41.16	1.5	9	38.38	44.06	40.36	1.38	12	37.04	42.1	39.11	1.59
Tmt BP	21	7.94	9.98	8.65	0.52	9	7.52	9.04	8.3	0.5	12	7.5	8.7	8.23	0.36
Tmt KB	19	2.94	3.98	3.24	0.24	9	3.1	3.6	3.3	0.14	12	2.96	3.34	3.17	0.12
Tmt BD	22	6.92	9.2	7.87	0.55	9	7.08	8.52	7.66	0.39	12	7.26	8.86	7.85	0.55
Tmt DD	21	7.68	9.64	8.79	0.49	9	8.36	9.8	8.98	0.4	12	8.34	9.54	8.9	0.38
Cor GL	23	40.86	46.88	44.25	1.78	7	43.24	46.2	45.09	1.2	12	41.46	47.2	43.75	1.84
Cor LM	24	37.66	43.28	41.17	1.56	7	39.7	42.8	41.67	1.27	12	41.56	43.16	40.44	1.55
Cor BD	24	14.42	18.06	16.45	0.88	7	15.62	18.48	16.99	0.88	12	14.32	18.12	16.45	1.19
Cor BB	24	10.9	13.94	12.39	0.73	6	11.6	14.02	13.04	0.81	11	10.34	13.48	12.41	1.2
Scap GL	8	60.36	67.42	63.57	2.39	5	61.96	66.9	64.42	2.21	11	59.3	67.24	63.17	2.72
Scap BC	9	9.76	11.28	10.74	0.46	7	10.28	11.36	10.94	0.37	11	9.34	11.52	10.4	0.64
Scap KC	9	4.76	5.5	4.7	0.37	7	3.92	5.04	4.42	0.34	11	3.8	4.86	4.3	0.28

Table 9.1. (Cont.). Measurement data of the modern *Lagopus lagopus* samples.

Lagopus lagopus samples							
	L. l. lagopus - Russia					*L. l. brevirostris*- Altai Mts.	*L. l. maior* - Kazakhstan
Measurements	**N**	**Min**	**Max**	**Mean**	**St. dev.**		
Hum GL	7	58.34	64.36	62.47	2.1	58.2	-
Hum TP	7	15.06	17.44	16.35	0.89	15.02	18.54
Hum KT	7	5.08	6.2	5.68	0.36	5.1	-
Hum KB	7	3.62	4.58	4.09	0.36	4.02	-
Hum TD	7	10.74	12.12	11.32	0.46	9.96	-
Uln GL	6	55.84	61.46	59.01	1.98	55.24	65.5
Uln DP	6	9.68	11.2	10.56	0.57	9.52	12.06
Uln BP	6	7.5	8.52	8.16	0.39	7.18	8.94
Uln TP	6	5.9	6.94	6.52	0.39	6.1	7.98
Uln KT	6	3.06	3.68	3.5	0.23	3.12	4.24
Uln DD	6	7.34	8.6	7.93	0.45	7.42	9.36
Rad GL	6	50.34	55.28	53.25	1.78	50.22	58.74
Rad TP	6	3.52	4.12	3.78	0.2	3.5	4.3
Rad BP	6	4.34	4.72	4.49	0.13	4.2	5
Rad KT	6	1.9	2.18	2.06	0.13	1.92	2.4
Rad TD	6	4.92	5.6	5.37	0.27	5.14	6.2
Cpt GL	6	32.48	36.1	34.51	1.33	32.32	37.42
Cpt TP	6	9.08	10.6	9.75	0.57	9.38	11.32
Cpt GT	6	5.02	5.68	5.36	0.28	4.42	5.96
Cpt DD	6	6.22	7.46	7.01	0.46	5.7	7.82
Cpt HS	6	3.92	7.98	5.0	1.5	3.42	4.4
Fem GL	8	59.04	63.34	61.50	1.66	56.82	69.52
Fem LM	8	55.5	60.2	57.97	1.7	53.64	65.4
Fem TP	7	6.6	7.4	7.01	0.36	6.9	7.92
Fem BP	8	10.6	12.98	11.67	0.74	11.04	13.06
Fem KB	8	4.26	4.92	4.63	0.26	4.3	5.5
Fem TD	8	7.54	8.92	8.22	0.55	7.7	9.66
Fem BD	8	9.16	10.64	9.97	0.51	9.72	11.64
Tib GL	6	73.26	81.92	78.68	3.23	70.34	88.06
Tib LA	6	72.4	79.8	77.21	2.87	69.36	86.66
Tib BP	7	7.96	9.4	8.75	0.5	-	-
Tib DP	7	12.1	13.98	13.16	0.6	12.14	14.94
Tib KB	6	3.3	4.06	3.74	0.28	3.46	4.44
Tib KT	6	3.2	7.8	4.99	2.14	3.28	4.1
Tib BD	6	6.9	7.82	7.54	0.34	7.1	8.62
Tmt GL	6	38.1	42.88	40.61	1.62	-	45.32
Tmt BP	6	7.82	8.74	8.31	0.3	-	9.02
Tmt KB	5	2.84	3.44	3.19	0.22	-	3.64
Tmt BD	6	7.02	8.96	7.99	0.67	-	8.84
Tmt DD	6	8.1	9.74	9.1	0.58	-	10.6
Cor GL	8	41.08	46.34	43.79	1.63	40.82	49.82
Cor LM	8	38.44	42.98	40.67	1.51	37.34	45.16
Cor BD	8	11.94	17.82	15.71	1.74	14.82	19.22
Cor BB	8	11.18	14.36	12.53	1.03	12.06	14.8
Scap GL	7	58.42	65.42	62.7	2.78	57.7	67.96
Scap BC	8	9.5	11.72	10.72	0.71	9.6	11.48
Scap KC	8	3.72	5.32	4.53	0.5	4.02	4.58

Table 9.2. Measurement data of the modern *Lagopus mutus* samples.

Lagopus mutus samples	*L. m. mutus* - Scandinavia					*L. m. millaisi* - Scotland					*L. m. islandorum* - Icelandorum				
Measurements	N	Min	Max	Mean	St. dev.	N	Min	Max	Mean	St. dev.	N	Min	Max	Mean	St. dev.
Hum GL	3	53.5	57.86	55.46	2.21	5	58.36	60.62	59.9	0.93	3	54.44	60.82	57.39	3.22
Hum TP	3	14.44	15.7	14.9	0.7	5	16.04	16.54	16.32	0.19	3	14.4	16.5	15.38	1.06
Hum KT	3	4.94	5.18	5.05	0.12	5	5.1	5.6	5.29	0.19	3	4.88	5.28	5.06	0.2
Hum KB	3	3.62	4.3	3.87	0.38	5	3.76	4.18	3.99	0.17	3	3.52	4.12	3.82	0.3
Hum TD	3	9.42	10.96	10.15	0.77	5	10.72	11.2	10.92	0.23	3	9.8	10.84	10.31	0.52
Uln GL	3	52.22	56.72	54.55	2.25	2	58.02	59.94	58.98	-	3	52	58.8	55.37	3.4
Uln DP	3	8.8	10.2	9.49	0.7	1	-	-	10.02	-	3	9.14	10.28	9.73	0.57
Uln BP	3	6.54	8.06	7.31	0.76	1	-	-	7.58	-	3	6.96	7.94	7.59	0.54
Uln TP	3	5.52	6.5	5.94	0.5	2	6.32	6.5	6.41	-	3	5.69	6.6	6.19	0.47
Uln KT	3	3.02	3.46	3.2	0.23	2	3.3	3.32	3.31	-	3	2.76	3.3	3.1	0.3
Uln DD	3	6.74	7.74	7.21	0.5	2	6.54	7.44	6.99	-	3	6.86	7.72	7.41	0.48
Rad GL	3	46.92	51.8	49.35	2.44	2	52.64	54.2	53.42	-	3	47.06	52.74	50.08	2.86
Rad TP	3	3.16	3.54	3.31	0.2	2	3.58	3.64	3.61	-	3	-	-	-	-
Rad BP	3	3.7	4.48	4.01	0.41	2	4.32	4.48	4.4	-	3	3.5	4.26	3.8	0.4
Rad KT	3	1.76	2.04	1.89	0.14	2	1.92	2.12	2.02	-	3	1.64	1.94	1.8	0.15
Rad TD	3	4.68	5.46	4.95	0.44	2	5.24	5.48	5.36	-	3	4.6	5.44	5.15	0.48
Cpt GL	3	30.9	33.3	31.91	1.24	2	32.06	34.34	33.2	-	3	29.82	34.54	32.47	2.4
Cpt TP	3	8.6	9.6	9.03	0.52	2	9.56	10	9.78	-	3	8.5	9.62	9.24	0.64
Cpt GT	3	4.8	5.6	5.17	0.4	2	5.4	5.42	5.41	-	3	4.74	5.38	5.15	0.36
Cpt DD	3	5.84	6.98	6.39	0.57	2	6.68	6.8	6.74	-	3	5.82	6.86	6.39	0.53
Cpt HS	3	3.86	4.94	4.47	0.55	2	4.5	4.54	4.52	-	3	3.9	4.72	4.21	0.44
Fem GL	2	51.92	52.6	52.26	0.35	2	58.32	69.58	58.95	-	3	52.24	57.82	54.97	2.79
Fem LM	3	48.52	51.8	49.74	0.16	2	55.3	5.42	55.36	-	3	49.22	54.74	51.11	3.15
Fem TP	3	4.96	6.5	5.85	1.24	2	6.82	7.12	6.97	-	3	6.2	6.88	6.52	0.34
Fem BP	3	9.36	10.66	9.85	0.48	2	11.16	11.96	11.56	-	3	10.1	11.06	10.52	0.49
Fem KB	3	3.82	4.12	4	0.13	2	4.34	4.4	4.37	-	3	3.94	4.42	4.13	0.25
Fem TD	3	6.62	8	7.21	0.68	2	9.44	9.74	9.59	-	3	7.56	9.18	8.38	0.81
Fem BD	3	7.84	9.02	8.54	0.56	2	7.82	8.34	8.08	-	3	7.46	8.96	8.27	0.76
Tib GL	2	64.82	65.44	65.13	0.48	2	73.58	74.58	73.99	-	3	63.74	71.34	66.96	3.93
Tib LA	2	63.4	63.8	63.6	1.79	2	72.4	72.94	72.67	-	3	62.68	70.24	65.97	3.87
Tib BP	2	7.04	7.5	7.27	0.8	1	8.6	-	8.6	-	3	7.1	8.1	7.53	0.51
Tib DP	2	10.5	11.34	10.92	0.7	2	-	12.02	11.8	-	3	10.74	??	13.1	3.63
Tib KB	2	2.7	3.16	2.93	0.15	2	3.18	3.4	3.29	-	3	2.7	3.4	3.05	0.35
Tib KT	2	2.7	3.2	2.95	0.79	2	2.14	3.32	2.73	-	3	2.6	3.26	2.89	0.34
Tib BD	2	6.28	6.5	6.39	0.59	2	7.1	7.3	7.2	-	3	6.14	7.08	6.61	0.66
Tmt GL	3	29.8	32.24	30.88	0.44	5	33.22	35.9	34.69	1.26	3	30.18	34.82	32.23	2.37
Tmt BP	2	6.58	7.26	6.92	0.28	5	7.24	7.86	7.55	0.26	3	7.28	7.66	7.45	0.19
Tmt KB	3	2.5	2.9	2.67	0.33	5	2.88	3.12	3.0	0.11	3	2.58	2.86	2.76	0.16
Tmt BD	3	6.44	7.44	6.8	0.59	5	7.12	7.84	7.5	0.26	2	6.54	7.44	6.83	0.51
Tmt DD	3	7.06	8.34	7.52	0.33	5	7.78	8.6	8.14	0.38	3	7.04	8.02	7.54	0.49
Cor GL	3	38.9	41.62	40.42	1.39	2	42.74	44.14	43.44	-	3	39.8	44.2	41.84	2.53
Cor LM	3	36.3	39.1	37.81	1.41	2	39.3	40.4	39.85	-	3	35.94	40.98	38.58	0.63
Cor BD	3	14.22	15.48	14.86	0.63	2	16.38	16.84	16.61	-	3	15.44	16.7	16.05	1.07
Cor BB	3	10.66	11.4	10.97	0.39	2	13.18	13.78	13.48	-	3	10.92	12.96	12.13	2.07
Scap GL	2	57.34	57.46	57.4	0.08	2	59.48	62.32	60.9	-	3	57.62	61.58	59.25	0.33
Scap BC	3	8.7	9.8	9.37	0.59	2	10.26	10.46	10.36	-	2	9.32	9.78	9.55	0.31
Scap KC	3	3.52	4.2	3.84	0.34	2	4.28	4.62	4.45	-	3	3.64	4.2	3.99	2.79

Table 9.2. (Cont.). Measurement data of the modern *Lagopus mutus* samples.

Lagopus mutus samples	*L. m. helveticus* - French Alps					*L. m. hyperboreus*	*L. m. nadezdae*	*L. m. kellogae*
Measurements	N	Min	Max	Mean	St. dev.	- Spitzbergen	- Altai Mts.	- Taimyr Peninsula
Hum GL	4	52.88	58.44	56.2	2.63	-	57.58	60.88
Hum TP	4	14.2	15.46	14.91	0.61	16.68	15.08	16.14
Hum KT	3	4.52	5.22	4.81	0.36	-	5.16	5.22
Hum KB	3	3.2	4.06	3.53	0.46	-	3.8	4.14
Hum TD	2	9.76	10.44	10.01	0.37	-	10.24	10.64
Uln GL	2	55.1	57.2	56.15	1.48	58.88	-	58.2
Uln DP	2	8.96	9.5	9.23	0.38	10.4	-	10.1
Uln BP	2	7.24	7.46	7.35	0.16	8.06	-	7.34
Uln TP	2	5.78	6.12	5.95	0.24	6.7	-	6.38
Uln KT	2	3.12	3.36	3.24	0.17	3.58	-	3.3
Uln DD	2	7.1	7.5	7.3	0.28	7.96	-	7.72
Rad GL	2	50.46	51.96	51.21	1.06	52.7	50.1	52.62
Rad TP	2	3.28	3.44	3.36	0.11	3.84	3.44	3.72
Rad BP	2	3.92	4.3	4.11	0.27	4.44	4.34	4.14
Rad KT	2	1.9	1.9	1.9	0	2.12	1.84	2.02
Rad TD	2	4.86	5.16	5.01	0.21	5.58	4.74	5.2
Cpt GL	2	31.14	33.04	32.09	1.34	35.84	-	34.92
Cpt TP	2	8.52	9.3	8.91	0.55	10.3	-	9.46
Cpt GT	2	4.9	4.98	4.94	0.06	5.3	4.42	5.32
Cpt DD	2	6.02	6.4	6.21	0.27	6.9	6.46	6.9
Cpt HS	2	3.7	4.28	4.17	0.16	5.42	3.64	-
Fem GL	2	53.48	55.92	54.7	1.73	-	-	55
Fem LM	2	50.5	53.1	51.8	1.84	-	-	52.86
Fem TP	2	5.52	6.66	6.09	0.81	-	-	7.14
Fem BP	2	9.56	11.08	10.32	1.07	11.7	-	10.94
Fem KB	2	3.9	4.16	4.03	0.18	7.2	-	4.16
Fem TD	2	7.42	8.36	7.89	0.66	-	-	7.94
Fem BD	2	6.94	9.2	8.07	1.6	-	-	8.72
Tib GL	2	68.44	71.48	69.96	2.15	-	-	70.34
Tib LA	2	67.64	70.92	69.28	2.32	-	69.22	68.94
Tib BP	2	7.16	7.76	7.46	0.42	-	-	7.9
Tib DP	2	4.16	11.64	11.4	0.34	-	-	11.8
Tib KB	2	2.96	3.36	3.16	0.28	-	3.2	3.18
Tib KT	2	2.8	3.14	2.97	0.24	-	3.1	3.04
Tib BD	2	6.4	6.74	6.57	0.24	-	7.4	6.64
Tmt GL	16	31.8	35.7	33.06	1.17	-	-	34.08
Tmt BP	16	6.62	7.56	7.11	0.35	-	-	7.3
Tmt KB	16	2.46	2.94	2.71	0.14	-	-	2.74
Tmt BD	16	6.34	7.24	6.67	0.26	-	-	6.9
Tmt DD	16	7.08	7.82	7.56	0.2	-	-	7.7
Cor GL	3	37	41.32	39.09	2.16	43.34	40.52	44.36
Cor LM	3	34.22	38.76	36.29	2.3	39.98	37.9	41.16
Cor BD	3	14.18	14.9	14.57	0.36	16.48	15.24	17.46
Cor BB	3	11	11.36	11.18	0.18	13.12	11.7	13.5
Scap GL	1	-	-	56.5	-	-		60.76
Scap BC	2	9.44	10.42	9.93	0.69	10.6		10.66
Scap KC	2	3.74	4.68	4.21	0.66	4.74		4.28

Table 9.3. *t*- tests between *Lagopus lagopus* and *L. mutus*.

Measurements	
Hum GL	***
Hum TP	***
Hum KT	***
Hum KB	**
Hum TD	***
Uln GL	***
Uln DP	***
Uln BP	***
Uln TP	***
Uln KT	***
Uln DD	***
Rad GL	***
Rad TP	*
Rad BP	*
Rad KT	***
Rad TD	*
Cpt GL	***
Cpt TP	***
Cpt GT	***
Cpt DD	***
Cpt HS	Ns
Fem GL	***
Fem LM	***
Fem TP	*
Fem BP	***
Fem KB	***
Fem TD	ns
Fem BD	***
Tib GL	***
Tib LA	***
Tib BP	***
Tib DP	***
Tib KB	***
Tib KT	***
Tib BD	***
Tmt GL	***
Tmt BP	***
Tmt KB	***
Tmt BD	***
Tmt DD	***
Cor GL	***
Cor LM	***
Cor BD	*
Cor BB	*
Scap GL	***
Scap BC	***
Scap KC	**

Key: *** p< 0.001, ** p< 0.01, * p< 0.05, s Not significant at the 5% level.

Table 9.4. *t*- tests between *Lagopus lagopus scoticus* and *L. l. lagopus*.

Measurements	
Hum GL	ns
Hum TP	ns
Hum KT	ns
Hum KB	ns
Hum TD	ns
Uln GL	**
Uln DP	ns
Uln BP	ns
Uln TP	ns
Uln KT	ns
Uln DD	ns
Rad GL	***
Rad TP	ns
Rad BP	ns
Rad KT	ns
Rad TD	ns
Cpt GL	ns
Cpt TP	ns
Cpt GT	**
Cpt DD	ns
Cpt HS	ns
Fem GL	**
Fem LM	**
Fem TP	ns
Fem BP	ns
Fem KB	ns
Fem TD	ns
Fem BD	ns
Tib GL	**
Tib LA	**
Tib BP	ns
Tib DP	ns
Tib KB	*
Tib KT	ns
Tib BD	ns
Tmt GL	**
Tmt BP	**
Tmt KB	ns
Tmt BD	ns
Tmt DD	ns
Cor GL	ns
Cor LM	ns
Cor BD	ns
Cor BB	ns
Scap GL	ns
Scap BC	ns
Scap KC	*

Key: *** p< 0.001, ** p< 0.01, * p< 0.05, ns Not significant at the 5% level.

Table 9.5. *t*- tests between *Lagopus lagopus lagopus* from Russia and *L. l. lagopus* from Scandinavia.

Measurements	
Hum GL	ns
Hum TP	ns
Hum KT	ns
Hum KB	ns
Hum TD	ns
Uln GL	ns
Uln DP	ns
Uln BP	ns
Uln TP	ns
Uln KT	ns
Uln DD	ns
Rad GL	ns
Rad TP	ns
Rad BP	ns
Rad KT	ns
Rad TD	ns
Cpt GL	ns
Cpt TP	ns
Cpt GT	ns
Cpt DD	ns
Cpt HS	ns
Fem GL	ns
Fem LM	ns
Fem TP	ns
Fem BP	ns
Fem KB	ns
Fem TD	ns
Fem BD	ns
Tib GL	ns
Tib LA	ns
Tib BP	ns
Tib DP	ns
Tib KB	ns
Tib KT	ns
Tib BD	ns
Tmt GL	ns
Tmt BP	ns
Tmt KB	ns
Tmt BD	ns
Tmt DD	ns
Cor GL	ns
Cor LM	ns
Cor BD	ns
Cor BB	ns
Scap GL	ns
Scap BC	ns
Scap KC	ns

Key: *** p< 0.001, ** p< 0.01, * p< 0.05, ns Not significant at the 5% level.

Table 9.6. *t*- tests between *Lagopus lagopus scoticus* from Scotland and *L. l. scoticus* from Derbyshire.

Measurements	
Hum GL	ns
Hum TP	ns
Hum KT	ns
Hum KB	ns
Hum TD	ns
Uln GL	ns
Uln DP	ns
Uln BP	ns
Uln TP	ns
Uln KT	ns
Uln DD	ns
Rad GL	ns
Rad TP	ns
Rad BP	ns
Rad KT	ns
Rad TD	ns
Cpt GL	ns
Cpt TP	ns
Cpt GT	ns
Cpt DD	ns
Cpt HS	ns
Fem GL	ns
Fem LM	ns
Fem TP	ns
Fem BP	ns
Fem KB	***
Fem TD	ns
Fem BD	ns
Tib GL	ns
Tib LA	ns
Tib BP	ns
Tib DP	ns
Tib KB	ns
Tib KT	ns
Tib BD	ns
Tmt GL	ns
Tmt BP	ns
Tmt KB	ns
Tmt BD	ns
Tmt DD	ns
Cor GL	ns
Cor LM	ns
Cor BD	ns
Cor BB	ns
Scap GL	ns
Scap BC	ns
Scap KC	*

Key: *** p< 0.001, ** p< 0.01, * p< 0.05, ns Not significant at the 5% level.

9.6. Sexual Dimorphism

Significant sexual dimorphism exists in both *L. lagopus* and *L. mutus* with males proving to be larger than females in both skeletal and traditional ornithological measurements (Kraft, 1972; Cramp, 1980). However, sample sizes of sexed birds in the present samples were too small to merit *t*-testing.

9.7. Fossil Samples

The Pleistocene fossil record of *Lagopus* in the Western Palaearctic has recently been reviewed (Tyrberg, 1995). This has confirmed the view already held that remains of *Lagopus* are the commonest bird fossils in the area during this period. The reason for this appears to be related to both taphonomic bias and a wider former distribution. The former distribution, as has already been mentioned, was far wider in the Pleistocene including low altitude areas in southern Europe where the birds are absent today. The taphonomic factor is the preponderance of bird fossils in caves, which appears to be the result of avian predation (Mourer-Chauviré, 1983 and Baales, 1992). This mode of accumulation for the bird remains has also had the effect of producing a pattern in the relative frequency of skeletal elements of the species. Tarsometatarsi and carpometacarpi are the most abundant elements to be represented in these faunas presumably because of the manner in which the predators handle their prey. This is fortuitous because the tarsometatarsus happens to be one of the bones which can most readily be identified to species within *Lagopus* (see previous section). Therefore, much of what follows deals with that skeletal element.

Holocene - Britain

Ossom's Eyrie Cave, Manifold Valley, Staffordshire

The sequence in this cave is described in detail in Chapter 10, and shall not be repeated here. The fossils of the genus *Lagopus* from Ossom's Eyrie Cave came from a part of the sequence broadly Mediaeval in age.

Table 9.7. *Lagopus* fossils and their measurements from Ossom's Eyrie Cave.

Specimen No.	Skeletal element	Measurements				
		GL	TP	KT	KB	TD
Sample 9. 9'' - 12''	R. humerus	61.6	16.52	5.7	4.38	13.2
		GL	BP	KB	BD	DD
Layer B. Section II	L. tarsometatarsus	41.36	8.28	7.6	3.5	9.04
Ledge 15" - 18" Section 3	Prox. tarsometatarsus	-	8.54	-	-	-

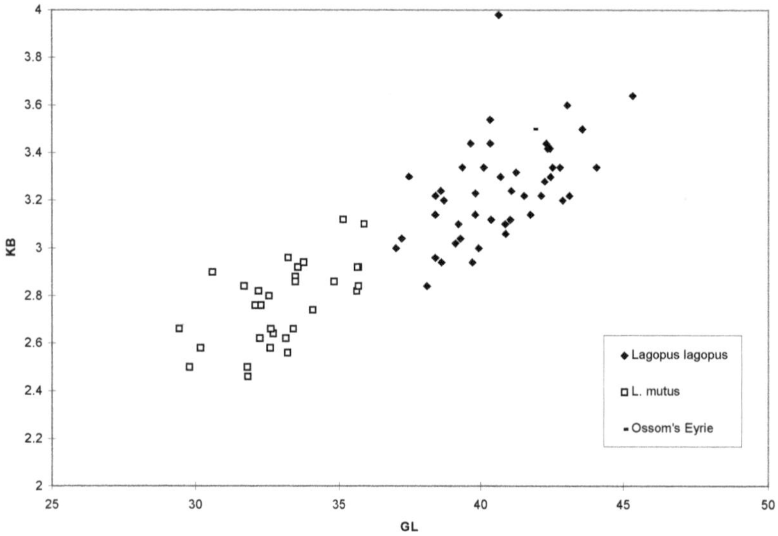

Figure 9.10. *Lagopus* sp. tarsometatarsus from Ossom's Eyrie plotted with modern *Lagopus* samples. Greatest length (GL) against shaft breadth (TP).

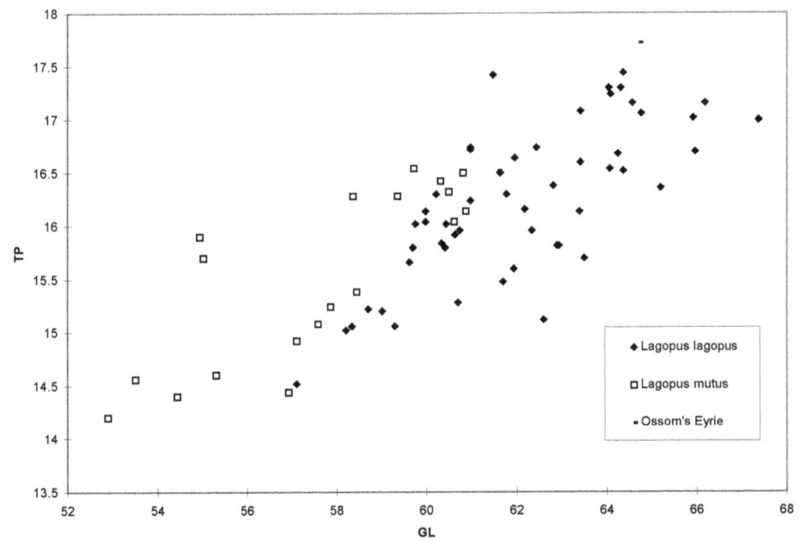

Figure 9.11. *Lagopus* sp. humerus from Ossom's Eyrie plotted with modern *Lagopus* samples. Greatest length (GL) against proximal breadth (KB).

The single complete *Lagopus* tarsometatarsus from Ossom's Eyrie plotted in Figure 9.10 is consistent in size and shape with *Lagopus lagopus*. Furthermore, the specimen is in the larger part of the species' distribution, similar to the red grouse *L. l. scoticus*, which still occurs in the general area today – c.f. the Derbyshire Peak District sample of red grouse (Figure 9.9 and Table 9.1).

The single humerus from Ossom's Eyrie is also large, consistent with *L .l .scoticus*. It posesses a greater proximal breadth (TP value) than all modern *Lagopus* specimens measured (Figure 9.11 and Table 9.1).

Late Glacial - Britain

Merlin's Cave (also known as Wye Valley Cave and Great Doward's Cave), Hereford and Worcester

The stratigraphy of Merlin's Cave was apparently disturbed during the early 19th century by iron miners (Hewer, 1924) and therefore the age of individual fossils, from all excavations, is difficult to know without direct dates. Despite this, there is clearly a Late Glacial component to the fauna, in addition to the later Holocene material; the latter including early Bronze Age to Romano-British artefacts. Recent investigations in the cave by the author as part of the Wye Valley Caves Project, coordinated by Dr. N. Barton of Oxford Brookes University, has confirmed that no *in situ* undisturbed sediment is left in the cave.

The faunal elements which would appear to be of Late Glacial age are pika *Ochotona pusilla*, collared lemming *Dicrostonyx torquatus*, norway lemming *Lemmus lemmus* and northern vole *Microtus oeconomus*. Furthermore, there is a radiocarbon date of 10 020 ± 120 BP on a pika bone from the cave (Housley, 1991). This would mean that at least part of the fauna is dated to the last, Younger Dryas phase of the Late Glacial.

Table 9.8. *Lagopus* fossils and their measurements from the Hewer excavation.

Specimen No.	Skeletal element	Measurements				
		GL	BP	KB	BD	DD
	R. tarsometatarsus	32.6	7.94	3.1	-	-
	R. tarsometatarsus	31.9	7.78	3.2	7.8	8.08
	R. tarsometatarsus	30.26	7.38	3.16	7.8	7.86
	R. tarsometatarsus	35.92	8.08	3.22	7.64	8.14
	R. tarsometatarsus	30.82	-	2.98	-	-
	R. tarsometatarsus	30.92	7.62	3	7.42	-
	R. tarsometatarsus	33.04	7.9	3.34	7.5	8.38
	L. tarsometatarsus	31.3	8.02	3.2	7.6	8.24
	L. tarsometatarsus	30	7.48	2.96	-	-
	L. tarsometatarsus	29.94	-	3.26	7.3	7.7
	R. tarsometatarsus	31.44	7.64	3.28	7.16	7.88
	R. tarsometatarsus	32.72	7.58	2.92	6.96	7.66
	R. tarsometatarsus	32.18	7.64	2.96	6.66	7.94
	R. tarsometatarsus	29.96	7.8	3.1	6.86	7.8
	R. tarsometatarsus	38.7	8.7	3.3	7.82	8.48
	R. tarsometatarsus	39.14	8.82	3.56	8.22	9.2
	Dist. R. tarsometatarsus	-	-	3.6	7.76	9.18

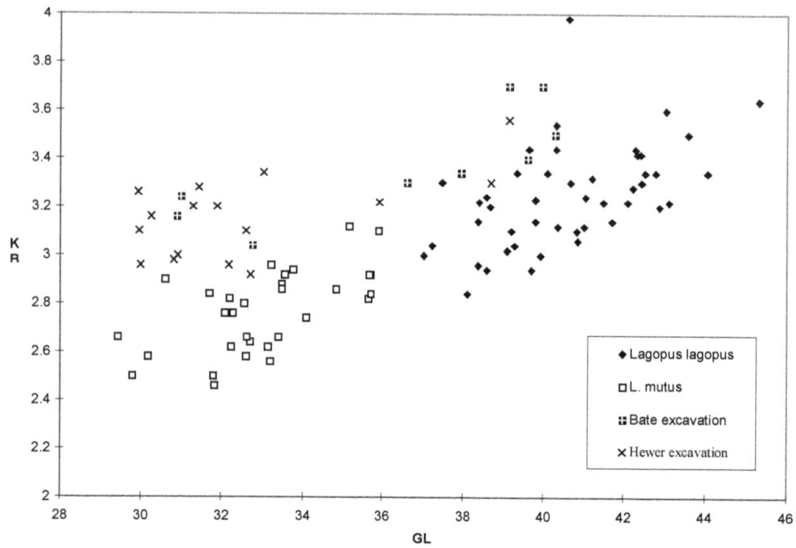

Figure 9.12. *Lagopus* sp. tarsometatarsi from Merlin's Cave / Wye Valley Cave plotted with modern *Lagopus* samples. Greatest length (GL) against shaft breadth (KB).

Table 9.9. *t*-tests between modern *Lagopus mutus* tarsometatarsi and those from Merlin's Cave.

Measurements	
Tmt GL	***
Tmt BP	***
Tmt KB	***
Tmt BD	**
Tmt DD	Ns

Key: *** p< 0.001, ** p< 0.01, * p< 0.05, ns Not significant at the 5% level.

Table 9.10. *t*-tests between modern *Lagopus lagopus* tarsometatarsi and those from Merlin's Cave.

Measurements	
Tmt GL	***
Tmt BP	ns
Tmt KB	ns
Tmt BD	ns
Tmt DD	ns

Key: *** p< 0.001, ** p< 0.01, * p< 0.05, ns Not significant at the 5% level.

The Merlin's Cave *Lagopus* tarsometatarsal dimensions (Table 9.8) plotted in Figure 9.12 indicate that both *L. lagopus* and *L. mutus* are present. The midshaft breadths (KB) are greater in relation to their lengths than the modern samples of both species measured (Figure 9.12), although only the *L. mutus* tarsometatarsi *t*-test as such (Tables 9.9 and 9.10). In length, both the *L. lagopus* and *L. mutus* tarsometatarsi fall in the lower part of the respective modern species size distributions. This means that both species had short legs on average in relation to the modern British representatives of the two species (Figure 9.9 and Table 9.2). This may explain why Newton (1924) believed that a "small ptarmigan" was present in the cave fauna. Newton only had access to reference material of the British subspecies of the two species.

The material below (Table 9.11) was collected by Dorothea Bate in 1901 and is in the collections of the Natural History Museum in South Kensington. This collection is smaller than that above and equally has no specific provenance data.

Table 9.11. *Lagopus* fossils and their measurements from the Bate excavation.

Specimen No.	Skeletal element	Measurements				
		GL	BP	KB	BD	DD
A1091	L. tarsometatarsus	32.76	7.22	3.04	7.14	7.62
	R. tarsometatarsus	30.9	7.3	3.16	7.52	7.94
	L. tarsometatarsus	31.0	-	3.24	7.18	7.7
	R. tarsometatarsus	-	7.4	3.18	-	-
	Dist. L. tarsometatarsus	-	-	-	7.04	7.64
A1090	R. tarsometatarsus	39.6	8.8	3.44	8.46	9.84
	R. tarsometatarsus	39.14	9.3	3.7	7.86	9.4
	R. tarsometatarsus	40.28	8.7	3.5	8.14	9.32
	R. tarsometatarsus	37.94	-	3.34	7.76	8.44
	L. tarsometatarsus	39.96	8.72	3.7	8.06	9.2
	L. tarsometatarsus	36.62	7.12	3.3	7.54	8.74

Figure 9.12 demonstrates that the *Lagopus* tarsometatarsi from the Bate excavation are similar to those of the later excavation.

Ossom's Cave, Manifold Valley, Staffordshire

Ossom's Cave is on the western slope of the Manifold Valley in Staffordshire. The cave was excavated by Bramwell between 1954 and 1956 after having been discovered by cavers (Bramwell, 1954, 1955, 1956). The stratigraphic sequence in the cave consisted of five layers which were named Layers A - E (Bramwell *et al.*, 1987). The top two layers (A and B) appear to have been Post-Glacial in origin and contained domestic animals, Romano-British pottery and recent debris. The next layer, Layer C, varied laterally in composition from a breccia towards the cave entrance to a sandy clay further in. Within Layer C there was a band of concentrated reindeer bones. Otherwise occasional large mammal and numerous small vertebrate remains were found throughout layer C. Bramwell *et al.* (1987) considered this deposit to be of Late Glacial age because of the several C^{14} dates taken from the deposit. Late Glacial dates fall between 10 590 ±70 and 12 220 ±320 on various reindeer bones. However, other dates gave later and earlier dates: two, a bovine tibia and a human tibia, dated to 2 030 ±80 and 4 860 ±80 radiocarbon years B.P respectively while a bovine humerus gave a date of 25 610 ±1500. The intrusive origin of the human bone is easily explained, as burials are often found within caves in Britian. The other date, however, causes more concern. The only taxon to produce Late Glacial ages are reindeer bone. It would be very interesting to further investigate the possibility of mixing of Post-Glacial material into Layer C, particularly given the occurrence of bank vole (*Clethrionomys glareolus*) in the deposit. This species is not normally expected in a Late Glacial deposit, although other instances (Stuart, 1983), suggest that it may have formed a genuine component of the fauna at these times. This would be an instance of a "disharmonious" fauna. Layer C has other taxa suggestive of locally-occuring woodland among the bird fauna, such as the mistle thrush (*Turdus viscivorous*) and the common buzzard (*Buteo* cf. *buteo*), although these could never by themselves be taken as categoric evidence for extensive deciduous woodlands. This is the kind of evidence that eventually led to the northern Refugia Hypothesis of Stewart and Lister (2001).

Table 9.12. *Lagopus* fossils and their measurements from Ossom's Cave.

Specimen No.	Skeletal element	Measurements				
		GL	BP	KB	BD	DD
Stratum C	R. tarsometatarsus	35.76	7.82	3.16	7.44	8.02
	R. tarsometatarsus	37.02	7.66	3.16	7.26	8.06
	Dist. R. tarsometatarsus	-	-	3.4	7.96	8.84
	Prox. L. tarsometatarsus	-	7.96	3.34	-	-
	Dist. R. tarsometatarsus	-	-	3.2	-	-
	Tarsometatarsus shaft	-	-	3.8	-	-
No stratum	Distal L. tarsometatarsus	-	-	2.8	6.74	7.32

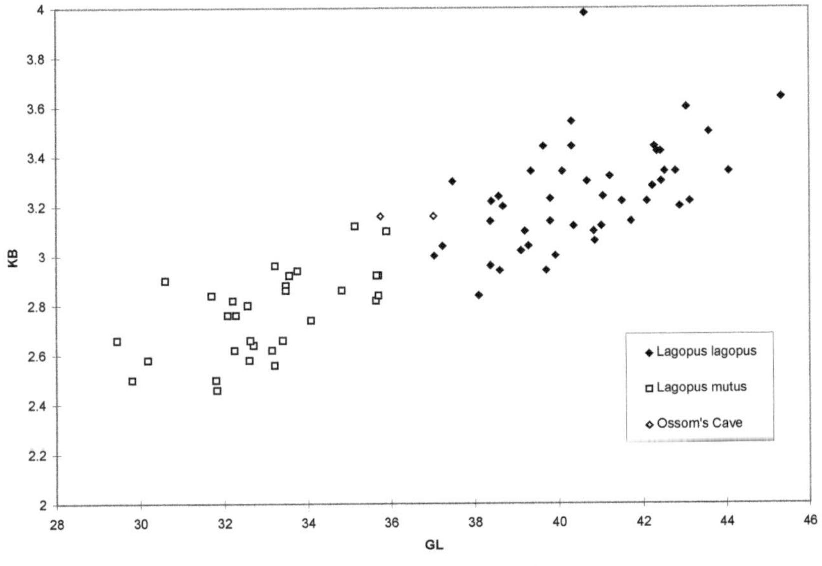

Figure 9.13. *Lagopus* sp. tarsometatarsi from Ossom's Cave plotted with modern *Lagopus* samples. Greatest length (GL) against shaft breadth (KB).

Table 9.13. *Lagopus* fossils and their measurements from Tornewton Cave.

Specimen No.	Skeletal element	Measurements				
		GL	BP	KB	BD	DD
A4106	R. tarsometatarsus	39.82	8.38	3.34	8.36	8.8
A4147	Prox. L. tarsometatarsus	-	-	2.82	-	-
	L. tarsometatarsus shaft	-	-	2.8	-	-
	R. tarsometatarsus frag.	-	-	3.06	-	-
	R. tarsometatarsus frag.	-	-	2.88	-	-

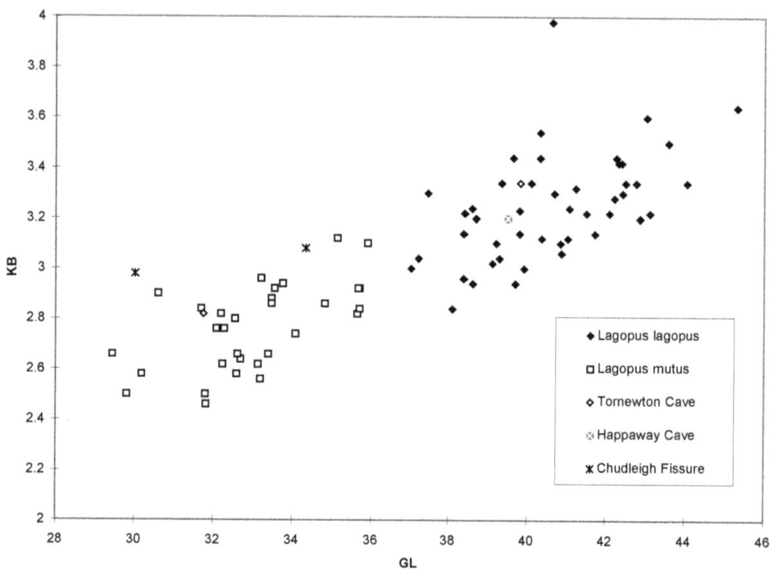

Figure 9.14. *Lagopus* sp. tarsometatarsi from Tornewton Cave, Happaway Cave and Chudleigh Fissure plotted with modern *Lagopus* samples. Greatest length (GL) against shaft breadth (KB).

The two complete *Lagopus* tarsometatarsi from Ossom's Cave fall in the narrow zone between *L. lagopus* and *L. mutus* (Figure 9.13). This makes it difficult to be sure of their identification. If the dimensions of the fragmentary material (Table 9.12) are compared to the modern data in Table 9.2 they can be seen to be similarly-shaped.

Tornewton Cave, Devon

The sequence in this cave is described in detail in Chapter 11. The fossils of the genus *Lagopus* came from the Reindeer Stratum, a part of the sequence which signifies that they probably date from the Late Glacial (OIS 2).

The single complete *Lagopus* tarsometatarsus (A4106, Table 9.13) from Tornewton Cave is no different from the modern dimensions of *L. lagopus* (Figure 9.14). The fragmentary tarsometatarsi (A4147 in Table 9.13) are not particularly robust and appear to belong to *L. mutus* judging from the lengths of the shafts that were present.

Inchnadamff, Sutherland

The bone cave near Inchnadamff in the Valley of Alt nan Uamh in Sutherland was investigated during the Geological Survey in the 1880's (Peach and Horne, 1917). The sequence within the cave started with a layer of rounded gravel including many rock types other than the limestone within which the cave was formed. This deposit represented a period when running water was present in the cave. Following this was a bed of limestone breccia with mammal and bird remains. This layer apparently formed during a cold and dry episode. Next came a fine grey clay with quartzite blocks followed by a red cave earth up to 3ft. in thickness with abundant mammal and bird bones and indications of human occupation. The cave earth signified that the climate had warmed, which was confirmed by the vertebrate taxa therein. Finally, there was a patchy deposit of marl covered by a layer of organic matter resulting from the accumulation of sheep manure. The *Lagopus* bones all came from the lower bone-rich deposit. It seems that this lower deposit was Late Glacial, while the upper one was probably Early Holocene in age (Stuart,1983). As pointed out by Stuart, a precise age for this fauna would be of interest because Scotland was affected by glaciation during the Late Glacial, the Loch Lomond readvance.

The taxa accompanying the *Lagopus* bones include stoat *Mustela ermina*, weasel *M. vulgaris*, bear *Ursus arctos*, hare *Lepus variabilis*, collared lemming *Dicrostonyx torquatus*, northern vole *Microtus oeconomus*, field vole *M. agrestis*, chaffinch? *Fringilla* sp., teal *Anas crecca*, wigeon *A. penelope*, tufted duck *Aythya fuligula*, long-tailed duck *Clangula hyemalis*, eider duck *Somateria mollissima*, common scotter *Melanitta nigra*, grey plover *Pluvialis squatarola*, little auk *Alle alle*, frog *Rana temporaria*, and salmon / trout Salmonidae (Newton, 1917). These identifications require checking although none of the taxa is unexpected for a Late Glacial assemblage. The presence of *D. torquatus* is not known after the Pleistocene and no other taxa suggest that mixing has taken place.

It is apparent that only *Lagopus mutus* remains were recovered from Inchnadamff (Figure 9.15). None of the tarsometatarsi reached the larger size of *L. lagopus*. The fact that only ptarmigan was present is contrary to Newton (1917) who identified *L. lagopus* as well. The specimens housed in the Geological Survey Museum still had labels reflecting these original identifications.

Table 9.14. *Lagopus* fossils and their measurements from Inchnadamff Cave.

Specimen No.	Skeletal element	Measurements				
		GL	BP	KB	BD	DD
27704	L. tarsometatarsus	31.18	6.8	6.7	7.52	3.02
27701	R. tarsometatarsus	33.9	7.14	7.3	8.08	3.12
27703	L. tarsometatarsus	32.4	7.56	7.18	7.88	3.02
27702	L. tarsometatarsus	33.52	7.56	6.92	7.76	3.22
27708	R. tarsometatarsus	29.7	7.06	6.52	7.22	3.1
27724	R. tarsometatarsus	30.72	7.72	6.82	7.72	3.16
27729	R. tarsometatarsus	31.08	7.22	7.16	7.56	2.74
27734	L. tarsometatarsus	32.72	7.26	7.14	7.96	2.96
27726	R. tarsometatarsus	31.54	7.44	6.84	-	2.84
27723	R. tarsometatarsus	31.82	7.06	6.82	-	2.76
27730	R. tarsometatarsus	30.54	7.46	7.14	7.8	2.98

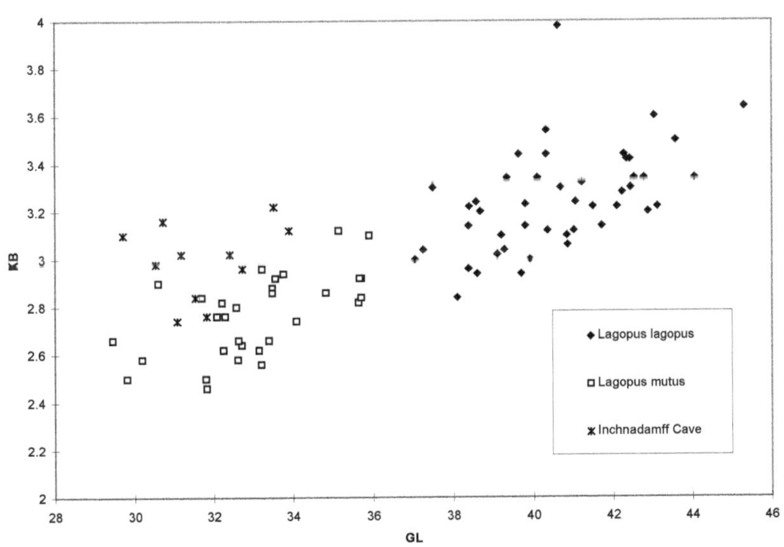

Figure 9.15. *Lagopus* sp. tarsometatarsi from Inchnadamff Cave plotted with modern *Lagopus* samples. Greatest length (GL) against shaft breadth (KB).

Table 9.15. *t*-tests between modern *Lagopus mutus* tarsometatarsi and those from Inchnadamff.

Measurements	
Tmt GL	*
Tmt BP	ns
Tmt KB	***
Tmt BD	**
Tmt DD	ns

Key: *** $p < 0.001$, ** $p < 0.01$, * $p < 0.05$, ns Not significant at the 5% level.

Furthermore, the tarsometatarsi are significantly broad in their midshafts measurements compared with modern *L. mutus* (Table 9.15). Their length dimensions are shorter on average than those in Scotland today (Table 9.2). Also different, although less significantly, are the distal breadths and the greatest lengths of the Inchnadamff *L. mutus* fossils (Table 9.15).

Gough's Old Cave, Cheddar, Somerset

Gough's Old Cave was excavated between 1954 and 1958 and the results were published by Tratman (1960). The vertebrate remains from the excavation were first described by Savage in Tratman (1960) although the birds were not identified. Harrison (1989) finally identified the birds, of which the *Lagopus* specimens are dealt with here.

The cave was excavated in foot layers and the specimens were labelled accordingly. Bird remains are available from the 6th foot to 10th foot layers. Carbon 14 dates have been obtained from three bones from the cave, 12 400 BP on a horse *Equus ferus* bone from the 6th ft. layer, 10 190 BP on a reindeer antler base from the 9th. ft. layer and 9 320 BP from a beaver mandible from the 10th ft. layer. It would therefore appear that there has been mixing of the cave sediments since the early Holocene at the earliest. The *Lagopus* bones were recovered from the 6th and 10th ft. layers. The *Lagopus* remains may well be Late Glacial in age but could be younger. Without absolute dates on the *Lagopus* bones themselves this question will remain open.

The other birds identified by Harrison (1989) were greylag goose *Anser anser*, mallard *Anas platyrhynchos*, tufted duck *Aythya fuligula*, goosander *Mergus merganser*, golden eagle *Aquila chrysaetos*, hobby *Falco subbuteo*, black grouse *Tetrao tetrix*, grey partridge *Perdix perdix*, great bustard *Otis tarda*, lapwing *Vanellus vanellus*, rock dove *Columba livia*, greater-spotted woodpecker *Dendrocops major*, blackbird / ring ouzel *Turdus merula* / *T. torquatus*, fieldfare *T. pilaris*, song thrush *T. philomelos*, redwing *T. iliacus*, stonechat / winchat *Saxicola torquata* / *S. rubetra*, snow bunting *Plectrophenax nivalis*, red-billed chough *Pyrrhocorax pyrrhocorax*, and jackdaw *Corvus monedula*.

Table 9.16. *Lagopus* fossils and their measurements from Gough's Old Cave.

Specimen No.	Skeletal element	Measurements				
		GL	TP	KT	KB	TD
GOC 0/10,11. 2nd ft of calcareous layer	R. humerus	58.5	16.1	5.42	10.88	4.2
		GL	BP	KB	BD	DD
GOC M17.N/11 10th. ft.	R. tarsometatarsus	38.72	8.54	3.34	7.58	9.12

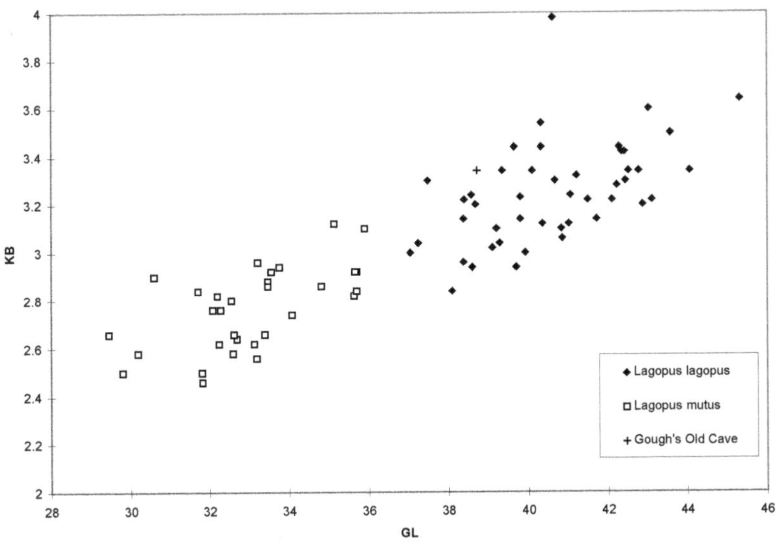

Figure 9.16. *Lagopus* sp. tarsometatarsus from Gough's Old Cave plotted with modern *Lagopus* samples. Greatest length (GL) against shaft breadth (KB).

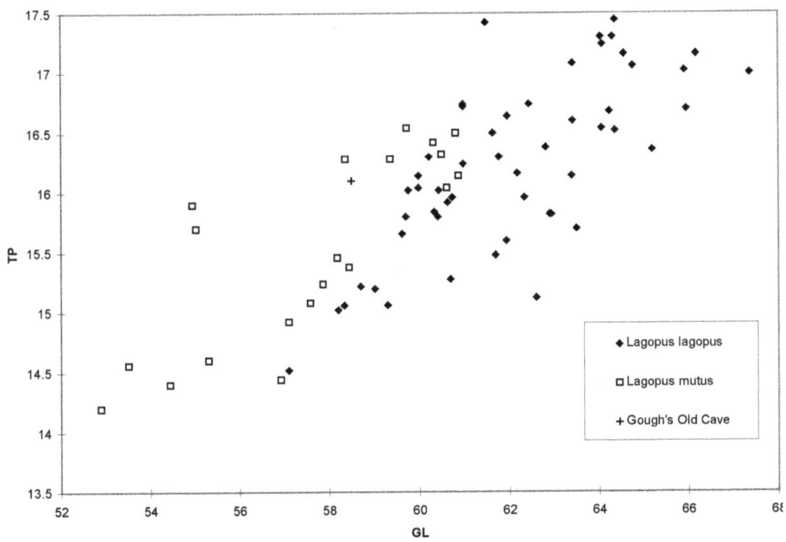

Figure 9.17. *Lagopus* sp. humerus from Gough's Old Cave plotted with modern *Lagopus* samples. Greatest length (GL) against proximal breadth (TP).

The *Lagopus* tarsometatarsus (Table 9.16) from Gough's Old Cave is that of the species *L. lagopus*, and it falls within the robust part of the modern distribution for that species (Figure 9.16). The humerus from Gough's Old Cave may be that of *L. mutus* whose humeri appear to be relatively broad in the proximal end compared to modern *L. lagopus* (Figure 9.17).

Pin Hole Cave, Creswell Crags, Derbyshire

The quaintly named Pin Hole Cave, named after an old local custom of visitors dropping a pin into a water-filled hollow and removing a pin left by a previous visitor, is in a cave-rich gorge, Creswell Crags, in Derbyshire. The cave extends 31 metres into the Magnesian limestone bedrock. The first published accounts of excavations in the cave are by Mello (1875, 1876), followed by Armstrong (1926, 1928 and 1929). The material examined here came from the Armstrong excavations, which will therefore be briefly described.

Armstrong recorded all material found *in situ* in two dimensions, and excavated sediment was sieved using 1/4 inch mesh screens. The two dimensions were defined on the basis of distance from the cave entrance and a depth measurement. The depth measurement, however, was not taken from a horizontal datum but from "the mid-point of the upper surface of the extensive flowstone stratum which capped the Devensian deposits" (Jenkinson, 1984). Because the flowstone varies vertically previous workers have found difficulties in reconstructing the cave's stratigraphy and interpreting finds therefrom (Jenkinson, 1984). Jenkinson claimed to have solved the problem by using the knowledge of Armstrong's recording system to put individual specimens into a correct stratigraphic position. However, it appears that there is a further problem because while Jenkinson's efforts put the specimens in their correct vertical position this did not necessarily mean that they were in correct stratigraphic position. Jacobi (per. comm.) has pointed out, on the basis of a number of C^{14} dates, that the strata, and the boundaries between them, were not horizontal, further complicating matters. He has reconstructed the stratigraphy and claims is able to accurately place specimens into their correct strata. Two assemblages are then recognised, a lower one of Middle Devensian age, and an upper one thought to be a mixture of Late Glacial and Holocene material. For the purposes of this study the assemblages have not been differentiated and are considered to be mixed, although it is conceded that a significant proportion of the *Lagopus* fossils are probably Pleistocene in age.

Table 9.17. *Lagopus* fossils and their measurements from Pin Hole Cave.

Specimen No.	Skeletal element	Measurements				
		GL	TP	KT	KB	TD
PH(F)1713	R. humerus	63.14	17.96	6.00	4.32	11.92
PH(F)99	L. humerus	66.88	18.46	6.02	4.66	12.36
		GL	BP	KB	BD	DD
PH(F)266	R. tmt	39.52	8.14	3.54	7.58	8.8
PH(F)980	R. tmt	40.22	8.46	3.6	8.22	-
1'-2'/35'-41'a	L. tmt	38.2	8.06	3.3	7.46	-
PH(F)504	R. tmt	37.82	8.7	3.54	8.56	-
PH(F)3	L. tmt	39.62	8.56	3.32	7.26	8.6
PH(F)1909	R. tmt	39.1	8.4	3.14	7.2	8.44
PH(F)304	L. tmt	37.04	8.22	3.22	7.24	8.74
PH(F)965	R. tmt	39.1	8	3.38	7.66	-
No no.a	L. tmt	37.26	8.04	3.28	7.3	-
PH(F)2934 - 2941 a	R. tmt	37.84	8.7	3.28	8.12	8.9
PH(F)2934 - 2941 b	L. tmt	38.58	8.16	3.46	7.7	9.26
PH(F)12566 -12568a	L. tmt	40.56	8.7	3.46	7.82	8.82
PH(F)8385-8392 a	R. tmt	38.24	8.3	3.44	7.92	-
PH(F)7589 - 7590	L. tmt	41.32	9.2	3.6	8.6	9.68
PH(F)7609 - 7611	Tmt	39.26	9.02	3.76	8.34	9.42
PH(F)9308	L. tmt	38.36	8.02	3.46	-	-
PH(F)294	L. tmt	40.26	8.6	3.68	-	-
PH(F)8034	Tmt	37.92	8.14	3.14	-	-
PH(F)7616 & PH(F)7617	R. tmt	39.02	8.58	3.88	-	-
PH(F)1064	L. tmt	31.24	6.98	3.02	6.52	7.14
PH(F)1039 or PH(F)784 or PH(F)586	R. tmt	32.68	7.02	3.12	6.82	7.4
PH(F)7 or PH(F)782 b	L. tmt	32.5	7.34	2.84	7.02	7.7
PH(F)7 or PH(F)782 a	L. tmt	31.1	6.7	2.86	6.24	-
PH(F)533	R. tmt	30.54	-	2.94	6.9	-
PH(F)581	R. tmt	32.02	7.42	3.22	7.06	7.56
PH(F)1137	L. tmt	31.44	-	2.88	6.38	-
PH(F)522	L. tmt	32.06	7.52	3.06	7.06	7.7
PH(F)9 i	R. tmt	32.36	7.4	3.18	7.36	8.02
PH(F)190	R. tmt	31.98	7.48	2.96	7.36	7.94
PH(F)9 ii	R. tmt	32.32	7.42	3.2	7.36	7.96
PH(F)542	L. tmt	32.8	7.24	3.02	6.56	-
PH(F)1039 or PH(F)784 or PH(F)586	R. tmt	31.2	7.06	2.9	7.2	7.46
PH(F)9314	R. tmt	31.54	7.94	3.48	7.52	-
(F)9316	L. tmt	31.2	-	3.04	-	-
PH(F)8515	L. tmt	30.72	7.34	2.94	6.74	7.58
PH(F)18551	L. tmt	31.96	7.12	3.04	-	-
PH(F) 2945 to PH(F)2947 a	R. tmt	31.34	-	2.94	-	-
PH(F)8434 to PH(F)8436 a	R. tmt	32.5	7.48	3.14	7.36	7.94

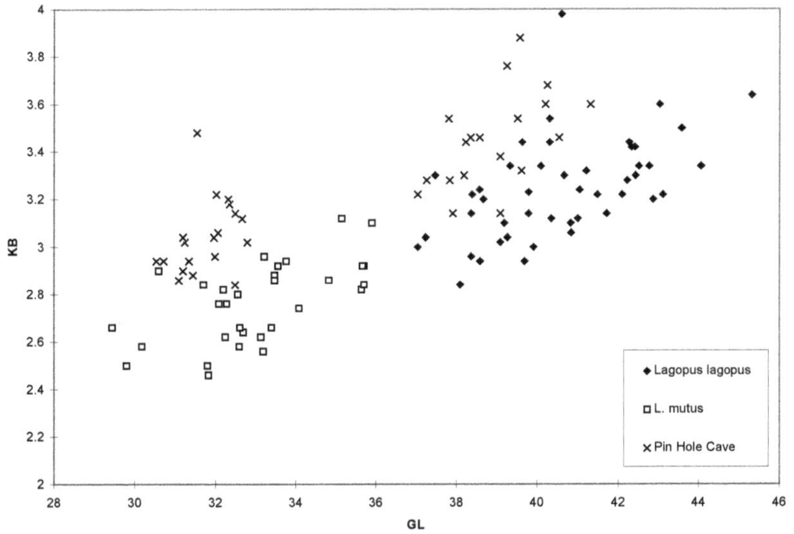

Figure 9.18. *Lagopus* sp. tarsometatarsi from Pin Hole Cave plotted with modern *Lagopus* samples. Greatest length (GL) against shaft breadth (KB).

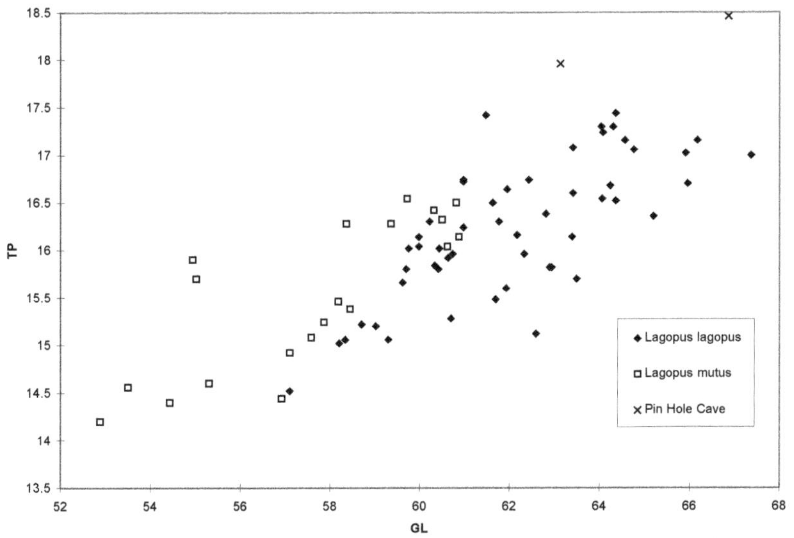

Figure 9.19. *Lagopus* sp. humeri from Pin Hole Cave plotted with modern *Lagopus* samples. Greatest length (GL) against proximal breadth (TP).

Table 9.18. *t*-tests between modern *Lagopus lagopus* tarsometatarsi and those from Pin Hole Cave.

Measurements	
Tmt GL	***
Tmt BP	ns
Tmt KB	***
Tmt BD	ns
Tmt DD	ns

Key: *** p< 0.001, ** p< 0.01, * p< 0.05, ns Not significant at the 5% level.

Table 9.19. *t*-tests between modern *Lagopus mutus* tarsometatarsi and those from Pin Hole Cave.

Measurements	
Tmt GL	***
Tmt BP	ns
Tmt KB	***
Tmt BD	ns
Tmt DD	ns

Key: *** p< 0.001, ** p< 0.01, * p< 0.05, ns Not significant at the 5% level.

Table 9.20. *t*-tests between modern *Lagopus lagopus* humeri and those from Pin Hole Cave.

Measurements	
Hum GL	ns
Hum TP	***
Hum KT	*
Hum KB	ns
Hum TD	**

Key: *** p< 0.001, ** p< 0.01, * p< 0.05, ns Not significant at the 5% level.

The *Lagopus* tarsometatarsi from Pin Hole Cave (Table 9.17) belong to both *L. lagopus* and *L. mutus*. Their midshaft measurements are large when compared to those of modern representatives of both species (Figure 9.18). These midshaft breadths together with the tarsometatarsal greatest lengths proved to be significantly different (Tables 9.18 and 9.19).

The two *Lagopus* humeri (Table 9.17) appear to belong to *L. lagopus* due to their gross size. They are more robust proximally in relation to their lengths than all modern

Lagopus specimens measured (Figure 9.19). Their shaft breadths and distal depths also exceed the dimentions of the modern members of *L. lagopus* (Table 9.20).

Undifferentiated Quaternary (Devensian?) – Britain

Happaway Cave, Devon

The sequence in this cave is detailed in Chapter 10.The *Lagopus* fossils from this cave are either Pleistocene or Holocene and cannot be more securely dated for the present.

The Happaway Cave tarsometatarsus (Table 9.21) falls near the middle of the range of modern *L. lagopus* tarsometatarsi (Figure 9.14).

Chudleigh Fissure, Devon

The sequence in this cave is detailed in Chapter 11. The *Lagopus* fossils from this cave are either Pleistocene or Holocene and cannot be more securely dated for the present.

The *Lagopus* tarsometatarsi from Chudleigh Fissure (Table 9.22) belong to *L. mutus* and have broad midshafts (Figure 9.14). Figure 9.20 shows that the complete humerus from Chudleigh is difficult to determine to species because its dimensions fall in the area of overlap between modern *L. mutus* and *L. lagopus*. However, it is possible, given the dimensions of other Late Pleistocene *Lagopus*, that the humerus belongs to *L. mutus*. This conclusion is only a tentative one, based on the fact that humeri from both species during that time were large relative to modern samples (Figure 9.32).

Table 9.21. *Lagopus* fossils and their measurements from Happaway Cave.

Specimen No.	Skeletal element	Measurements				
		GL	BP	KB	BD	DD
A1126	L. tarsometatarsus	39.5	8.6	3.2	7.8	8.84

Table 9.22. *Lagopus* fossils and their measurements from Chudleigh Fissure.

Specimen No.	Skeletal element	Measurements				
		GL	TP	KT	KB	TD
A1267	L. humerus	60.4	16.48	6.2	-	10.7
	Prox. L. humerus	-	17.36	-	-	-
		GL	BP	KB	BD	DD
A3512	R. tarsometatarsus	30.02	-	2.98	-	-
	L. tarsometatarsus	-	-	3.32	-	-
	Prox. R. tarsometatarsus	-	7.5	-	-	-
A3511	R. tarsometatarsus	34.34	7.52	3.08	7.2	8.2
	Dist L. tarsometatarsus	-	-	-	7.84	8.24

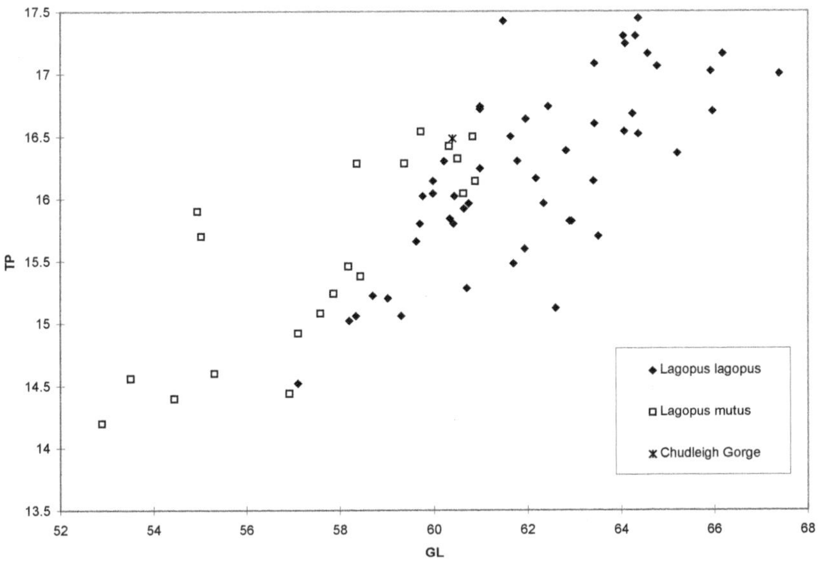

Figure 9.20. *Lagopus* sp. humerus from Chudleigh Gorge plotted with modern *Lagopus* samples. Greatest length (GL) against proximal breadth (TP).

Wetton Mill Rock Shelter, Manifold Valley, Staffordshire

This site is in a low position relative to the Manifold Valley and included three stratigraphic layers, Layers A to C. These layers were dated by their contained archaeology (Kelly, 1976). Layer A dated from the historical period, possessing a fauna which included butchered domestic animals, many rabbit bones and even the brass ends of shotgun cartridge cases (Bramwell, 1976). Layer B contained archaeology from the Neolithic to Iron Age. The fauna from this deposit included wild and domestic cattle, red and roe deer, wild and domestic pig, brown hare and extant voles. Layer C, meanwhile, seems to have spanned the Mesolithic to Late Glacial periods. Evidence for mixing of Late Pleistocene and early Holocene material comes from the composition of the vertebrates of this deposit. Animals like the reindeer *Rangifer tarandus*, collared lemming *Dicrostonyx torquatus*, arctic fox *Alopex lagopus*, and wolverine *Gulo gulo* were present with wild boar *Sus scrofa*, wood mouse *Apodemus sylvaticus* and bank vole *Clethrionomys glareolus*. The former are usually associated with tundra environments, the latter, today, with temperate woodlands. However, Bramwell (1976) points out that the apparently anomalous rodents are a common element of the Late Glacial faunas in Britain and that the bank vole remains may belong to the tundra species *C. rufocanus*, the grey-sided vole, which is similar in morphology. The remains of *Lagopus* from Wetton Mill came from Layer C and are therefore either Late Glacial or Early post-glacial in age. They were identified as representing *L. lagopus* (4 bones) with certainty, and possibly *L. mutus* (1 bone) (Bramwell, 1976).

The present study was only able to locate two bones of *Lagopus* in the Stoke-on-Trent Museum, which were, however, labelled as coming from Wetton Mill Minor rather than Wetton Mill Rock Shelter. It is likely that this locality is the same as Wetton Mill Rock Shelter (Yalden, pers. comm.). A further cause for concern is the fact that one of the bones at the museum was labeled as comming from Layer B (Table 9.23).

Table 9.23. *Lagopus* fossils and their measurements from Wetton Mill Rock Shelter.

Specimen No.	Skeletal element	Measurements				
		GL	BP	KB	BD	DD
Layer C, 2528	Tarsometatarsus	36.18	7.7	3.36	-	-
Layer B	L. tarsometatarsus	31.4	7.2	3.14	6.44	-

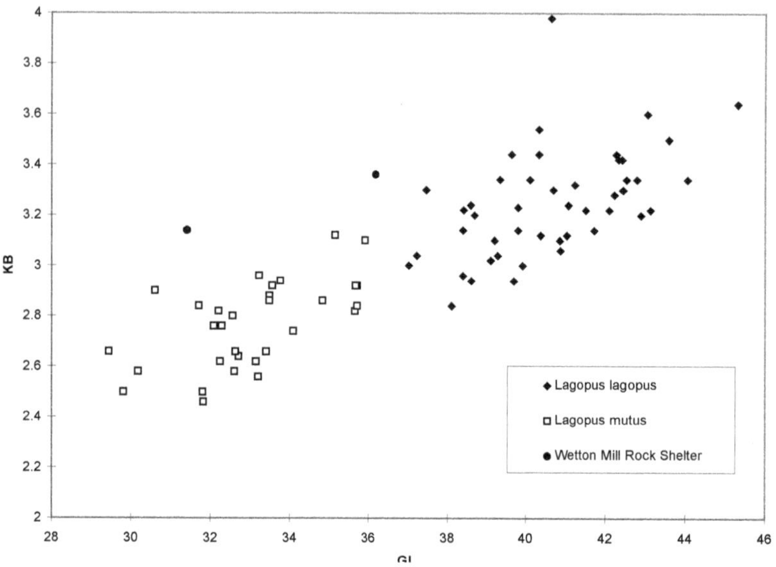

Figure 9.21. *Lagopus* sp. tarsometatarsi from Wetton Mill Rock Shelter plotted with modern *Lagopus* samples. Greatest length (GL) against shaft breadth (KB).

Of the two tarsometatarsi from Wetton Mill Rock Shelter one is clearly *L. mutus*, while the other's length falls in between the two species of *Lagopus* today (Figure 9.21). Both the specimens are relatively broad in the midshaft compared to the modern specimens.

Early Middle Pleistocene - Britain

Westbury-sub-Mendip, England

The cavern fills at Westbury-Sub-Mendip were first studied and excavated by Bishop during 1969 (Bishop, 1974, 1982). Bishop recognised three faunas, an Early Pleistocene one and two from the unnamed post-Cromerian, pre-Anglian interglacial. Subsequently, excavations were carried out by the Natural History Museum between 1976 and 1982. The more recent excavations have revealed that the sequence is more complex than at first thought (Stringer *et al.*, 1996). Nineteen statigraphic units were eventually recognised, which yielded evidence for two warm stages and two cold stages. This signifies that the possibility that the whole of the Westbury fauna might correlate with a late part (Substage IV) of the type Cromerian interglacial is impossible (Stringer *et al.*, 1996). So while Westbury-Sub-Mendip has assumed a role in elucidating the correlation of British Quaternary terrestrial sites with the oxygen isotope curve (Chapter 4), the situation is yet more complex than at first thought.

Westbury-Sub-Mendip is the earliest site in Britain to include levels with a steppe-tundra fauna and it is from the unit with a particularly cold fauna, the rodent earth (Unit 15/8 of the Western sequence and Unit 18 of the Eastern Sequence), that the *Lagopus* fossils from the cave came (Andrews, 1990). It is important to stress that Westbury-Sub-Mendip contained faunal material other than from the putative post-Cromerian, pre-Anglian interglacial, because there has been a tendency to consider material from the cave as representing a single climatic stage (Stringer *et al.*, 1996).

Although not plotted, the single measurements taken on the tarsometatarsus and humerus (table 9.24) are large in relation to modern representatives of the two modern species of *Lagopus* (Tables 9.1 and 9.2). It seems likely that the bones represent a large member of what may well be *L. lagopus*. This is tentative, however, because the fossils are more than twice the age of the next youngest fossils of the genus examined in this study. Harrison, in Andrews (1990), had however identified both *L. lagopus* and *L. mutus* in the fauna. The conclusion that only *L. lagopus* is present would, however, fit with the findings of Mourer-Chauviré, (1975a), who considers that *L. lagopus* was already present in the earlier part of the Middle Pleistocene.

Table 9.24. *Lagopus* fossils and their measurements from Westbury-Sub-Mendip.

Specimen No.	Skeletal element	Measurements				
		GL	TP	KT	KB	TD
A5023	Prox. L. humerus	-	17.28	-	-	-
		GL	BP	KB	BD	DD
A5027	Prox. L. tarsometatarsus	-	-	3.42	-	-

Table 9.25. *Lagopus* fossils and their measurements from the Grotte de Remouchamps.

Specimen No.	Skeletal element	Measurements				
		GL	BP	KB	BD	DD
MRAH 115A	Tarsometatarsus	39.44	8.52	3.24	8.52	9.06
MRAH 115B	Tarsometatarsus	41.64	8.6	3.4	8.12	9.1
MRAH 117	Tarsometatarsus	39.5	8.9	3.2	8.92	9.38

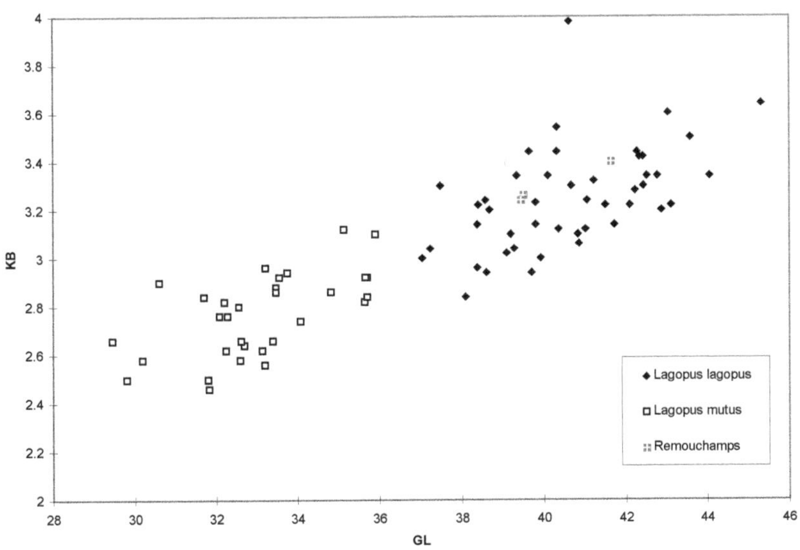

Figure 9.22. *Lagopus* sp. tarsometatarsi from Remouchamps plotted with modern *Lagopus* samples. Greatest length (GL) against shaft breadth (KB).

Late Glacial - Belgium

La Grotte de Remouchamps, Belgium

The Grotte de Remouchamps is in the valley of the Amblève in the Belgian Ardennes. According to Cordy (1991) the fauna from this cave dates to the Younger Dryas at the end of the Late Glacial. Radiocarbon dates for the site give ages of 10,330 ± 110 BP and 10,800 ± 110 BP (Hedges *et al.* 1994).

Table 9.26. *t*-tests between modern *Lagopus lagopus* tarsometatarsi and those from Remouchamps.

Measurements	
Tmt GL	ns
Tmt BP	ns
Tmt KB	ns
Tmt BD	ns
Tmt DD	ns

Key: *** p< 0.001, ** p< 0.01, * p< 0.05, ns Not significant at the 5% level.

The three *Lagopus* tarsometatarsi from Remouchamps (Table 9.25) belong to *L. lagopus* and fall within the size and shape range of modern tarsometatarsi (Figure 9.22; Table 9.26).

Late Glacial - France

La Balme-Les-Grottes, France

La Balme-Les-Grottes is situated at the northern extremity of Crémieu Island, on the left bank of the Rhône. The site was excavated in 1866 by E. Chantre, then in 1895 by E. Jacquemet, and finally by L. Chapuis between 1934 and 1937. The material from the Jacquemet excavation is housed in the Earth Science Department of Claude Bernard University of Lyon. Two principle faunal layers were distinguished by Jacquemet. A superficial one, between 50 and 90 cm in thickness, contained mammals such as reindeer *Rangifer tarandus*, sheep *Ovis aries*, Ibex *Capra ibex* and small cattle *Bos* sp., as well as the bones of both species of *Lagopus*. The second faunal layer was below a 1.5 m sterile layer, and possessed a fauna dominated by large bovid and equid bones, together with a single flint artefact. The upper deposit's archaeology was described as "Romanello-azilian" and believed to date to the end of the "Wurm IV (Dryas 1 c)" (Mourer-Chauviré, 1975a).

Other bird taxa identified by Mourer-Chauviré (1975) from the upper layer include teal *Anas crecca*, black grouse *Lyrurus tetrix*, rock partridge *Alectoris graeca*, quail *Cortunix cortunix*, corncrake *Crex crex*, woodcock *Scolopax rusticola*, stock dove? *Columba cf. oenas*, woodpigeon *C. palumbus*, alpine swift *Apus melba*, swallow *Hirundo rustica*, pied / white wagtail *Motacilla*

Table 9.27. *Lagopus* fossils and their measurements from La Balme-Les-Grottes.

Specimen No.	Skeletal element	Measurements				
		GL	BP	KB	BD	DD
No no.	R. tarsometatarsus	32.30	8.24	3.20	7.82	7.94
No no.	L. tarsometatarsus	32.24	7.32	3.20	-	-
No no.	Tarsometatarsus	-	-	2.96	-	-
No no.	Tarsometatarsus	-	-	2.9	-	-
No no.	L. tarsometatarsus	38.32	8.18	3.36	7.18	8.64
No no.	R. tarsometatarsus	37.36	8.30	3.04	6.98	8.84
No no.	L. tarsometatarsus	38.80	-	3.26	7.64	-
No no.	Tarsometatarsus	38.60	8.50	3.42	7.56	8.84
No no.	L. tarsometatarsus	36.96	8.30	3.12	7.82	9.02
No no.	L. tarsometatarsus	37.80	8.40	3.48	7.92	9.10
No no.	L. tarsometatarsus	37.04	-	3.04	-	-
No no.	R. tarsometatarsus	36.36	-	3.00	6.80	-
No no.	L. tarsometatarsus	38.30	-	3.30	7.84	-
No no.	L. tarsometatarsus	37.96	-	3.18	7.42	-

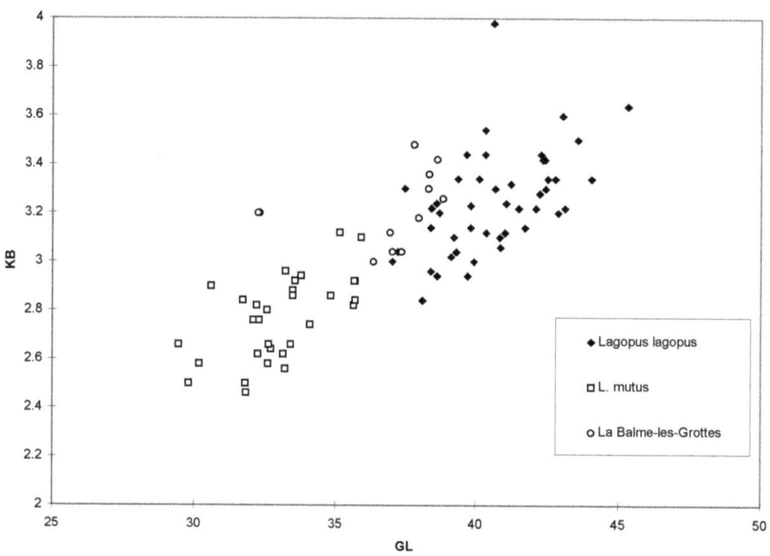

Figure 9.23. *Lagopus* sp. tarsometatarsi from La Balme-les-Grottes plotted with modern *Lagopus* samples. Greatest length (GL) against shaft breadth (KB).

alba, blackbird *Turdus merula*, chaffinch *Fringilla coelebs*, bullfinch *Pyrrhula pyrrhula*, alpine chough *Pyrrhocorax graculus*, jackdaw *Covus monedula*, and raven *C. corax*.

The presence of small cattle and sheep would seem to signify that the upper layer has a more modern component and that it is not all contemporary with the Upper Palaeolithic artefacts. The presence of woodland birds such as woodpigeon may also be of concern. Nevertheless, until such time as bones of *Lagopus* from the site are directly dated, it is probably safe to assume that they are Late Glacial in age.

The majority of the La-Balme-les-Grottes *Lagopus* tarsometatarsi (Table 9.27) are the length of small modern specimens of *L. lagopus* (Figure 9.23), although a large sized population of *L. mutus* cannot be ruled out. Two specimens, however, appear to belong to *L. mutus* and are robust in their midshaft compared to modern *L. mutus* measured (Figure 9.23). Assuming the majority belong to *L. lagopus*, on the basis of the presence of the two discretely smaller tarsometatarsi, they are very significantly shorter on average than modern individuals (Table 9.27). The *L. lagopus* tarsometatarsi are also significantly broader distally.

Table 9.28. *t*-tests between modern *Lagopus lagopus* tarsometatarsi and those from La Balme les Grottes.

Measurements	
Tmt GL	***
Tmt BP	ns
Tmt KB	ns
Tmt BD	*
Tmt DD	ns

Key: *** $p< 0.001$, ** $p< 0.01$, * $p< 0.05$, ns Not significant at the 5% level.

L' Abri de La Colombière à Neuville sur Ain (Ain), France

The rock shelter of La Colombière is large (46 m long), and is situated by the river Ain. The first excavations were conducted in 1913 and 1914 by L. Mayet and J. Pissot and subsequent ones by H.L. Movius in 1948 (Mourer-Chauviré, 1975a).

The sequence was described by Mayet and Pissot as follows: "at the base of the sequence was a sterile stratified sand unit, followed by levels D, D^{I} and D^{II} considered to be Aurignacian, followed by another sterile sand unit and finally level B of the Magdalanian. In the area of the drop-zone at the entrance of the shelter all the deposits are vertically interrupted (truncated or a facies change of long duration?) by a very rich deposit of small mammal bones. These bones no doubt came from the pellets of birds of prey which have been moved by water and deposited there." The bird fossils from the first excavations were originally studied by E.T. Newton (Mourer-Chauviré, 1975a).

The second excavation found bird fossils in all levels including those formerly described as sterile, with the exception of the Magdalenian level. All the levels were considered by Movius as having an Upper Perigordian industry. C^{14} dates on charcoal from hearths at these levels gave the following results: 15 500 ± 700 BP, 13 400 ± 400 BP and 11 650 ± 600 BP. The Perigordian is considered to be *ca.* 24 000 years old, so the C^{14} was considered to have been contaminated by modern organic matter. Therefore, a date was sought from mammoth collagen from level D, which gave a date of 13 390 ± 300 BP, corresponding to the age generally believed for the Middle Magdalenian and not the Upper Perigordian (Mourer-Chauviré, 1975a).

The material from both excavations unfortunately no longer has any provenance data, therefore necessitating that the fauna be considered as one homogenous collection (Mourer-Chauviré, 1975a). Mourer-Chauviré (1975a) considered this collection to be from "Dryas 1c".

Table 9.29. *Lagopus* fossils and their measurements from La Colombière.

Specimen No.	Skeletal element	Measurements				
		GL	TP	KT	KB	TD
No no.	L. tarsometatarsus	31.38	7.1	2.92	6.96	7.58
No no.	L. tarsometatarsus	32.7	7.54	2.94	6.96	7.94
No no.	L. tarsometatarsus	31.48	7.56	3.12	7.06	7.72
No no.	L. tarsometatarsus	30.94	7.62	3	7.18	7.62
No no.	L. tarsometatarsus	31.34	6.84	2.84	6.52	7.58
No no.	R tarsometatarsus	31.92	7.88	2.92	7.06	7.62
No no.	L. tarsometatarsus	32.76	-	2.86	7.02	-
No no.	R. tarsometatarsus	33.16	-	2.6	6.48	7.44
No no.	R. tarsometatarsus	29.58	7.46	3.04	6.92	7.46
No no.	R. tarsometatarsus	31.08	7.6	3.32	7.26	7.96
No no.	R. tarsometatarsus	31.48	7.68	3.02	6.92	-
No no.	L. tarsometatarsus	30.36	-	3	6.96	7.48
No no.	R. tarsometatarsus	30.06	7.22	3.04	6.88	7.28
No no.	R. tarsometatarsus	30.86	7.58	3.36	7.14	7.54
No no.	L. tarsometatarsus	31.32	-	2.9	6.84	7.78
No no.	L. tarsometatarsus	29.6	7.4	2.78	6.76	7.5
No no.	R. tarsometatarsus	33.44	-	2.98	6.84	7.78
No no.	R. tarsometatarsus	32.12	7.54	3.12	7.08	7.82
No no.	L. tarsometatarsus	32.4	7.44	3.12	7.14	7.6
No no.	R. tarsometatarsus	31	7.24	2.81	6.72	7.66
No no.	L. tarsometatarsus	32.96	7.94	3.16	7.88	-
No no.	R. tarsometatarsus	29.26	7.2	2.96	6.86	7.3
No no.	L. tarsometatarsus	32.22	7.46	2.88	6.66	7.58
No no.	R. tarsometatarsus	29.94	7.4	2.78	7.12	7.66
No no.	L. tarsometatarsus	31.6	-	3.14	7.14	7.74
No no.	L. tarsometatarsus	30.8	7.6	3.14	7	7.54
No no.	L. tarsometatarsus	30.88	6.7	2.9	6.8	-
No no.	L. tarsometatarsus	30.04	6.84	2.92	6.8	7.5
No no.	R. tarsometatarsus	31.28	7.46	2.9	7.3	7.76
No no.	L. tarsometatarsus	33.72	-	3.02	7.2	-
No no.	R. tarsometatarsus	39.1	-	3.5	7.94	8.6
No no.	L. tarsometatarsus	38.84	8.5	3.74	8.24	-
No no.	R. tarsometatarsus	38.88	8.02	3.26	7.7	-
No no.	R. tarsometatarsus	39.76	-	3.3	7.7	8.38
No no.	L. tarsometatarsus	36.88	8.42	3.38	7.68	-

Table 9.29 (Cont.). *Lagopus* fossils and their measurements from La Colombière.

Specimen No.	Skeletal element	Measurements				
		GL	TP	KT	KB	TD
No no.	Tarsometatarsus	38.14	-	3.38	7.88	8.52
No no.	L. tarsometatarsus	37.22	8.42	3.6	8.02	9.04
No no.	R. tarsometatarsus	38.94	8.7	3.34	7.8	9.1
No no.	R. tarsometatarsus	37.68	-	3.26	7.8	-
No no.	L. tarsometatarsus	38.7	8.58	3.24	7.66	9.14
No no.	R. tarsometatarsus	35.46	8.38	3.16	7.32	8.24
No no.	R. tarsometatarsus	36.7	8.64	3.38	7	-
No no.	R. tarsometatarsus	39.2	8.68	3.4	7.9	9.2
No no.	R. tarsometatarsus	37.94	8.7	3.48	7.96	9.04
No no.	Tarsometatarsus	39.02	-	3.42	7.94	9.2
No no.	R. tarsometatarsus	37.74	-	3.42	7.74	8.96
No no.	L. tarsometatarsus	38.7	8.6	3.34	7.74	9.28
No no.	L. tarsometatarsus	38.06	8.28	3.24	7.56	8.82
No no.	L. tarsometatarsus	38.72	8.6	3.34	7.78	-
No no.	R. tarsometatarsus	37.88	8.42	3.24	8.14	9.2
No no.	R. tarsometatarsus	37.7	-	3.3	-	-
No no.	L. tarsometatarsus	36.32	-	3.04	7.2	7.96
No no.	L. tarsometatarsus	38.86	8.76	3.64	8.3	9.14
No no.	R. tarsometatarsus	40.5	8.72	3.6	8.3	9.56
No no.	L. tarsometatarsus	37.92	8.1	3.32	7.8	8.82
No no.	R. tarsometatarsus	38.56	8.66	3.7	7.94	8.96
No no.	L. tarsometatarsus	35.2	8.18	3.28	7.56	8.5
No no.	R. tarsometatarsus	40.6	8.6	3.2	8.3	9.3
No no.	L. tarsometatarsus	38.3	8.24	4.08	7.52	7.52
No no.	R. tarsometatarsus	38.64	8.16	3.4	7.32	8.7

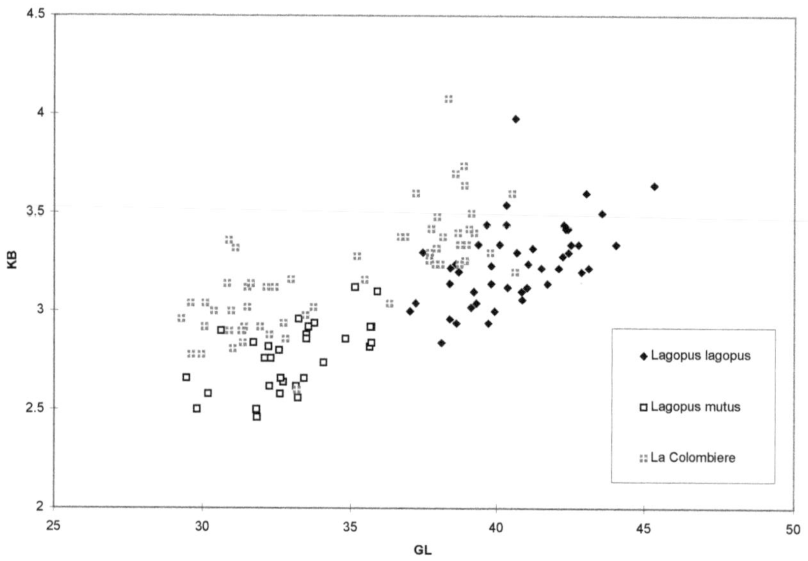

Figure 9.24. *Lagopus* sp. tarsometatarsi from La Colombière plotted with modern *Lagopus* samples. Greatest length (GL) against shaft breadth (KB).

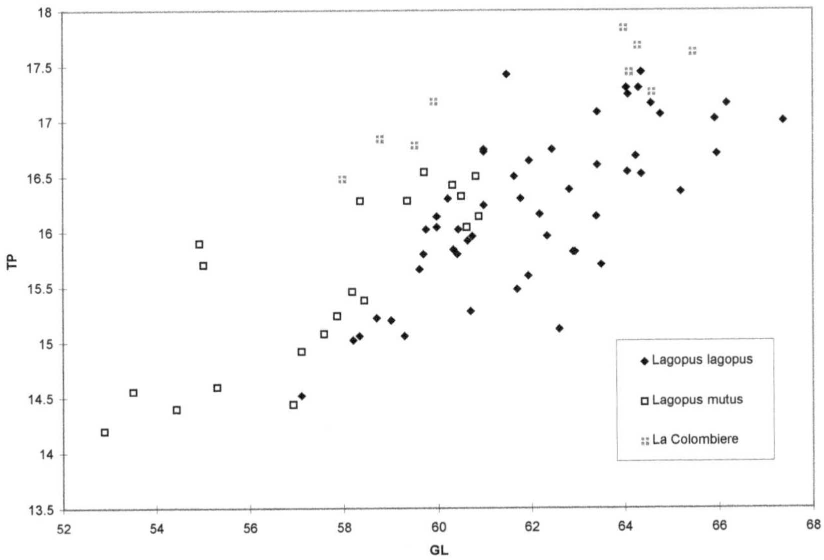

Figure 9.25. *Lagopus* sp. humeri from La Colombière plotted with modern *Lagopus* samples. Greatest length (GL) against proximal breadth (TP).

Table 9.30. *t*-tests between modern *Lagopus mutus* tarsometatarsi and those from La Colombière.

Measurements	
Tmt GL	***
Tmt BP	-
Tmt KB	***
Tmt BD	-
Tmt DD	-

Key: *** p< 0.001, ** p< 0.01, * p< 0.05, ns Not significant at the 5% level, - not tested.

Table 9.31. *t*-tests between modern *Lagopus lagopus* tarsometatarsi and those from La Colombière.

Measurements	
Tmt GL	***
Tmt BP	-
Tmt KB	**
Tmt BD	-
Tmt DD	-

Key: *** p< 0.001, ** p< 0.01, * p< 0.05, ns Not significant at the 5% level, - not tested.

Table 9.32. *t*-tests between modern *Lagopus lagopus* humeri and those from La Colombière.

Measurements	
Hum GL	*
Hum TP	***
Hum KT	ns
Hum KB	*
Hum TD	***

Key: *** p< 0.001, ** p< 0.01, * p< 0.05, ns Not significant at the 5% level.

Table 9.33. *t*-tests between modern *Lagopus mutus* humeri and those from La Colombière.

Measurements	
Hum GL	ns
Hum TP	**
Hum KT	*
Hum KB	ns
Hum TD	*

Key: *** p< 0.001, ** p< 0.01, * p< 0.05, ns Not significant at the 5% level.

The *Lagopus* tarsometatarsi from La Colombière (Table 9.29) clearly represent both species. These bones are significantly broader in the midshaft, when compared with modern individuals of the two species (Figure 9.24; Table 9.30 and 9.31). The tarsometatarsi of both species from La Colombière are also shorter than the total modern sample measured. The *Lagopus* humeri appear to include *L. lagopus* and *L. mutus*, judging from the two clusters in Figure 9.25. These humeri possess wider proximal extremities compared to modern specimens of the two species measured (Table 9.32 and 9.33). The two clusters also appear to be at the upper end of range of the respective modern species, suggesting relatively long humeri, however only *L. lagopus* proved to be significantly longer than the modern sample. Also different are the distal breadth and depth of *L. lagopus* and the shaft breadth and distal depth *of L. mutus*.

Late Middle Pleistocene - France

La Fage, France

The sequence in this cave is described in detail in Chapter 10. The fossils of the genus *Lagopus* from the cave are "Rissian" or broadly late Middle Pleistocene in age. In Table 9.34 the specimens attributed by Mourer-Chauvire (1975a) to the two extinct subspecies are listed together with the measurements taken in the present study.

Table 9.34. *Lagopus* fossils and their measurements from La Fage.

Specimen No.	Skeletal element	Measurements				
		GL	BP	KB	BD	DD
L. lagopus noaillensis						
41.594	**R. tarsometatarsus**	**39.3**	**8.94**	**3.4**	**7.82**	**8.6**
41.595	L. tarsometatarsus	37.86	7.78	3.34	7.7	8.26
41.596	L. tarsometatarsus	37.4	8.5	3.44	7.82	9.02
41.597	L. tarsometatarsus	39.04	7.92	3.34	7.4	8.8
41.598	L. tarsometatarsus	37.94	7.62	3.22	7.26	8.58
41.599	L. tarsometatarsus	38.04	8.3	3.34	7.2	8.56
41.600	R. tarsometatarsus	36.52	-	3.22	7.6	8.2
L. mutus correzensis						
41.601	**R. tarsometatarsus**	**34.12**	**7.56**	**3.1**	**6.96**	**7.8**
41.602	L. tarsometatarsus	34.8	7.6	2.96	6.8	-
No no.	L. tarsometatarsus	-	7.52	3.1	-	-
41.603	L. tarsometatarsus	33.24	7.16	3.4	-	-
41.604	L. tarsometatarsus	33.6	-	3.42	-	-
41.606	L. tarsometatarsus	33.86	7.64	3.28	7.42	7.84
41.605	L. tarsometatarsus	33.14	-	2.74	-	-
41.607	L. tarsometatarsus	33.64	7.3	3.02	7.08	7.44
41.608	R. tarsometatarsus	32.3	-	2.9	6.86	-
41.610	L. tarsometatarsus	31.7	-	2.96	6.64	7.84
41.609	L. tarsometatarsus	33.14	7.46	-	-	-
41.611	L. tarsometatarsus	33.9	7.2	2.9	6.64	7.3
41.612	L. tarsometatarsus	32.56	-	2.96	6.44	-
41.613	R. tarsometatarsus	-	-	2.88	7.18	7.8
41.615	L. tarsometatarsus	31.42	7.32	3.14	6.54	7.06
41.616	L. tarsometatarsus	35.26	7.5	2.82	7.22	-
41.614	Tarsometatarsus	34.5	-	2.98	-	-

Note: Holotypes in bold.

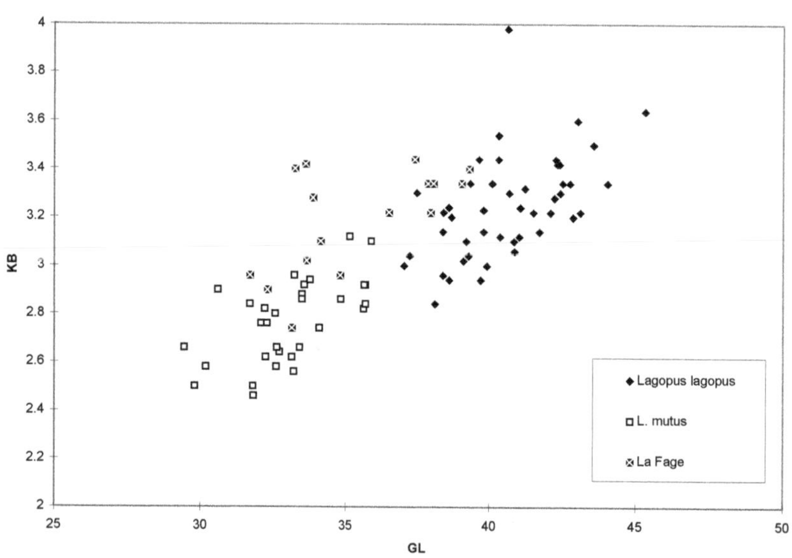

Figure 9.26. *Lagopus* sp. tarsometatarsi from La Fage plotted with modern *Lagopus* samples. Greatest length (GL) against shaft breadth (KB).

Table 9.35. *t*-tests between modern *Lagopus mutus* tarsometatarsi and those from La Fage.

Measurements	
Tmt GL	ns
Tmt BP	-
Tmt KB	***
Tmt BD	-
Tmt DD	-

Key: *** p< 0.001, ** p< 0.01, * p< 0.05, ns Not significant at the 5% level, - not tested.

Table 9.36. *t*-tests between modern *Lagopus lagopus* tarsometatarsi and those from La Fage.

Measurements	
Tmt GL	**
Tmt BP	-
Tmt KB	ns
Tmt BD	-
Tmt DD	-

Key: *** p< 0.001, ** p< 0.01, * p< 0.05, ns Not significant at the 5% level, - not tested.

The *Lagopus* tarsometatarsi from La Fage are more difficult to confidently assign to species than many younger samples. This is because their lengths are not bimodal to the same extent as later fossils of the genus (Figure 9.26). Figure 9.26 appears to confirm the finding of Mourer-Chauviré (1975a) that the La Fage *Lagopus* tarsometatarsi have broader midshafts although this is because they are broad in relation their lengths. *T*-tests were performed on the tarsometatarsal measurements despite the lesser degree of bimodality indicating that the division between the two species is less certain. Fossils were assigned according to prior descions as to identity made by Mourer-Chauviré in her original work. The results of *t*-testing (Tables 9.32 and 9.33) confirm only *L. mutus* as differing in shaft breadth from modern specimens measured in this study. *L. lagopus*, on the other hand, proves to differ significantly in length, being shorter than all modern samples measured here, in contrast the results of Mourer-Chauviré's analysis. The samples of fossils from La Fage analysed are the same and the means of Mourer-Chauviré's measurements are very similar to those of the present study. The modern samples of each species are, however, not the same and have clearly affected the results of the comparison. This highlights the problems inherent in doing such comparative work when the modern representatives of a species are widespread and consequent polytypic with varying osteological size and shape.

The analysis of the La Fage Lagopus fossils illustrates another issue, which is that the *t*-test is a univariate test and will not pick up proportional shape differences. Therefore while the La Fage *L. lagopus* tarsometatarsi are clearly seen to be relatively robust in relation to the modern data on the scattergram (Figure 9.26), the shaft breadth is not statistically different (Table 9.36).

Vistulian (Late Pleniglacial) - Poland

Mamutowa Cave, Poland

Mamutowa Cave is 20 Km from Krákow in the south of Poland. The cave has a number of deposits most of which date from the Vistulian glaciation, equivalent to the Weischelian (Bochenski, 1985). The fossils dealt with here are likely to date from that time.

Table 9.37. *Lagopus* fossils and their measurements from Mamutowa Cave.

Specimen No.	Skeletal element	Measurements				
		GL	BP	KB	BD	DD
Layer 2						
Sample 11/61	R. tarsometatarsus	39.84	8.5	3.56	8.56	-
	L. tarsometatarsus	38.44	8.38	3.22	-	-
	R. tarsometatarsus	37.3	7.88	3.22	7.76	8.14
	R. tarsometatarsus	35.74	7.62	2.84	7.24	8.1
	L. tarsometatarsus	36.84	8.6	3.28	7.58	8.14
	L. tarsometatarsus	34.78	6.96	2.74	-	-
	R. tarsometatarsus	32.1	7.46	2.82	6.84	7.74
Sample 4/62	R. tarsometatarsus	39.6	9.02	3.6	8.06	9.06
	R. tarsometatarsus	39.42	8.88	3.34	8.16	9.46
	R. tarsometatarsus	38.7	8.3	3.32	8.04	8.5
	R. tarsometatarsus	37.64	8.54	3.4	8.1	8.9
	R. tarsometatarsus	36.34	8.18	3.44	7.86	8.66

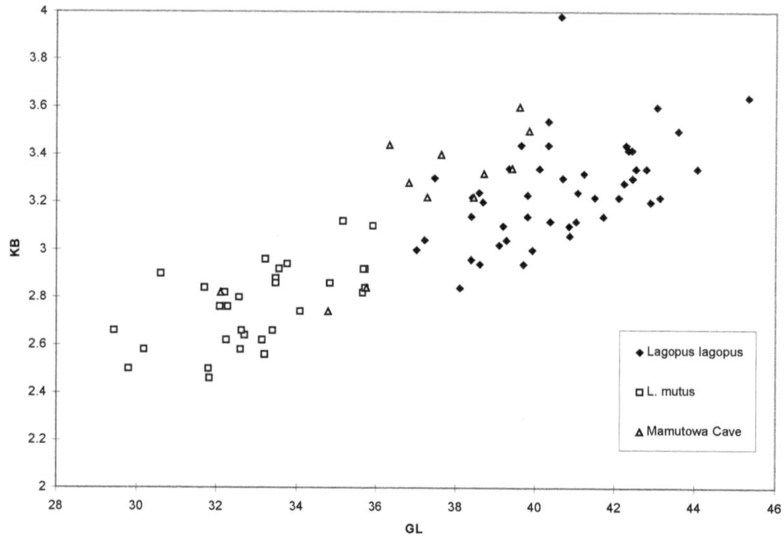

Figure 9.27. *Lagopus* sp. tarsometatarsi from Mamutowa Cave plotted with modern *Lagopus* samples. Greatest length (GL) against shaft breadth (KB).

Table 9.38. *t*-tests between modern *Lagopus lagopus* tarsometatarsi and those from Mamutowa Cave.

Measurements	
Tmt GL	***
Tmt BP	ns
Tmt KB	ns
Tmt BD	ns
Tmt DD	ns

Key: *** $p < 0.001$, ** $p < 0.01$, * $p < 0.05$, ns Not significant at the 5% level.

The measurement on tarsometatarsi (Table 9.37) from Mamutowa Cave show that both *Lagopus lagopus* and *L. mutus* are present (Figure 9.27). The larger tarsometatarsi which belong to *L. lagopus* are on the whole more broad in their shaft in relation to their lengths than modern specimens. However, the *L. lagopus* fossils could only be seen to differ significantly from modern specimens in length (Table 9.38). Those belonging to *L. mutus*, however, were similar in all dimensions to modern samples.

9.8. Conclusions

9.8.1. What is the pattern of morphology in *Lagopus lagopus* and *L. mutus* today throughout their geographical range, and what is the explanation for that variation?

In *L. mutus*, there is a noticeable difference in the mean size of the tarsometatarsi among modern subspecies in Europe (Figures 9.1 - 9.10). The Scottish subspecies (*Lagopus mutus millaisi*) is larger than populations from Scandinavia (*L. m. mutus*) or the Alps (*L. m. helveticus*). This difference is perhaps best seen in the length, although the bones appear to be isometric. This confirms the findings of Bochenski (1974), although slight differences in the means in the two studies exist. Kraft (1972) alleged differences between nominate *L. m. mutus* from Scandinavia and *L. m. helveticus* from the Alps, but this could not be confirmed in the present study due to small sample sizes.

With *Lagopus lagopus* there appears to be greater overlap between the samples, although *L. l. scoticus* is signigicantly larger on average than nominate *L. l. lagopus*. The single specimen of *L. l. maior* is very much larger, although sample size prevents reliable consideration of this subspecies, which is, however, regarded as larger by ornithologists (Dement'ev and Gladkov, 1967).

The patterns of size amongst the different measured populations of the two species show the British populations to be the largest in Western Europe. This is interesting although not easily explained. Other large subspecies are *L. l. maior* and *L. m. hyperboreas*. Bergmann's Rule cannot be invoked as causal, as the distribution of the different sized subspecies, within each species, do not follow a latitudinal or altitudinal pattern. Work by Moss (Moss, 1967; Moss, 1968; Gardarsson and Moss, 1970) has demonstrated that both species are strongly selective in their choice of food plants, selecting for the most nutritious species at different times of the year and in different regions. This may therefore, be a better explanation of the geographic size variation within the two species. Differences in vegetation in the areas from which the samples of both *Lagopus* species emanated may be contributing to their size differences. A detailed field study would be required to corroborate this.

Finally, the longer leg dimensions of *L. L. scoticus* than *L. l. lagopus* may relate to shorter periods of snow cover in Scotland than in mainland Europe, hence the lack of white winter plumage.

9.8.2. What is the pattern of morphology in the fossil record of *Lagopus lagopus* and *L. mutus* throughout their geographical range and what does it signify?

The youngest fossil sample of *Lagopus* analysed in the present study is that from Ossom's Eyrie which is

Medieval in age. Only *Lagopus lagopus* was found in the cave which might have been predicted if the presence of *L. mutus* in lowland situations was solely a Pleistocene phenomenon, although the sample is small. Unfortunately, the sample of *L. lagopus* is also too small to draw any reliable conclusions regarding any change in its morphology in the area since that time. The few fossils that are present are consistent in size and shape with *L. l. scoticus* today (Figure 9.10 and 9.11).

The next youngest British samples are those from Merlin's / Wye Valley Cave in Hereford and Worcestor, Inchnadamff in Sutherland, Tornewton Cave in Devon, Ossom's Cave in Staffordshire, and Gough's Old Cave in Somerset, which are all probably Late Glacial in age. Merlin's Cave or Wye Valley Cave includes both *Lagopus lagopus* and *L. mutus*, although no evidence was found for the third species, or "small ptarmigan", said to be present by E.T. Newton (1924). The source of Newton's conclusion probably lies in the fact that the ptarmigan skeletons he had for comparative purposes were Scottish *L. m. millaisi* which possess comparatively long tarsometatarsi. The fossils from the Wye Valley included specimens which were the length of modern Scottish *L. mutus*, as well as a majority which were noticeably shorter leading him to conclude that another species of *Lagopus* had existed (Figure 9.12). Newton (1924) did not name these, but simply referred to them as "small ptarmigan". When the comparison is made between the *Lagopus mutus* fossils from Merlin's / Wye Valley Cave and a more inclusive sample of the species from Europe today, most of the fossils can be seen to be within the range of those from Scandinavia, Russia and Iceland (Table 9.2). It is interesting that the *L. lagopus* fossils from the cave are also short, compared to those of modern British members of that species (Table 9.1). In addition to smaller average lengths, both species of *Lagopus* in the cave have broader tarsometatarsal shafts.

Inchnadamff in Sutherland, Scotland yielded only *L. mutus*, contrary to Newton (1917), who claimed that both species were present. The reason for the misdiagnosis is puzzling. It does not appear to be because some of the remains seen by Newton have gone missing. The *L. mutus* remains seen in the Geological Survey Museum are labelled either as *L. mutus* or as *L. lagopus*. Some material, not seen during this study, was housed in the National Museum of Scotland in Edingburgh and may include the specimens correctly identified by Newton as *L. lagopus*. The labels on the material at the Geological Survey Museum may have been made later, and so may not relate directly to Newton's original identification. Newton commented that *L. mutus* was present in greater numbers than *L. lagopus*, which he put down to the ptarmigan's greater abundance in Scotland at the time, which may be true. However, the absence of *L. lagopus* bones, if correct, may be due the site's altitude, at between 1000 and 2000 ft. (*ca.* 300 - 600 m). Today, *Lagopus mutus* replaces *L. lagopus* altitudially, which is thought to reflect a greater tolerance of cold temperatures. In Scotland *L. lagopus* is currently found at this altitude (Cramp, 1980), although during a colder climate, such as that of the colder episodes of the Late Glacial, it may be that only *L. mutus* could exist at such a height. Alternatively, the absence or smaller numbers of *L. lagopus* may be because the predator responsible for accumulating the assemblage was preferentially feeding on the ptarmigan. The Inchnadamff *L. mutus* possessed greater tarsometatarsal shaft breadths in relation to their lengths than do modern birds (Figure 9.15). They are also possibly slightly shorter than the modern Scottish subspecies.

The identity of the two tarsometatarsi from Ossom's Cave is difficult to determine with confidence as they fall between the smaller modern *L. lagopus* and the larger *L. mutus* (Figure 9.13).

The single tarsometatarsus and humerus from Gough's Old Cave are within the size range of modern *L. lagopus* and *L. mutus* respectively, although they are in the upper part of the modern size distributions of those species (Figure 9.16 and 9.17).

Tornewton Cave's Reindeer Stratum has been dated to Oxygen Isotope Stage 2 (Late Glacial) by Currant (1996). Both *L. lagopus* and *L. mutus* are present in small numbers, although neither of them can be shown to differ morphologically from the modern *Lagopus* populations (Figure 9.14). This may, however, be due to the small sample size.

Pin Hole Cave, whose assemblage is probably of mixed age, contained a large number of *Lagopus* fossils and clearly included the two species, both of which had broader tarsometatarsal shafts (Figure 9.18). The *Lagopus* humeri at Pin Hole Cave also possessed greater proximal breadths than modern populations (Figure 9.19).

Other samples of probable Late Pleistocene age whose precise ages are not known include Chudleigh Fissure, Happaway Cave and Wetton Mill Rock Shelter. These sites have very small samples, inhibiting reliable interpretation. However, at Chudleigh Fissure where only *L. mutus* can be confirmed as present, the two tarsometatarsi are relatively robust compared to modern specimens of the species (Figure 9.14). The humerus from Chudleigh Gorge is difficult to interpret because it may belong to either a large *L. mutus* or a middle sized *L. lagopus* (Figure 9.20). At Happaway Cave, where only *L. lagopus* is present, the tarsometatarsus is of the size and shape of modern birds (Figure 9.14). At Wetton Mill Rock Shelter, a site which may be Late Glacial in age, the *Lagopus* assemblage is represented by two robust tarsometatarsi, one of which is *L. mutus* while the other falls in the overlap range between the two species (Figure 9.21)

Westbury-sub-Mendip is the oldest sample of *Lagopus* examined during the present study, from an unnamed part

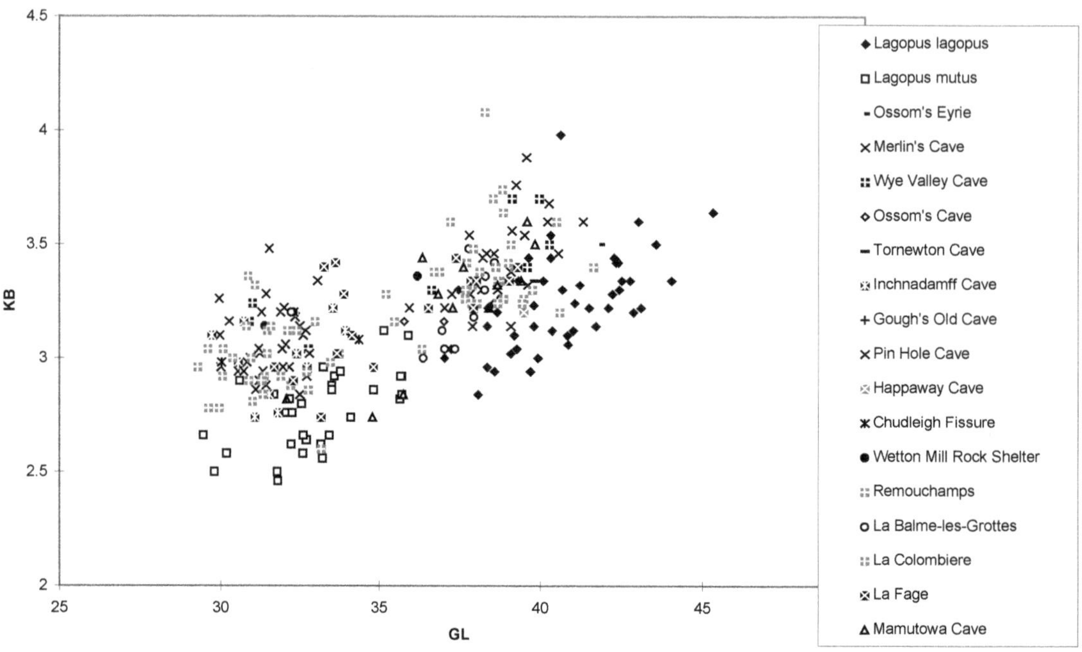

Figure 9.28. Summary diagram of fossil and modern *Lagopus* sp. tarsometatarsi. Greatest length (GL) plotted against shaft breadth (KB).

of the Cromerian Complex (*ca.* 500 Kyrs). The extremely fragmentary nature of the fossils confirmed as belonging *Lagopus* hindered specific identification. However, the measurements taken would suggest a large species slightly exceeding the size of most modern *L. lagopus*. This may be further evidence for early *L. lagopus* (Mourer-Chauviré, 1993).

Remouchamps in Belgium yielded three *L. lagopus* tarsometatarsi which were not different from modern samples of that species and were probably Late Glacial in age (Figure 9.22, Table 9.25).

In France the youngest sites from which *Lagopus* fossils were examined were La Balme-les-Grottes and La Colombière. At La Balme-les-Grottes both *L. lagopus* and *L. mutus* appear to be present, the latter in smaller numbers (2) than the former (9), though this would need to be confirmed by a larger sample size. The tarsometatarsi which appear to belong to *L. lagopus* were quite small, falling close to the modern divide between the two species (Figure 9.23). The two *L. mutus* tarsometatarsi, unlike those of *L. lagopus*, appear to be significantly robust (Figure 9.23). The difference between the two species at La Balme-les-Grottes shows further that not all Late Glacial populations of *Lagopus* were robust. It is possible that this relates to the climatic oscilations which are known to occur during this period although at present this is difficult to test due to the inadequacy of the dating for this and other sites.

The La Colombière sample of *Lagopus* demonstrates that there can be fossil tarsometatarsi which are difficult to identify to species with confidence (Table 9.25). This could be because the sample is large and therefore includes a greater number of outliers. Alternatively, it is due to time averaging, the differently-aged fossils giving the appearance of a greater range of morphology than would a synchronous sample. Finally, the explanation could lie in Newton's suggestion that a third species had existed in the Late Pleistocene. The last explanation is perhaps the least likely. The tarsometatarsi were broad in relation to their lengths (Figure 9.24, Table 9.29 and 9.30). Futhermore, the humeri had wide proximal ends in relation to modern samples of both species (Table 9.26).

The final French site to receive attention was La Fage, where distinct subspecies have been named (Mourer-Chauviré, 1975). The La Fage tarsometatarsi, which from their dimensions seem to include both *L. lagopus* and *L. mutus* are, however, closer in size to each other than in other fossil or modern situations. Both species possess robust tarsometatarsi but this is less pronounced than in certain samples from the last glacial in France and elsewhere.

At Mamutowa Cave in Poland, the assemblage is largely Vistulian in age, which is equivalent to the Weichselian. Both *L. mutus* and *L. lagopus* appear to be present. The interesting phenomenon is that, whereas the willow grouse possessed greater breadths in their tarsometatarsal midshafts, the ptarmigan did not.

In conclusion, most Pleistocene samples of *L. lagopus* and *L. mutus* tarsometatarsi were found to be broader in their midshafts than modern samples. Exceptions to this are three *L. mutus* from Mamutowa Cave in Poland, the two specimens of *Lagopus* sp. from Ossom's Cave, the single *L. lagopus* specimens from Tornewton Cave's Reindeer Stratum and Happaway Cave and the three *L.*

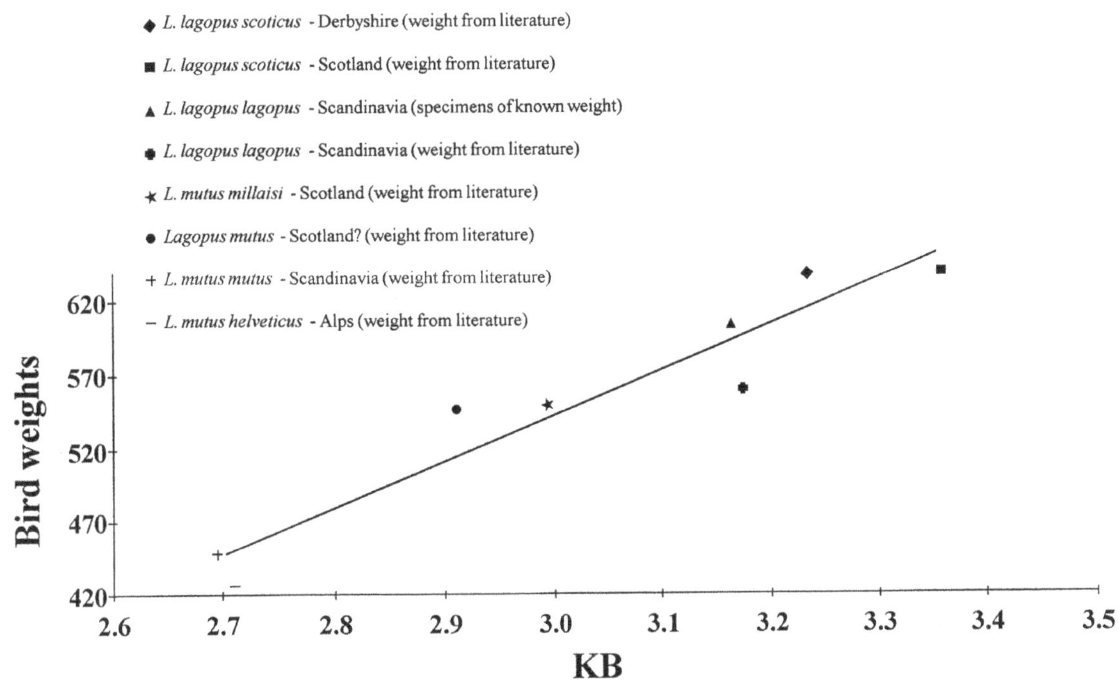

Figure 9.29. Linear regression of mean weights of *Lagopus* versus mean shaft breadth of tarsometatarsi. The data plotted are based on mean weights of subspecies, taken from Cramp (1980), against mean tarsometatarsal shaft breadths and lengths of the corresponding subspecies measured in the present study. Means weights were taken from the literature in all but one instance, as in most cases skeletons in collections had no such data recorded for individual birds. The one exception was the sample of *Lagopus lagopus* from Scandinavia where weights were recorded.

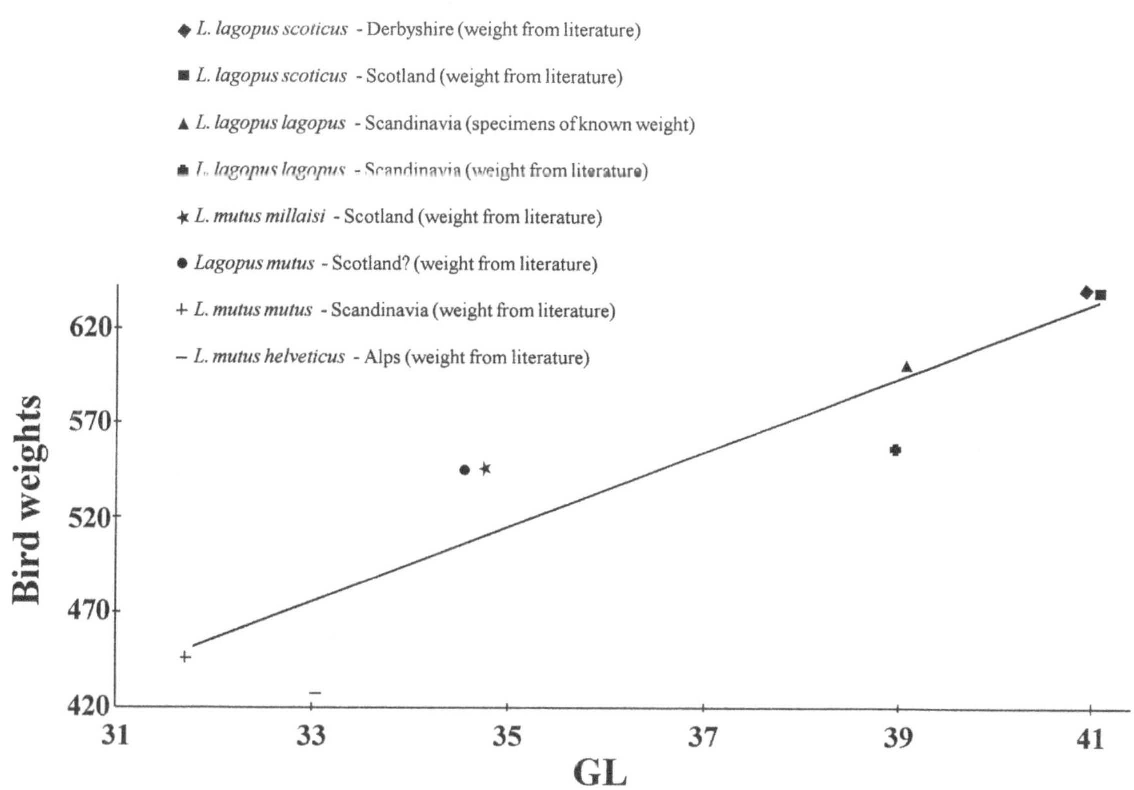

Figure 9.30. Linear regression of mean weights of *Lagopus* versus their mean length of tarsometatarsus. See legend to Figure 9.29 for source of data.

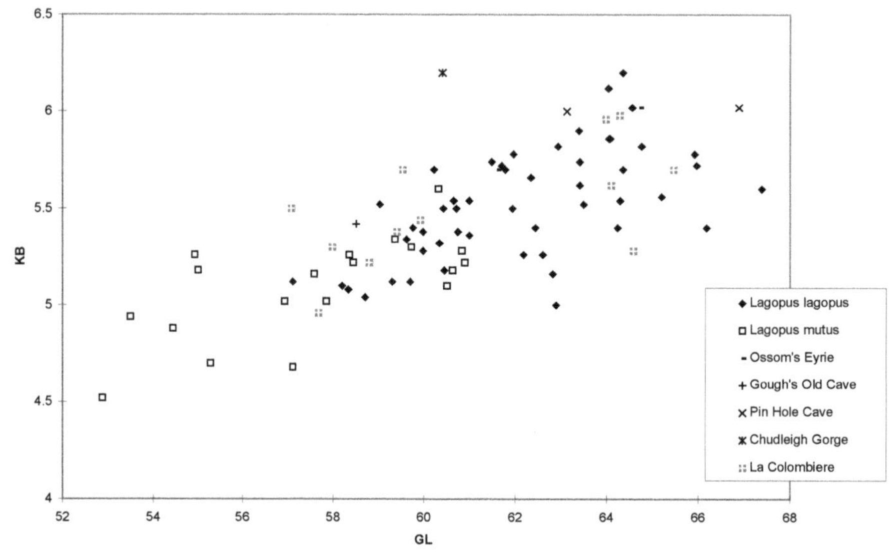

Figure 9.31. Summary diagram of fossil and modern *Lagopus* sp. humeri. Greatest Length plotted (GL) against shaft breadth (KB).

lagopus tarsometatarsi from Remouchamps. It is important to note, however, that all of these constitute very small samples.

The almost ubiquitous robusticity of the tarsometatarsi is important because the modern samples are geographically widely spaced and the fossils come from sites of significantly different ages as well as being widely spaced geographically. In addition, there is a tendency for fossils of both taxa from the last glaciation to have shorter tarsometatarsi, thus making these bones very robust (Figure 9.28). An exception to this pattern is provided by the study made by Mourer-Chauviré (1975a) on a sample from Gigny in France which, although from the last ("Würm") glaciation, had relatively long tarsometatarsi. Tarsometatarsi from La Fage ("Riss") also show this tendency (Mourer-Chauviré, 1975a and Figure 9.28). Therefore, length is probably more variable amongst fossil populations than is shaft breadth. It is possible, as discussed in the modern morphometric section, that the variation in leg length relates to winter snow cover variation through both time and space; however, this is difficult to test.

Given the consistency of greater robustness in Pleistocene *Lagopus* tarsometatarsi, an explanation should be sought. Although it is conceivable that the one species may have changed into another through evolution or replacement, both *L. lagopus* and *L. mutus* can be traced from the Pleistocene, suggesting intraspecific adaptational changes.

Research into bovid limbs and the variables affecting their morphology has shown that bone shaft breadth is closely correlated with body mass (Scott, 1985) because of weight-bearing constraints. This is particularly true of the more distal leg bones in bovids. Therefore, the greater robustness of the fossil tarsometatarsi of *Lagopus* may reflect greater mean body weight. To test this hypothesis an attempt was made to assess the degree to which tarsometatarsal dimensions are correlated to body weight in modern *Lagopus*. Estimates of bird weight have been calculated by Campbell and Marcus (1992) but in that case femoral and tibiotarsal circumference was used.

Both species of *Lagopus* are included in this analysis to augment the size and range of samples. This approach is considered valid as the relationship should be strictly mechanical and comparable between closely related taxa; it would seem unlikely that *L. lagopus* and *L. mutus* bones possess significantly different mechanical properties. Figure 9.29 shows that there is a close positive correlation between bird weights and their tarsometatarsal shaft breadths (correlation coefficient [r] = 0.951). Due to the positive correlation between the shaft breadths and lengths, a regression has also been performed for the tarsometatarsal lengths against bird weights. In this instance (Figure 9.30) a positive correlation is again revealed (r = 0.914). It may be significant that although high, the r value for the tarsometatarsus length is lower than that for the tarsometatarsal shaft breadth. This is not completely unexpected as similar trends were observed by Scott (1985) for the bovid family, lengths being free to vary with other adaptive constraints e.g. locomotion. The relationship between length and breadth may be tighter within a genus, such as *Lagopus*, where isometric relationships between skeletal dimensions are more likely.

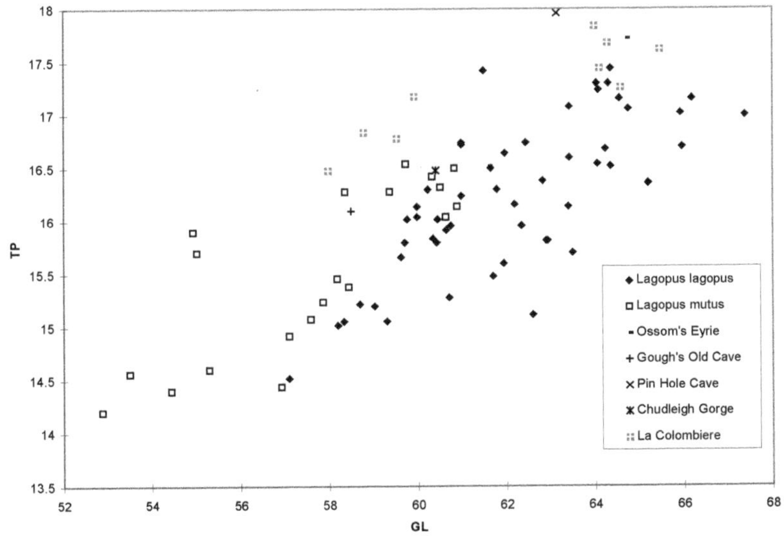

Figure 9.32. Summary diagram of fossil and modern *Lagopus* sp. humeri. Greatest Length plotted (GL) against proximal breadth (TP).

Because both tarsometatarsal breadth and length are positively correlated with body weight the interpretation that the more robust Pleistocene tarsometatarsi represent heavier birds becomes less certain, although not unexpected from the work of Scott (1985) on bovids. The other skeletal element analysed, although more difficult to interpret due to greater overlap between the two taxa, proves that the humerus was different in the Pleistocene. The humeri from La Colombière and Pin Hole Cave are particularly important in this respect. A plot of humeral length versus humeral diaphysial breadth showed either that the Pleistocene birds conformed in these dimensions with the modern birds of the two species, or that there may be some hint of stoutness (Pin Hole, Chudleigh, Ossom's) (Figure 9.31). However, when greatest length was plotted against proximal breadth, the fossil humeri proved to have wider proximal extremities than their modern counterparts (Figure 9.32). It was found that despite the great overlap between the two species of *Lagopus* today, the fossils plot as distinct clusters indicating the presence of *L. lagopus* and *L. mutus* for the larger and smaller humeri respectively. The wider proximal extremities of the fossils are likely to reflect larger areas of muscle attachment, as this measurement takes in the degree of development of both the crista pectoralis and the crista bicipitalis. This in turn implies that there may have been a greater development of the pectoral and bicipital muscles which are needed for flight, supporting the hypothesis of the animals' greater body weight.

The hypothesis that both *Lagopus lagopus* and *L. mutus* were birds of greater body weight during the Pleistocene is supported when compared to all modern populations and subspecies examined, except perhaps for *L. lagopus maior*. Unfortunately, the exact timing of this change in mean body weight cannot be ascertained due to the lack of early Holocene fossils available. If, as likely, body size declined at the Pleistocene / Holocene boundary, a number of possible causes may be suggested, such as climatic change, vegetational changes or interspecific competition.

Interspecific competition in the form of character displacement can be invoked as an explanation of size variation within certain species. Studies such as that on the pygmy shrew *Sorex minutus*, in northern Europe, have shown that where two ecologically similar taxa occur in sympatry, their sizes will be more divergent than when in allopatry (Malmquist, 1985). This is does not appear to affect *Lagopus* today, and could not affect the change in size seen through time, as these changes are independent of sympatry/allopatry i.e. the two species are mostly sympatric throughout the time covered by the study. *L. lagopus* and *L. mutus* are presumably not ecologically similar enough for character displacement to take place. In any case, they seem to change in the same direction, not to diverge.

An often quoted hypothesis to account for change in body size during the Quaternary is that of climate and in particular temperature - the mechanism often invoked to account for Bergmann's Rule (Chapter 2). Many Pleistocene mammals from glacial episodes were larger than today and certain authors have suggested that thermoregulation is the causal mechanism (Davis, 1981). However, other palaeontologists and biologists have agreed that this mechanism has been applied where it may not be appropriate and that the subject is a much more complex one (Lister, 1992). A counter-argument proposed by Guthrie (1984, 1990) and Geist (1986) is that it is not the climate that directly affects an animal's size, but the consequences of the length and quality of the plant growing season, which themselves are affected by climate. The Late Quaternary vegetational environment, called steppe-tundra or mammoth-steppe, has been described as very productive on the basis of the large herbivores it supported (Guthrie,1990). The vegetation was a mosaic of high diversity, although dominated by grassland. It should be noted, however, that some palynologists have disagreed with the concept of the mammoth-steppe. They believe the vegetation was poor, a barren tundra, based on the apparently low pollen influx at the time. The idea that the vegetational environment was a rich steppe-tundra has recently been expanded by Lister and Sher (1995) who have suggested that the steppe-tundra vegetation relied on a climatic regime that has vanished. They pointed out that the detailed climatic records, such as studies of the Greenland the ice cores, have shown that the Holocene is distinct from the Late Pleistocene in having unusually stable conditions. Pleistocene climatic instability may have allowed the mosaic vegetation of the steppe-tundra to persist. Once this climatic regime ceased to exist, the megafauna, which relied so heavily on the vegetation type it supported, changed along with it. Some animals became extinct, like the giant deer *Megaloceros giganteus* and the woolly rhinoceros *Coelodonta antiquitatis*, or locally extinct, like the lion *Panthera leo* and spotted hyena *Crocuta crocuta* (Stuart, 1991). Others underwent a reduction in body size in response to Pleistocene / Holocene vegetational change, such as the wolf (*Canis lupus*) and brown bear (*Ursus arctos*) (Kurtén, 1968). It is, therefore, an attractive hypothesis that certain birds such as the two species of *Lagopus*, which abounded in the steppe-tundra environment, also underwent changes upon its demise, such as reduced geographic ranges and body size. The specifics of the dietry difference which presumably accounts for the change are difficult to fathom. However, the fact that the kind of vegetational environment which *Lagopus* enjoy today was present at lower latitutes probably signifies that the productivity of that vegetation was greater in the steppe-tundra than the nearest analogues present today in northern latitudes and on higher elevated grounds further south (Guthrie, 1990). Interestingly, the largest subspecies of the genus, *L. l. maior*, lives today on the wooded steppes of Kazakhstan.

In support of this idea is the fact that all the fossil populations examined come from deposits considered to belong to cold phases of the Pleistocene. The oldest members of the genus *Lagopus* examined come from a cold horizon above the interglacial at Westbury-sub-Mendip in England which is early Middle Pleistocene (Oxygen Isotope Stage 12). Unfortunately, the fossils are few in number and fragmentary, which makes it difficult to assess to which species they belong. They do, however, possess relatively robust tarsometatarsi, so it may be that *Lagopus* was already by this time adapted to the steppe-tundra and was larger in relation to today's birds. The next youngest assemblage examined in this survey is that from La Fage, which is late Middle Pleistocene. Here both species are definitely present.

The relatively greater areas of the crista pectoralis and crista bicipitalis may indicate that both *L. lagopus* and *L. mutus* were heavier in the Pleistocene. If there were a primary selective force for large body size, so that birds were heavier, they would require relatively greater muscle bulk to fly, which in turn further adds to body weight. Alternatively, the birds may have become larger because of selection for better-developed flight muscles under a climatic regime when the birds were less sedentary. This hypothesis of greater mobility would be bolstered by the findings of Bochenski (1974, 1985) and Bochenski and Tomek (1994) who demonstrated that the distal wing elements of *Lagopus lagopus* and *L. mutus* during the Pleistocene of Poland and Austria were relatively longer than in present-day birds, implying greater flying power. This conclusion, however, was not confirmed by the samples analysed in the present study because it was found that *L. lagopus* and *L.mutus* carpometacarpi could not be differentiated from each other or with reliability from other similarly sized galliformes.

Concerning tarsometatarsus length, Bochenski (1974) claimed on the basis of modern samples that it was clearly positively correlated with temperature, although he pointed out that local vegetation type was also influential. The nature of the variability seen in the tarsometatarsal length over both time and space implies that, unlike the shaft breadths, local factors may be having an influence. This seems more likely than the variation reflecting the other thermoregulatory biogeographic rule (Allen's Rule) which would produce more uniform clines across the birds' former geographic ranges. It may, therefore, be that influences such as the local terrain are more important, since locomotion is generally regarded as important in determining leg length in mammals (Scott, 1985). Therefore, it is possible that the extent of snow cover in winter determines the tarsometatarsus length of the members of the genus *Lagopus*. Certainly today the subspecies of *Lagopus lagopus* without a white Winter phase in its plumage, the red grouse (*L. l. scoticus*), has longer legs than the nominate subspecies the willow grouse (*L. l. lagopus*). The lack of white plumage in the British birds is usually explained by lower levels of snow precipitation. Other aspects of local terrain may not be relevant due to the small magnitude of the differences involved and the small size of the birds in relation to the ground relief, and are difficult to test.

Lagopus lagopus and *L. mutus* differ allometrically between the Pleistocene and the present, a finding which is almost completely consistent even when fossils from widely distributed areas and times are compared with greatly dispersed modern samples from across Europe. This difference is most readily identified in the shaft breadth of the tarsometatarsus. The tarsometatarsal length changes however, appear to be reactions to variations in local conditions which remain unknown. An analogous situation might be a local adaptation similar to that proposed for the Holocene origin of the red grouse (*L. l. scoticus*) in the British Isles (Voous, 1960; Tyrberg, 1991). Conversely, the uniformity of tarsometatarsal breadth change across Europe implies regional rather than local effects. Global events which occurred at the Pleistocene / Holocene boundary seem to be a likely explanation. However, three problems arise. First, pinpointing the exact timing of the changes; second, untangling cause and effect between climatic and vegetational factors, which are closely linked; and third, the possibility that a change in the birds' vagility occurred.

It is perhaps more easy to conceive that the two *Lagopus* species changed size due to climate change and its effect on vegetation directly, rather than to changing their degree of sedentism. The birds appear to have reacted in much the same way as many mammals which survived the Holocene / Pleistocene boundary, which also became smaller. It is suggested here that the change in length of the growing season and vegetation type was the primary reason for this, and not temperature. The birds today, in the northern areas of Europe, and particularly in mountainous regions further south in the case of *L. mutus*, probably exist at much the same temperatures as they did in the Pleistocene in Europe south of their present range. This would eliminate temperature as the primary causal factor in the genus' size decrease over this period because birds of comparative sizes to those of the Pleistocene are not present in northern Europe today. Dietary shifts caused by vegetational changes are proposed here as the most significant factor leading to their size reduction. The fact that various modern studies have shown that both *L. lagopus* and *L. mutus* are highly selective in their choice of food (Moss, 1967, 1968; Gardarsson and Moss, 1970) may imply that they are quick to respond to changes in available resources, which may have given further impetus to the size change proposed. Their preferences for the taxa and parts of plants consumed, varying seasonally and regionally, relate directly to the nutritional value of those vegetable foods.

It therefore seems reasonable to suggest that the birds in the genus *Lagopus* in the Palaearctic today are refugial populations which originated as steppe-tundra species.

9.8.3. Is there any evidence for the timing of divergence of *Lagopus lagopus* and *L. mutus*?

Mourer-Chauviré (1993) suggested that *L. lagopus* and *L. mutus* diverged shortly before the "Rissian", or the late Middle Pleistocene period represented by the site of La Fage. The oldest site where one of the two species has been identified in Europe is the late Early Pleistocene sites of Valerot in France and Kozi Grzbiet in Poland where *L. lagopus* was recognised (Mourer-Chauviré, 1993).

At Westbury-Sub-Mendip in England, which is early Middle Pleistocene, there is a deposit with cold-adapted taxa probably from Oxygen Isotope Stage 12. Within this fauna are included the bones of *Lagopus*. Harrison, in Andrews (1990), had considered both species to be present at Westbury-Sub-Mendip, although this could not be confirmed here as the skeletal elements are inadequate for species-level diagnosis.

The next youngest assemblage examined in this survey is that from La Fage. Here both species are definitely present, although it may be significant that they appear to be less divergent from each other in their tarsometatarsal lengths than modern birds (Figure 9.28). This may be support for the suggestion (Mourer-Chauviré, 1993) that the species had diverged not long before. However, an alternative explanation might be that two species were closer in their tarsometarsal morphology due to ecological adaptive reasons. There is evidence at later sites, such as La Colombière, for a blurring of the morphological boundary between *Lagopus lagopus* and *L. mutus*. Unfortunately, the comparative lack of fossils of the genus *Lagopus* in the Middle Pleistocene (Tyrberg, 1995) will prove a hindrance to any study of the possible divergence of the two neospecies. This scarcity of fossil *Lagopus* may in part be due to the nature of avian assemblages generally from that time period. In Britain for example, there are relatively few large assemblages of this age, and most of these are open air sites which are less likely to yield *Lagopus* fossils as they are dominated by water birds. (Only one occurance of a *Lagopus* find from an open site is known to the author, that at Belchatów in Central Poland (Wiszniowska, 1991). This find, assuming it was correctly identified, is from the Ferdynandovian Interglacial, making it Middle Pleistocene and therefore one of the oldest European Pleistocene records for the species.) Westbury-Sub-Mendip is one of only two avian cave assemblages before the Ipswichian and it produced *Lagopus*, albeit in low numbers. (The other is the Otter Stratum in Tornewton Cave which is completely dominated by large numbers of anatids.) It may be significant that this particular Westbury-Sub-Mendip assemblage is a cold fauna (Stringer et al, 1986). *Lagopus* fossils are predominantely found in the Late Pleistocene and particularly in the Late Glacial. The fact that this is also the case with cold adapted mammals (Tyrberg, 1991 and Kurtén, 1968) need not be confirmation that steppe-tundra animals only became common during the Late Pleistocene. Instead, it may simply be that both are an indication of the scarcity of earlier Pleistocene cold assemblages. Therefore, it is not possible to confirm at present when the two *Lagopus* species diverged.

9.8.4. Are the subspecies of both *L. lagopus* and *L. mutus* named by Mourer-Chauviré valid when compared to modern subspecies?

The diagnosis of *L. lagopus noaillensis* and *L. mutus correzensis* was that they possessed greater breadths of their tarsometatarsi. The present study has confirmed the morphology which characterises the two species at La Fage. However, it is not solely these sites whose *Lagopus* species possessed such broad tarsometatarsi. While Mourer-Chauviré (1975a) was aware of this, she only included the Middle Pleistocene material under the names *L. lagopus noaillensis* and *L. mutus correzensis*. It would appear that the age of the various assemblages influenced the decision to allocate fossils to these subspecies. Therefore, the subspecies names should either be dropped from use, or applied to all Pleistocene *Lagopus* samples, regardless of age, whose tarsometatarsi are broader in their midshaft than modern populations.

However, the question arises whether a subspecies should be used to describe fossils spread widely in time and space. It is possible that in the Late Glacial the *Lagopus* species may each have been monotypic and so the use of a single different new subspecies is theoretically legitimate to describe all the Late Glacial populations of each of the two species. The reason that the two *Lagopus* species may have been monotypic in the Late Glacial is the more continuous spread of the two taxa than today. Today the European distribution of *L. lagopus*, and particularly *L. mutus*, are discontinuous, with isolated populations, often in more elevated areas. These populations are distinct subspecies, and it is the isolation which is believed to have caused their differentiation since the end of the Pleistocene (*ca.* 10 Kyr ago). However, the legitimacy of using the subspecies names given to the populations of *Lagopus* at La Fage for later populations, although similar in terms of the character used in their diagnosis, cannot be sustained. This is because they cannot realistically be considered a homogeneous population in time and space.

The variability in the tarsometatarsal lengths of the *Lagopus* species, both through time and space, causes one to question even the homogeneity of the Late Glacial sample. The mean lengths of tarsometatarsi from La Fage ("Rissian") and Gigny (Late Glacial) (Mourer-Chauvire, 1975a) exceed those of other Pleistocene samples. There is also variation in the tarsometatarsal breadths, as seen by the La Balme les Grottes sample of *L. lagopus* whose tarsometatarsi are similar in breadth to modern populations, unlike all other Late Glacial samples. In conclusion it is perhaps wise to abandon the use of the subspecies named at La Fage (*L. lagopus noaillensis* and *L. mutus correzensis)* because their diagnosis has not been considered in relation to other Pleistocene populations and modern subspecies of the two species.

9.8.5. Are there any further conclusions to be made on the basis of the present study of *Lagopus* in the Quaternary of Europe?

The study of the two *Lagopus* species through time and space has implications for the taxonomic use of metric-shape differences in skeletal elements for palaeontology. Often such metric-shape differences are given greater significance than a mere size difference. The *Lagopus* example clearly shows that a change in size may not be conferred to all elements isometrically. Therefore, dimensions thought to be significant should be considered in terms of their relationship to size before taxonomic decisions are taken. Furthermore, metric-shape changes have been observed by neontologists to occur over short periods of time e.g. the increased bill size observed in *Geospiza fortis* on the island of Daphne Major, Galapagos as a result of strong selection through differential survival of large seeded plants in response to a drought (Grant, 1986) (Chapter 2). These metric-shape changes have no taxonomic significance but simply represent organisms adapting to the complex pattern of natural selective pressures. Thus caution should be used when interpreting metric-shape differences seen in fossils, where the differences seen between Pleistocene and modern *Lagopus* are of the type which have sometimes been considered to have greater taxonomic significance than merely at the subspecies level.

Chapter 10

Corvus corax (the Raven)

10.1. Introduction

The raven (*Corvus corax*) is a member of the crow family (Corvidae) and is congeneric with other Palaearctic species such as the carrion / hooded crow (*Corvus corone corone* / *C. c. cornix*), the rook (*Corvus frugilegus*) and the jackdaw (*C. monedula*). Ravens have a widespread distribution, much reduced by human agency in Western Europe, with little in the way of specific habitat requirements and are found in most terrestrial situations. Across their range they vary morphologically and several subspecies have been recognised. They are relatively common in the Pleistocene fossil record and in one instance, La Fage in France, the reputed Middle Pleistocene ancestor (*Corvus antecorax*) of the present species forms the dominant part of an assemblage consisting of nearly 30 000 bones (Mourer-Chauviré, 1975a). Other extinct members of this genus have been named in Europe such as *Corvus pliocaenus janossyi* and *C. praecorax*, both of which are older than *C. antecorax*. During the Holocene they are quite often present on archaeological sites, particularly those dating to Iron Age and Roman times.

The main objective of this chapter is to consider both modern and fossil raven osteological diversity and the determinants of that variation. Where appropriate, the other larger members of the genus (the crow, *C. corone* and rook, *C. frugilegus*) shall receive limited attention to help in the interpretation of the raven osteological patterns.

The most immediate questions which arise from a preliminary survey of the geographical and temporal distribution of the taxon and the past literature dealing with ravens are as follows:

- What is the geographical variation in the osteology of modern Western Palaearctic ravens and what is the explanation for this variation?
- What variation can be seen in the fossil record and are there any temporal and geographical patterns in that record which may be explained by the modern patterns?
- Is *Corvus antecorax* a genuine species or should it be considered conspecific with *Corvus corax* as claimed by Pierce Brodkorb (1978)?
- Is there any evidence in the fossil record for the timing of the evolution of *Corvus corax*?
- Is there any evidence in the fossil record for the timing of the evolution of the modern subspecies of *Corvus corax*?

10.2. Systematics

Family Corvidae

This family includes medium to very large oscine passerines which are some of the most adaptable and successful birds of the order Passeriformes. They can be found in most habitat types and have an almost world-wide distribution excepting the Arctic, the Antarctic and most oceanic islands. Their diets are mostly generalised, the birds taking a range of both animal and plant foods, and often scavenging. They have a tendency to being sedentary or partial migrants, irruptive during adverse conditions. There are 115 species world-wide.

Genus *Corvus* Linnaeus 1758

The genus *Corvus* is widespread globally and in Europe includes four breeding species. They are the raven *C. corax*, the carrion crow or hooded crow *C. corone*, the rook *C. frugilegus* and the jackdaw *C. monedula*.

Corvus corax (Linnaeus, 1758). The raven.

The raven is the largest Western Palaearctic member of the family Corvidae and as such is the region's largest passeriforme. The species has a Holarctic distribution and also a wide-ranging choice of habitat. Ravens are essentially sedentary although northern populations are more prone to making southward movements in winter.

The raven is a polytypic species with seven subspecies being recognised in the Western Palaearctic and four others in Asia and North America. The geographical variation is partly in the form of plumage differences - intensity of gloss, colour of feather bases and lengths of throat feathers. Size is also known to vary, as are the relative length of the tarsus and relative length and breadth of the bill. These subspecies can be difficult to pinpoint in the field because they intergrade clinally. The seven Western Palaearctic subspecies are (Cramp and Perrins, 1994):-

1/ *Corvus corax corax* - Distribution: Europe and Asia, from Scandinavia, Britain and France east to the Yenisey Basin.

2/ *Corvus corax kamtschaticus* - Distribution: Between the Yenisey and Lena Basins and in the area around Lake Baikal, south to the Pyrenees, mainland Italy, Balkan countries, western Greece, north-east Turkey, Transcaucasia, northern Iran, basin of the Volga, southern Urals and Kazakhstan south to 51° N.

3/ *Corvus corax hispanus* - Distribution: Iberia and the Balearic Islands.

4/ *Corvus corax laurencei* (Syn. *C. c. subcorax*) - Distribution: Eastern Greece, Cyprus, Turkey, Levant, Iraq and Lower Volga to eastern Kazakhstan, China (Sinkiang), Afghanistan, Pakistan and north-west India.

5/ *Corvus corax tingitanus* - Distribution: North Africa (Morocco to north-west Egypt).

6/ *Corvus corax canarius* - Distribution: Canary Islands.

7/ *Corvus corax varius* - Distribution: Faroes and Iceland.

The four subspecies outside the Western Palaearctic are:
8 *Corvus corax principalis* - Distribution: Northern and eastern North America.
9 *Corvus corax kamtchaticus* - Distribution: Eastern Siberia, south to northern and western Mongolia, northern Manchuria, Armurland and Hokkaido (Japan).
10/ *Corvus corax tibetanus* - Distribution: The mountains of central Asia from Ten Shan and Pamir-Alay system to Himalayas and Tibet.
11/ *Corvus corax sinatus* - Distribution: Rockies (North America) to Nicaragua.

Corvus corone **(Linnaeus, 1758). The carrion crow / hooded crow.**

The species *Corvus corone* is polytypic, with two "groups" of sub-species: the *corone* group or carrion crows and the *cornix* group or hooded crows. These two groups were at one time classed as full species but because they have hybrid zones in the areas where the two meet they have been relegated to subspecies. They are large, heavy-billed crows and are either all black in the case of the carrion crows or with contrasting grey backs and underbids in the hooded crows.

Their distribution is over much of the Palaearctic and they are found in many different habitat types. *C. c. cornix* breeds in the Western Palaearctic from the sub-arctic and boreal zones to the Mediterranean, steppe and desert zones. In Scotland they are more likely to be found at higher altitudes than the nominate form and are more likely to nest on rock ledges or even on the ground than are *C. c. corone*. However, in other areas the subspecies is no less arboreal than the nominate, *C. c. corone*. *C. c. corone* is found in the west and south of Europe and is sedentary. *C. c. cornix* varies in its migratory behaviour between populations, from sedentary to fully migratory, being more migratory towards the north of its range.

The food of these crows is varied and includes principally invertebrates and cereal grain, although carrion is a significant element in their diets.

Corvus frugilegus **(Linnaeus, 1758). The rook.**

The rook is a slightly smaller bird than the carrion crow, with a more slender bill and a seemingly deeper body due to the loose feathers flanking the thighs. As adults their faces are bare of feathers and their plumage has a characteristic gloss. Their habitats include only boreal to temperate zones in the breeding season although they will move further south in the winter. They like tall trees adjacent to open spaces for breeding, avoiding dense woodland, dry or rocky environments and wetlands.

They vary from sedentary to migratory and the proportions of birds with each behaviour can vary depending on the climate. Their food is mainly beetles, although cereals and other invertebrates, as well as carrion, are taken regularly.

They are have two subspecies in the Palaearctic (*Corvus frugilegus frugilegus* and *C. f. pastinator*) although only the nominate lives in the Western Palaearctic.

Corvus antecorax **(Mourer-Chauviré, 1975).**

This fossil species was described and named by Mourer-Chauviré as part of her Ph.D. publication (Mourer-Chauviré, 1975a) on the basis of a very large collection, nearly 30 000 bones, from the cave site of La Fage at Noailles in France. The holotype is a complete left humerus in the collection of the Museum d'Histoire naturelle de Lyon. All skeletal elements are represented in this assemblage and on the basis of an osteometric comparison of most of the major limb bones with extant ravens a new species was diagnosed. *Corvus antecorax* is described as intermediate in size between *Corvus praecorax* from the Pliocene and *Corvus corax* of the present day. Furthermore, the species was described as morphologically indistinguishable from *C. corax* excepting its smaller size. The modern skeletal material of *C. corax* used by Mourer-Chauviré in the comparative study of the La Fage assemblage came from the collections of the Paris Museum, the Geneva Museum, the Budapest Museum, the Natural History Museum, the Institute of Human Palaeontology in Paris and the Department of Earth Sciences of Lyon.

The following sites have yielded bones attributed to *Corvus antecorax*: Aven I of the Abîmes de La Fage, Les Valerots at Nuits-Saint-George, Mas Rambault at Frontgnan, Lunel-Viel (Grotte I), Orgnac 3 at Orgnac-l'Aven, L' Arago at Tautavel, Lazaret (Locus VIII) at Nice, Cap de la Biehle at Nestier, Pech de l'Azé II at Carsac, all of which are in France (Mourer-Chauviré, 1975a); Atapuerca in Spain (Sanchez, 1987a, 1987b, 1992). All of these are Middle Pleistocene in age. Furthermore, there are possible occurrences from the last glaciation at Pouas, Eivissa, Ibiza (Florit *et al.*, 1989) and Cueva Nerja, Malaga, Spain (Hernández Carrasquilla, 1995).

Corvus pliocaenus **(Portis, 1889).**

This species was first described from the Val d' Arno in Italy on the basis of a distal ulna by Portis who originally identified the bone as a new species of *Numenius*, the genus to which modern curlew (*Numenius arquata*) and whimbrel *(N. phaeopus*) belong (Mourer-Chauviré, 1975a). Later Regalia (1902) placed the specimen in the genus *Corvus*. The diagnosis of this species is that it is a member of the genus *Corvus*, intermediate in size between *Corvus corone* and *C. frugilegus* on the one hand and *C. corax* on the other (Regalia, 1902).

The subspecies *Corvus pliocaenus janossyi* (Mourer-Chauviré, 1975a) was named by Mourer-Chauviré on the basis of material from Saint-Estève-Janson, France. The diagnosis of this subspecies is that it is slightly larger than the original specimen from the Val d' Arno. Jànossy, inexplicably, in his review of fossil birds from the Plio-Pleistocene of the Carpathian Basin, elevates the subspecies to a specific level, referring to it as *Corvus janossyi* (Jànossy, 1979a).

C. p. janossyi has been also identified at Aven I of the Abîmes de La Fage, Lunel-Viel (Grotte I). All the reported localities are Middle Pleistocene in age (Mourer-Chauviré, 1975a).

***Corvus praecorax* (Depéret, 1890).**

This species was named by Depéret on the basis of material from the Pliocene of Fort de Serrat-d'en-Vaquer, Perpignan. Depéret described the species as being a quarter smaller than the modern raven and possessing a few unimportant osteologically distinct features. He also stated that it was undoubtedly the ancestor of the modern raven.

10.3. Osteology of the genus *Corvus* and its constituent species.

The dominant theme in any discussion of the osteology of members of the genus *Corvus* is that certainly in the postcranium the species are comparatively conservative in morphology. This is borne out by the discussions of both Harrison (1987a) regarding the distinction between *C. corone* and *C. frugilegus* and Mourer-Chauviré (1975a) regarding the distinction between *C. corone* and *C. corax*. Difficulties were noted as early as 1960 by Bramwell (1960a). Therefore any identification of skeletal remains must rely on the size and proportions of the different species. This signifies that certain species which overlap greatly in size and proportions such as the *C. corone* and *C. frugilegus* cannot be reliably identified.

10.4. The modern *Corvus corax* samples

Of the subspecies in the Western Palaearctic, skeletal examples were available, in the various institutions visited, of *Corvus corax corax*, *C. c. hispanus* and *C. c. varius* in addition to a single specimen of *C. c. principalis* from Greenland. This assumes that their provenance can be taken to indicate the relevant subspecies, as these details are never given in museum specimen registers.

Corvus corax - Unprovenanced.

Raven material in this category are those skeletons whose data were limited, indicating only the species they represented. Many of these specimens were collected fifty years ago and more. It was decided not to ignore these data altogether as at least the species was known and they might add to the variation seen within ravens. However, the lack of geographical provenance does not allow the subspecies to be inferred for the specimens. Included here are five specimens in the Natural History Museum, Tring; a specimen in the Archaeology Department of Sheffield (formerly in the D. Bramwell collection), a specimen in the Liverpool Museum, and another in the collections of the Musée d'Histoire Naturelle in Brussels.

Corvus corax corax - Britain.

The British raven skeletal material all derives from localities were they occur today. Their range in Britain has been dramatically reduced in historical times and they are now limited to the less populated and largely upland parts of these isles such as south-west Britain, Wales, the Lake District, Scotland and Ireland. These are thus the areas from which the skeletons measured during this study derive. They include a specimen from southern Devon, two from Wales, one from Cumberland and one from South Uist. These specimens came from Natural History Museum, Tring; the Archaeology Department of Sheffield (formerly in the D. Bramwell collection) and the Liverpool Museum. Another specimen which was included was from Dr. Yalden's private collection at Manchester University and originated from the Southern Uplands in Scotland.

Corvus corax corax - Switzerland.

This small sample of three raven skeletons is housed in the collections of the Musée d'Histoire Naturelle in Brussels. The birds formed part of a programme to reintroduce the species back into Belgium from were they have been eliminated by the activities of humans. They were bred in captivity from birds which came from Switzerland where, conversely, their forestry commission is trying to keep numbers of the species down. The three specimens were collected in the wild in Buzenol in the province of Luxembourg, Belgium but are clearly not true Belgian ravens. They have been considered to be a sample of Swiss ravens for the purposes of this study, although this could be questioned as they are one generation removed from that country. A question would therefore remain if any dramatic conclusions were to made on the basis of their bones, particularly given the experiments on transplanting eggs of the red-winged blackbird (*Agelaius phoeniceus*) in North America (James, 1983) (Chapter 2). None of the specimens is sexed.

Corvus corax corax - Scandinavia.

The modern skeletal material from Scandinavia came from the following localities: 2 from Jamtland, 1 from Jinterkvarteret, 1 from Jylland, 1 from Hagested in Holboek, 2 from Helsinge, 2 from Steenstrup in Denmark; 2 from Wedellsborg Jyen, 3 from Skaane, 1 from Svenstrup, 1 from Engelholm in Sweden; 1 from Trondheim, 1 from Kirumoden, 2 from Malmberget, 1 from Utan Fynddata, 1 from Frijsenborg, 1 from Lindenvborg, 1 from Florö in Noway; 1 from Sweden unspecified, 2 from Denmark unspecified, 3 from Scandinavia unspecified. They are housed in the Zoological Museum, Copenhagen University and the Swedish Museum of Natural History, Stockholm. Of these, 20 specimens possessed only the skeletal elements which would be left after the preparation of skins: the humeri, femora, coracoids and scapulae. These did, however, generally possesss better sexing data, possibly due to the method by which skins are prepared, requiring the examination of the animals' viscera. There were 12 unsexed, 8 females and 10 males. They had been collected between 1845 and 1987.

Corvus corax corax - Poland.

The sample from Poland is from the collection of the Institute for Evolution and Systematics in Krakow and totals 12. Five of these were from the north east of the country, one from the east, one from the south, three from Krakow Zoo which were probably local birds (Z. Bochenski, pers. comm.) and two were unspecified. The skeletons were collected between 1966 and 1990 and the sexed specimens included four of each sex. Five of the specimens showed signs of being immature in their osteology, although this information was not included when the specimens were collected and catalogued.

Corvus corax hispanus - Spain

The skeletal specimens measured came from the collections of both the Universidad Autonoma de Madrid and the Museo Nacionale de Cienca Naturale de Madrid. None of these had their subspecific identities recorded although it is very likely that they all belong to the Iberian subspecies *C. c. hispanus.*

The modern skeletal material from Spain came from the following localities: 1 from Badajos, 2 from Merida, 1 from Huelua, 1 from Santader, 5 from Madrid, 1 from Toledo, 1 from Avila, 2 from Cuenca, 1 from Asturias, 1 from Segovia. They included 1 females, 2 males and 13 unsexed individuals.

Corvus corax varius - Iceland.

Four specimens from Iceland were available for study in the collections of the Zoological Museum in Copenhagen. They were collected between 1898 and 1909. The two specimens that were sexed were both males. It is probably reasonable to assume that these specimens belong to the subspecies present on Iceland today, *C. c. varius*.

C. c. principalis - Greenland.

A single specimen of raven from Greenland is present in the collection at Tring and was collected in or before 1843. No other information was available for this specimen and it was included for completeness despite the fact that it came from beyond the area covered by the present study.

10.5. Morphometric comparison of modern populations and subspecies of *Corvus corax*

It seemed reasonable to split up the nominate *C. c. corax* sample geographically because the sample was large enough to do so without creating too many sub-samples which were too small. Also, because the geographical spread of these was large both latitudinally and longitudinally, differences may well be present between them. Consideration was given to further dividing the Scandinavian sample because of the very large area this represents; however, this proved impractical because the localities were too evenly spread across the area. The measurements taken on the raven skeletons are described in Chapter 6.

The humerus of the *Corvus corax* populations examined from Europe shows marked variation in size which is only partially related to their subspecific affinities (Figure 10.1). There appears to be some suggestion of a cline from south to north with size increasing in that direction. The largest birds are from Scandinavia, while the smallest are from Spain. This trend is reflected in the means of the measurements taken on the humerus although there is considerable overlap (Table 10.1). The humeri from Britain, Poland and Switzerland are intermediate in size. The Icelandic subspecies *C. c. varius*, and *C. c. principalis* from Greenland, although not represented by large samples, have a similar large size to the birds from Scandinavia. One British specimen has a relatively short and therefore robust humerus.

The ulna measurements, while maintaining the order of size exhibited by the humeri for the different populations, demonstrate a greater degree of overlap among the populations of smaller birds (Figure 10.2). The shaft breadth aligns the samples to the greatest extent. The Scandinavian birds are the largest although the Spanish, Polish and British populations appear to overlap considerably. Again *C. c. varius* and *C. c. principalis* appear to be most like the Scandinavian birds. *C. c. hispanus* appear to have relatively wide proximal ends indicating a possible shape difference (Figure 10.3).

The plot of radii is similar to those of the ulnae, with greater length overlap between the smaller birds. However, the distal breadth of several of the Polish individuals was relatively small compared to the radii of similar length of the British, Swiss and Spanish birds (Figure 10.4). There was one radius from Britain which was markedly long. Compared to the Scandinavian birds the British and Spanish birds had slightly more robust distal ends, although not proximal ends.

The carpometacarpus also exhibits overlap among the samples of populations of smaller size, particularly as far as the distal part of the bone is concerned. The largest average size belongs to the Scandinavian birds. The British and Spanish populations differ in the dimension of the proximal carpometacarpus, the British birds exceeding the mean of the Spanish birds, while the Polish birds overlap with both (Figure 10.5). The Icelandic population is again within the general range of the Scandinavian population.

The largest coracoids belong to *C. corax corax* from

Scandinavia, *C. c. principalis* from Greenland and *C. c. varius* from Iceland, although the Scandinavian population overlaps a great deal with the smaller populations (Figure 10.6). Again, as with many of the fore-limb bones there is considerable overlap in the smaller populations. The great amount of variability among the coracoid shaft breadths (SC) is difficult to account for and is most clearly seen in the Scandinavian sample.

The situation with the scapula is almost identical to that in the coracoid measurements. There is greater overlap between measurements from localities generally, while the largest specimens and highest means belong to the Scandinavian and Icelandic birds, the lowest means to Spanish and Polish birds (Figure 10.7).

The pattern for the femora is similar to that seen in the fore-limb skeletal elements and the samples fall roughly into two groups, with the northern Scandinavian sample, the Icelandic subspecies (*C. c. varius*) and the single individual *C. c. principalis* from Greenland as the larger birds, the other samples being on average smaller. There is, however, considerable overlap (Figure 10.8).

There are less data for the tibiotarsus than for previously described elements. The general pattern remains the same with the largest individuals belonging to the Scandinavian and Icelandic birds and the smallest to the Spanish and Polish birds, while the British ravens appear to be intermediate (Figure 10.9).

The tarsometatarsi differ from all other skeletal elements in exhibiting a different pattern across Europe. In this case the order of size of the raven populations is similar to that of other elements except for the Spanish subspecies *C. c. hispanus* which can be seen to have long tarsometatarsi (Figure 10.10). This length is not, however, matched by greater breadth. The comparison is best made between the Spanish and Polish birds which appear to be similar in size for other elements, while Table 10.1 shows little overlap between the two populations in their tarsometatarsal greatest length.

T-tests were performed to examine the difference between the total nominate subspecies sample and the Spanish subspecies (Table 10.2). Testing of the subsamples of the *C. c. corax* was not worthwhile because the samples are too small to produce meaningful results. The results seen in Table 10.2 show that there are differences in most measurements taken on the two subspecies. In the humerus and ulna the greatest differences are to be found, *C. c. corax* being very significantly larger. Most other dimensions are significantly larger in *C. c. corax* than in *C. c. hispanus*. The exceptions are the distal breadth of the radius which is not significantly different and the tarsometatarsal greatest length which is significantly longer in *C. c. hispanus* than *C. c. corax*.

Therefore, while overlap exists there is a general size cline in ravens in Europe from north to south with the smallest populations being in Spain and the largest from Scandinavia and Iceland. The Polish sample is relatively small in relation to its geographical position and represents a reverse in the general pattern from south to north. The other geographical variation of morphology is the relatively long tarsometatarsi in the Spanish birds relative to the more northern populations. These two morphological trends appear to agree with Bergmann's and Allen's rules respectively. However, there is no evidence for a general cline in the tarsometatarsal length from north to south: it is merely that the Spanish population differs greatly from all others. Unlike the tarsometatarsal length difference there is no necessary correspondence between the subspecific attribution of populations with size. This is clear from the variation present within the nominate *C. c. corax* which is of the same order as between subspecies.

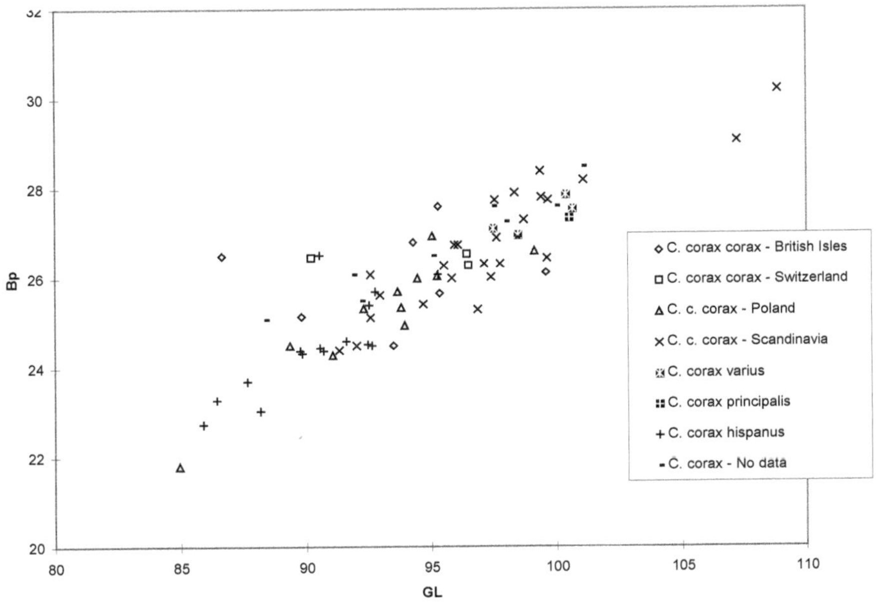

Figure 10. 1. Scattergram of European *Corvus corax* population humeral greatest length (GL) plotted against proximal breadth (Bp).

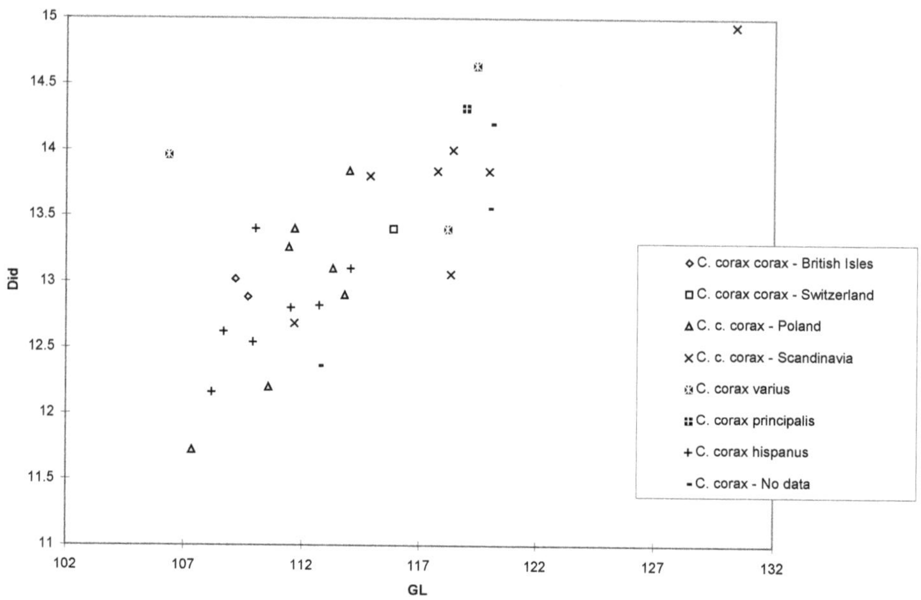

Figure 10. 2. Scattergram of European *Corvus corax* population ulnar greatest lengths (GL) plotted against distal diagonal depths (Did).

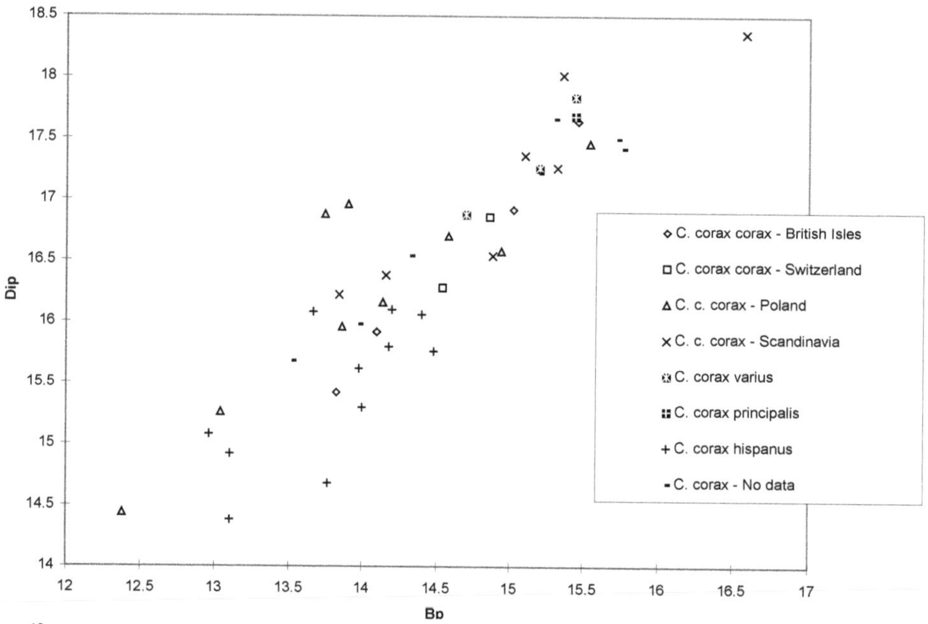

Figure 10. 3. Scattergram of European *Corvus corax* population ulnar proximal diagonal measurements (Dip) plotted against proximal breadth (Bp).

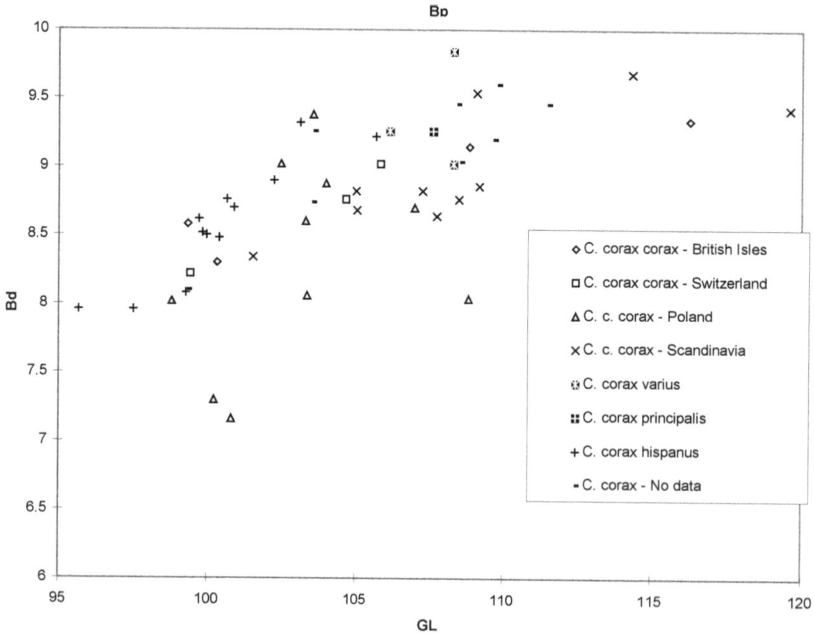

Figure 10. 4. Scattergram of European *Corvus corax* population radius Greatest length (GL) plotted against distal breadth (Bd).

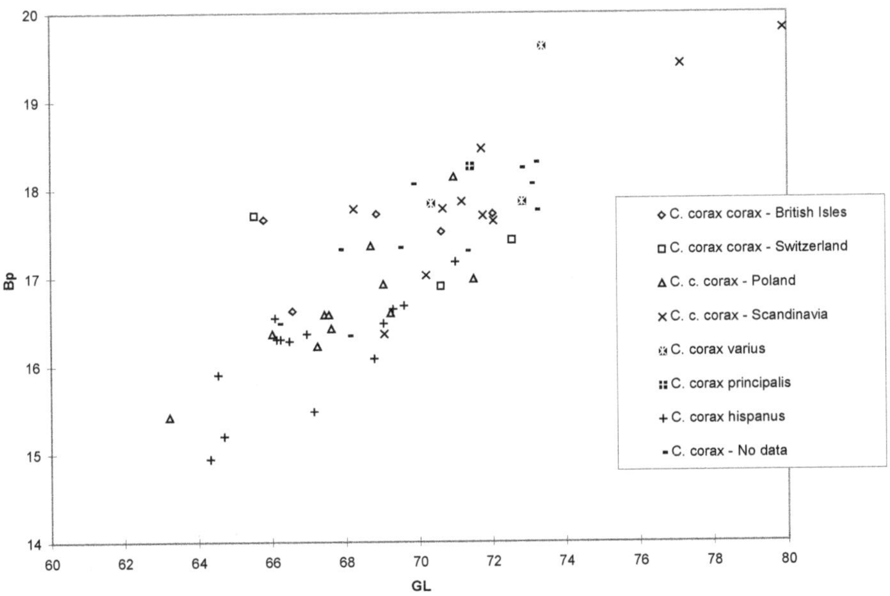

Figure 10. 5. Scattergram of European *Corvus corax* population carpometacarpal greatest length (GL) plotted against proximal breadth (Bp).

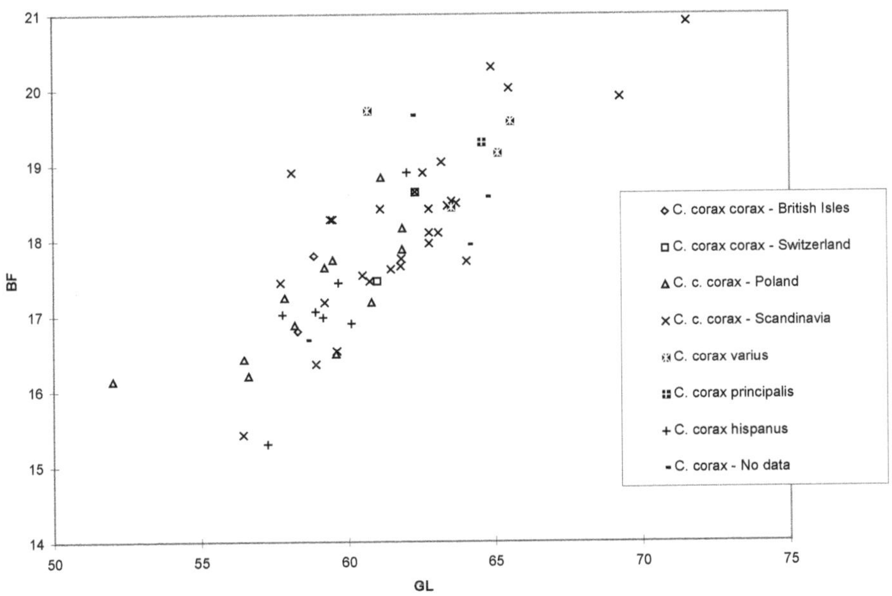

Figure 10. 6. Scattergram of European *Corvus corax* population coracoideal greatest length (GL) plotted against basis articularis sternalis (BF).

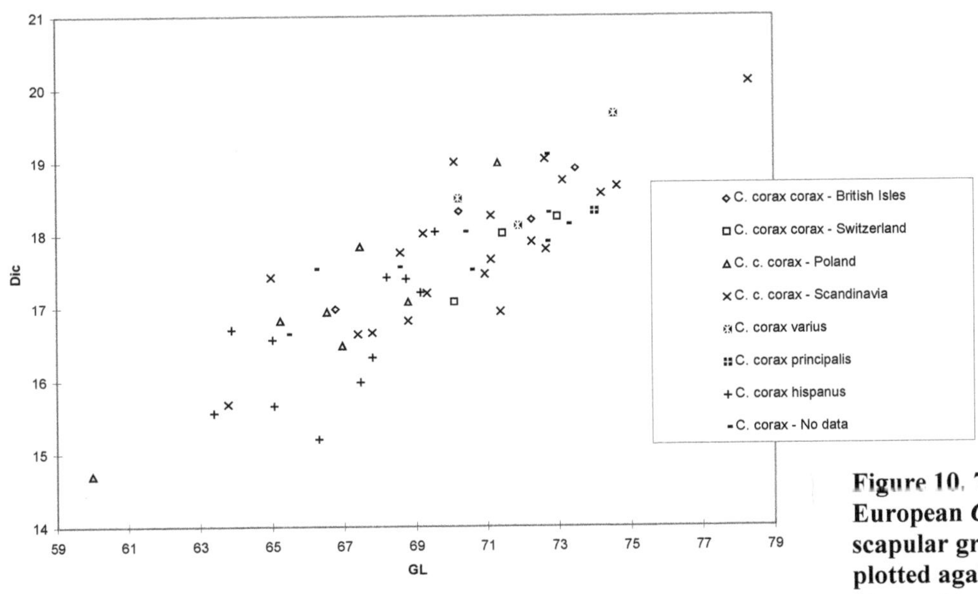

Figure 10. 7. Scattergram of European *Corvus corax* population scapular greatest length (GL) plotted against the greatest cranial diagonal measurement (Dic).

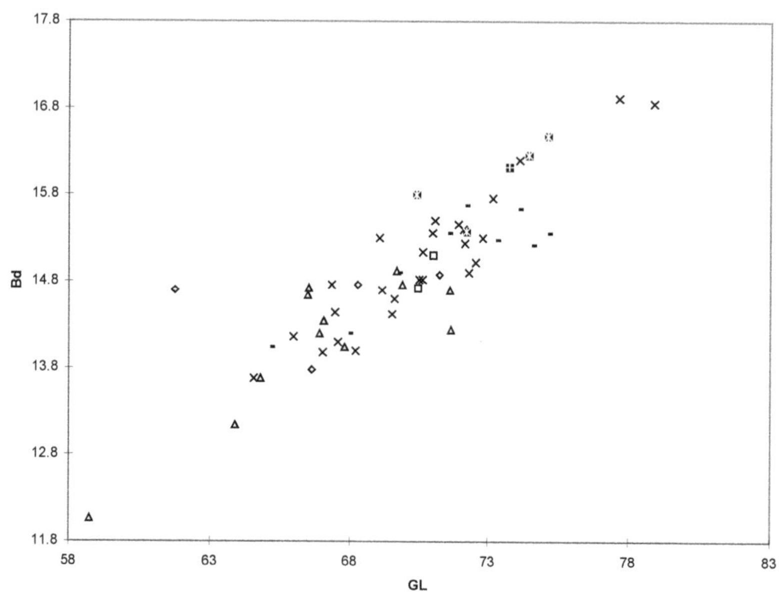

Figure 10. 8. Scattergram of European *Corvus corax* population femoral greatest length (GL) plotted against distal breadth (Bd).

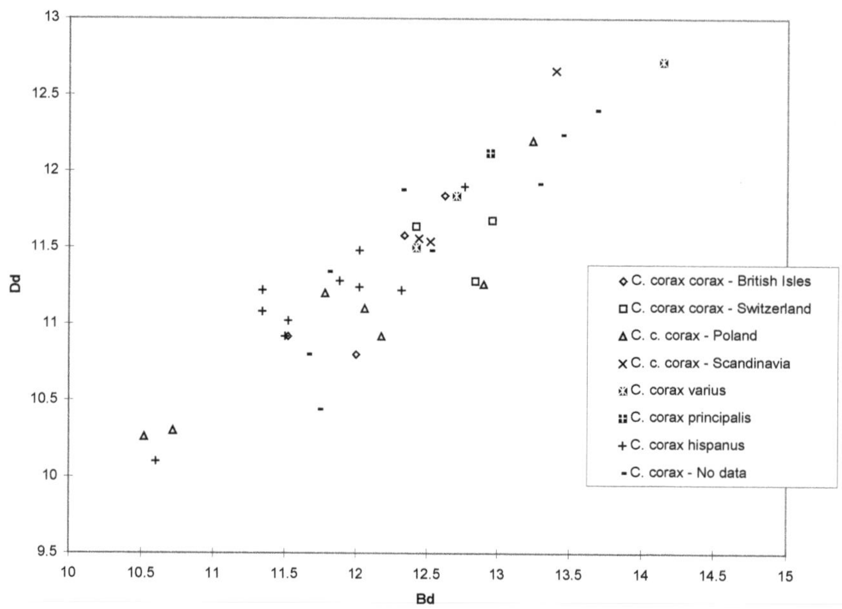

Figure 10. 9. Scattergram of European *Corvus corax* population tibiotarsal distal breadth (Bd) plotted against distal depth (Dd).

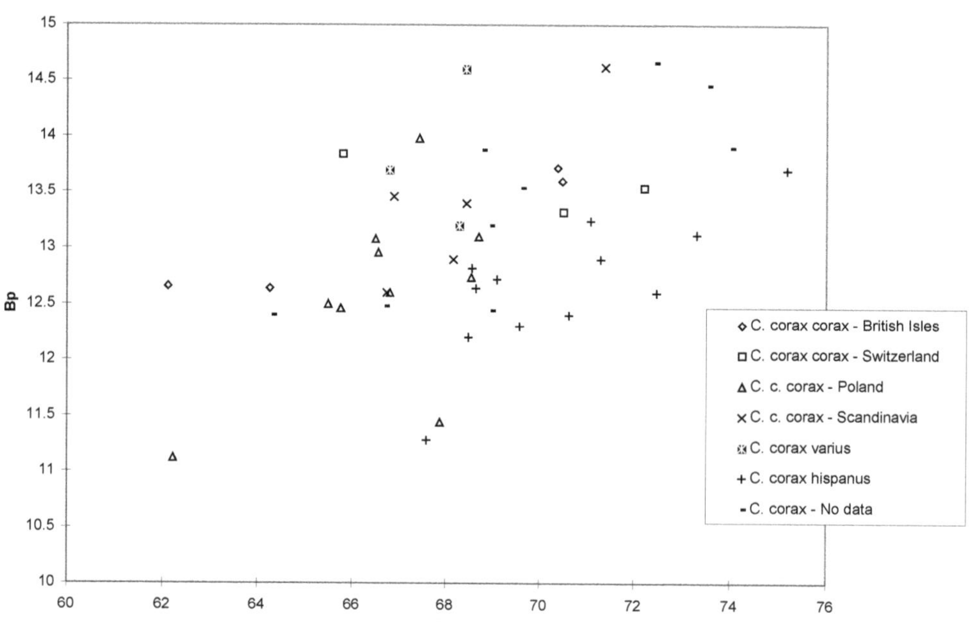

Figure 10. 10. Scattergram of European *Corvus corax* population tarsometatarsal greatest length (GL) plotted against proximal breadth (Bp).

Table 10.1. Mensural data for the populations or *Corvus corax* measured.

	Corvus corax samples														
	C. c. corax - Britain					*C. c. corax* - Scandinavia					*C. c. corax* - Poland				
Measurements	**N**	**Min**	**Max**	**Mean**	**St. dev.**	**N**	**Min**	**Max**	**Mean**	**St. dev.**	**N**	**Min**	**Max**	**Mean**	**St. dev.**
Hum GL	7	86.64	99.6	93.5	4.19	26	91.32	108.84	97.4	4.07	13	84.96	99.14	93.17	3.36
Hum Bp	7	24.5	27.6	26.05	1.05	28	24.4	30.2	26.66	1.4	14	21.8	26.94	25.41	1.32
Hum Bd	7	22.1	24.52	23.43	0.76	26	21.84	26.42	24.12	1.08	13	20.12	24.64	22.92	1.13
Hum SC	7	7.7	9.02	8.51	0.5	26	7.44	10.08	8.89	0.53	13	7.24	9.1	8.26	0.49
Uln GL	6	109.16	118.82	113.82	3.86	9	111.66	130.4	119.27	5.79	10	107.32	119.58	112.98	3.16
Uln Bp	6	13.82	15.46	14.6	0.67	10	13.84	16.86	15.14	0.96	10	12.38	15.54	14.02	0.9
Uln Dip	6	15.42	17.64	16.61	0.85	7	16.22	18.36	17.16	0.82	10	14.44	17.46	16.26	0.88
Uln Did	6	12.88	14.02	13.34	0.43	10	12.68	15.1	13.8	0.78	10	11.72	13.84	12.92	0.61
Uln SC	5	5.96	6.68	6.47	0.3	10	5.92	7.4	6.67	0.41	10	5.58	6.8	6.12	0.37
Rad GL	6	99.34	116.28	105.72	6.34	10	101.52	119.6	108.73	5.09	12	98.8	108.84	103.17	2.73
Rad Bp	6	4.74	7.26	6.43	0.88	10	5.9	7.08	6.4	0.39	12	5.56	6.82	6.19	0.39
Rad Bd	6	8.3	9.7	8.97	0.52	10	8.34	9.68	8.96	0.44	13	7.16	9.38	8.37	0.67
Cpt GL	5	65.8	72.02	68.78	2.63	10	68.24	79.9	72.19	3.6	13	63.22	71.5	68.06	2.1
Cpt Bp	5	16.62	17.72	17.45	0.47	10	16.36	19.82	17.98	1.03	13	15.42	18.14	16.73	0.66
Cpt Did	5	15.44	17.7	17.02	0.91	10	16.16	18.96	17.17	0.93	13	14.76	17.42	16.11	0.66
Fem GL	4	61.74	71.22	66.97	3.97	28	64.56	78.88	70.54	3.18	14	58.74	71.64	67.09	3.32
Fem Bp	4	14.46	15.58	15.23	0.52	27	13.9	18.6	16.05	1.04	14	12.3	16.28	14.84	1.02
Fem Dp	4	8.1	9.18	8.79	0.47	28	8.22	10.3	9.05	0.54	14	7.12	9.28	8.58	0.6
Fem Bd	4	13.78	14.88	14.53	0.51	29	13.68	16.92	14.98	0.78	14	12.06	14.92	14.15	0.77
Fem SC	4	5.96	6.7	6.35	0.37	29	5.9	7.36	6.34	0.32	14	5.32	6.64	6.17	0.36
Tib GL	4	104	118.5	112.31	6.29	8	117.02	127.88	118.08	4.95	8	106.48	119.54	111.97	4.1
Tib Dip	4	18.4	20.2	19.25	0.75	8	20.02	22.54	20.58	1.13	8	16.58	20.26	19.08	1.2
Tib Bd	4	11.52	12.62	12.12	0.47	7	12.44	16.94	13.21	1.71	8	10.52	13.24	12	0.97
Tib Dd	4	10.8	11.84	11.29	0.5	7	11.56	12.66	11.8	0.66	8	10.26	12.2	11.05	0.64
Tmt GL	4	62.12	70.48	66.81	4.27	6	66.74	71.38	68.31	1.67	12	62.72	68.7	66.79	1.8
Tmt Bp	4	12.64	13.72	13.16	0.59	5	12.6	14.62	13.4	0.77	12	11.12	13.98	12.64	0.76
Tmt Bd	4	9.1	9.82	9.54	0.33	6	9.1	10.3	9.5	0.41	13	7.78	9.7	8.95	0.62
Tmt SC	4	4.8	5.44	5.14	0.27	5	4.7	5.98	5.15	0.57	13	3.8	5.66	4.82	0.53
Cor GL	6	58.3	62.86	60.55	1.74	29	56.44	71.54	62.13	3.19	14	52.02	61.86	58.77	2.6
Cor Lm	6	53.24	57.86	55.56	1.79	29	52.24	65.22	56.79	2.78	14	48.1	56.62	53.65	2.28
Cor BF	6	16.3	18.2	17.35	0.79	29	15.42	20.9	18.22	1.16	14	16.14	18.84	17.25	0.8
Cor SC	6	4.26	5	4.64	0.37	29	3.38	5.36	4.6	0.4	14	3.94	4.9	4.38	0.27
Scap GL	4	66.78	73.5	70.69	2.94	22	63.78	78.32	70.78	3.29	9	60	71.32	66.56	3.08
Scap Dic	5	16.98	18.9	17.91	0.82	26	15.68	20.34	17.4	1.08	14	14.7	18.98	16.97	1.07

Table 10.1. (Cont.). Mensural data for the populations or *Corvus corax* measured.

Corvus corax samples	*C. c. corax* - Switzerland					*C. c. hispanus* - Spain					*C. c. varius* - Iceland				
Measurements	**N**	**Min**	**Max**	**Mean**	**St. dev.**	**N**	**Min**	**Max**	**Mean**	**St. dev.**	**N**	**Min**	**Max**	**Mean**	**St. dev.**
Hum GL	3	90.2	96.52	94.39	3.63	15	85.88	95.28	90.45	2.59	4	97.5	100.66	99.26	1.52
Hum Bp	3	26.28	26.54	26.43	0.13	15	22.74	26.52	24.51	1.08	4	27.1	27.54	27.37	0.41
Hum Bd	3	23.54	23.78	23.68	0.12	15	21.4	23.24	22.22	0.62	4	24.28	24.92	24.7	0.64
Hum SC	3	8.5	9.06	8.78	0.28	15	7.18	8.94	8.09	0.49	4	8.86	9.3	9.15	0.21
Uln GL	3	108.9	115.88	113.43	3.93	14	105.3	114.08	109.74	2.54	3	106.32	119.42	114.65	7.24
Uln Bp	3	14.54	14.86	14.74	0.17	16	12.84	14.48	13.73	0.54	3	15.2	15.44	15.11	0.38
Uln Dip	2	16.28	16.86	16.57	0.41	15	14.38	16.14	15.41	0.59	3	17.26	17.84	17.33	0.48
Uln Did	3	13.2	13.4	13.32	0.11	16	11.72	13.4	12.59	0.48	3	13.96	14.64	14	0.62
Uln SC	3	6.22	6.4	6.32	0.09	16	5.1	6.52	5.94	0.38	3	6.6	6.5	6.56	0.05
Rad GL	3	99.42	105.86	103.31	3.42	15	95.68	105.7	100.21	2.34	3	106.16	108.3	107.59	1.24
Rad Bp	2	6.66	6.68	6.67	0.01	15	5.54	6.5	5.94	0.3	3	6.44	6.8	6.5	0.27
Rad Bd	3	8.22	9.02	8.67	0.41	15	7.96	9.32	8.6	0.45	3	9.26	9.84	9.37	0.42
Cpt GL	3	65.54	72.54	69.56	3.61	15	64.32	71	67.29	2.06	3	70.34	73.38	72.18	1.62
Cpt Bp	3	16.9	17.7	17.34	0.41	14	14.94	17.18	16.17	0.61	3	17.84	19.62	18.44	1.02
Cpt Did	3	16.66	17.66	17.09	0.52	15	14.96	17.8	16.05	0.73	3	17.2	18.36	17.65	0.62
Fem GL	3	67.22	71	69.56	2.04	14	62.96	70.28	66.86	2.09	4	70.4	75.12	73.04	2.16
Fem Bp	3	15.24	15.76	15.47	0.26	14	12.96	15.64	14.22	0.65	4	16.44	16.44	16.45	0.11
Fem Dp	2	9.26	9.42	9.34	0.11	14	7.72	9.44	8.44	0.44	4	9.62	10.14	9.77	0.38
Fem Bd	3	14.72	15.1	14.9	0.19	13	12	14.8	13.9	0.71	4	15.8	16.48	15.98	0.49
Fem SC	3	6.2	6.42	6.28	0.12	13	5.16	6.48	6.02	0.38	4	6.34	6.88	6.59	0.23
Tib GL	3	110.88	118.24	115.44	3.98	12	106.6	120.76	113.35	4.17	3	114.82	121.1	117.97	3.14
Tib Dip		19.6	20.64	119.98	0.57	12	16.82	19.96	18.5	0.9	3	20.84	21.16	20.85	0.31
Tib Bd		12.42	12.96	12.74	0.28	13	10.6	12.76	11.74	0.53	3	12.7	14.14	13.09	0.92
Tib Dd		11.28	11.68	11.53	0.22	13	10.1	11.9	11.19	0.44	3	11.84	12.72	12.02	0.63
Tmt GL		65.8	72.2	69.5	3.32	13	67.6	75.18	70.73	2.34	3	66.8	68.42	67.83	0.9
Tmt Bp		13.32	13.84	13.57	0.26	13	11.25	13.7	12.65	0.58	3	13.7	14.6	13.83	0.71
Tmt Bd		9.5	9.58	9.55	0.04	14	7.44	9.76	8.95	0.61	3	9.64	9.9	9.86	0.2
Tmt SC		4.88	5.38	4.77	0.26	14	4.12	5.28	4.77	0.34	3	5.16	5.5	5.33	0.17
Cor GL	3	59.56	62.3	60.95	1.37	14	56.54	62.02	58.65	1.57	4	63.52	65.56	63.73	2.2
Cor Lm	3	54.3	56.5	55.55	1.13	14	51.1	56.46	53.64	1.5	4	57.84	59.38	58.87	0.77
Cor BF	2	17.46	18.64	18.05	0.83	14	15	18.9	16.75	0.04	4	18.44	19.58	19.23	0.57
Cor SC	3	4.28	4.6	4.43	0.16	14	3.7	5.08	4.32	0.34	4	4.98	4.86	4.79	0.17
Scap GL	3	70.08	72.98	71.5	1.45	11	63.38	69.56	66.78	2.17	4	70.2	75.1	72.94	2.3
Scap Dic	3	17.08	18.24	17.78	0.62	14	15.2	18.04	16.48	0.82	3	18.5	19.66	18.76	0.8

Table 10.1. (Cont.). Mensural data for the populations or *Corvus corax* measured.

Corvus corax samples						
	C.corax - No provenance data					*C. c. Principalis* - Greenland
Measurements	**N**	**Min**	**Max**	**Mean**	**St. dev.**	
Hum GL	9	88.38	101.08	95.84	4.24	100.54
Hum Bp	9	25.08	28.48	26.78	1.09	27.34
Hum Bd	9	22.32	25.78	24.02	1.22	24.9
Hum SC	9	8.04	9.22	8.76	0.44	8.68
Uln GL	9	108.5	120.06	115.74	4.09	118.98
Uln Bp	9	13.52	15.76	14.7	0.84	15.44
Uln Dip	9	15.68	17.66	16.82	0.74	17.68
Uln Did	9	10.7	14.2	12.98	1.05	14.32
Uln SC	9	5.44	6.84	6.32	0.42	6.4
Rad GL	10	99.34	111.5	106.57	3.96	107.6
Rad Bp	10	5.82	7.06	6.45	0.43	6.78
Rad Bd	10	8.1	9.6	9.06	0.47	9.26
Cpt GL	10	66.2	73.2	70.5	2.57	71.4
Cpt Bp	10	16.34	18.3	17.52	0.7	18.26
Cpt Did	10	16.56	17.32	16.92	0.31	17.48
Fem GL	10	65.16	75.14	71.16	3.35	73.74
Fem Bp	10	14.58	16.98	15.79	0.73	17.24
Fem Dp	10	8.26	10.02	8.99	0.48	10.06
Fem Bd	10	14.04	15.68	15	0.6	16.12
Fem SC	10	5.96	6.64	6.41	0.21	6.74
Tib GL	10	102.5	124.36	115.18	7.71	116.3
Tib Dip	10	18.94	20.88	19.98	0.73	21.02
Tib Bd	10	9	13.68	12.18	1.33	12.94
Tib Dd	10	10.44	12.4	11.59	0.62	12.12
Tmt GL	9	64.32	74.02	69.71	3.17	-
Tmt Bp	10	12.4	14.66	13.4	0.83	-
Tmt Bd	8	10.32	10	9.57	0.47	-
Tmt SC	10	4.76	5.72	5.15	0.31	-
Cor GL	10	58.62	66.16	62.42	2.19	64.58
Cor Lm	10	53.2	59.9	56.87	2.22	58.8
Cor BF	10	16.68	19.66	18.26	0.89	19.3
Cor SC	3	4.7	4.86	4.76	0.09	-
Scap GL	9	65.44	73.28	70.28	2.95	74.02
Scap Dic	10	16.64	19.1	17.87	0.64	18.32

Table 10.2. ***t*- tests between two modern subspecies *Corvus corax corax* and *C. c. hispanus*.**

	Corvus corax corax vs. *C. c. hispanus*
Hum GL	***
Hum Bp	***
Hum Bd	***
Hum SC	***
Uln GL	***
Uln Bp	**
Uln Dip	***
Uln Did	***
Uln SC	***
Rad GL	***
Rad Bp	**
Rad Bd	ns
Cpt GL	*
Cpt Bp	***
Cpt Did	*
Fem GL	*
Fem Bp	***
Fem Dp	**
Fem Bd	**
Fem SC	*
Tib GL	*
Tib Dip	**
Tib Bd	*
Tib Dd	*
Tmt GL	***
Tmt Bp	*
Tmt Bd	*
Tmt SC	*
Cor GL	**
Cor Lm	**
Cor BF	**
Cor SC	*
Scap GL	**
Scap Dic	***

Key: *** $p< 0.001$, ** $p< 0.01$, * $p< 0.05$,
ns Not significant at the 5% level.

10.6. Sexual dimorphism in *Corvus corax*?

Sexual dimorphism, while present in the raven, is relatively slight compared to birds normally regarded as being sexually dimorphic, such as birds of prey and certain Galliformes. The figures quoted by Cramp and Perrins (1994) demonstrate that female wing lengths in ravens are never less than 95% those for males from corresponding areas, and the equivalent figures for the tarsus are lowest in the sample from Greenland (*C. c. principalis*) where it is 94.9%. While a difference between the sexes exists it does not affect the present study to any extent. Due to the size of the overlap between the sexes, the small amount of sexual dimorphism will not cause any problems for the interpretation of any statistically significant results.

10.7. The other larger extant corvids.

During this study a large number of other corvid specimens were measured. The intention of this exercise was to investigate the magnitude of the gap between *Corvus corax* and the *C. corone* / *C. frugilegus* size groups. This was deemed important because of the extinct *Corvus* species which had been named in Europe and diagnosed as intermediate in size between *C. corax* and *C. corone* / *C. frugilegus*. It was felt that, as with *C. corax*, the range of specimens examined by earlier authors to gain an accurate view of the of the variability within the smaller species was too small.

Table 10.2 gives the mensural data for the three respective corvids (the two subspecies of *C. corone* and *C. frugilegus*). Measurements of the individual specimens of these samples are plotted in Figures 10.42 – 10.49 together with regression lines. This shows that there is no overlap between them and the ravens.

Table 10.3. Mensural data for the other larger extant Western Palaearctic corvids.

Measurement	*Corvus corone corone* N	Mean	*C. c. cornix* N	Mean	*C. frugilegus* N	Mean
Hum GL	50	66.16	42	66.48	26	64.15
Hum Bp	50	18.58	42	18.48	26	17.53
Hum Bd	50	17.43	40	17.29	26	16.08
Hum SC	49	5.99	42	6.09	26	5.79
Uln GL	48	80.16	31	81.36	25	78.48
Uln Bp	49	10.4	31	10.6	25	9.72
Uln Dip	48	11.66	31	11.85	25	5.79
Uln Did	49	9.72	31	9.79	25	9.22
Uln SC	49	4.76	31	4.8	24	4.43
Rad GL	49	72.41	30	73.19	25	71.52
Rad Bp	50	4.8	31	4.86	26	4.23
Rad Bd	50	6.26	31	6.37	25	5.97
Cpt GL	50	49.51	32	49.93	25	47.89
Cpt Bp	50	12.44	32	12.63	25	11.72
Cpt Did	49	11.87	32	11.93	25	11.17
Fem GL	51	52.67	33	53.2	25	49.56
Fem Bp	51	10.55	33	10.77	24	9.85
Fem Dp	47	6.09	33	8.82	24	5.83
Fem Bd	51	10.65	32	8.06	25	9.84
Fem SC	52	4.45	33	4.44	24	4.2
Tib GL	50	88.24	26	88.49	25	85.49
Tib Dip	49	14.08	28	14.19	25	12.97
Tib Bd	51	8.89	29	8.87	25	8.38
Tib Dd	51	8.31	28	8.54	25	7.95
Tmt GL	46	58.52	27	57.83	24	55.2
Tmt Bp	46	9.7	27	9.69	24	9.21
Tmt Bd	46	7.1	26	7.24	24	6.56
Tmt SC	45	4.25	27	3.57	24	3.34
Cor GL	51	44.21	35	44.84	26	42.71
Cor Lm	51	40.95	34	41.31	25	39.54
Cor BF	49	12.67	34	12.82	25	11.89
Cor SC	51	3.27	35	3.31	24	3.11
Scap GL	43	50.48	24	52.14	21	48.2
Scap Dic	48	12.25	26	12.61	25	11.92

10.8. Morphometric comparison of fossil samples.

The fossil material here examined is mostly from Britain although important collections were also studied from France as well as a single bone from Gibraltar. The fossils range in age from Medieval to Middle Pleistocene. The latter sample is that from La Fage and was named as an ancestral species to *C. corax*, *C. antecorax*. This extinct species was the primary reason for the inclusion of ravens in the present study.

Britain - Medieval

Ossom's Eyrie, Manifold Valley, Staffordshire

Ossom's Eyrie is a small cave in the Manifold Valley in Staffordshire. It lies 30m up a limestone cliff-face which reaches a total of 40m in height. The cave opens onto a small ledge and overlooks a deep gorge. The present material was excavated by D. Bramwell and the Peakland Archaeological Society (Bramwell, 1956, 1958a, 1958b) The deposit in the cave was largely homogeneous, except for a layer of charcoal concentrated near the front of the cave, and therefore the deposit was arbitrarily divided into spits during the excavation and later these were called layers 1 - 8.

Despite its inaccessible location, human occupation evidence was found in the form of a charcoal spread and a limited number or artefacts including a sherd of Romano-British pottery. This pottery came from near the base of the excavation (layer 6), therefore giving a date of approximately 300 AD for the maximum age of deposition within the cave. Further up in the sequence was found a mass of bones which was interpreted as evidence for the occupation of the cave by eagles and probably the site of a nest, hence Ossom's Eyrie (Bramwell *et al.*, 1987, 1990). Due to the finding of bones of immature golden eagle, *Aquila chrysaetos*, it would appear that this is the likely species using the site. The age of the mass of bones which is in layers 4 and 5 is thought to be no younger than the medieval period due to the presumed occurrence of wild *Sus scrofa*, extinct in the area in the 15th Century. The virtual absence of rabbit, *Oryctolagus cuniculus*, may suggest an earlier date than that as they are a favoured food of *A. chrysaetos* (Bramwell *et al.*, 1990). Most of the raven bones came from the layer in which the mass of bones was present, so it is likely that they date from some time in the medieval period.

Table 10.4. *Corvus corax* fossils from Ossom's Eyrie with their measurements.

Specimen No.	Skeletal element	Measurements			
		GL	Bp	Bd	SC
E35	R. humerus	94.4	-	23.82	8.7
Section I, 9" - 12"	R. humerus	91.78	24.46	22.72	7
	R. humerus	86.94	22.6	20.54	7.54
	L. humerus	90.54	-	-	7.38
Ledge 9" - 12, Sample 35	Dist. L. humerus	-	-	22.3	8.5
	Dist. R. humerus	-	-	22.66	-
No no. (Drawer 3)	R. humerus	91.18	24.48	21.94	8.1
Layer B, Section II, No. 1	L. humerus	92.92	24.82	23.32	7.2
Ledge 12" - 15", Sample 25	Prox. L. humerus	-	25.5	-	-
Ledge 12" - 15", Sample 27	L. humerus	-	-	-	8.1
Ledge 12" - 15", Sample 34	Dist. R. humerus	-	-	22.12	8.72
Ledge 18" - 21"	Humerus	92.4	25.58	-	8.62

Specimen No.	Skeletal element	GL	Bp	Dip	Did	SC
Section I, 9" - 12"	R. ulna	110.6	13.66	14.8	12.54	5.96
Ledge 9" - 12, Sample 35	Prox. L. ulna	-	13.88	15.9	-	-
	L. ulna	-	14.14	13.08	-	6.84
Layer B, Section II, No. 1	R. ulna	110.6	13.8	16.32	12.56	6.22
Ledge 15" - 18"	Prox. L ulna	-	12.2	14.34	-	5.44
	R. ulna	109.08	13.06	14.88	12.48	6.06
Ledge 12" - 15", Sample 25	Dist. ulna	-	-	-	13.2	-
Ledge 12" - 15", Sample 37	Ulna	-	-	-	13.1	6.12

Table 10.4. (Cont.) *Corvus corax* fossils from Ossom's Eyrie with their measurements.

		GL	Bp	Bd
Ledge 9" - 12, Sample 35	Radius	97.3	5.96	7.64
Layer B, Section II, No. 1	Radius	103.06	5.8	8.18
Ledge 15" - 18"	Radius	-	6.26	-
Ledge 12" - 15", Sample 37	Radius	-	-	8.46
Section 2, 3" - 6"	R. carpometacarpus	66.2	16.82	16.46
Ledge 9" - 12, Sample 35	L. carpometacarpus	67.52	16.88	16.3
Section 1B, Sample 14, 12" - 15"	L. carpometacarpus	69.6	17.2	16.72
	L. carpometacarpus	68.6	16.84	16.74

		GL	Bp	Dp	Bd	SC
Section I, 9" - 12"	L. femur	65.94	14.52	7.76	13.6	5.7
Ledge 9" - 12, Sample 35	R. femur	65.54	14.74	8.3	10.1	5.88
Layer B, Section II, No. 1	L. femur	64.18	14.08	7.86	13.28	5.82

		GL	Dip	Bd	Dd
Section I, 9" - 12"	Distal tibiotarsus	-	-	11.32	10.22
Ledge 15" - 18"	L. tibiotarsus	-	-	11.86	-
Ledge 12" - 15", Sample 37	Tibiotarsus	-	-	11.38	10.28

		GL	Bp	Bd	SC
E1A/B	R. tarsometarsus	66.08	12.02	-	4.24
Section I, 9" - 12"	Tarsometatarsus	~ 65.9	11.8	8.96	-

		GL	Lm	BF	SC
Section I, 9" - 12"	R. coracoid	57.36	52.66	16.52	4.46
	R. coracoid	-	52.3	-	4.6
Ledge 9" - 12, Sample 35	L. coracoid	58.8	54.16	16.6	4.48
	L. coracoid	59.58	54.8	17.16	4.88
	R. coracoid	59.8	55.2	17.1	4.8
Layer B, Section II, No. 1	R. coracoid	58.64	52.7	17.66	-
Ledge 15" - 18"	L. coracoid	58.54	54.48	17.22	4.88
	R. coracoid	57.44	52.9	16.2	4.42
Sample 16, 15" - 18"	L. coracoid	58.5	53.1	-	4.54
	R. coracoid	54.58	50.6	14.6	4.42
Section I, 18" - 20", Sample 17	L. coracoid	-	-	-	4.96
Ledge 12" - 15", Sample 25	R. coracoid	59.06	55.92	16.32	4.8
	R. coracoid	54.24	52.04	14.48	4.62
Ledge 12" - 15", Sample 23	Prox. Coracoid	-	-	15.6	-

Table 10.4. (Cont.) *Corvus corax* fossils from Ossom's Eyrie with their measurements.

		GL	Lm	BF	SC
Ledge 9" - 12", Sample 31	R. coracoid	60.52	55.3	18.4	5.06
Ledge 12" - 15", Sample 37	L. coracoid	-	-	-	4.28
Ledge 12" - 15", Sample 38	R. coracoid	56.34	52.14	16.34	4.7
		GL		Dic	
Section 2, 6" - 9"	L. scapula	-		16.74	
Section I, 9" - 12"	R. scapula	-		15.7	
No no. (Drawer 3)	R. scapula	-		15.9	
Layer B, Section II, No. 1	R. scapula	-		16.34	
Layer A, 15" - 18"	R. scapula	-		15.8	
Ledge 12" - 15"	Scapula	-		16.08	
Ledge 12" - 15", Sample 25	R. scapula	-		16.38	
	R. scapula	-		17.12	
	R. scapula	-		16.32	
Ledge 12" - 15", Sample 23	R. scapula	-		15.54	
	R. scapula	-		17.32	
Ledge 12" - 15", Sample 27	L. scapula	-		16.6	
Ledge 12" - 15", Sample 37	L. scapula	-		13.84	
	L. scapula	-		16.18	
Ledge 9" - 12", Sample 39	R. scapula	-		17.06	
Ledge 12" - 15", Sample 38	R. scapula	-		15.32	

The six raven humeri from Ossom's Eyrie are similar in size and shape to those from present-day Poland and close to those from Britain (Figure 10.11).

The three ulnae are in the size and shape range of the samples measured from Poland and Spain. This means they are relatively small compared to the modern British sample (Figure 10.12). The same holds for the Ossom's Eyrie radii which are small. The radii compare best in size and shape with those of the birds from Poland. This is particularly apparent in the distal breadth i.e. they are relatively slender, though the sample is very small (Figure 10.13).

The carpometacarpi from Ossom's Eyrie are consistent in size and shape with the modern ones from Poland, Spain and Britain (Figure 10.14). The sample of raven coracoids from the cave is relatively large, numbering twelve. When their dimensions are compared with those from the modern European survey they are found to have similar medial lengths to those from Poland and Britain but with relatively high values of their shaft breadths i.e. they are stouter than any modern samples (Figure 10.15).

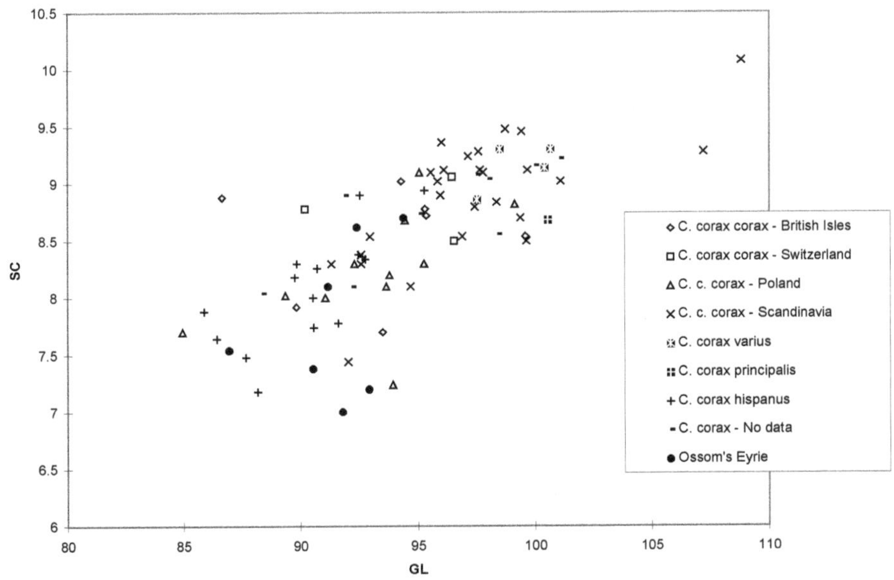

Figure 10.11. *Corvus corax* humeri from Ossom's Eyrie plotted with modern European samples. Greatest length (GL) plotted against shaft breadth (SC).

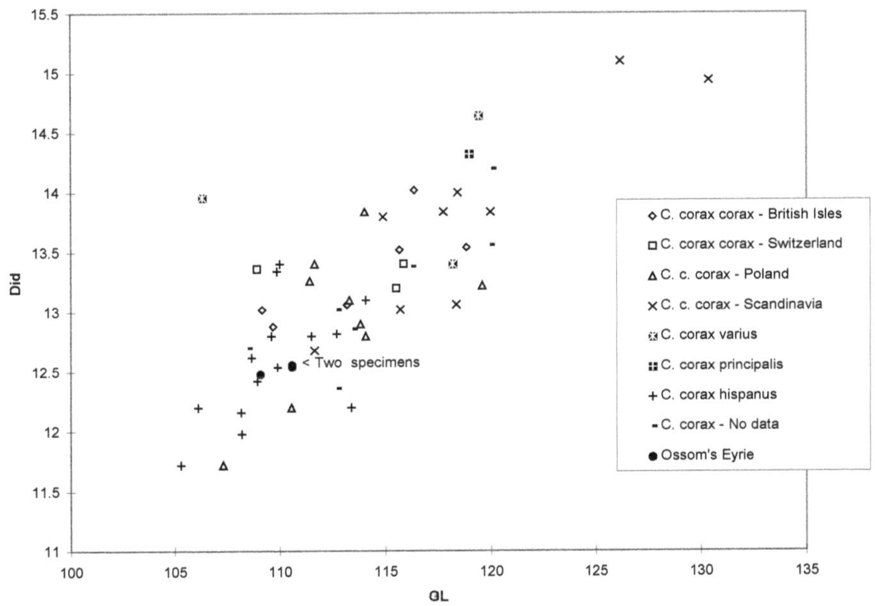

Figure 10.12. *Corvus corax* ulnae from Ossom's Eyrie plotted with modern European samples. Greatest length (GL) plotted against distal diagonal depth (Did).

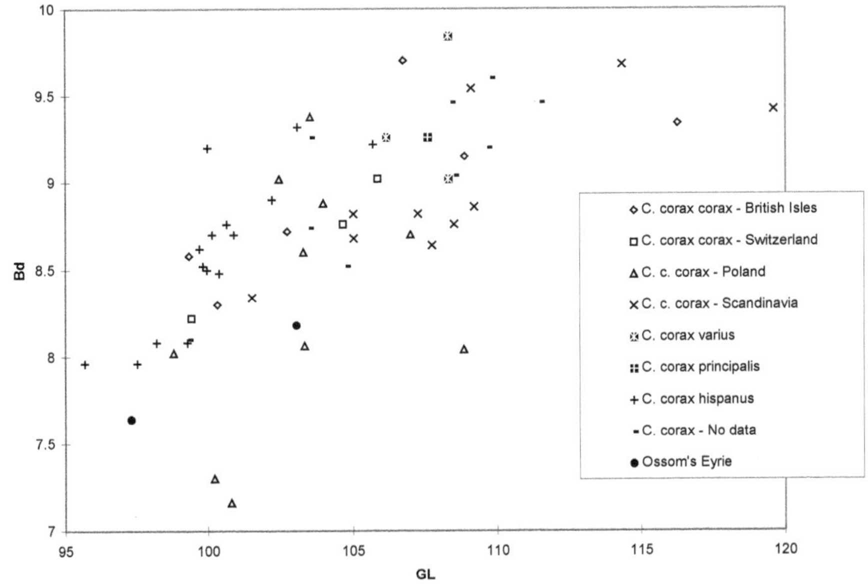

Figure 10.13. *Corvus corax* radii from Ossom's Eyrie plotted with modern European samples. Greatest length (GL) plotted against distal breadth (Bd).

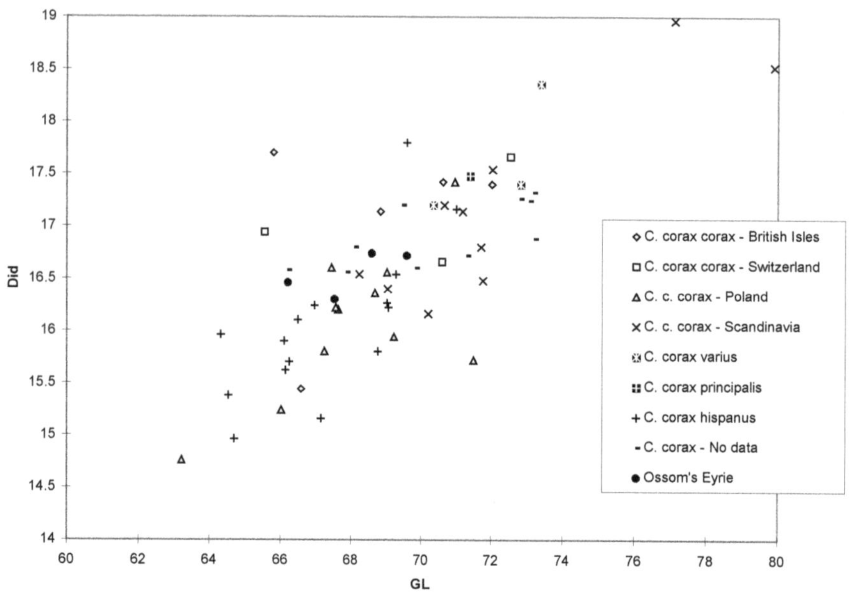

Figure 10.14. *Corvus corax* carpometacarpi from Ossom's Eyrie plotted with modern European samples. Greatest length (GL) plotted against distal diagonal depth (Did).

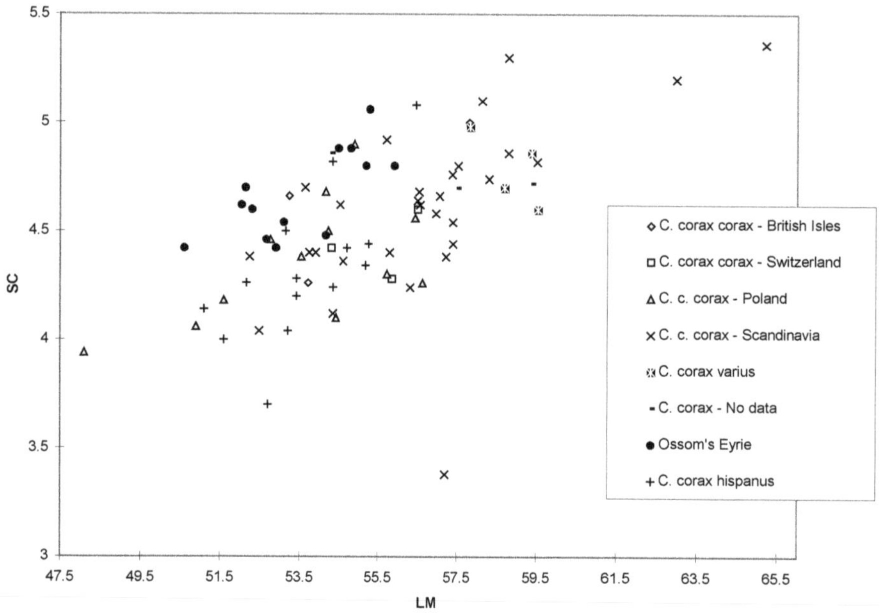

Figure 10.15. *Corvus corax* coracoids from Ossom's Eyrie plotted with modern European samples. Medial length (LM) plotted against shaft breadth (SC).

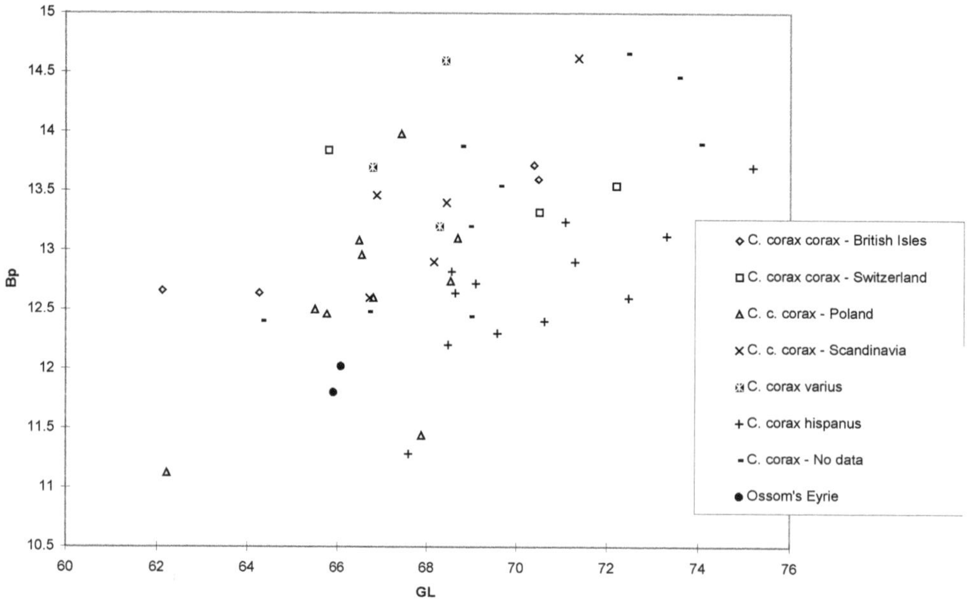

Figure 10.16. *Corvus corax* tarsometatarsi from Ossom's Eyrie plotted with modern European samples. Greatest length (GL) plotted against proximal breadth (Bp).

Table 10.5. *Corvus corax* fossils from London Wall with their measurements.

Specimen No.	Skeletal element	Measurements		
		GL	Bp	Bd
A 1028	R. carpometacarpus	65.28	16.62	16.24

Table 10.6. *Corvus corax* fossils from Witcombe Villa with their measurements.

Specimen No.	Skeletal element	Measurements				
		GL	Bp	Dip	Did	SC
A5213	Prox. R. ulna	-	13.36	14.9	-	-
		GL	Bp	Bd		
A5213	R. carpometacarpus	65.48	16.48	15.5		
		GL	Bp	Bd	SC	
A5213	R. tarsometatarsus	64.5	12.28	8.72	4.58	

The two tarsometatarsi from Ossom's Eyrie are small and fall in the range of those modern specimens from Poland. The breadth of the bones is within the range of the Spanish subspecies although their greatest lengths are significantly shorter (Fig. 10.16).

In summary, therefore, the Ossom's Eyrie ravens appear to have been relatively small compared to ravens in Britain today with more gracile dimentions in many instances with the exception of the coracoid.

Britain - Roman

London Wall, London

The raven carpometacarpus from this site was donated to the Natural History Museum by A.S. Kennard in 1915 and appears to have come from an excavation of part of the ancient city wall of London (Harrison, 1980a). The remains from this site are considered to be Roman in age and were part of what was probably food debris. Other taxa identified by Harrison from the site include items which no doubt were food items such as greylag goose (*Anser anser*), teal (*Anas crecca*), mallard (*A. platyrhynchos*), wigeon (*A. penelope*), common crane (*Grus grus*), woodcock (*Scolopax rusticola*), lapwing (*Vanellus vanellus*), domestic pigeon or woodpigeon (*Columba livia / palumbus*), song thrush (*Turdus philomelos*), starling (*Sturnus vulgaris*) and, unusually, night heron (*Nycticorax nycticorax*).

In size, the London Wall specimen is in the "small" part of the range of the modern carpometacarpal distribution. It is closest to those from Poland and Spain (Figure 10.19).

Witcombe Villa, Gloucestershire

The bird remains from this villa were described by Colin Harrison (Harrison, 1980a) although very little appears to be known of the site otherwise. The Roman age for these bones is presumably reliable and they are therefore included here. Apart from the raven remains there appears to be only a single tarsometatarsus identified as a barn owl (*Tyto alba*). The site was listed in Parker's review of British archaeological birds (Parker, 1988).

Both the carpometacarpus and tarsometatarsus dimensions from Witcombe fall towards the lower end of those for the modern British population measured (Figure 10.19 and 10.20).

Dunstable, Bedfordshire

Associated with a Romano-British cemetery in Dunstable was a deep cess-pit which yielded animal remains among which were a number of raven remains. These were identified by Graham Cowles (Cowles, 1981) as part of the site report (Matthews, 1981). Other vertebrates recovered were white-tailed sea eagle (*Haliaeetus albicilla*), frog (*Rana temporaria*) and water vole (*Arvicola terrestris*). The raven remains here described are housed in the sub-department of Ornithology at Tring in Bedfordshire. They comprise two boxes, one containing an assortment of raven bones which represent more than one individual (Minimum Number of Individuals = 2, based on two right humeri), the other containing what appears to be an associated skeleton.

A date for the raven remains has been derived from associated pottery. The date is pre- 150 AD and the assemblage is probably about 50 years older than that

(Matthews, 1981). It is confusing, however, that along with the specimens which appear to be an associated skeleton, there is a note that their age is Roman, with an age of 310 - 330 AD. It is possible that two groups of raven bones were recovered, that from the cess pit with an age of pre- 150 AD and the skeleton which was from a different archaeological context on the site, explaining the different ages quoted.

Table 10.7. *Corvus corax* fossils from Dunstable with their measurements.

Specimen No.	Skeletal element	Measurements				
		GL	Bp	Bd	SC	
No no.	R. humerus	86.7	24.7	21.9	7.82	
No no.	Prox. L. humerus	-	25.7	-	-	
No no.	Prox. R. humerus	-	25.52	-	-	
		GL	Bp	Dip	Did	SC
No no.	Dist. R. ulna	-	-	-	13.24	-
No no.	Dist. L. ulna	-	-	-	13.1	-
No no.	L. ulna	114.66	15.1	16.82	14.26	6.56
		GL	Bp	Bd		
No no.	Dist. radius	-	-	9.88		
No no.	Dist. radius	-	-	9.78		
		GL	Bp	Did		
No no.	L. carpometacarpus	68.38	17.74	16.82		
No no.	L. carpometacarpus	69.44	17.82	17.36		
		GL	Bp	Dp	Bd	SC
No no.	L. femur	68.7	15.6	-	14.4	6.42
No no.	L. femur	69.72	16.12	9.42	15.5	6.38
		GL	Dip	Bd	Dd	
No no.	L. tibiotarsus	110.3	19.06	12.46	11.28	
No no.	L. tibiotarsus	118.26	20.1	13.42	11.9	
		GL	Bp	Bd	SC	
No no.	L. tarsometatarsus	68.16	13.14	9.54	5.3	
No no.	R. tarsometatarsus	68.32	13.3	9.62	5.3	
No no.	L. tarsometatarsus	69.6	13.94	10.4	5.12	
		GL	Lm	BF	SC	
No no.	R. coracoid	60.18	54.74	18.9	-	
	L. coracoid	60.8	56.16	18.9	-	
		GL	Dic			
No no.	L. scapula	70.52	18.42			

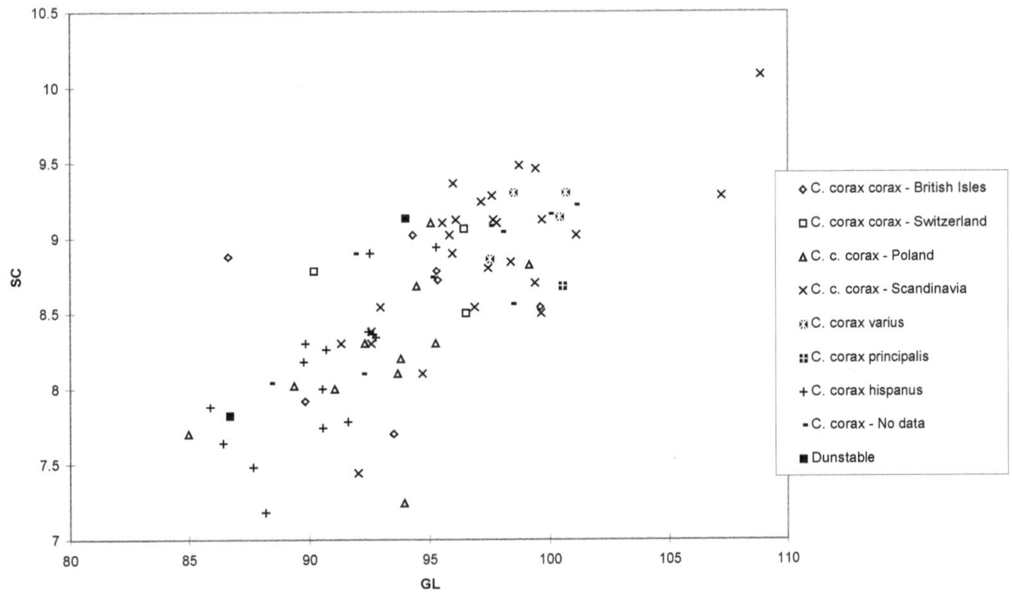

Figure 10.17. Scattergram of the humeri from Dunstable with modern European *Corvus corax* populations. Humeral greatest length (GL) plotted against shaft breadth (SC).

The two humeri from Dunstable are of the same order of size as those of modern British ravens as well as those from Poland. One of the two falls within the small part of the size range for modern raven humeri, thus being closer to both modern Spanish and Polish populations (Figure 10. 17).

The ulna is close in size to the modern British sample although slightly more robust distally (Figure 10.18). The only other birds to have similar distal dimensions are those from Iceland (*C. c. varius*).

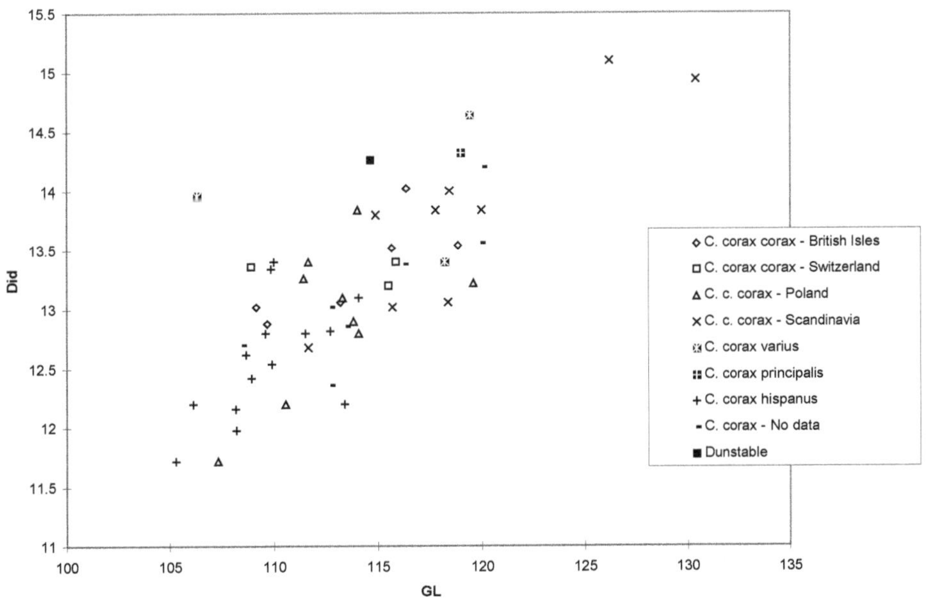

Figure 10. 18. Scattergram of the ulna from Dunstable with modern European *Corvus corax* populations. Ulnar greatest length (GL) plotted against shaft breadth (SC).

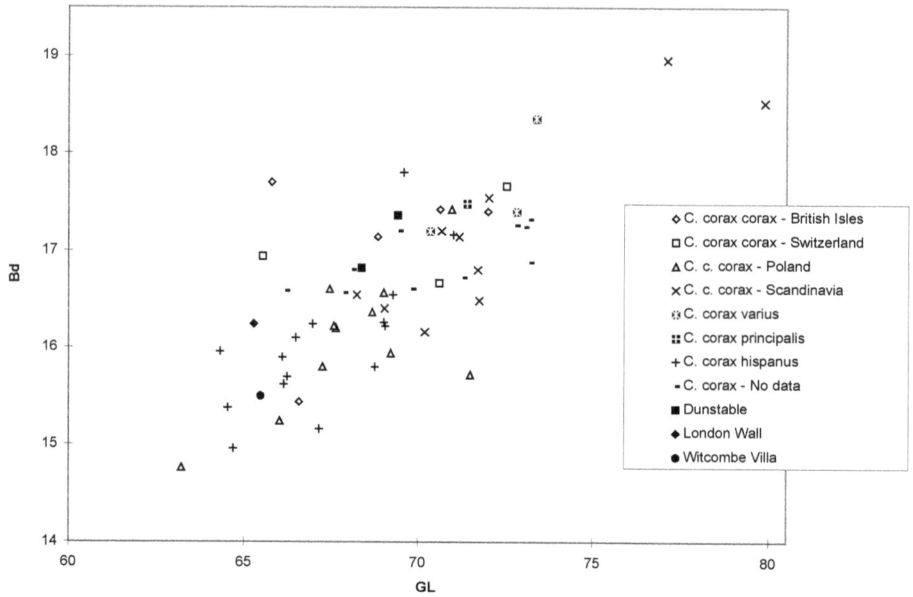

Figure 10. 19. Scattergram of the carpometacarpi from Roman aged sites with modern European *Corvus corax* populations. Carpometacarpal greatest length (GL) plotted against distal breadth (Bd).

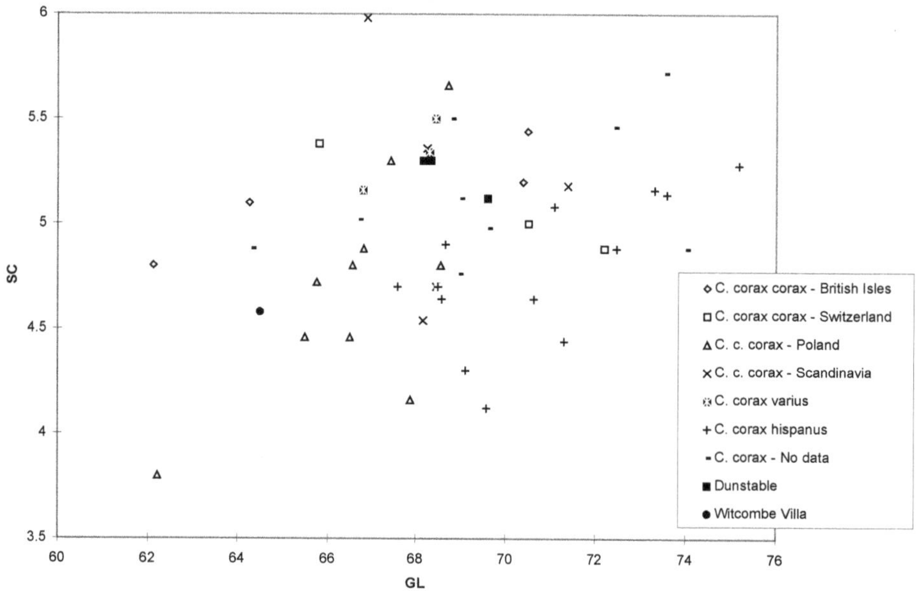

Figure 10. 20. Scattergram of the tarsometatarsi from Roman sites with modern European *Corvus corax* populations. Tarsometatarsal greatest length (GL) plotted against shaft breadth (SC).

The two carpometacarpi from Dunstable are in the range of distribution of modern British, Swiss and Spanish ravens and close to that of Poland (Figure 10.19).

The three raven tarsometatarsi from Dunstable, two of which probably came from one individual, are within the range of variation seen in the limited sample of modern British birds measured. They also fall in the range of Scandinavian birds, partially agreeing with those from Poland, and are close to those from Spain (Figure 10.20).

In sum the ravens from Dunstable are approximately the same size as those from Britain today measured.

Britain - Iron Age

Danebury Hillfort, Hampshire

A number of articulated skeletons of raven were found at Danebury Hillfort. Given their quantity and the fact that they were found as isolated burials, it is possible that they represent some kind of ritual behaviour by people as ravens were sacred to the Celtic god Lugus (Serjeantson, 1991). This is fortuitous as it enables one to work with known associated skeletons rather than dispersed, isolated elements. Of further interest, although not strictly relevant to the present study, this ritual behaviour may have spanned a considerable amount of time. This is evidenced by the range of dates available from the deposits in which they were found. These include early to late ceramic phases (pottery dates) from 550 - 450 BC through to 300 - 100/50 BC (Cunliffe, 1984). All the ravens can confidently be dated to the Iron Age with C^{14}-calibrated ceramic dates between the figures quoted (550 - 50 BC) (Cunliffe, 1984). The measurements which it was possible to take on these ravens are given in Table 10.9.

Table 10.8. *Corvus corax* fossils from Danebury with their measurements.

Specimen No.	Skeletal element	Measurements				
		GL	Bp	Bd	SC	
DA83 P2183 (6)	L. humerus	92.26	24.64	22.8	8.68	
DA83 P2218 (1)	Prox. L. humerus	-	26.86	-	-	
DA88 P2609 (5)	Prox. L. humerus	-	24.42	-	-	
DA86 P2568 (3)	L. humerus	90.24	24.84	24.22	8.3	
DA83 P2178 (4)	Prox R. humerus	-	24.54	-	-	
DA83 P2286 (5)	R. humerus	88.58	24.5	22.3	8.04	
DA82 P2030 (4)	L. humerus	91.42	25.32	23.34	8.48	
DA88 P2587 (5)	R. humerus	94.64	-	23.68	8.54	
		GL	Bp	Dip	Did	SC
DA83 P2183 (6)	L. ulna	108.22	14.2	15.62	12.54	6.24
DA86 P2568 (3)	L. ulna	108.32	13.86	15.9	12.32	6.28
DA88 4140141 (1)	L. ulna	101.86	13.9	16.02	12.5	6.38
DA83 P2286 (5)	R. ulna	107.3	13.4	15.48	12.56	5.98
DA82 P2030 (4)	L. ulna	110.8	14.34	16.48	13.02	6.3
DA88 P2587 (5)	R. ulna	114.24	13.9	16.82	12.84	6
		GL	Bp	Bd		
DA83 P2183 (6)	Radius	99.7	6.48	8.7		
DA86 P2568 (3)	Radius	-	6.42	8.48		
DA88 4140141 (1)	Radius	101.68	-	8.68		
DA83 P2286 (5)	Prox. radius	-	6.04	-		
DA82 P2030 (4)	Radius	100.62	6.16	8.84		
DA88 P2587 (5)	Dist. radius	-	-	9.02		
		GL	Bp	Did		
DA83 P2183 (6)	R. carpometacarpus	68.58	-	15.84		
DA86 P2568 (3)	R. carpometacarpus	66.9	16.62	16.14		
DA83 P2178 (4)	L. carpometacarpus	69.28	16.82	16.22		
DA83 P2286 (5)	R. carpometacarpus	65.82	16.58	15.68		
DA82 P2030 (4)	L. carpometacarpus	68.1	17.1	16.08		
DA88 P2587 (5)	R. carpometacarpus	70.62	17.82	-		

Table 10.8. (Cont.). *Corvus corax* fossils from Danebury with their measurements.

		GL	Bp	Dp	Bd	SC
DA83 P2183 (6)	R. femur	-	14.5	9.18	-	-
DA83 P2218 (1)	L. femur	-	15.02	8.82	14.72	-
DA88 P2609 (5)	L. femur	64.96	14.4	8.66	13.6	6.08
DA83 P2178 (4)	L. femur	67.02	14.38	8.1	13.52	5.74
DA83 P2286 (5)	R. femur	63.58	13.62	-	13.44	5.92
DA82 P2030 (4)	L. femur	67.02	14.36	8.66	-	-

		GL	Dip	Bd	Dd
DA83 P2183 (6)	Prox. Tibiotarsus	-	18.26	-	-
DA83 P2218 (1)	Tibiotarsus	-	18.32	11.86	11.52
DA88 P2609 (5)	L. tibiotarsus	107.68	17.86	11.42	10.56
DA83 P2286 (5)	R. tibiotarsus	107.86	18.12	11.02	10.28
DA82 P2030 (4)	L. tibiotarsus	112.36	19.48	12.44	10.88

		GL	Bp	Bd	SC
DA83 P2183 (6)	L. tarsometatarsus	68.3	13.18	-	5.1
DA88 P2609 (5)	L. tarsometatarsus	65.06	-	8.94	5.02
DA83 P2286 (5)	L. tarsometatarsus	64.26	11.98	-	5.02
DA82 P2030 (4)	L. tarsometatarsus	66.5	13.22	8.92	4.94

		GL	Lm	BF	SC
DA83 P2183 (6)	R. coracoid	-	54.56	-	5.04
DA83 P2218 (1)	L. coracoid	-	55.54	-	4.84
DA88 P2609 (5)	R. coracoid	54.7	51.84	-	4.44
DA86 P2568 (3)	L. coracoid	-	-	-	4.62
DA83 P2286 (5)	L. coracoid	-	-	-	4.56
DA82 P2030 (4)	L. coracoid	57.82	52.7	17.28	4.56
DA88 P2587 (5)	R. coracoid	58.4	54.16	16.58	4.54

		GL	Dic
DA83 P2183 (6)	L. scapula	-	17.36
DA83 P2218 (1)	L. scapula	-	17.42
DA88 P2609 (5)	L. scapula	-	16.54
DA86 P2568 (3)	L. scapula	-	16.96
DA83 P2286 (5)	R. scapula	64.6	16.14
DA82 P2030 (4)	R. scapula	67.66	17.66
DA88 P2587 (5)	R. scapula	-	17.26

The raven humeri from Danebury are within the distribution of the modern British birds measured. They also overlap with the Spanish and Polish populations and to some extent with those from Scandinavia (Figure 10.21).

The ulnae from Danebury are mostly smaller than modern British specimens measured. They are most like those from Spain and to a lesser extent Poland (Figure 10. 22).

The four coracoids from Danebury form two sets of two which are close in size to each other. They appear to be quite robust compared to all modern populations (Figure 10. 23). The tibiotarsi meanwhile are small in both length and distal depth dimensions, comparing best to Spanish and Polish samples (Figure 10. 24).

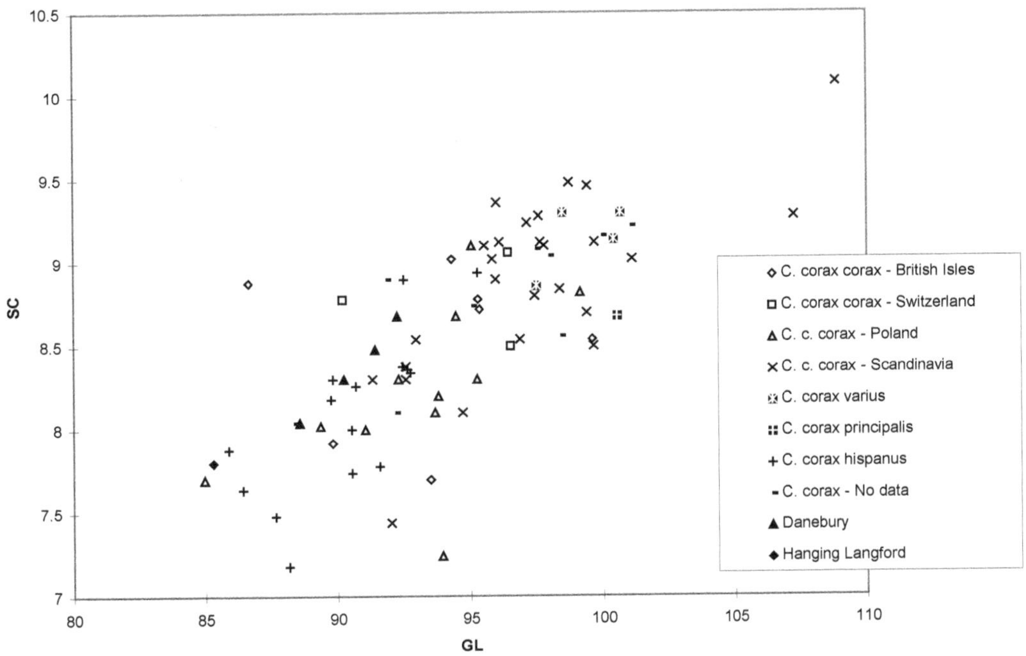

Figure 10.21. *Corvus corax* humeri from Iron Age sites plotted with modern European samples. Greatest length (GL) plotted against shaft breadth (SC).

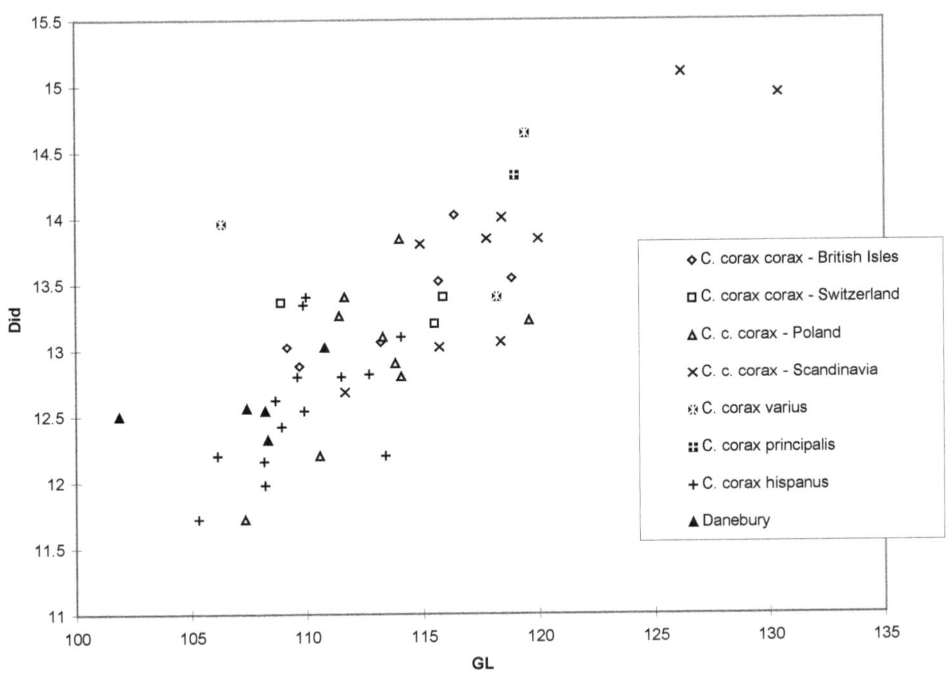

Figure 10.22. *Corvus corax* ulnae from Iron Age sites plotted with modern European samples. Greatest length (GL) plotted against distal diagonal depths (Did).

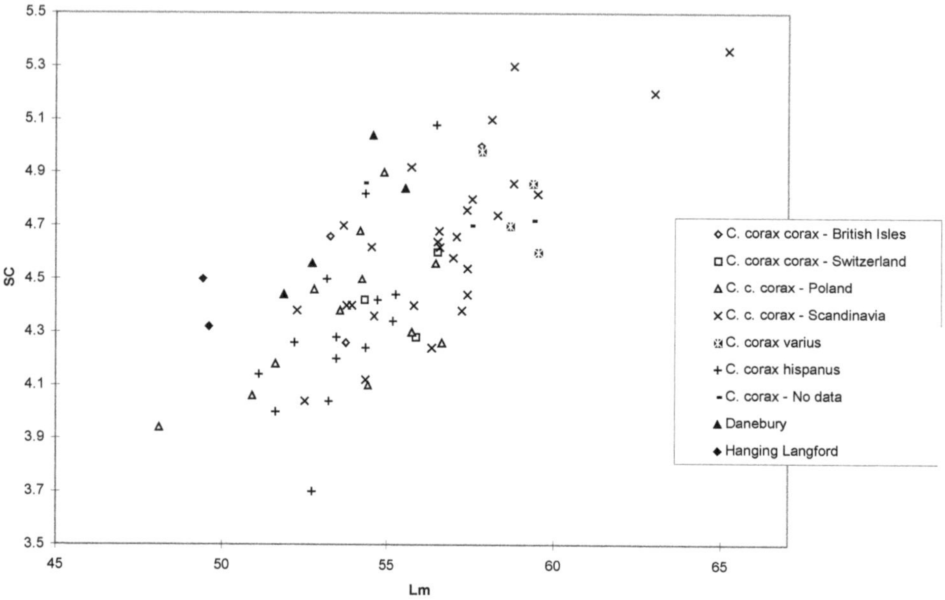

Figure 10.23. ***Corvus corax*** **coracoids from Iron Age sites plotted with modern European samples. Medial length (Lm) plotted against shaft breadth (S**

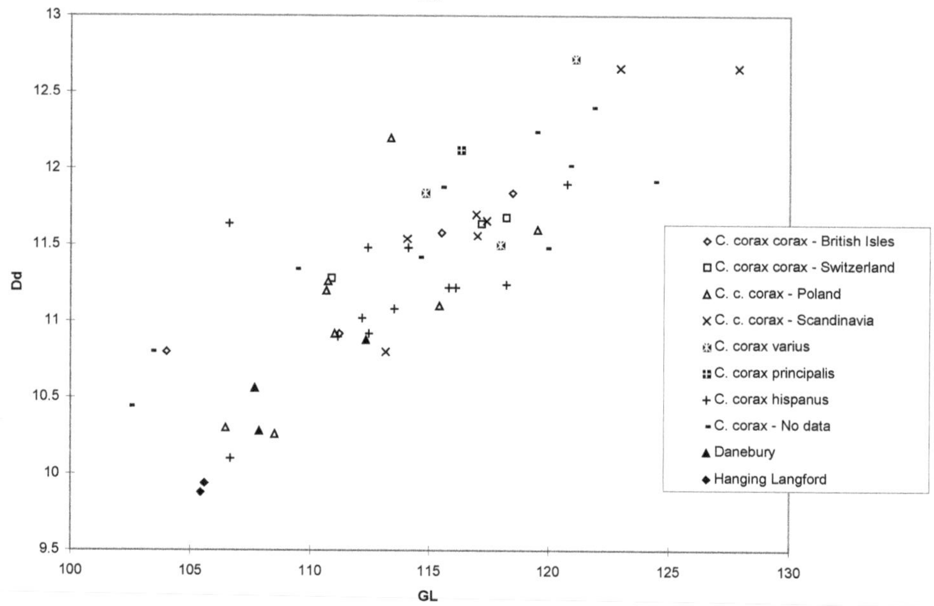

Figure 10.24. ***Corvus corax*** **tibiotarsi from Iron Age sites plotted with modern European samples. Greatest length (GL) plotted against distal depths (Dd**

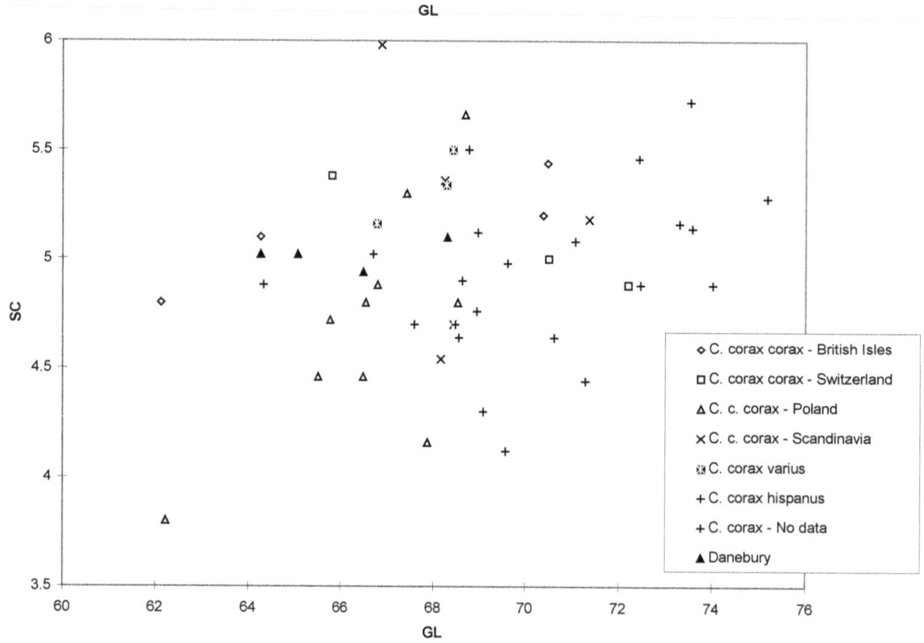

Figure 10.25. ***Corvus corax*** **tarsometatarsi from Iron Age si plotted with modern European samples. Greatest length (GL) plotted against shaft breadth (SC**

Table 10.9. *Corvus corax* fossils from Hanging Langford Camp with their measurements.

Specimen No.	Skeletal element	Measurements				
		GL	Bp	Bd	SC	
A1584	R. humerus	85.3	24.02	21.38	7.8	
		GL	Bp	Dp	Bd	SC
A1584	L. femur	63.2	13.78	8.3	12.92	5.58
		GL	Dip	Bd	Dd	
A1584	R. tibiotarsus	105.6	17.32	10.42	9.94	
	L. tibiotarsus	105.44	17.16	10.44	9.88	
		GL	Lm	BF	SC	
A1584	R. coracoid	53.54	49.42	-	4.5	
	L. coracoid	-	49.6	-	4.32	
		GL	Dic			
A1584	R. scapula	-	15.08			

The tarsometatarsi are in the range of the modern British sample and do not appear to be particularly small. They show no sign of being small like the tibiotarsi from the site (Figure 10. 25). This implies that there is a difference in limb bone proportions between these samples and the modern British birds.

The Danebury birds appear to be relatively small compared to modern British specimens measured although this is not apparent in all skeletal elements implying that there may be slight metric-shape differences whose significance is dificult to explain.

Hanging Langford Camp, Wylye, Wiltshire

Harrison (1980a) first described the bird assemblage from this site, which was presented to the Natural History Museum in 1922 by R.S. Newell. The fauna is labelled Iron Age, although no other information is available, such as the nature of the site or the means by which it was dated.

The humerus is relatively small, similar to modern Spanish and Polish ravens (Figure 10. 21). The two coracoids are quite robust in the shaft and short unlike modern populations (Figure 10. 23). The tibiotarsi are smaller than those of all modern ravens measured (Figure 10. 24).

The results of the comparison of the Hanging Langford ravens with modern comparative samples shows that British Iron age raven humeri and tibiotarsi are relatively small and that they have robust coracoids.

Britain - Devensian

Pin Hole Cave, Derbyshire

Pin Hole Cave is described in detail in Chapter 10 on *Lagopus*. The fossils from Pin Hole are likely to be a mixture of Late Pleistocene and Holocene in age.

The proximal ulna from Pin Hole Cave is similar in the dimensions representing the distance between the olecranon and the cranial edge of the cotyla dorsalis, but has a smaller breadth than modern British ravens (Figure 10.26).

The carpometacarpus from Pin Hole Cave is very close to those of modern British birds and within the range of those from Spain, and Poland (Figure 10.27). The two distal tarsometatarsi meanwhile, appear to be relatively large compared to most modern European samples measured. They are, however, close in size to the British sample (Figure 10.28).

Table 10.10. *Corvus corax* fossils from Pin Hole Cave with their measurements.

Specimen No.	Skeletal element	Measurements				
		GL	Bp	Dip	Did	SC
PH(F) 1122	Dist L. ulna	-	-	-	11.88	-
No no.	Prox. L. ulna	-	13.34	16.16	-	6.18
		GL	Bp	Bd		
PH(F) 639	Dist. radius	-	-	8.64		
		GL	Bp	Bd		
PH(F) 13075 - 13076	R. carpometacarpus	67.46	16.18	15.86		
		GL	LM	BF	SC	
PH(F) 7840	Dist. R. coracoid	-	-	17.86	-	
No no.	L. coracoid (Immature)	-	-	-	4.26	
		GL	Bp	Dp	Bd	SC
No no.	Dist. L. Femur	-	-	-	14.42	6.18
		GL	Dip	Bd	Dd	
PH (F) 1119 and PH(F) 6	Dist. L. tibiotarsus	-	-	12.94	11.8	
No no.	L. tibiotarsus	-	18.8	-	-	
		GL	Bp	Bd	SC	
PH (F) 1119 and PH(F) 6	Dist. L. tarsometatarsus	-	-	9.94	5.62	
No no.	Dist. L. tarsometatarsus	-	-	9.38	5.6	

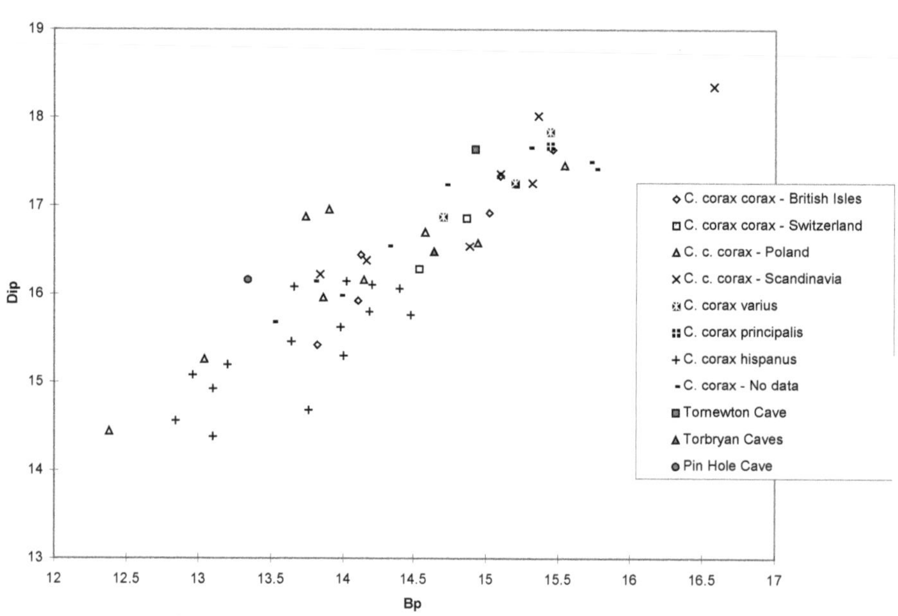

Figure 10.26. *Corvus corax* ulnae from British Devensian sites plotted with modern European samples. Proximal breadths (Bp) plotted against proximal diagonal depths (Dip).

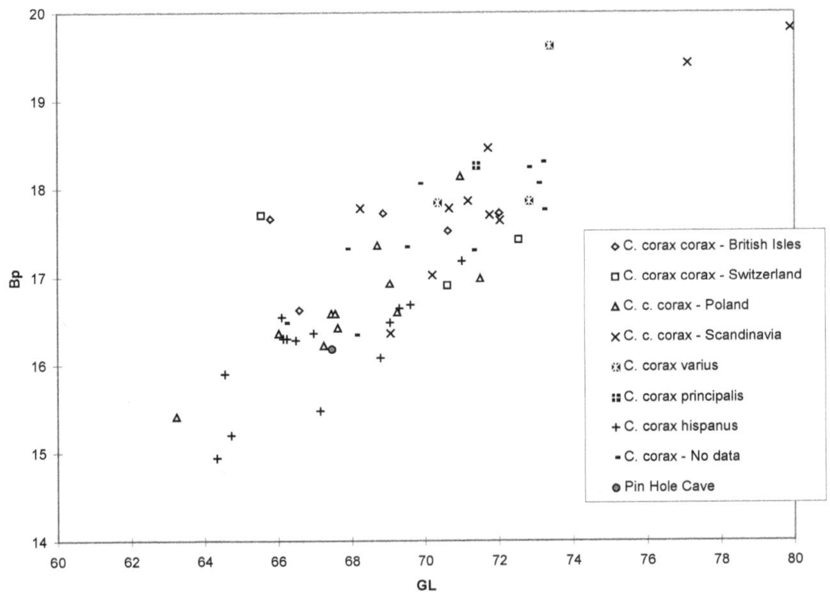

Figure 10.27. *Corvus corax* carpometacarpus from Pin Hole Cave plotted with modern European samples. Greatest lengths (Bp) plotted against proximal breadths (Bp).

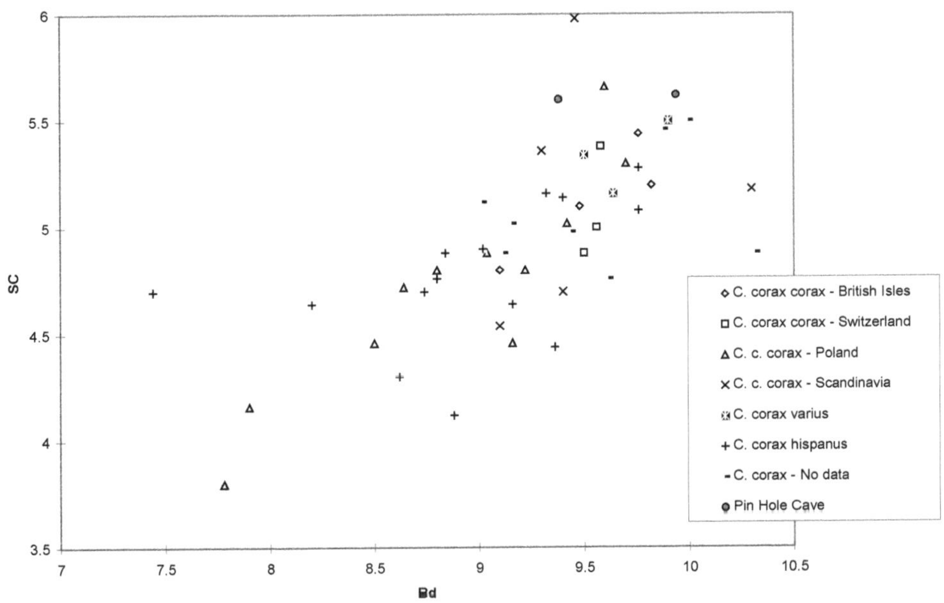

Figure 10.28. *Corvus corax* tarsometatarsi from Pin Hole Cave plotted with modern European samples. Distal breadths (Bp) plotted against shaft breadths (SC).

Happaway Cave, Devon

The Happaway Cave collection was presented to the museum Natural History Museum by Mrs. Pengelly in 1896 upon Pengelly's death.

The age of the single raven bone from Happaway Cave is difficult to gauge accurately because the excavation during which it was found is an old one and very little is known about the provenance of any of the fossils from the cave. Judging from the other taxa, and particularly the mammals, it would appear that a mixture of Pleistocene and Holocene fossils were excavated from the cave (Jacobi, pers. comm.). This makes it difficult to ascertain the age of the tarsometatarsus below without directly dating the specimen itself.

Table 10.11. *Corvus corax* fossils from Happaway Cave with their measurements.

Specimen No.	Skeletal element	Measurements			
		GL	Bp	Bd	SC
A1106	R. tarsometatarsus	61.78	12.92	-	5.14

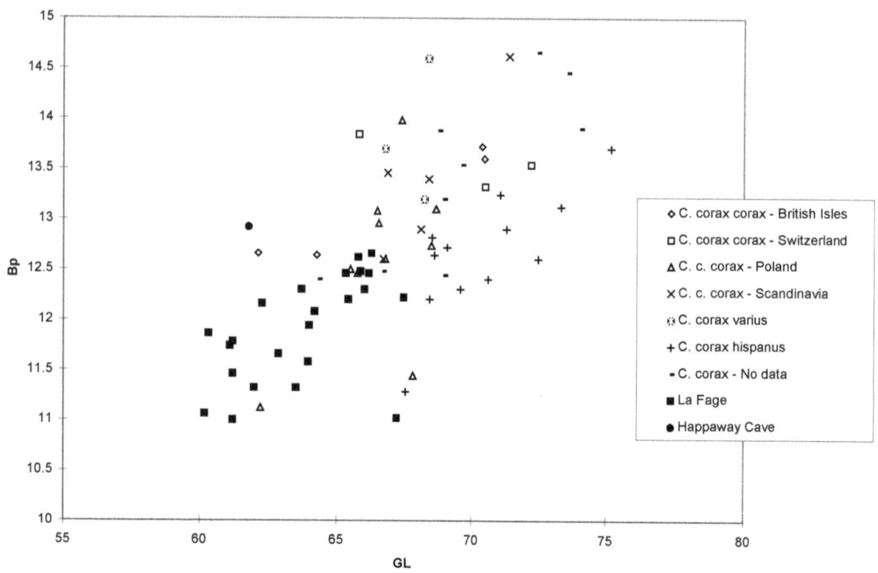

Figure 10.29. *Corvus corax* tarsometatarsi from Happaway Cave plotted with modern European samples as well as a sample of "*Corvus antecorax*" from La Fage. Distal breadths (Bp) plotted against proximal breadths (Bp).

Colin Harrison originally identified the specimen and added that the specimen was small on a note with the specimen. The recognition of small size was not followed up or indeed mentioned in any of the publications noting the species presence at the Happaway Cave (Harrison, 1980b and 1987a). This remark highlighted the possibility that the specimen was similar in proportions to the raven at La Fage (*C. antecorax*).

Figure 10. 29 demonstrates that the tarsometatarsus was indeed relatively short compared to most extant ravens measured in Europe in this study. However, the specimen falls outside the range of those specimens measured from La Fage. Specifically, it is more robust and is most similar in size to one of the modern British specimens. Harrison's observation, therefore, does not hold true after further investigation.

Ravencliffe Cave, Derbyshire

The material from Ravencliffe Cave, Cressbrook Dale in Derbyshire was presented to the Natural History Museum in 1907 by W. Storrs Fox. Little is known about this excavation although both large Pleistocene mammals and artifacts of the Neolithic and Bronze and Roman periods were recovered from the cave (Harrison, 1980a). This implies that little stratigraphic control can be assumed for these fossils, although they are often assumed to be Devensian in age.

Table 10.12. *Corvus corax* fossils from Ravencliffe Cave with their measurements.

Specimen No.	Skeletal element	Measurements			
		GL	Bp	Bd	SC
A1166	R. humerus	91.16	25.42	22.9	8.32
A1165	R. humerus	95.52	26.44	23.74	8.58
	Prox. L. humerus	_	25.36	_	8.28

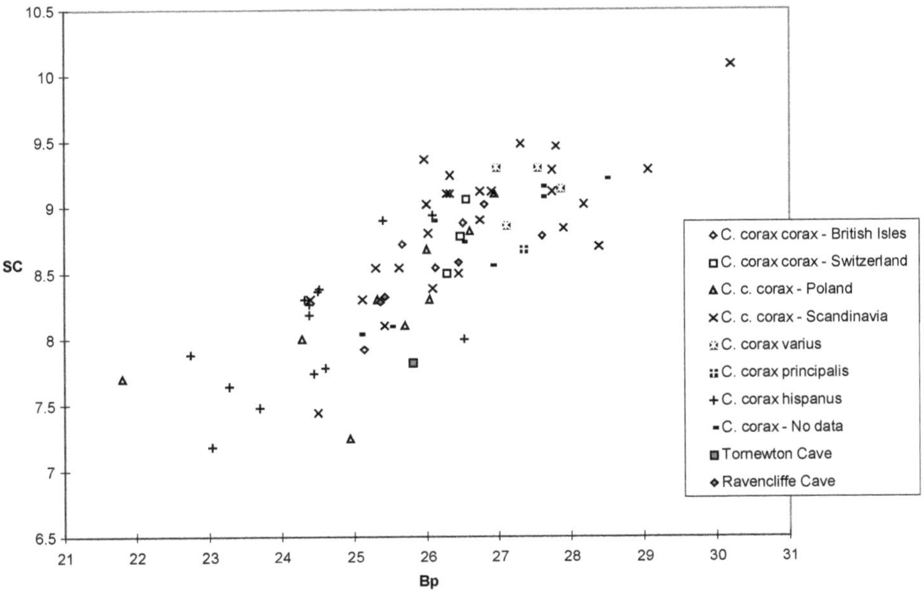

Figure 10.30. *Corvus corax* humeri from British Devensian sites plotted with modern European samples. Proximal breadths (Bp) plotted against shaft breadths (SC).

The three humeri are the same size as those from Britain today. They also fall within the upper part of the range of the distribution for Spain and Poland and the lower part from Scandinavia (Figure 10.30).

Tornewton Cave, Devon

All the raven fossils from Tornewton Cave were recovered during the excavations by A. Sutcliffe from 1944 - 1964 (Sutcliffe and Zeuner, 1962). They derive from both the Lower Hyaena and Sub-Bear (the original name for the Glutton Stratum) Strata. As detailed further in the section on the *Sturnus* fossils from the site (Chapter 12) there has been a significant reinterpretation of the cave sediments recently, both in terms of their ages and stratigraphic relationships (Currant, 1996).

The age of the Hyaena Stratum raven fossils are considered here to be from Oxygen Isotope Stage 5 in its broadest sense. This age can not be readily refined at the present moment due to problems with the biostratigraphic dating of assemblages of this date (see further, Chapter 12). The Sub-Bear or Glutton Stratum has also caused controversy in terms of dating. Currant (1996) quotes the fact that a mixture of boreal and temperate taxa are present among the smaller mammals as well as evidence that the deposit may be intrusive and of mass movement origin. These features suggest that this deposit is not immediately pre-Ipswichian in age (Oxygen Isotope Stage 6) as originally proposed. The best evidence, however, came from the AMS C^{14} date of 22 160 +/- 460 BP (OxA-4587) derived from a well preserved glutton (*Gullo gullo*) maxillary. This suggests that the slump deposit was active during Oxygen Isotope Stage 2 and may well have been active prior to and since that time. Therefore any fossils in the Glutton Stratum could be from Oxygen Isotope Stage 5 onwards.

The size and proportions of all the raven fossils from Tornewton Cave fall within the range of modern European ravens.

Table 10.13. *Corvus corax* fossils from Tornewton Cave with their measurements.

Specimen No.	Skeletal element	Measurements				
		GL	Bp	Bd	SC	
Sub-Bear Stratum A4161	Prox. L. humerus	-	26.12	-	-	
	Dist. L. humerus	-	-	24.12	-	
Lower Hyeana Stratum A4142	Prox. R. humerus	-	25.8	-	7.82	
		GL	Bp	Dip	Did	SC
No stratum recorded A4094	L. ulna	118.52	14.92	17.64	13.42	6.56

Table 10.13 (contd.). *Corvus corax* fossils from Tornewton Cave with their measurements.

		GL	Dip	Bd	Dd
Sub-Bear Stratum A4161	Dist. L. tibiotarsus	-	-	12.14	11.02
No stratum recorded A4119	Prox. L. tibiotarsus (Immature)	-	19.48	-	-

		GL	Bp	Bd	SC
No stratum recorded A420	L. tarsometatarsus	-	12.42	-	4.74
No stratum recorded A4160	Dist. L tarsometatarsus	-	-	9.7	-
	Dist. R. tarsometatarsus	-	-	9.38	-

		GL	Lm	BF	SC
Lower Hyeana Stratum A4131	R. coracoid	-	-	-	4.32
	L. coracoid	-	-	-	4.54
No stratum recorded A4132	L. coracoid	-	58.28	-	4.74

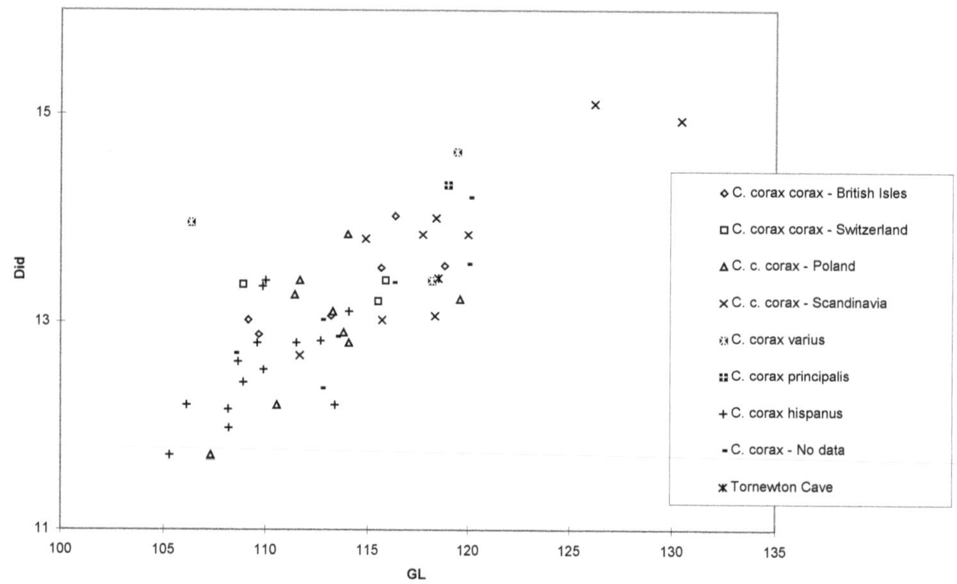

Figure 10.31. Scattergram of the ulna from Tornewton Cave with modern European *Corvus corax* populations. Ulnar greatest length (GL) plotted against distal diagonal breadth (Did).

The humeri are within or close to the modern British size distribution of ravens measured, although only one specimen was complete enough to allow two measurements to be taken and hence plotted (Figure 10.30). The ulna once again is inside the distribution of modern British birds (Figure 10.31), although relatively robust proximally (Figure 10.26).

France - Weischelian

La Grotte d'Isturitz, France

The cave locality of Isturitz is in the French Pyrenees and therefore on the northern edge of Iberia and hence on the boundary between *C. c. corax* and *C. c. hispanus* today. The site is 150 m above sea level, just north of the French border. There are three main chambers to the cave, the North Chamber, the South Chamber and the Large Chamber. The site is best known for the Upper Palaeolithic musical instruments, pierced bird ulnae which represent flutes.

The material examined in the Université Claude Bernard in Lyon was unlabelled except for the fact that the bones came from Isturitz, therefore detailed provenancing is impossible. However, the deposits in the main are late Pleistocene with archaeological industries which span the Mousterian to the Magdalenian and Azilian (Davies, pers. comm.).

Table 10.14. *Corvus corax* fossils from Isturitz with their measurements.

Specimen No.	Skeletal element	Measurements				
		GL	Bp	Bd	SC	
No no.	L humerus	92.34	25.6	23.4	8.72	
	L. humerus	100.86	28.00	-	9.3	
		GL	Bp	Did		
No no.	R. carpometacarpus	70.02	16.98	16.82		
No no.	L. carpometacarpus	68.88	17.44	16.76		
		GL	Bp	Dp	Bd	SC
No no.	L. femur	67.74	15.12	8.3	14.5	6.4
		GL	Dip	Bd	Dd	
No no.	R. tibiotarsus	112.66	18.18	11.48	10.3	
		GL	Bp	Bd	SC	
No no.	L. tarsometarsus	64.62	12.04	8.86	5.1	
	R. tarsometarsus	64.28	-	-	4.86	
		GL	Lm	BF	SC	
No no.	R. coracoid	56.64	52.92	15.62	4.54	

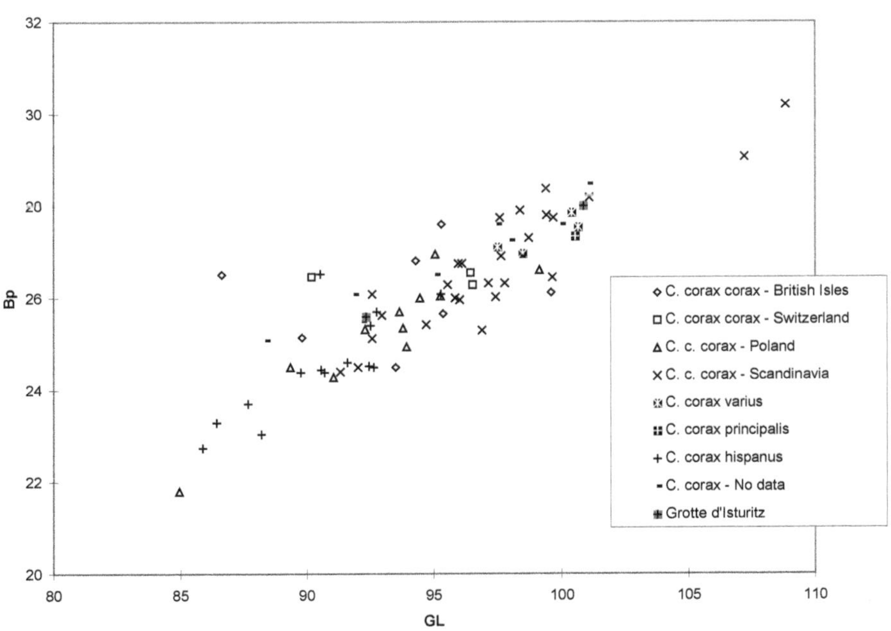

Figure 10.32. *Corvus corax* humeri from the Grotte d'Isturitz plotted with modern European samples. Greatest lengths (GL) plotted against proximal breadths (Bp).

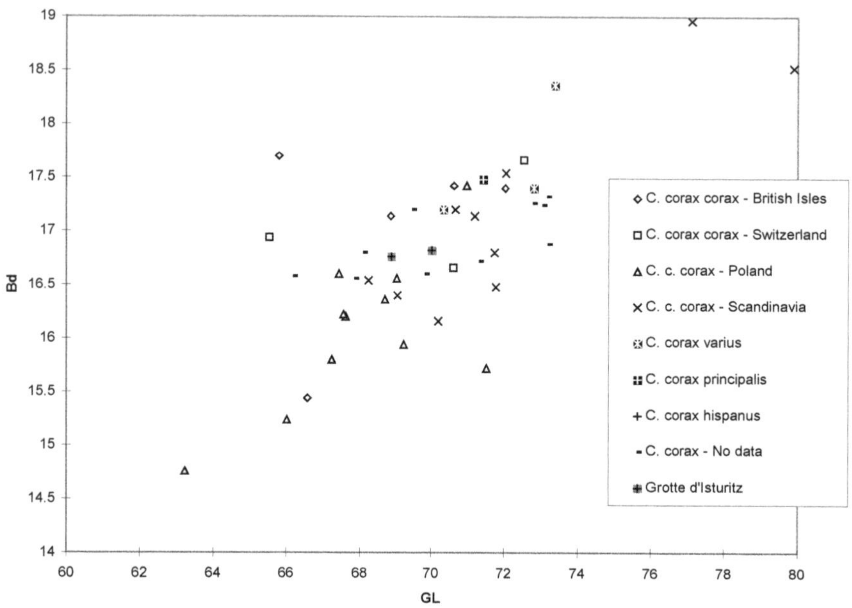

Figure 10.33. *Corvus corax* carpometacarpi from the Grotte d'Isturitz plotted with modern European samples. Greatest lengths (GL) plotted against distal breadths (Bd).

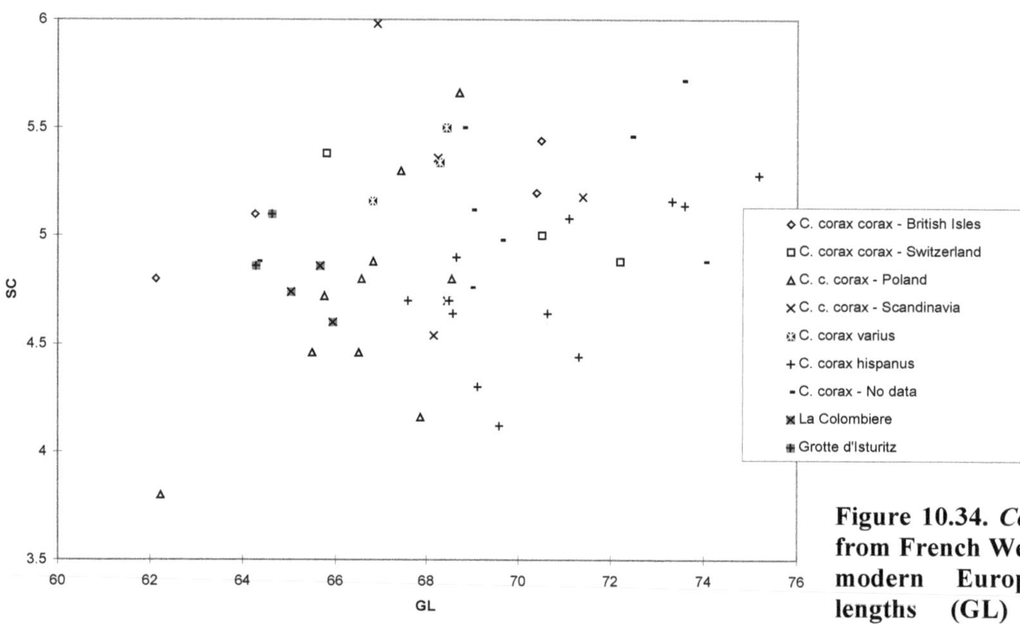

Figure 10.34. *Corvus corax* tarsometatarsi from French Weischelian sites plotted with modern European samples. Greatest lengths (GL) plotted against shaft breadths (SC).

The two humeri comprise one which is large, coinciding with the larger Scandinavian and Icelandic modern specimens and another which is more towards the middle of the modern European size range (Figure 10. 32). Both the carpometacarpi meanwhile fall in the overlap zone in the middle of the present size distribution of ravens (Figure 10. 33).

The tarsometatarsi from the Grotte d'Isturitz are relatively short and robust when compared to those of most modern European populations. They most resemble those from Britain today (Figure 10.34).

Given that Isturitz is in the Pyrenees and in close proximity to the modern Iberian subspecies *C. c. hispanus* with its long and hence gracile legs these tarsometatarsi are of interest in elucidating the Pleistocene patterns of morphology in ravens.

L' Abri de La Colombière a Neuville sur Ain (Ain), France

L' Abri de La Colombière is described in detail in Chapter 10 on *Lagopus*.

Table 10.15. *Corvus corax* fossils from La Colombière with their measurements.

Specimen No.	Skeletal element	Measurements			
		GL	Bp	Bd	SC
	R. tarsometatarsus	65.94	10.5	-	4.6
	L. tarsometatarsus	65.04	-	9.42	4.74
	L. tarsometatarsus	65.66	13.14	9.66	4.86

The three tarsometatarsi from La Colombière are similar in size and proportions, although slightly less robust than those from La Grotte d' Isturitz (Figure 10.34). The modern specimens which plot closest to them, come from Poland.

France - Middle Pleistocene

La Fage, France

The cave (Aven 1) of La Fage, at Noailles in Corrèze, is a swallow hole which opens up onto the Martel plateau. The swallow hole was filled with sediments containing many small vertebrates, including birds, as well as large mammals.

The sediments are thought to be referable to the "Rissian" Alpine stage of the Pleistocene, by Chaline (1975), on the basis of the rodent assemblage recovered. The original assignation of the site to the "Würmian" was contradicted by the presence of *Pliomys lenki* and *Allocricetus bursae*. Furthermore, comparing the rodents from La Fage with the post "Mindel" sites of Orgnac 3, Caune de l'Arago, Lazaret, Aldène and Saint-Estève Janson, has confirmed that age. The reason Chaline gave for this, was the lack of *Mimomys* and *Allophaiomys,* the respective ancestors of *Arvicola* and *Microtus* and *Pitymys*, in the La Fage fauna. This reasoning is now invalid because the *Mimomys* - *Arvicola* transition is now thought to have occurred before the Mindel (Anglian / Elsterian) Glaciation (van Kolfschoten, 1996) and the latest date attributed to *Allophaiomys* greatly exceeds the age of the Mindel Glaciation (Lister, 1998), rendering its absence unsurprising. The advanced morphology of *Pliomys lenki* at La Fage, *P. l. relictus*, together with the absence of *P. episcopalis*, apparently supports a post "Mindel-Riss" age, as does the advanced form of *Allocricetus bursae, A. b. correzensis* Chaline (1975).

The age of the La Fage deposits has been given as "Rissian" of the Alpine scheme, although this is difficult to correlate with the chronologies in North Western Europe (Chapter 5). It is perhaps safest to consider the fauna to be post Anglian / Elsterian and pre Ipswichian / Eemian in age, or upper Middle Pleistocene in its broadest sense. Whilst the site's stratigraphy has been divided into a number of units (Couche 5 - 2), and the deposits are of a considerable depth, the deposits appear to have been interpreted as having accumulated relatively rapidly and hence do not span more than one stage of the Alpine scheme. However, a diagram in the Chaline paper shows the Rissian as a "complex" rather than a simple stage, and La Fage is seen to cover more than one climatic section of that "complex". This is not important for the present study as most raven fossils studied here are from Couche (Layer) 5, unless they possessed no provenance data. Couche 5 is the second cold phase within the scheme Chaline presents.

The palaeoenvironmental information derived from faunal elements of Couche 5 all indicate a boreal fauna. Both the collared lemming and the Norway lemming are present although in small numbers, indicating the presence of a cold steppe habitat possibly at some distance from the cave. Confirmation of this is given by the occurrence of both *Microtus gregalis* and *Ochotona pusilla* in the assemblage.

Table 10.16. Measurements of a sample of raven fossils from La Fage.

Measurement	N	Min Max	Mean	Stand. dev.
Hum GL	22	81.32 - 93.4	86.89	2.95
Hum Bp	12	21.92 - 25.7	23.72	1.16
Hum Bd	18	20.04 - 22.04	21.23	0.7
Hum SC	22	7.26 - 8.9	7.96	0.44
Uln GL	22	96.56 - 109.92	103.89	3.42
Uln Bp	19	12.08 - 13.64	12.98	0.45
Uln Dip	18	13.02 - 15.56	14.63	0.67
Uln Did	22	10.82 - 12.8	12.1	0.5
Uln SC	22	4 - 6.88	5.67	0.54

Table 10.16 (contd.) Measurements of a sample of raven fossils from La Fage.

Rad GL	1	-	97.1	-
Rad Bp	1	-	5.86	-
Rad Bd	1	-	7.41	-
Cpt GL	22	58.72 - 66.98	62.3	1.97
Cpt Bp	22	14.3 - 16.88	15.13	0.61
Cpt Did	21	13.7 - 16.4	14.92	0.62
Fem GL	22	60.02 - 66.12	63.7	1.84
Fem Bp	21	12.24 - 15.02	13.83	0.62
Fem Dp	20	7.08 - 8.54	7.87	0.35
Fem Bd	21	9.08 - 14.2	13.38	1.03
Fem SC	22	5.2 - 6.98	5.98	0.27
Tib GL	20	98.24 - 109.82	105.5	3.33
Tib Dip	18	15.68 - 18.54	17.67	0.8
Tib Bd	19	9.68 - 11.68	10.74	0.58
Tib Dd	17	9.42 - 11.22	10.13	0.39
Tmt GL	27	60.18 - 67.52	63.71	2.21
Tmt Bp	25	11 - 12.66	11.89	0.5
Tmt Bd	23	8.22 - 9.44	8.8	0.32
Tmt SC	27	4.18 - 4.98	4.59	0.22
Cor GL	13	52.8 - 58.8	55.32	1.66
Cor Lm	21	48.2 - 54.582	51.44	1.93
Cor BF	12	14.22 - 17.98	15.61	0.98
Cor SC	21	3.08 - 4.68	4.16	0.35

The raven material at La Fage has been described as a new species by Mourer-Chauviré (1975a). However, as mentioned above the assemblage from that site was compared with modern specimens within a relatively limited geographic range. The museums of Paris, Geneva, Budapest, London and Lyon are unlikely to hold specimens from localities much outside their national boundaries. This means that the fossils were mainly compared with modern skeletons of the nominate *Corvus corax corax*. Clearly, further comparative work was necessary. Material from the La Fage collection was measured at the Museum d'Histoire Naturelle of Lyon and compared with the larger spread of data from my metric survey. The results of this comparison can be seen in the scatters (Figures 10.35 - 10. 41) and the results of a statistical comparison are laid out in Table 10.20.

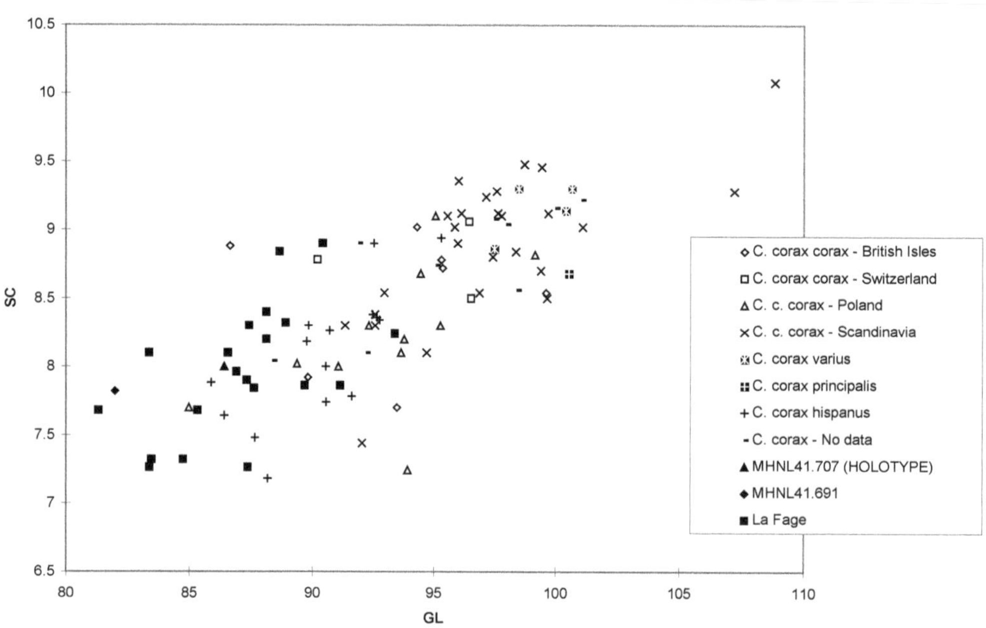

Figure 11.35. Scattergram of the humeri of large corvids from La Fage assigned to *Corvus antecorax* with modern European *Corvus corax* populations. Humeral greatest length (GL) plotted against shaft breadth (SC).

The La Fage humeri are more robust than modern populations. Their breadths are similar to the modern population measured from Spain *C. c. hispanus* and Poland, although their lengths are on average shorter (Figure 10.35). The ulnae, carpometacarpi, coracoids, femora and tibiotarsi on the other hand, are simply smaller (Figures 10.36 - 10.40). Figure 10.41 shows that the tarsometatarsi from La Fage are robust, although more noticeably so than the humeri, because of the length of the tarsi of the similarly sized modern Spanish subspecies. The La Fage tarsometatarsi do not overlap on the scatter with any of the Spanish birds.

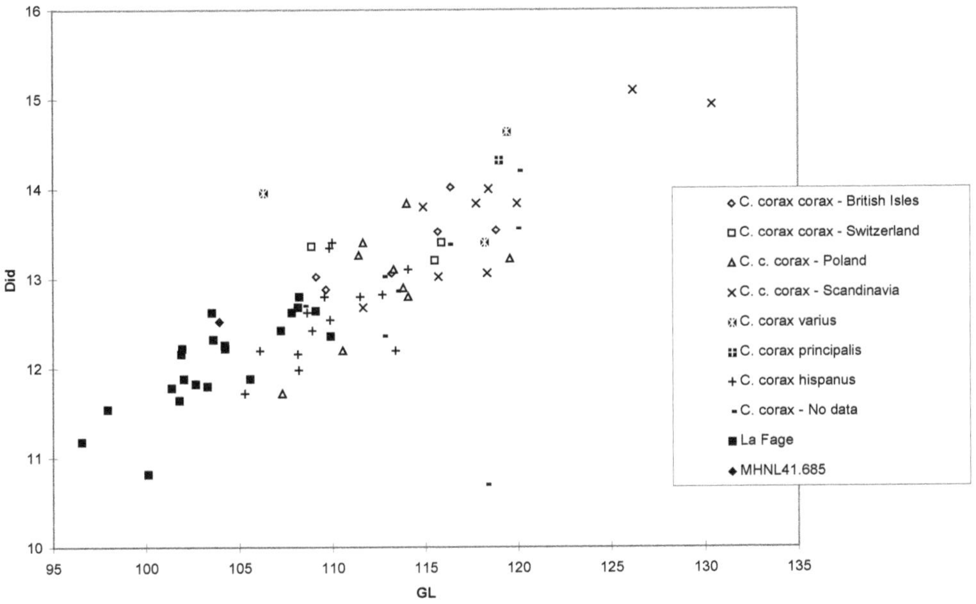

Figure 10.36. Scattergram of the ulnae of large corvids from La Fage assigned to *Corvus antecorax* with modern European *Corvus corax* populations. Ulnar greatest length (GL) plotted against distal diagonal depth (Did).

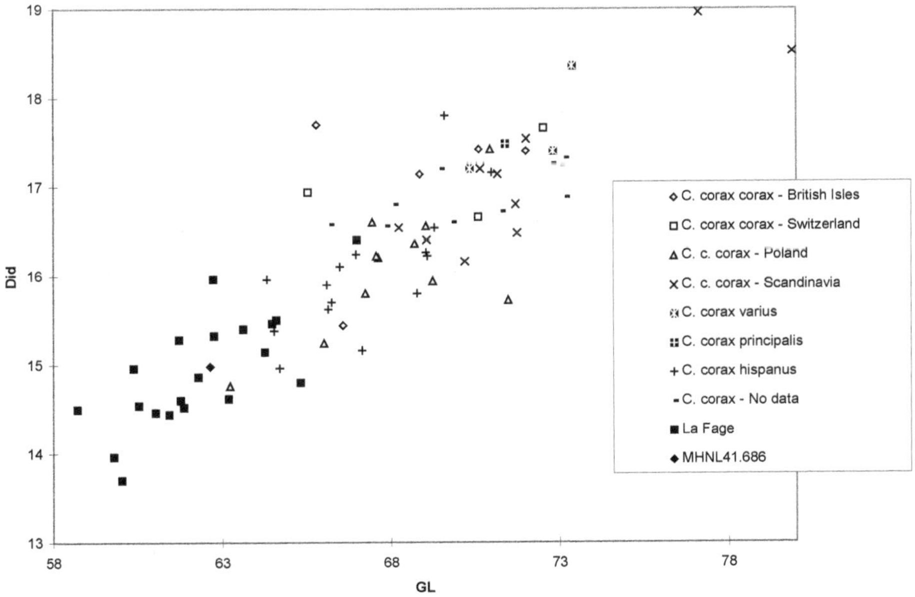

Figure 10.37. Scattergram of the carpometacarpi of large corvids from La Fage assigned to *Corvus antecorax* with modern European *Corvus corax* populations. Carpometacarpal greatest length (GL) plotted against distal diagonal depth (Did).

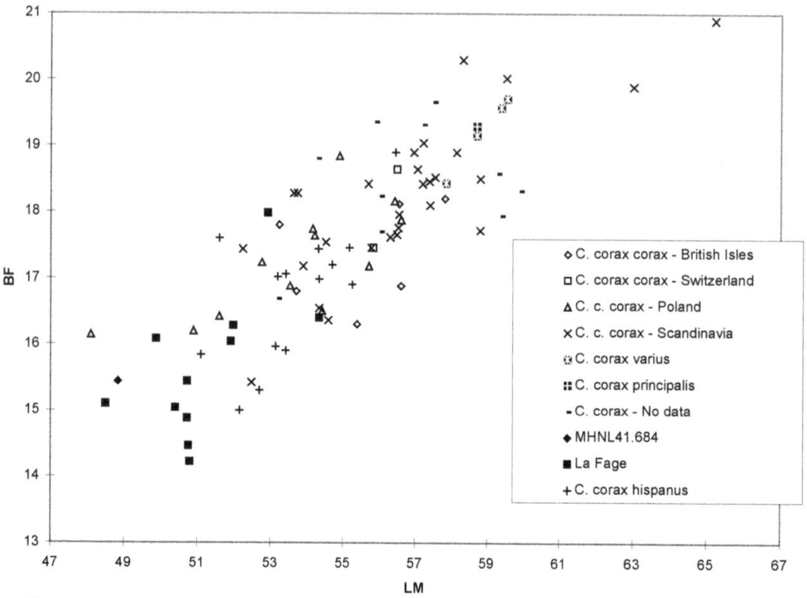

Figure 10.38. Scattergram of the coracoids of large corvids from La Fage assigned to *Corvus antecorax* with modern European *Corvus corax* populations. Coracoideal medial length (LM) plotted against the breadth of the basis articularis sternalis (BF).

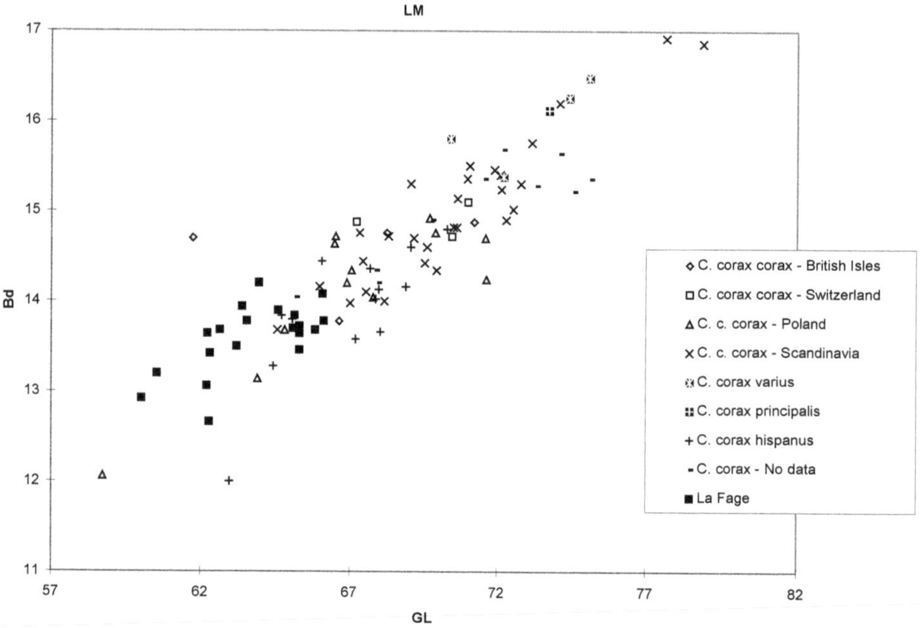

Figure 11.39. Scattergram of the femora of large corvids from La Fage assigned to *Corvus antecorax* with modern European *Corvus corax* populations. Femoral greatest length (GL) plotted against distal breadth (Bd).

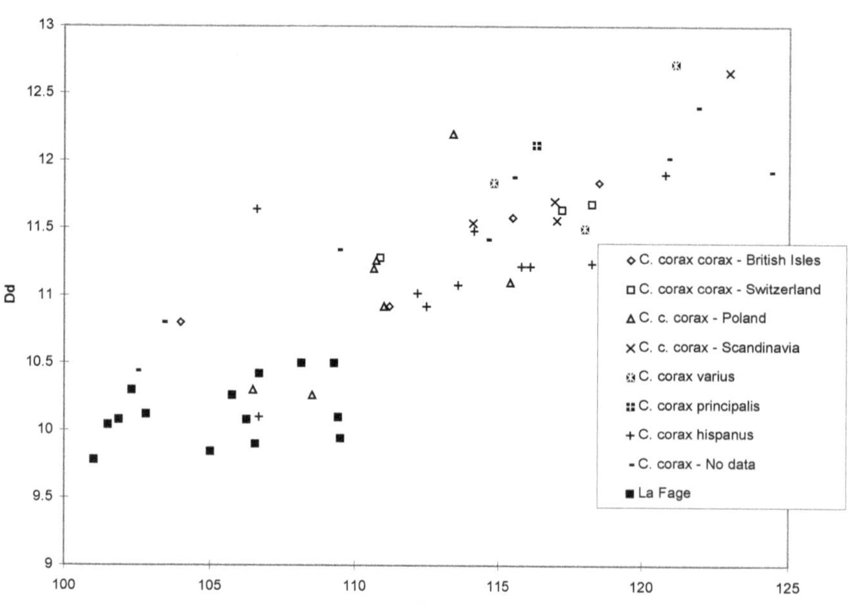

Figure 11.40. Scattergram of the tibiotarsi of large corvids from La Fage assigned to *Corvus antecorax* with modern European *Corvus corax* populations. Tibiotarsal greatest length (GL) plotted against distal depth (Dd).

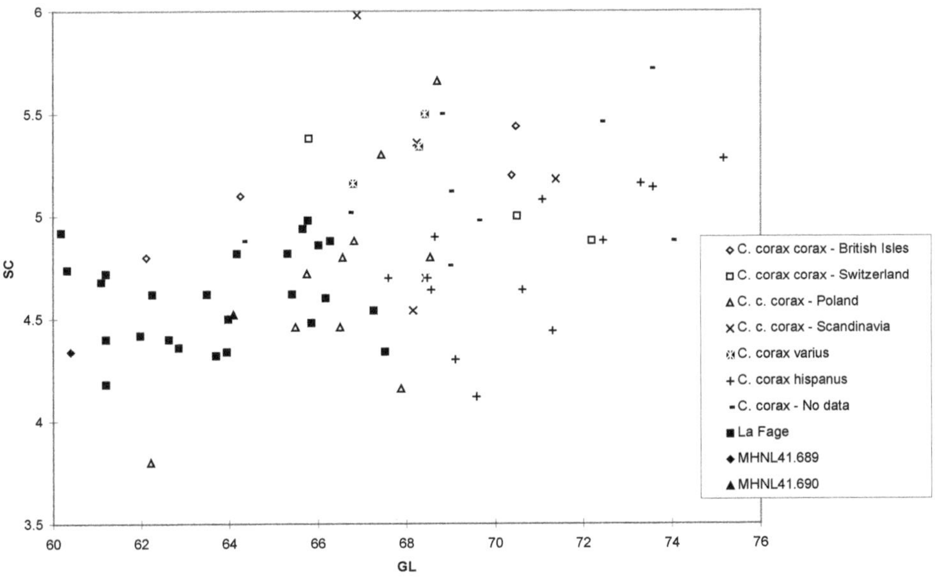

Figure 10.41. Scattergram of the tarsometatarsi of large corvids from La Fage assigned to *Corvus antecorax* with modern European *Corvus corax* populations. Tarsometatarsal greatest length (GL) plotted against shaft breadth (SC).

The diagnosis of *C. antecorax* was based on size. The species was said to be intermediary in size between *C. corax*, the extant raven, and *C. praecorax*, its supposed Pliocene ancestor. *C. praecorax* in turn was described as a quarter smaller than the modern raven. This means that the size was beginning to approach that of the next smallest modern corvids, the carrion/hooded crows (*Corvus corone corone* / *C. c. cornix*) and rooks (*Corvus frugilegus*). Therefore, it was felt that all of these taxa, together with *C. pliocaenus janossyi*, which is also in this general size range, should be considered during the evaluation of the corvids from La Fage named *C. antecorax*.

Figures 10.42 to 10.49 show regression lines transposed onto scatterplots of the various relevant taxa. These are *Corvus corax*, *C. corone corone*, *C. c. cornix*, *C. frugilegus* and the material from La Fage named *C. antecorax*. Below each of the graphs are the R^2 values for the relationships between two variables plotted for each skeletal element measured. In addition to the plots of the modern corvid samples and that from La Fage, the mean values quoted by Mourer-Chauviré (1975a) for *C. praecorax* and *C. pliocaenus janossyi* were plotted on the relevant scatters.

The R^2 Values for all the different samples of modern taxa are lower than that for the La Fage corvids. More importantly, the La Fage sample was the one to give the highest R^2 values to signifying positive relationships between the variables plotted. This implies that proportions, more than size, of individuals is tightly controlled by intra-populational factors. This may be because the La Fage corvids ("*C. antecorax*") are a more genotypically homogenous sample than the others. Alternatively, it may mean that natural selection was exerting greater than usual pressure on the La Fage population to have very specific body proportions. It was not unexpected, that while generally high, the R^2 values for the La Fage corvid samples are lowest for the leg elements. This is particularly true of the tarsometatarsus, which has an R^2 values ($R^2 = 0.0419$) even lower than the modern samples of other taxa. Intra-population metric-shape variation tends to be highest in the tarsometatarsus.

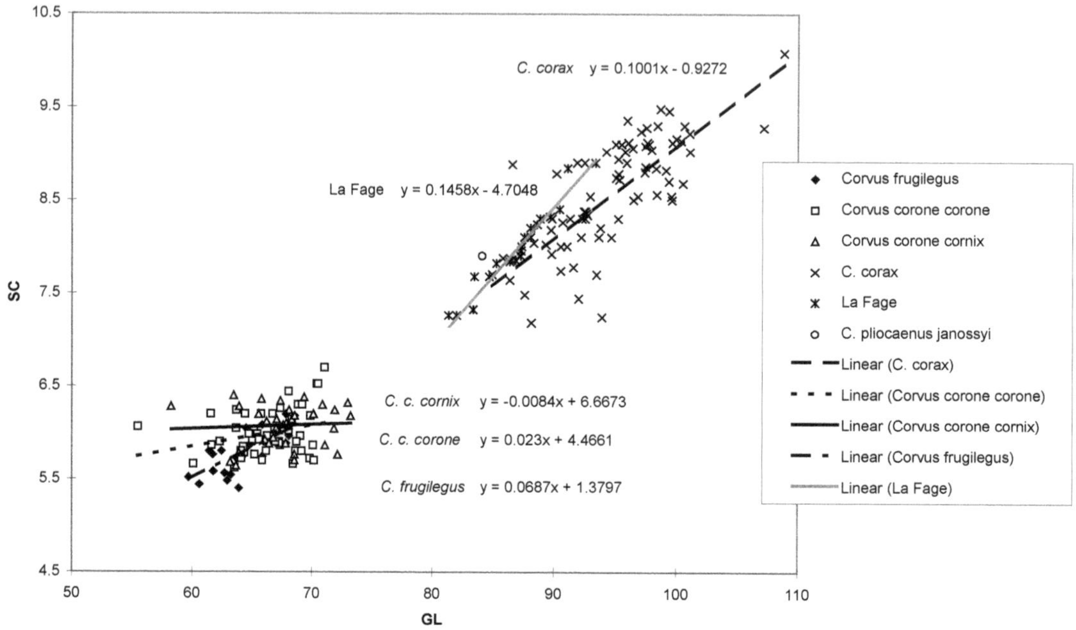

Figure 10.42. Scattergram with regression lines of humeri of the La Fage corvids together with all other large species of the genus in Europe. Humeral greatest length (GL) plotted against shaft breadth (SC). R^2 Values: *C. frugilegus* 0.5813, *C. corone corone* 0.082, *C. corone cornix* 0.0172, *C. corax* 0.5447, La Fage 0.9474.

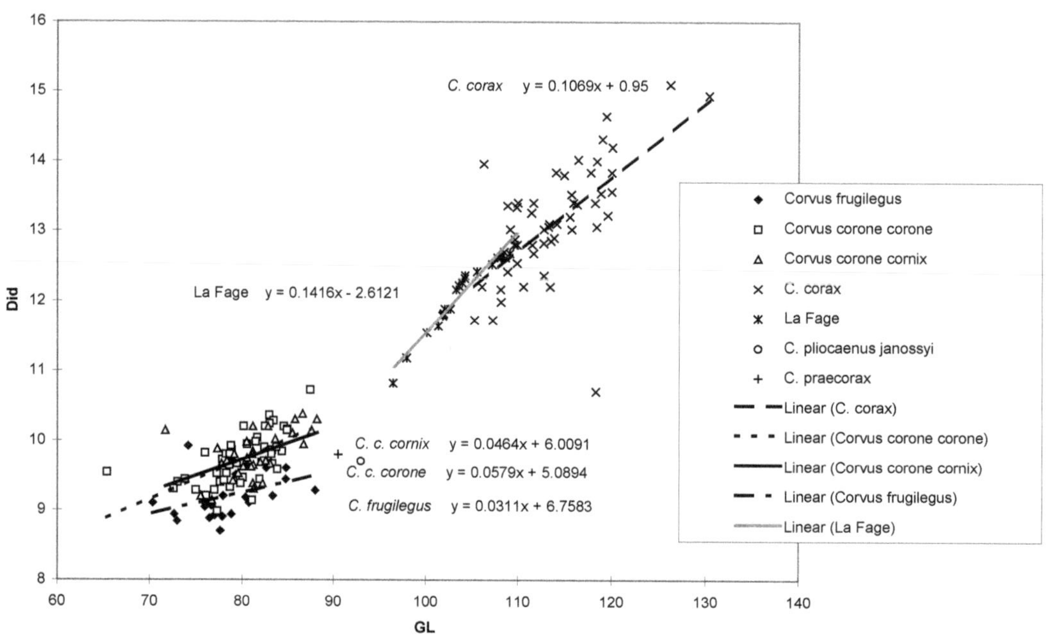

Figure 10.43. Scattergram with regression lines of ulnae of the La Fage corvids together with all other large species of the genus in Europe. Ulnar greatest length (GL) plotted against distal diagonal depth (Did). R^2 Values: *C. frugilegus* 0.1821, *C. corone corone* 0.3769, *C. corone cornix* 0.2568, *C. corax* 0.4124, La Fage 0.9412.

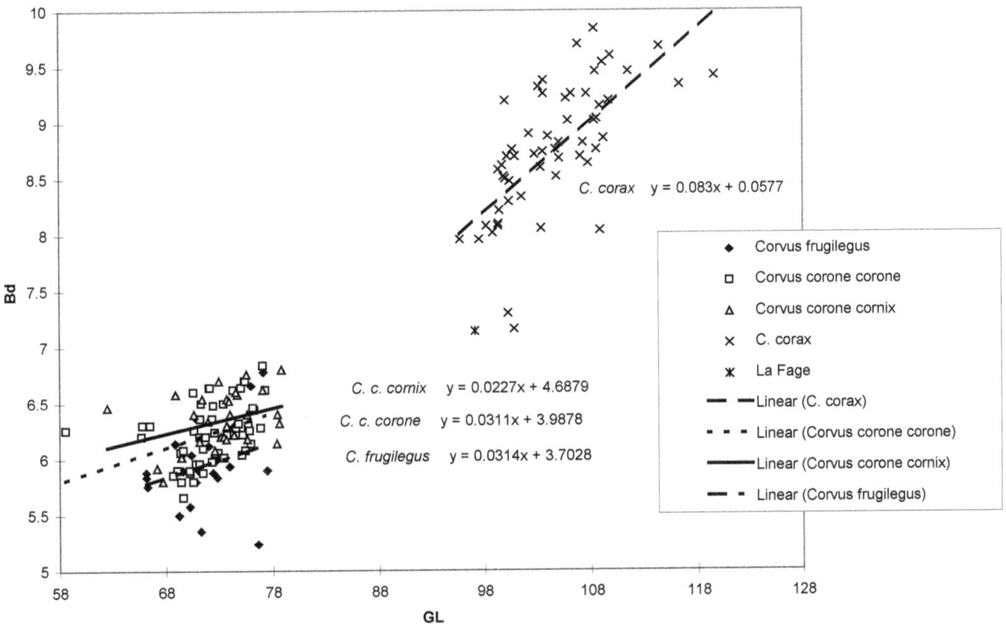

Figure 10.44. Scattergram with regression lines of radii of the European large corvids and one radius from La Fage. Radial greatest length (GL) plotted against distal breadth (Bd). R^2 Values: *C. frugilegus* 0.0864, *C. corone corone* 0.1446, *C. corone cornix* 0.1096, *C. corax* 0.4324.

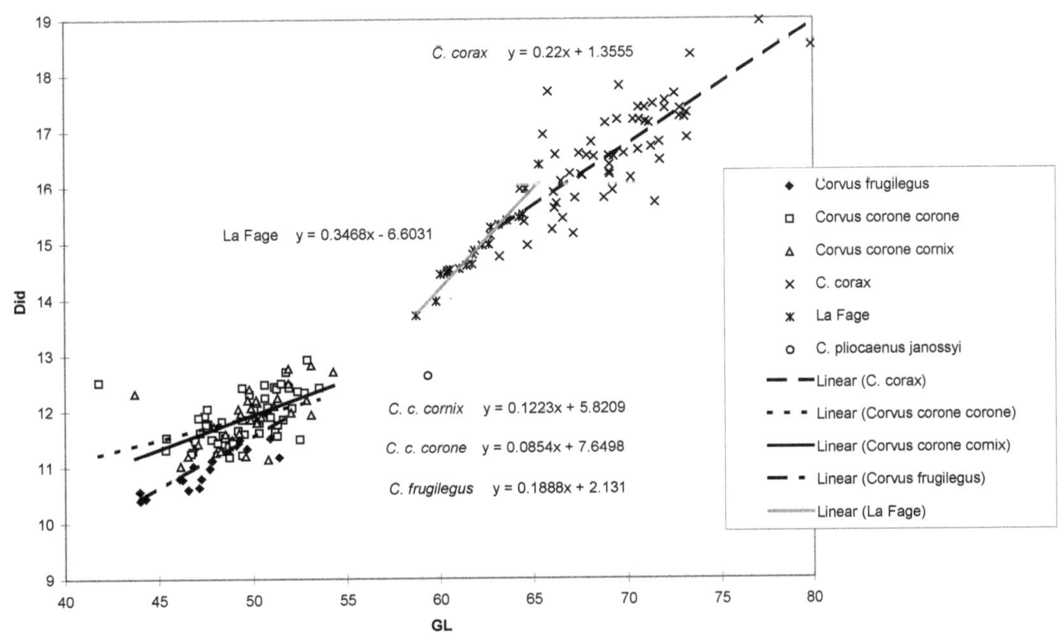

Figure 10.45. Scattergram with regression lines of carpometacarpi of the La Fage corvids together with all other large species of the genus in Europe. Carpometacarpal greatest length (GL) plotted against distal diagonal depth (Did). R^2 Values: *C. frugilegus* 0.6832, *C. corone corone* 0.1999, *C. corone cornix* 0.3165, *C. corax* 0.5916, La Fage 0.9391.

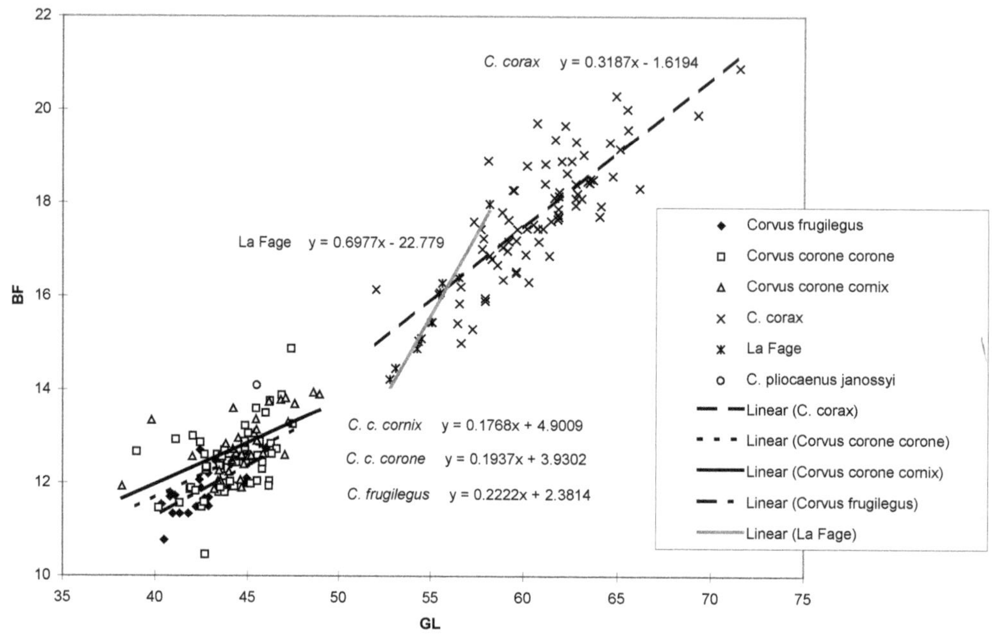

Figure 10.46. Scattergram with regression lines of coracoids of the La Fage corvids together with all other large species of the genus in Europe. Coracoideal greatest length plotted against the breadth of the basis articularis sternalis (BF). R^2 Values: *C. frugilegus* 0.624, *C. corone corone* 0.2545, *C. corone cornix* 0.3583, *C. corax* 0.4987, La Fage 0.9652.

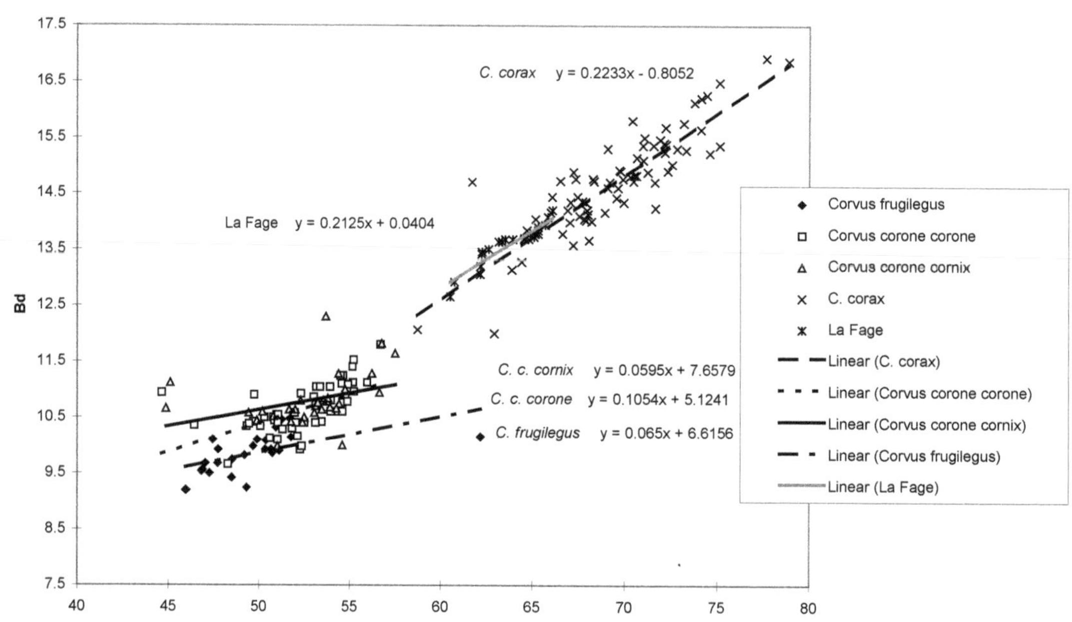

Figure 10.47. Scattergram with regression lines of femora of the La Fage corvids together with all other large species of the genus in Europe. Femoral greatest length (GL) plotted against distal depth breadth (Bd). R^2 Values: *C. frugilegus* 0.283, *C. corone corone* 0.0015, *C. corone cornix* 0.1576, *C. corax* 0.1576, La Fage 0.8829.

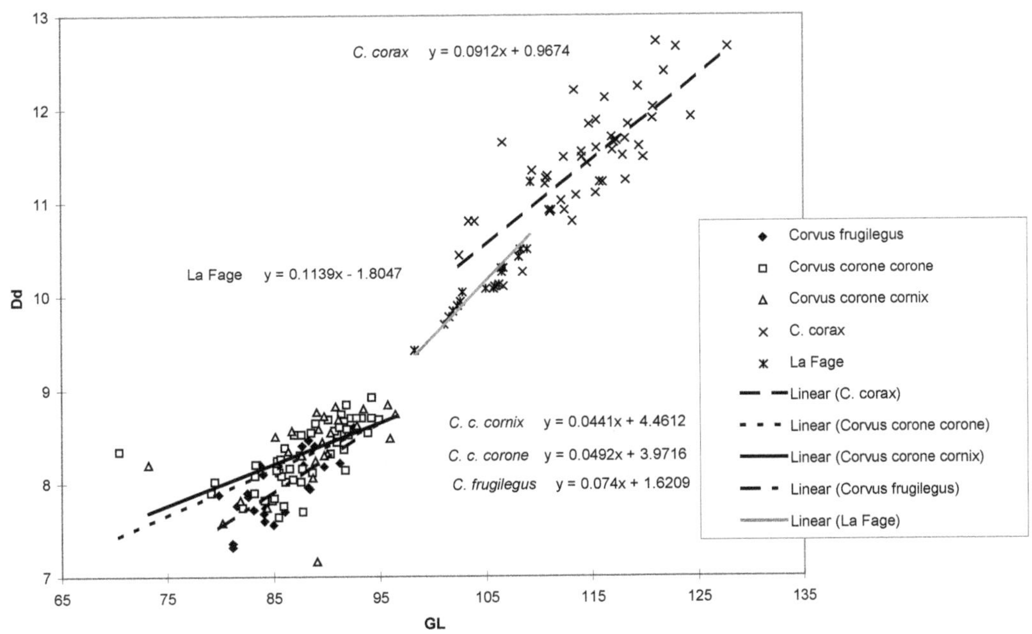

Figure 10.48. Scattergram with regression lines of tibiotarsi of the La Fage corvids together with all other large species of the genus in Europe. Tibiotarsal greatest length (GL) plotted against disatl depth (Dd). R^2 Values: *C. frugilegus* 0.5472, *C. corone corone* 0.4125, *C. corone cornix* 0.2816, *C. corax* 0.0755, La Fage 0.8163.

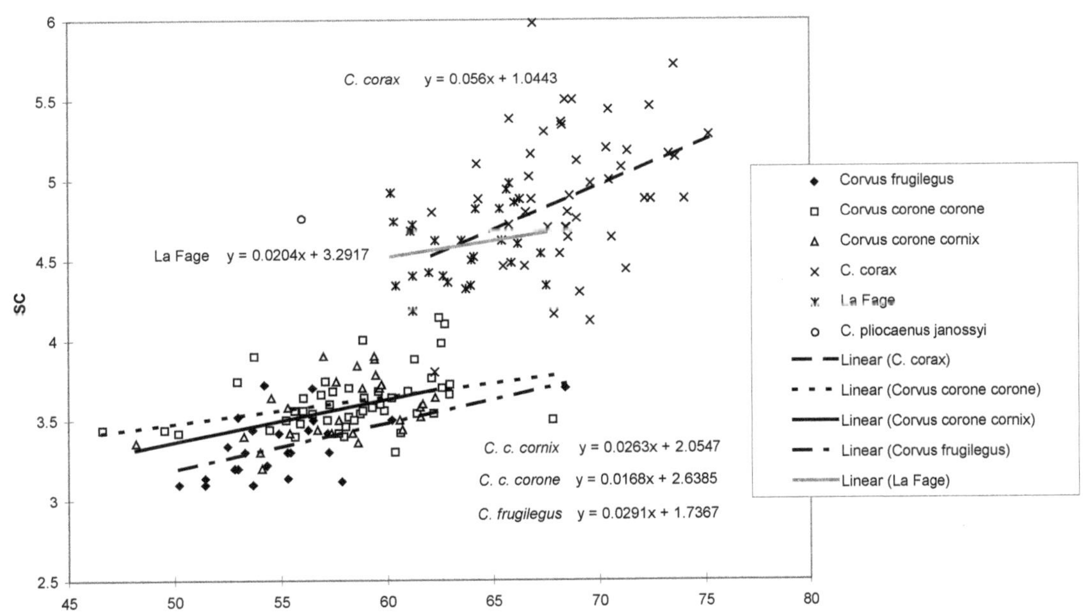

Figure 10.49. Scattergram with regression lines of tarsometatarsi of the La Fage corvids together with all other large species of the genus in Europe and *C. pliocaenus janossyi.* Tarsometatarsal length (GL) plotted against shaft breadth (SC). R^2 Values: *C. frugilegus* 0.1869, *C. corone corone* 0.1207, *C. corone cornix* 0.1869, *C. corax* 0.1869, La Fage 0.0419.

In summary, the ravens from La Fage named *C. antecorax* may be said to be small in relation to the modern populations examined in the present study. In limb proportions they were similar to the nominate subspecies *C. c. corax*.

Iberia - Weischelian

Ibex Cave, Gibraltar

Ibex Cave was excavated recently by a team including Chris Stringer and Andy Currant from the Natural History Museum and Nick Barton from Oxford Brooks University. It is during this excavation that the raven fossils here described were recovered. Unfortunately the fossils were found in debris sand and hence not *in situ*. Despite this it appears that the bones belonged to a single individual and are probably Late Pleistocene in age (J. Cooper, in press).

Table 10.17. *Corvus corax* fossils from Ibex Cave with their measurements.

Specimen No.	Skeletal element	Measurements				
		GL	Bp	Dip	Did	SC
No. 162	Dist L. ulna	-	-	-	12.2	5.9
		GL	Bp	Bd		
No. 149	Dist. L. carpometacarpus	-	-	14.38		
		GL	Dip	Bd	Dd	
No. 164	Dist. R. tibiotarsus	-	-	10.4	10.48	
		GL	Bp	Bd	SC	
No. 132	Dist. R. tarsometatarsus	-	-	8.36	-	

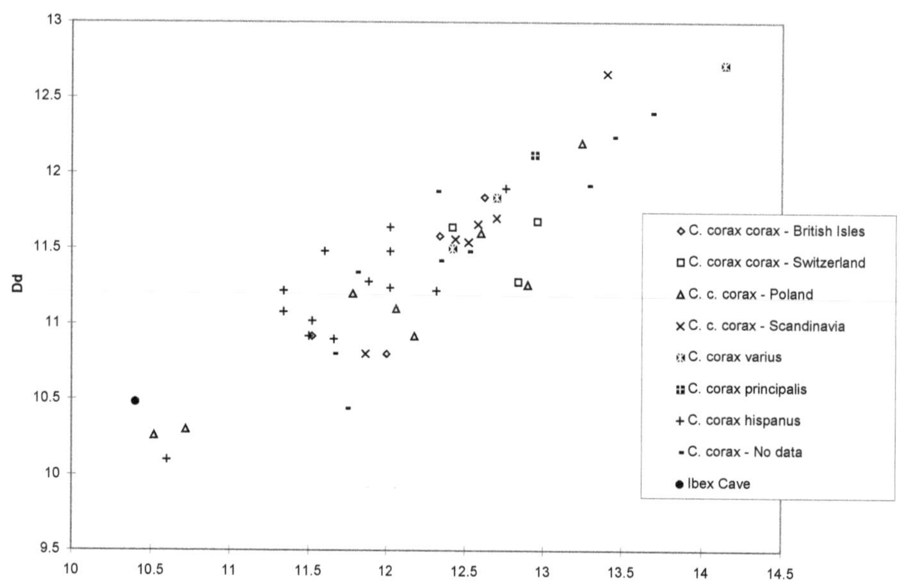

Figure 10.50. *Corvus corax* distal tibiotarsus from Ibex Cave plotted with modern European samples. Distal breadth (Bd) plotted against distal depth (Dd).

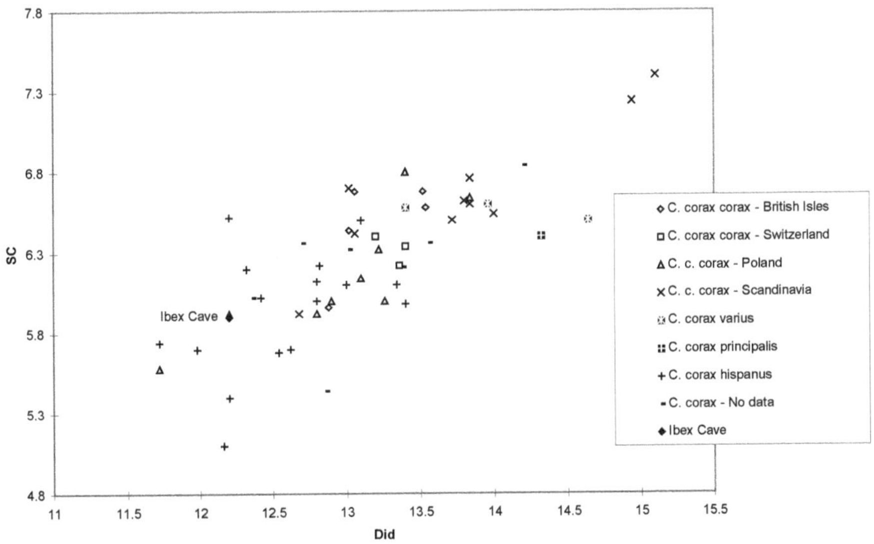

Figure 10.51. *Corvus corax* distal ulna from Ibex Cave plotted with modern European samples. Distal diagonal depth (Did) plotted against shaft breadth (SC).

The Ibex Cave raven fossils proved to be small in keeping with the dimensions of modern birds in the area today (Figures 10.50 and 10.51). Unfortunalely, the other characteristic of modern Iberian ravens, the relatively long tarsometatarsi, could not be investigated for the fossils because the only tarsometatarsus was incomplete.

10.9. Conclusions

10.9.1. What is the geographical variation in the osteology of modern Western Palaearctic ravens and what is the explanation for this variation?

Size variation has been recognised as being a significant aspect of the variation within modern ravens (Cramp and Perrins, 1994). Cramp and Perrins (1994) state that there are three main size groups, composing a large *principalis* and *tibetanus* group from North America and Greenland, and two other groups from the mountains of Central Asia respectively. Next there is a group of small ravens including both *tingitanus* from North Africa and *canariensis* from the Canary Islands, and finally there is an intermediate group including all other ravens.

The present study has confirmed that there is a strong geographical component to the size of ravens. Broadly speaking, the more northern the distribution of raven subspecies or populations, the larger they are. This therefore appear to broadly conform to Bergmann's Rule. The material which was available for study has further elucidated the degree of variation that can be present within subspecies, such as the widespread *C. c. corax*. The small size of the Polish population of *C. c. corax*, contravenes Bergmann's rule and is a new observation. The reason for this may lie in a local factor, which for the present remains obscure.

The most noticeable metric-shape variable exhibited by the raven's osteology today is that of the relative tarsometatarsus length (Table 10.18). The phenomenon is documented by Cramp and Perrins (1994), although in this instance it is inferred from the relative tarsus to wing length percentages (Table 10.19). While the data from Cramp and Perrins (1994) is less accurate, being based on ornithological measurements, it never-the-less allows for a more thorough treatment of the subject as skeletal material was not seen during the present study of all subspecies in the Palaearctic. Both sets of data suggest that there are two groups of ravens, those with relatively short tarsometatarsi (with a value around 70% in Table 10.18) and those with relatively long tarsometatarsi (with a value nearer 80% in Table 10.18). These two groups consist of more northerly and more southerly distributed populations respectively. A similar pattern is seen in Table 10.19, where again the more southerly subspecies are those with the longest tarsi, while the most northerly distributed populations possess the shortest tarsi. Table 10.19, being more geographically inclusive, shows that the tarsal proportions are clinal and do not readily split into two groups.

Table 10.18. Geographical variation in *Corvus corax* humerus to tarsometatarsus proportions, based means in Table 10.1.

Populations	Tarsometatarsus GL % of humerus GL
Scandinavia - *C. c. corax*	70.13%
Iceland - *C. c. varius*	68.33%
Poland - *C. c. corax*	72.08%
Britain - *C. c. corax*	71.45%
Switzerland - *C. c. corax*	73.63%
Spain - *C. c. hispanus*	78.19%
Balearic Islands - *C. c. corax**	78.71%
Canary Islands - *C. c. tingitanus**	80.44%
La Fage - *C. antecorax*	73.32%

* Data from Hernandez *et al.* (1993). The identity of the subspecies from these areas in this publication is suspect, as it is different from those designated for these islands in *Birds of the Western Palaearctic* (Cramp and Perrins, 1994). The Balaearic Islands should have *C. c. hispanus*, the Canary Islands *C. c. canariensis*.

Table 10.19. Geographical variation in *Corvus corax* wing to tarsus proportions, based on means from the literature (Cramp and Perrins, 1994).

Population	Tarsus length % of wing length
Northern and eastern N. America and Greenland - *C. c. principalis*	15 - 16%
Scandinavia, Britain, Pyrenees east to Yenisey basin - *C. c. corax*	15.9 - 16.6%
Iceland - *C. c. varius*	15.9 - 16.6%
Mountains of central Asia - *C. c. tibetanus*	15 - 16%
East Mediterranean and lower Volga east to central Asia - *C. c. laurenci*	16.9 - 17.4%
Spain - *C. c. hispanus*	16.9 - 17.4%
Canary Islands - *C. c.canariensis*	17.3%
North Africa - *C. c. tingitanus*	18.1%

An explanation for this pattern might be that the ecogeographic rule, Allen's Rule, is being adhered to. This rule, which was discussed in detail in Chapter 2 predicts that populations of endothermic animals will demonstrate a negative correlation between latitude and body extremity lengths. This correlation reflects heat conservation by more northern endotherms. In other words, a bird might be expected to have longer legs in the more southern part of its distribution, compared to its more northerly distributed populations. Other correlates of variations in limb lengths may be associated with locomotion, such as the speed with which an animal requires to run, or a response to the nature of the terrain over which it is moving (Chapter 2). While these two possibilities cannot be eliminated, it is difficult to verify exactly what is taking place. This is particularly true given the nature and circumstances of the locomotion of the raven, walking relatively slowly when feeding and if alarmed or in danger taking to the wing. Animals which may vary in leg length due to the requirements of speed are usually ones which run to escape from predators or alternatively chase after prey, neither of which applies to the raven. The second correlate of leg length, the nature of the terrain, may also be of little importance as much of the raven's movements are achieved on the wing, during which time they scout for food. When they alight, they usually do so on the ground, at some distance from the carcass they may wish to scavenge. They will then generally hop, using a few wing beats, towards their quarry (Cramp and Perrins, 1994). Little information appears to exist concerning the relative degree to which the southern and northern European ravens alight on trees, which could be significant. The prediction, based on birds which forage on different degrees of perch stability, is that birds using unstable perches will tend to have shorter tarsi, while those using more stable perches will have longer tarsi (Grant, 1969). Ground feeding birds will presumably have the longest tarsi in this model. The relative use of trees as opposed to rock faces for nesting is another possible explanation. Another hypothesis which has been considered for relative tarsus lengths is based on the work on growth of domestic fowl, where it was found that there was a high correlation between nasal length, including the bill, and the length of the tarsus of birds (Cock, 1969). This led Grant (1971) to consider the possibility that this may be the source of variation he observed, in tarsus length. He suggested that natural selection need not actually favour long tarsi, but if long bills were selected for then the tarsus might follow as a necessary genetic consequence, i.e. a pleiotropic effect. This hypothesis is invalid in this instance, as the relationship among raven populations is opposite to that found in fowl by Cock (1971). It would therefore appear that Allen's rule is the most plausible explanation for the metric-shape variation in the raven today.

If the explanation for relative tarsometatarsal length is to be found in the climatic rule, it may follow that the birds should follow a cline in relation to latitude. The present data, and indeed that in the literature (Cramp and Perrins, 1994), appear to confirm this to a certain extent, although there are problems due to a lack of data from the transitional area between the more southern long legged forms and those with shorter legs. The variation within the nominate subspecies, however, would appear to confirm the suggestion of a cline, as the birds appear to gradually decrease in relative tarsometatarsus length as one goes north. (Table 10.18).

10.9.2. What variation can be seen in the fossil record and are there any temporal and geographical patterns in that record which may be explained by the modern patterns?

It is possible that ravens in Medieval times in Britain were small in relation to the modern British population. In addition to a generally smaller size compared to modern British birds the Ossom's Eyrie ravens possesed narrow proximal tarsometatarsi, narrow distal ulnar diagonal depths and radial breadths as well as more robust coracoids. This is based on one site, Ossom's Eyrie, and the number of modern British specimens is relatively small (N=6), so the conclusion should be treated with caution.

The British Roman material is more diverse in origin, although the number of specimens is relatively small. There is no evidence for a difference between Roman ravens and present day British birds.

Certain skeletal elements from the British Iron Age, give an indication of smaller size than modern British birds, such as the ulnae and tibiotarsi from Danebury and the single humerus from Hanging Langford, whilst other bones appear to be consistent in size with today's ravens. The Danebury and Hanging Langford coracoidea appear to be relatively robust.

There are problems with the dating of some of the sites which may be from the last glaciation (Devensian) in Britain. Pin Hole Cave, Ravencliffe Cave and Happaway Cave are therefore unreliable. Their measurements, however, indicate that the ravens may have possessed robust tarsometatarsi (Pin Hole Cave and Happaway Cave), whilst other elements were possibly similar to modern British individuals. Tornewton Cave in Devon, whose fossils are better provenanced, include raven bones from both Oxygen Isotope Stage 5 and 2, although their small number prevent comparisons through time. They do show that in Oxygen Isotope Stage 2, the British ravens were probably similar in size to those in the area today.

In France, ravens from the last Glaciation (Weischelian) were the size of mid-European latitude (Britain and Switzerland) ravens today. The tarsometatarsi were relatively robust, reminiscent of the specimen from Happaway Cave in Britain. These tarsometatarsi may represent evidence for ravens with relatively short and stout tarsi in a cold period.

Earlier in the Pleistocene of France (late Middle Pleistocene), there was a population of ravens in the Corrèze region, which "came to rest" in the cave of La Fage and were subsequently named *Corvus antecorax* (Mourer-Chauviré, 1975a). The ravens are small, with isometrically proportioned limb bones (including their tarsometatarsi) in relation to the modern nominate subspecies *C. c. corax*. They have tarsometatarsi of similar length to those of the last Glaciation in France, although less robust. This presumably reflects a lower mean body mass for these birds. The climate at the time is difficult to judge accurately because of the apparent mixture of boreal and temperate mammals in the cave's fauna. It is possible that it was not as cold as the climate occupied by the ravens at La Colombière and Isturitz in the last glacial episode.

Finally, the few specimens examined from Ibex Cave in Gibraltar date to the last Glaciation and indicate that small individuals of this species inhabited the area during that time, as now. This, together with the larger birds to the North, during this period of generally colder temperatures, seems to indicate that clinal variation existed during the Pleistocene. However, without more accurate dates for the various fossils, allowing for close correlation with the climatic fluctuations of that period, this must remain speculation.

10.9.3. Is *Corvus antecorax* a genuine species or should it be considered conspecific with *Corvus corax* as claimed by Pierce Brodkorb (1978)?

As described above, *C. antecorax* was named by Mourer-Chauviré (1975a) on the basis of its intermediary size between *C. praecorax* and *Corvus corax*. No distinct non-metric morphological features were found. This diagnosis was challenged by Pierce Brodkorb and the form was dropped to a subspecific level *C. corax antecorax* (Brodkorb, 1978). Brodkorb did not explain his decision, although it seems likely that behind it lay an appreciation of the polytypic nature of the species *Corvus corax* and that size variation was a large component of that polytypy. The large amount of size variation within the species in Europe today has been confirmed by the present study and greater detail added to our knowledge of their osteological proportions. However, while there is some overlap, *Corvus antecorax* cannot be totally subsumed within the range of morphology found in *Corvus corax* today, although it is likely that if osteological material had been available for *C. c. canariensis*, the difference would have been eroded still further (Hernandez *et al.*, 1993). It could be argued that because *C. antecorax* falls outside the metrical range seen in *C. corax* today that this strengthens the case for *C. antecorax*'s validity as a species. However, such an assertion would require that the range of morphologies found in a species in the present, sets strict limits on the possible variation in that species in the past. This is clearly not supported by what we know of other species in the fossil record such as mammals like red deer (*Cervus elaphus*) and spotted hyaena (*Crocuta crocuta*) (Lister, 1987; Scott and Klein, 1989) and birds like the red / willow grouse *Lagopus lagopus* and ptarmigan *Lagopus mutus* (Stewart, 1999a, 1999b and Chapter 9).

An alternative way to tackle this problem is to see whether *Corvus antecorax* is more different in size and shape from extant raven subspecies than they are from each other. This is not supported by the present study, for while the La Fage material proved to be significantly smaller than extant subspecies including the smallest *C. corax hispanus* (Table 10.20), they do not add any more variation within the species than do some of the Scandinavian specimens. Table 10.20 gives the results of *t*-tests between two modern subpecies of raven (*C. corax hispanus* and *Corvus corax corax*) and each with *C. antecorax*. This demonstrates that the difference between *C. c. corax* and *C. c. hispanus* is of a similar order of magnitude to that between *C. c. hispanus* and *C. antecorax*. In Mourer-Chauviré's (1975a) analysis, where only *C. c. corax* is compared to *C. antecorax*, the case appears stronger for distinct identity than if a more wide-ranging comparison is made. The present analysis thus demonstrates that the La Fage fossils cannot be considered significantly more different from modern ravens, than different modern subspecies are from each other.

Furthermore, the La Fage fossils, although small, are merely isometrically different from the nominate subspecies *C. c. corax,* while *C. c. hispanus* exhibits a degree of metric-shape difference. The result is a relatively long tarsometatarsi for their body size. In her original diagnosis of the species, Mourer-Chauviré had access to only a single subspecies of the raven, the nominate, and she only examined samples of these from a limited range of latitudes (Mourer-Chauviré, 1975a). This did not allow her to fully appreciate the polytypic nature of the species and hence to assess whether the fossils from La Fage she called *C. antecorax*, merited distinct specific status.

The demonstration that the La Fage fossil ravens have a different metrical character than modern birds, is not sufficient reason for the erection of a new species. This is not an isolated instance where a species diagnosis has been based purely on size, as other examples exist in the literature such as *Mergus connectens* (Jánossy, 1972) (Chapter 1). This approach appears to be arbitrary because there are no reasons why in some circumstances size differences are considered to be subspecific, in others specific, and others still are not given any distinct taxonomic status. The practice of using size as a taxonomic tool has all but completely been abandoned amongst Quaternary mammalian palaeontologists, but has been slow to be accepted among avian workers. Another reason which no doubt lies behind the naming of the La Fage material is geological age, although this has not been expressed. Certainly this would appear to be the case with the two subspecies of *Lagopus* (*Lagopus lagopus noaillensis* and *L. mutus correzensis*) which Mourer-Chauviré named from the site. In this case, although different to modern populations, they do not differ from some later Pleistocene populations which apparently do not merit distinct names (Chapter 9).

There is evidence that *C. antecorax*, or at least similarly small ravens, occurred at other localities synchronous with La Fage, as well as in earlier times. This does not, however, give support to the validity of the species, as most of these sites are in the southern half of France, with one occurrence in Spain at Atapuerca (Mourer- Chauviré, 1975a and Sanchez Marco, 1987a, 1987b, 1992). We therefore, have very few ravens of the same ages in markedly differently geographical areas. This means that we know nothing of the geographical variation of ravens in the Middle Pleistocene. Second, even if there is a trend, as seems possible, towards all ravens becoming larger over time, that does not necessarily imply a species-level change. Equally, however, this argument cannot rule out the possibility that there was a species level event.

The systematic relationship between the fossils called *C. antecorax* and *C. corax* is difficult to clarify. The complete lack of morphological differences, beyond size, which distinguish the two taxa and other members of the genus, further signify that *C. antecorax* could even, in principle be more closely related and ancestral to *C. corone* and *C. frugilegus*, rather than to *C. corax*. The fundamental problem is the lack of distinctive non-metric traits which may serve as shared derived characters and which extend to corvid osteology generally. Despite this it may be more reasonable to assume until shown otherwise, that the fossils attributed to *C. antecorax* are ancestral to *C. corax* for two reasons. First, because it may be more parsimonious to attribute the larger *C. antecorax* to a phylogenetic position ancestral to *C. corax* and the sympatric (at La Fage), smaller *C. pliocaenus janossyi* to the ancestry of *C. corone* or *C. frugilegus* or both. Second, this may be supported by the possible continuity of *C. corax* sized fossils back to the late Middle Pleistocene, and possibly the Pliocene, which are likely to represent ancestral ravens. In other words, it is more probable that the order of sizes of these taxa has remained the same, than that they swapped over.

Table 10.20. *t*- tests between two modern subspecies of Corvus corax and the La Fage corvids named *C. antecorax* (Mourer-Chauviré, 1975).

	Corvus corax corax vs. "*C. antecorax*"	*C. c. hispanus* vs. "*C. antecorax*"	*Corvus corax corax* vs. *C. c. hispanus*
Hum GL	***	***	***
Hum Bp	***	*	***
Hum Bd	***	***	***
Hum SC	***	*	***
Uln GL	***	***	***
Uln Bp	***	***	**
Uln Dip	***	**	***
Uln Did	***	**	***
Uln SC	***	**	***
Rad GL	?	?	***
Rad Bp	?	?	**
Rad Bd	?	?	ns
Cpt GL	***	***	*
Cpt Bp	***	***	***
Cpt Did	***	***	*

Table 10.20 (contd.) *t*- tests between two modern subspecies of Corvus corax and the La Fage corvids named *C. antecorax* (Mourer-Chauviré, 1975).

Fem GL	***	***	*
Fem Bp	***	*	***
Fem Dp	***	***	**
Fem Bd	***	*	**
Fem SC	**	Ns	*
Tib GL	***	***	*
Tib Dip	***	**	**
Tib Bd	***	***	*
Tib Dd	***	***	*
Tmt GL	***	***	***
Tmt Bp	***	***	*
Tmt Bd	**	*	*
Tmt SC	**	*	*
Cor GL	***	***	**
Cor Lm	***	***	**
Cor BF	***	**	**
Cor SC	***	*	*
Scap GL	?	?	**
Scap Dic	?	?	***

Key: *** p< 0.001, ** p< 0.01, * p< 0.05, ns Not significant at the 5% level, question mark indicates that no test was possible due to inadequate sample size.

It is, therefore, best to consider the species *C. antecorax* as a junior synonym of *Corvus corax,* until shown otherwise. There seems little point in supporting a subspecific rank for the name as so little is known about the geographic variation of ravens at the time.

10.9.4. Is there any evidence in the fossil record for the timing of the evolution of *Corvus corax*?

As has been stated above, a number of species of large corvids have been described from the Plio-Pleistocene of Europe. These species are all purported to be intermediary in size between modern *Corvus corax* and *C. corone*. The relationships between these species are difficult to ascertain. It is tempting to stack them chronologically as a lineage, however, considering the general lack of morphological differences between present day corvid species, this ignores the conservative osteological nature of the group. The *Corvus* species today vary primarily in size. Indeed, it would appear that a solution to this problem is unlikely to be found as the identity of the fossils and their relationships to each other will no doubt always be marred by a lack of distinctive features. Even if more fossils were to be found, in large sample sizes and from diverse geological ages, it is unlikely that this situation would change. The best that can be hoped for, given a great many more samples, is for an investigation of the upper and lower limits in size and metric shape of *Corvus* fossils through time. This may then confirm certain overall trends, such as the supposition from our present state of knowledge that the very largest birds have appeared most recently.

10.9.5. Is there any evidence in the fossil record for the timing of the evolution of the modern subspecies of *Corvus corax*?

There is some suggestion from the fossil record, including material examined here, that clinal variation in the raven's size existed in the Pleistocene, certainly in the last glaciation. The smallest ravens from the last glaciation came from Spain: Ibex Cave (this study) and Cueva Nerja (Hernández Carrasquilla, 1995). Furthermore, those from Britain of that period were larger than in Spain and of the same order of size as the modern British population. However, the existence of a cline does not necessarily signify that the individual fossil populations belonged to the same subspecies as are present in those areas today. The relative sizes are presumably the product of local natural selective pressures which existed during the Late Pleistocene. These pressures include such factors as the relative climatic conditions responsible for the observations which led to the recognition of Bergmann's rule. It is interesting to note that despite the assumed colder climatic regime at the time there appears to be concordance in size between the populations in these areas through time. This could be argued as supporting the case for continuity of populations in these areas through time and that the climatic rule was less important in determining the birds' size and shape than their genetic origins.

The other cline seen in today's birds, that of relative tarsometatarsal length, from the limited evidence available (one specimen from Cueva Nerja: Hernández Carrasqilla, 1995) may imply that the populations in the Late Pleistocene were in fact different. The specimen discussed by Hernández Carrasquilla (1995) is short when compared to the modern Iberian raven *C. corax hispanus*. This led him to consider whether the specimen represented a relict form of *C. antecorax*, which are similarly small with relatively short tarsometatarsi. However, as with size, the length of tarsometatarsus is probably a plastic feature of ravens, which could rapidly change given the appropriate environmental conditions. This means that the short tarsometatarsus of the Cueva Nerja may well have evolved independently form those of *C. antecorax*. It will be interesting to examine more Late Pleistocene ravens from Iberia and elsewhere to further investigate this matter.

Chapter 11

The genus *Sturnus* (Starlings)

11.1. Introduction

The genus *Sturnus* is in the family Sturnidae of the Order Passeriformes. They are medium-sized birds and include three resident species in the Western Palaearctic. These are the common starling (*Sturnus vulgaris*), the spotless starling (*S. unicolor*) and the rosy pastor or rose-coloured starling (*S. roseus*). The common starling is the best known in the British Isles, being the only resident there, although rose-coloured starling has been known as a vagrant (Whitlock, 1953). The common starling (*Sturnus vulgaris*) is the most widespread member of the genus, having been introduced around the world to North America, Southern Africa, Australia and New Zealand (Feare, 1984). The rose-coloured starling normally resides in summer in Eastern Europe and winters in India, while the spotless starling is confined to Iberia, certain Mediterranean islands and North Africa.

Starlings have been recognised in the fossil record of both Britain and mainland Europe from deposits as old as Cromerian through to Holocene. All three modern European species have been identified, although problems probably exist with these identifications both as species and in some cases as genera. The difficulties lie in the morphological overlap among the species within the genus *Sturnus* as well as with members of similarly-sized passerine genera and especially *Turdus* (Ballman, 1978; Stewart, 1992).

Bearing in mind the identification difficulties, two examples of possible evolution in *Sturnus* have been recognised in the fossil record of the Western Palaearctic. The first is by Tchernov (1962) who mentioned the occurrence in Palestine of smaller members of the genus *Sturnus* in Levalloiso-Mousterian levels than occur in that region today. The other is Harrison (1980b) who described some large starling bones from Pleistocene deposits of Tornewton Cave, Devon, though he regarded these as being conspecific with the extant *S. vulgaris* present in that area today.

A number of questions regarding the genus *Sturnus* arise:

- What is the pattern of size and shape in modern members of the genus *Sturnus* both inter- and intraspecifically and what does the pattern signify?
- What is the pattern of size and shape in fossil *Sturnus* and what does the pattern signify?
- Are there any extinct *Sturnus* species in the fossil record?
- Can the present study of *Sturnus* in the Quaternary of Europe shed any light on the historical biogeography of the genus *Sturnus*?

11.2. Systematics

Family Sturnidae

There has been much debate regarding the relationships of the species and genera within the family Sturnidae. It appears that the taxon can be subdivided in different ways according to the criteria used. The family was at first split into 50 genera, 33 of which were monotypic. Amadon (1943), however, followed this by synonymising a number of these genera resulting in a total of 24. He stated that many of the genera had "become largely redundant equivalents of the specific names, rather than denoting groups of related species". Amadon and earlier workers based their classification on the study of skins collected by others and deposited in museums. Other studies have used biomolecular information (Sibley and Ahlquist, 1972) or details of the musculature and osteology of the jaw (Beecher, 1978) to divide up the family into genera. The different classifications arise because certain characteristics are shared derived traits while others are parallellisms (possibly for adaptive reasons) and opinions differ as to which are which.

Beecher's (1978) use of the feeding apparatus focuses on the varying degree of adaptation to a prying or spreading action of the bill that different species possess. This behaviour is reflected in the jaw musculature and osteology. The prying action allows the birds to push apart the substrate, by an upwards movement of the maxilla, to reach invertebrates within. In all other passerines this action is not possible due to the lack of development of the relevant muscles. The skull in these birds is also adapted to this behaviour by having more forward-pointing eye sockets. The adaptation is at its most highly developed in the common starling *Sturnus vulgaris*, and intermediates between this extreme and non prying taxa exist. The non prying taxa (e.g. *Aplonis* and *Lamprotornis*) are thought to be primitive and are arboreal, eat predominately fruit and also tend to live in more tropical situations. The prying genera feed mostly on the ground, probing for insects in short turf, and it is thought that this enabled them to spread to dryer, more temperate regions.

Genus *Sturnus* Linnaeus, 1758

As with the classification of the family Sturnidae, there is controversy over the relationships within the genus *Sturnus*. Debate exists as to how many species belong in the genus. Amadon (1956) has suggested 16 species while Beecher (1978) believed that there was justification in reviving genera which had previously been disbanded. Beecher left the genus *Sturnus* with only four species (*S. vulgaris*, *S. contra*, *S. unicolor* and *S. cineraceus*).

Beecher's classification has however not achieved the degree of acceptance as has that of Amadon which is most often used (e.g. in Feare, 1984 and Howard and Moore, 1991).

Sturnus is believed to have its origins in south-east Asia as seen by the greatest concentration of species of that genus which occur in the Oriental Region. Fourteen of the sixteen species spend at least part of their lives there. In addition to this, closely related genera such as *Acridotheres*, *Ampeliceps*, *Mino* and *Gracula* are Oriental. It appears that *Sturnus* spread out from that region both to the north into northern China and to the west into Europe and North Africa. The more northerly breeding members of the genus are migratory. Of the migratory species within *Sturnus*, *S. vulgaris* is the only one to have an east-west migratory route; all the others have a west-east or north-south route. The only starling to have completely lost contact with the east is the sedentary spotless starling (*S. unicolor*) (Feare, 1984).

***Sturnus vulgaris* Linnaeus, 1758 (Common Starling)**

The common starling is a species close in size to the blackbird (*Turdus merula*), although it has a pointed bill, sloping head and triangular wings. They appear black at a distance but are iridescent and quite colourful close up, and have white spots in winter.

The starling (*Sturnus vulgaris*) is a polytypic species with a disputed number of subspecies. Vaurie (1962) and Amadon (1962) accept eleven subspecies while Feare (1984) recognises a twelfth, and others authors still have considered giving populations such as that from Britain subspecific status (Bullough, 1942). The subspecies are diagnosed chiefly by differences in their plumage sheens, although some differ in their sizes. *S. vulgaris* forms a super-species with *S. unicolor* (see below).

In the Western Palaearctic starlings breed within the ca.10 - 30 degrees centigrade July isotherms, encompassing the Arctic fringe though to the Mediterranean climatic zones. They nest in holes or crevices in rocks or trees and will travel considerable distances to feed in open grasslands and floodlands. After fledging, birds frequent open country such as the upper parts of salt marshes, heaths and rocky shorelines. They will, however, avoid low dense vegetation and dense forests, especially conifers and extensive wetlands, although they require trees or reed-beds for roosting.

Sturnus vulgaris is migratory in the north and east of its range and a partial migrant or resident in the south and west. Subspecies such as *zetlandicus*, *faroensis* and *minor* are residents.

Food consists of both plant and animal material all year round although the latter is predominately fed to the young. Plant material predominates in the autumn and winter. Animal food consists of insects and their larvae, while plant matter includes soft fruits and seeds (Cramp and Perrins, 1994).

***Sturnus unicolor* Temminck, 1820 (Spotless starling)**

The spotless starling forms a super-species with *S. vulgaris* and sometimes is combined with it as a single species, as similarities exist between the two in habits, morphology, moult and eggs (Cramp and Perrins, 1994). However, the races (subspecies) of the starling *S. vulgaris* grade morphologically into one another clinally while *S. unicolor* replaces *S. vulgaris vulgaris* abruptly. *S. unicolor* also differs in morphology from *S. vulgaris vulgaris* to a greater extent than do any individual races of *S. vulgaris*. No hybrids are yet known despite slight geographical overlap in their breeding range. Post-mating barriers to reproduction are evident from captive breeding experiments. Also the timing of gonad readiness for reproduction differs, *S. unicolor* nesting later than *S. vulgaris*.

S. unicolor is slightly larger than *S. vulgaris* and differs in having more elongated feathers on its head and throat as well as having black, evenly glossed plumage with little in the way of spots.

The spotless starling is monotypic although it may decline in size from East to West; however, the sample size from which this was deduced was small (Cramp and Perrins, 1994).

The species is a resident or a partial short-distance migrant and subject to nomadic dispersal. The spotless starling has recently spread in Northern Spain and Southern France probably due to changing agricultural practices. They were formerly confined to Southern and central Iberia, Corsica, Sardinia, Sicily and parts of Tunisia and Morocco.

Their habitat preference is similar to that of *S. vulgaris*, nesting in buildings and tree holes. They like open woodland near short grasslands. Their diet consists of invertebrates from early spring to summer and seeds and fruit during the rest of the year. The species has a strong association with cattle, taking plant remains from their dung, and flushed insects as well as parasites from the bodies of animals. They eat more adult beetles than do *S. vulgaris*; this may be a more localised source of prey than are soil dwelling insect larvae, preferred by *S. vulgaris*. This may cause *S. unicolor* to have smaller populations and more restricted ranges (Cramp and Perrins, 1994).

***Sturnus roseus* Linnaeus, 1758 (Rose-coloured starling)**

The rose-coloured starling is a monotypic species and is not thought to vary geographically. They are appoximately the same size as *S. vulgaris* with a shorter bill. They have a crest and orange-pink legs. The adult has a pink jacket on the back, flanks and belly, which is spotted in fresh plumage. The rest of the plumage is blue to purple-black. They breed in the east of the Western Palaearctic and migrate to winter in India and Sri Lanka.

S. roseus's habitat in the breeding season is over lower middle latitudes, mainly in steppe, semi-desert and Mediterranean lowland zones, dry, open spaces of rocky terrain or grassland. They also track ephemeral localised insects (particularly Orthoptera). In the breeding season their diet is mainly insects (locusts and grasshoppers), although after the young fledge, fruit, nectar and seeds are mostly eaten (Cramp and Perrins, 1994).

11.3. Studies in Historical Biogeography

Feare (1984) considered the distribution of *S. vulgaris* in relation to Pleistocene climates. He stated that whatever time *S. vulgaris* first evolved it would not have spread to fill its present distribution until after the Weichselian glaciation. In relation to *S. unicolor* he stated that it "obviously evolved from a *vulgaris* -like ancestor and is presumably a relic of a former invasion of Europe by *Sturnus* starlings: If this occured before the last glaciation, the present geographical distribution of this species may represent the ice free refuge in which it was able to survive".

Feare also discussed subspecies of *Sturnus vulgaris* in the Western Palaearctic and referred to work on Quaternary starlings in so doing. He suggested that the isolated location of *S. v. grantii* in the Azores in relation to other breeding *S. vulgaris* in the Western Palaearctic, which is less apparent today because of the recent expansion of starlings, is indicative of a former wider range of *S. vulgaris*. He quotes Pleistocene occurences of *S. vulgaris* in the Mediterranean (Fischer and Stephan, 1974) as evidence of this former wider distribution.

Feare hypothesised that given the isolated, western-most nature of *S. unicolor*'s distribution in relation to other members of the genus, and its morphological distinctiveness from all *S. vulgaris* subspecies, it is attractive to regard it as the earliest arrival in the Western Palaearctic of the *S. unicolor / vulgaris* super-species. If this is the case they would have arrived from the East, the supposed place of origin of the genus. The next arrival in the area could well have been *S. vulgaris* and this may have happened in two or more waves. This may have been after the northwards retreat of cold climates at a glacial-interglacial boundary, allowing for the westwards expansion of *S. vulgaris* from a refugium in the East. According to this model, the isolation of sedentary subspecies of *S. vulgaris* on islands such as the Faroes (*S. v. faroensis*), the Shetlands (*S. v. zetlandicus*) and the Azores (*S. v. granti*) would suggest that there had been two pulses of westerly spread by *S. vulgaris*. The first pulse resulted in the isolation of the island endemic races, while the second resulted in the recent Western range of the nominate subspecies *S. v. vulgaris* (Feare, 1984).

As regards the British starlings, Feare (1984) notes that they were comparatively rare prior to the middle of the last century when they increased their population size greatly. This increase in numbers is thought to be part of the trend of population increase and westerly expansion by starlings in Europe as a whole. The cause for this has been attributed to climatic amelioration which has also triggered mild winters and hence reduced migratory distances (Berthold, 1968). Despite this suggestion, the activities of man may have had a large role in enabling the increase in population size and range due to increased forest clearance. This may also have encouraged a more sedentary habit (Feare, 1984).

The history of *S. roseus* in this regard was not discused by Feare (1984) although it may always have had a more easterly Western Palaearctic distribution indicated by the area in which it winters (India and Sri Lanka).

There are, however, problems with the interpretation of modern geographic ranges in relation to origination of taxa. As already stated, the modern ranges may not be very old and clearly will be influenced by changes of the environment by man. In fact the increased use of certain agricultural practices has been suggested by Feare to be the reason for the recent expansions by both *S. vulgaris* and *S. unicolor*. It is also possible that more direct interventions by humans may have had a role in the modern European distribution of *Sturnus* species as evidenced by the introduction of *S. vulgaris* to many continents by Europeans since the nineteenth century. This clearly complicates the interpretation of modern ranges and potentially enhances the importance of the fossil record. However, the fossil record should be used with great caution because the identification of the starling bones to species is not without problems. There is little value in dipping into the palaeontological literature without questioning identifications.

The timing of the three proposed geographical range expansions into Europe by the *S. vulgaris / unicolor* superspecies is also unknown at present. It may be possible to shed light on this by means of a cautious use of the fossil record.

11.4. Identification of the genus *Sturnus*

It has long been considered difficult to identify fossil passerine bones (Lydekker, 1891b; Newton, 1924a). Past studies to remedy this have mainly concentrated on the cranium (Moreno, 1985, 1986, 1987; Cuisin, 1981, 1982, 1983) and humerus (Jánossy, 1983; Ashley, 1941) which have been perceived as possessing the most reliable taxonomic characters. The first anatomical work on these two skeletal elements was undertaken to unravel the interrelationships between the families and not directly for use in palaeontology. Work on the cranium includes Beecher (1953), Bock (1960, 1966), Jollie (1958), Tordoff (1954) and Wetmore (1957) amongst many others, while that on the humerus is more limited (Bock, 1962; Wetmore, 1957; Berger, 1957). The anatomical differences of the cranium are of little use in palaeontology because they are rarely preserved and if they are the important details of the palate, formed of a separate bone which easily becomes dissociated, are generally lost.

Postcranial bones are far more numerous in the fossil record and are also more often preserved whole. The work on the humerus is therefore of much greater use. It is the state of the pneumatic fossa of the humerus in the Passeriformes which exhibits much variation among families and genera, hence lending itself to be used as an identification criterion. Ashley (1941) was the first to stress the significance of the pneumatic fossa, although the feature's potential was known as early as the late nineteenth century by Shuffeld (1888). In Ashley's work the single fossa was considered to be a primitive state, the double fossa an advanced state. Argument has existed, however, over the value of such characters to phylogenetic reconstructions (Delacour and Vaurie, 1957; Mayr, 1957); Bock (1962) reviewed the situation and was equivocal about the pneumatic fossa's use in systematics. This does not necessarily diminish its usefulness for identification of extant genera and families in the Quaternary. Hence the identification key of Jánossy (1983), which was influenced by Bock's paper, remains of great value. In this key the Sturnidae are said to fall into a size category with *Turdus*, *Monticola* and *Pinicola* and can be distinguished from these largely on details of the pneumatic fossa. He also states that the *Sturnus* humerus differs from those of *Monticolla*, *Turdus* and *Pinicola* by being proximally and distally narrower and by the fact that the *crus dorsale fossae* completely separates the deep *fossa anconaea.*

Sturnus vs. *Turdus*

Stewart (1992) found that *Turdus* presented the greatest problem when it came to the identification of *Sturnus* postcranial bones generally.

Bock (1962), Ballman (1978) and Jánossy (1983) described the means of distinguishing the humeri of *Sturnus* species from those of the genus *Turdus*, and Stewart (1992) additionally considered other major postcranial skeletal elements. During the present study it was felt that the reliability of these diagnostic criteria was in need of testing. To that end the characters originally described were re-evaluated and others were sought, all of which were tested for reliability in the manner described by Lister (1996) *Sturnus* versus *Turdus* characters are assessed in Chapter 7.

11.5. Material

Modern samples

Sturnus vulgaris - British sample

Most of the 25 specimens of common starling from Britain were from southern Britain and particularly the home counties (Surrey, Essex and particularly Hertfordshire, where Tring is located). The majority of the specimens came from the Natural History Museum collections in Tring (20) while four came from the Ancient Monuments Laboratory collections in London and one from the Environmental Archaeology Unit in York.

An attempt was made to include equal numbers of males and females in the sample but this was not possible because many were unsexed specimens and the males outnumbered the females by eight to two among the sexed specimens. Another problem encountered with the specimens was the age or maturity of the individual birds. Some specimens were recorded as juveniles although there was no indication of this from their bones, while others appeared to be juvenile osteologically (porous grainy texture of bone and incompletely fused proximal end of tarsometatarsi) but were not documented as such. Two reasons may account for this. The most likely is that mistakes or omissions have been made during the recording of this information. Secondly, however, it is possible that the state of the birds' plumage (adult vs. juvenile) may not be synchronous with the birds' bone maturity. This cannot be adequately solved in this study but might warrant further investigation. The British sample has some juveniles among it and so should be viewed from this perspective when the mensural data are interpreted.

The dates recorded for the British sample are mostly between 8/7/1970 and 25/5/1988. Most of the collection dates (all but two) are outside the season during which the British starling population is inflated by migratory birds. This signifies that the British starling sample can be regarded as derived from a breeding population. However, not all the specimens have data on the dates of collection; one specimen from E.T. Newton's collection has a 1930 date recorded. This is the date when the specimen was accessioned into the Natural History Museum's collection, as all Newton's specimens are recorded as 1930.

Sturnus vulgaris - Belgian sample

The Belgian sample of *S. vulgaris* is housed in the Musée d'Histoire Naturelle in Brussels. The total was 20, 12 of which were from the province of Brabant, 1 from the province of Luxembourg, 4 from the province of West Flanders, 1 from the province of Liege and 2 from the province of Namur. The sexes were almost equally represented: 8 females, 11 males and one unsexed specimen. There were also three specimens recorded as juveniles, which was not apparent osteologically. The material was collected probably between 9/3/1938 and 18/6/1963 as seen from the dates on the specimens. However, these dates are probably accession dates rather than collection dates so cannot be used to determine the season during which they were collected, and hence it is not possible to know if they represent Belgian breeding birds or migrants. That aside the sample is a reasonably adequate sample of starlings from Belgium, which being a small country is a relatively confined area.

Sturnus vulgaris - Scandinavian sample

The Scandinavian sample is housed in the Swedish Museum of Natural History. Many of these (14) were collected in Ottenby in Sweden, which is a staging post for migratory birds going south (Ericson, pers. comm.). The others were collected in Ängfallet (1 specimen), Lögla (1 specimen), Dalköpinge (2 specimens), Bjurfors (1 specimen) and Djursholm (1 specimen). They were accessioned between 1979 and 1992 and were probably collected over that period of time. The ratio of sexes in the Scandinavian sample is 10 males, 6 females and 4 unsexed. One specimen was recorded as being a juvenile which was not visible from the bones, while conversely two were immature-looking osteologically although not listed as juvenile. The Lögla specimen, interestingly from the point of view of the British sample, had a British Museum ring. This further highlights the potential inclusion of non-British breeding birds in the British sample. The Scandinavian sample, however, would have bred in Scandinavia and so reflects the starlings there more accurately. However, some of the individuals may be non-migrants living in, for example, Southern Sweden, so there may still be a mixture of migrants and non-migrants, although all the birds from Ottenby are almost certainly migratory individuals.

Sturnus vulgaris - Polish sample

The Polish sample is housed in the collections of the Institute for Evolution and Systematics in Krakow and totals 12. Most of these were collected on the Baltic coast (10), although one specimen came from near Krakow and another from the Polish Carpathians. They were collected between 1969 and 1976. Only two of the specimens were sexed, one male and one female. Due to lack of time the birds from Poland only had the humerus, carpometacarpus and tibiotarsus measured. This was thought to be sensible because these are the elements which can be reliably identified in the fossil record. It is not known wether the Polish birds were breeding or passage migrants.

Sturnus vulgaris - Other samples

In the collection of the Institute for Evolution and Systematics in Krakow are five specimens from outside Poland. They comprise two birds from near the Black Sea in Bulgaria, two from northern Iraq and one from Algeria. Although these do not form adequate samples, they were measured for completeness. Again, however, only the humerus, carpometacarpus and tibiotarsus were measured from these specimens. None of these specimens was sexed.

Sturnus unicolor

The sample of spotless starlings totals 17 and are mostly specimens in the collections of the Universidad Autonoma de Madrid and the Museo Nacional de Cienca Naturale de Madrid. Of these specimens, most that were documented with collection localities came from Madrid and its environs. Other specimens came from Avila, Guadalajara, Pontevada and La Coreña, also in Spain. The last three, as well as two others, consisted only of trunk skeletons which would have been the result of preparing skins. These therefore only made coracoids, femora and scapulae available, the other limb bones retained by the prepared skins. The last locality from which a specimen was available was Sicily, a skeleton in the Tring collection which had been aquired during the early part of the 19th century. Unfortunately there was little information as to the sex of the specimens - a single female and a male were the only recorded with this information. The Spanish birds form an adequate sample of the species.

Sturnus roseus

Eight specimen of rose-coloured starling were available. Three were from the Musée d'Histoire Naturelle in Brussels, a male and a female and an unknown, which were ex-captive birds (one from Antwerp Zoo). Two were from the collection of the Institute for Evolution and Systematics in Krakow, one of which had no provenance and the other coming from Tadzhikistan. A further two males came from the Smithsonian Institution in Washington (both ex-captive birds). The last specimen is from Tring and is a trunk skeleton and only the femur, coracoid and scapula could be measured. The ratio of known sexes amongst these specimens was 4 males and 1 female.

11.6. Morphometric comparison of *Sturnus vulgaris*, *S. unicolor* and *S. roseus*

The following comparisons are undertaken to establish the interspecific variation within the three European *Sturnus* species.

In general, *S. unicolor* is the largest of the three species, with *S. vulgaris* and *S. roseus* being smaller (Table 11.1; Figures 11.1 - 11.11). The humerus of *S. roseus*, however, while being of a similar size to *S. vulgaris*, differs from it in having a significantly narrower proximal end (Figure 11.1; Table 11.1 and 11.2). The distal dimensions taken for *S. roseus* align the species to a greater extent with that of *S. vulgaris* (Table 11.2).

Sturnus unicolor have, on average, the longest ulnae (Figure 11.2). There is considerable overlap between *S. vulgaris* and *S. roseus* (Table 11.1 and 11.2). The apparent tendency for the ulnae of *S. roseus* to have small distal dimensions (Figure 11.2) is not statistically significant at the 95% level in the small sample available (Table 11.2).

While *S. unicolor* is maintained as having the largest radius dimensions, its overlap with the other two species is greater than with the two previous skeletal elements described (Figure 11.3).

The order of size of the species is maintained in the carpometacarpus, with *S. unicolor* being on average larger than either *S. vulgaris* or *S. roseus*, which overlap considerably with each other (Figure 11.4; Tables 11.1 and 11.2). It may be important that the range of proximal breadths is more tightly correlated in relation to the greatest length in *S. unicolor* when compared to *S. vulgaris* and maybe *S. roseus* (R^2 values in Figure 11.4).

The measurements of the coracoid reveal that *S. unicolor*, while having a greater average length, have relatively low value for the facies articularis sternalis (BF) i.e. the dimension overlaps greatly with *S. vulgaris* (Table 11.1 and 11.2). The coracoid of *S. roseus* appears to be the shortest on average with *S. vulgaris* falling in between it and *S. unicolor* (Table 11.1). The coracoid shaft width of *S. vulgaris* is significantly smaller than that of *S. unicolor* (Figure 11.5; Table 11.2).

S. unicolor has the largest scapular dimensions and again *S. vulgaris* and *S. roseus* overlap considerably (Figure 11.6).

In the femur, the mean greatest length is greater in *S. unicolor* than it is in *S. vulgaris*. *S. roseus*, however, also has relatively long legs, although in absolute dimensions they overlap greatly with other species. There is also a great deal of overlap between *S. unicolor* and *S. vulgaris* (Figure 11.7). *S. roseus* has long and thin femora compared to either *S. vulgaris* or *S. unicolor* (Table 11.1 and 11.2; Figure 11.7).

The tibiotarsus produces the greatest amount of overlap between the species (Figure 11.8). The longer legs of *S. roseus* are not expressed in the tibiotarsus (Table 11.1 and 11.2; Figure 11.8). There are, however, significant differences in mean of certain breadth measurements between the three species (Table 11.2).

The tarsometatarsus shows the least correlation between dimensions (Figure 11.9). The width measurements are more widely spread in relation to the length than all other post-cranial elements measured. *S. roseus* overlaps much with *S. unicolor* indicating that it has relatively long legs for its size, while there are many long gracile tarsometatarsi belonging to *S. vulgaris* (Table 11.1 and 11.2; Figure 11.9)

Analysis of relative bone dimensions can reveal interesting patterns. In Figure 11.10, the European representatives of the genus *Sturnus* have their humerus greatest lengths plotted against their carpometacarpus greatest lengths. The three species allign along approximately the same axis in this case, suggesting that they have the same proportions.

To conclude this comparison of the three European *Sturnus* species it is clear that *S. unicolor* is on average larger than both *S. roseus* and *S. vulgaris*. Furthermore, there are some metric-shape patterns which have been illustrated. The humeri of *S. roseus*, for instance, are narrower proximally than in *S. vulgaris*, which is otherwise similarly sized. Furthermore, the femora of *S. roseus* are relatively long and thin and their tarsometatarsi are relatively long, overlaping with *S. unicolor*.

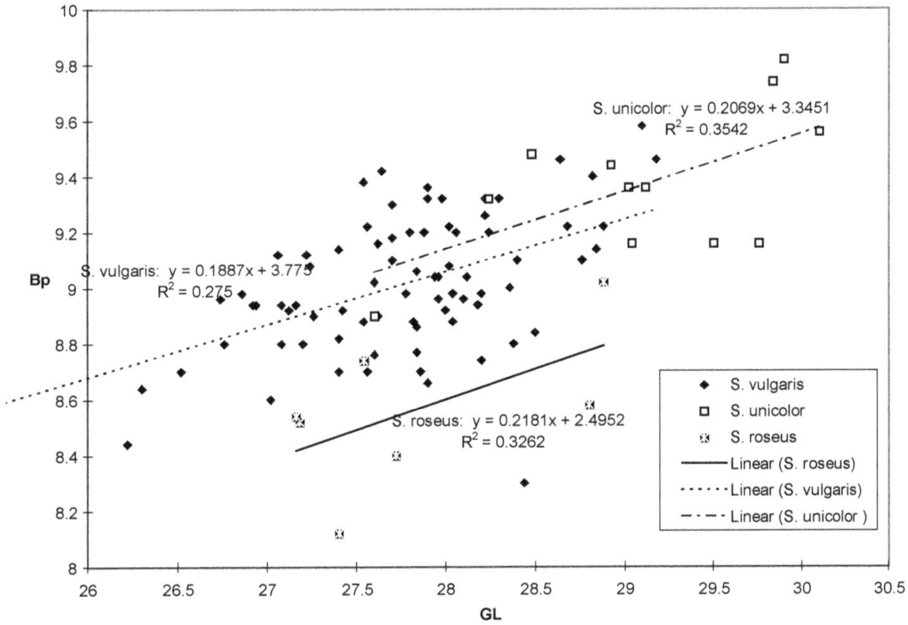

Figure 11.1. Scattergram of European *Sturnus* humeri. Greatest length (GL) plotted against proximal breadth (Bp).

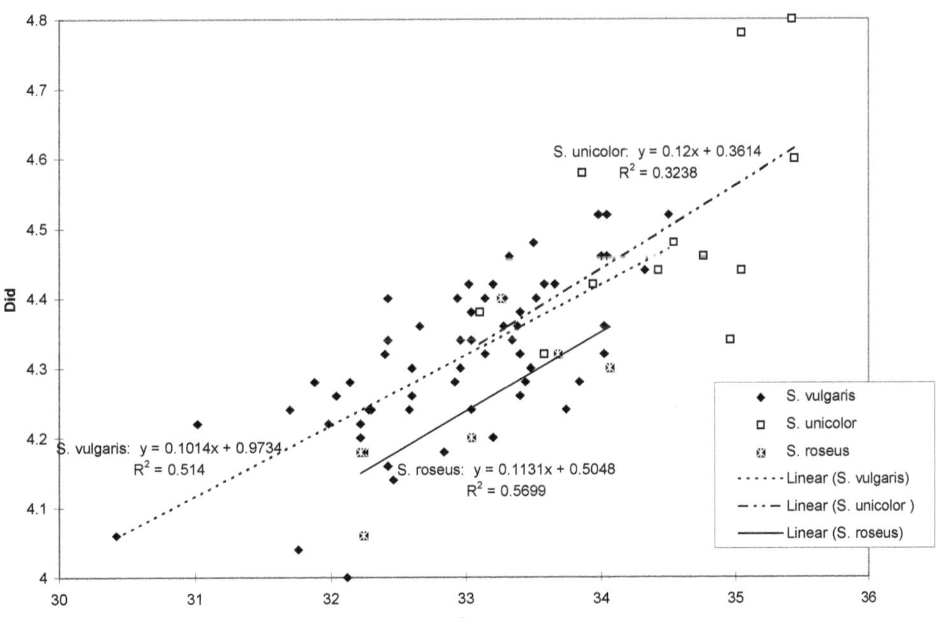

Figure 11.2. Scattergram of European *Sturnus* species ulnae. Greatest length (GL) plotted against distal diagonal depth (Did).

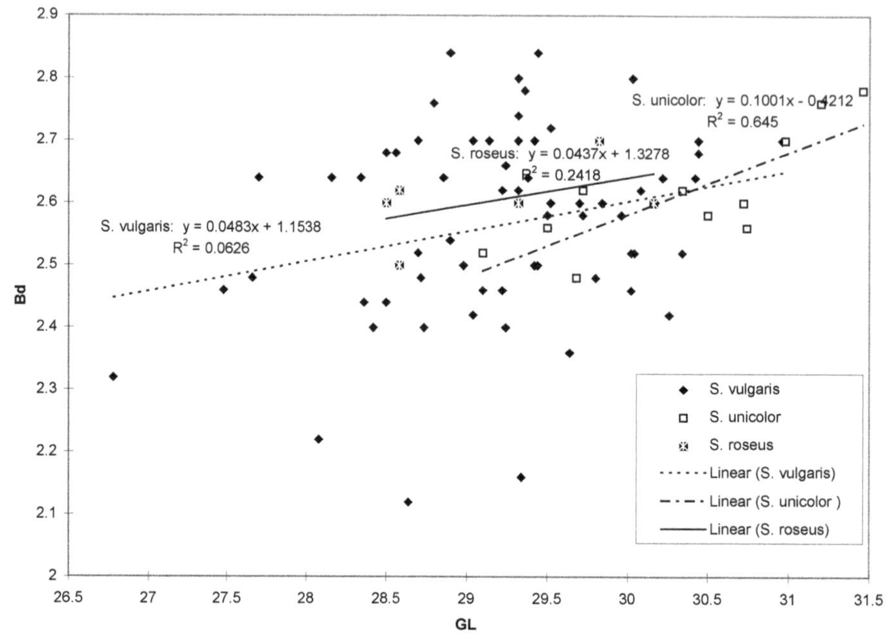

Figure 11.3. Scattergram of European *Sturnus* species radii. Greatest length (GL) plotted against distal breadth (Bd).

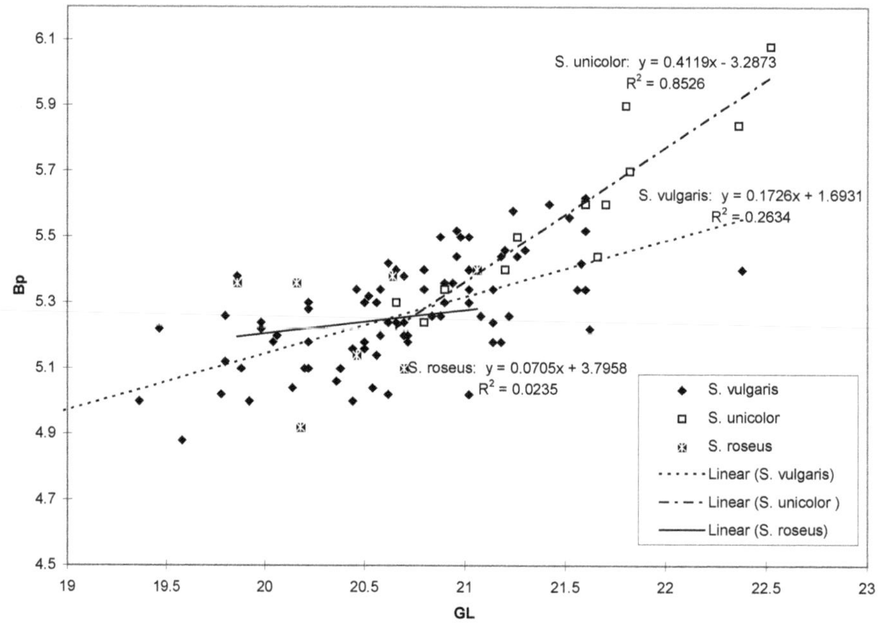

Figure 11.4. Scattergram of European *Sturnus* carpometacarpi. Greatest length (GL) plotted against proximal breadth (Bp).

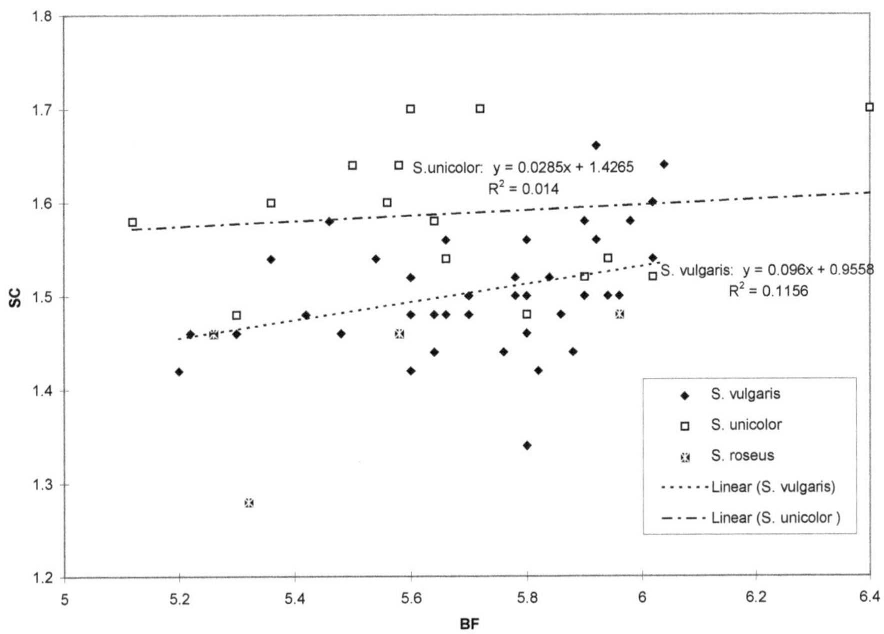

Figure 11.5. Scattergram of European *Sturnus* coracoidea. Breadth of the facies articularis sternalis (BF) plotted against shaft breadth (SC).

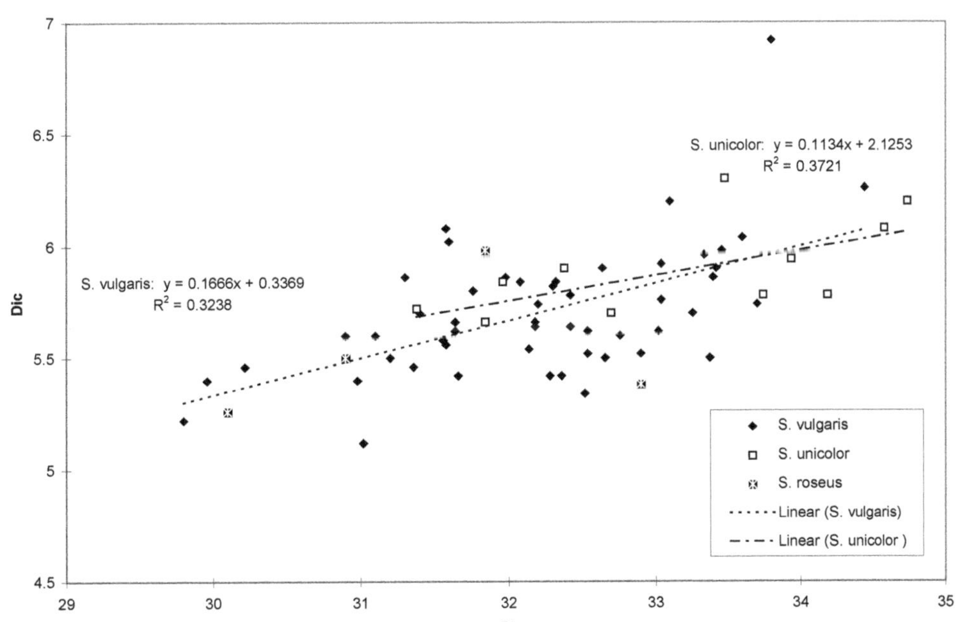

Figure 11.6. Scattergram of European *Sturnus* scapulae. Greatest length (GL) plotted against cranial diagonal breadth (Dic).

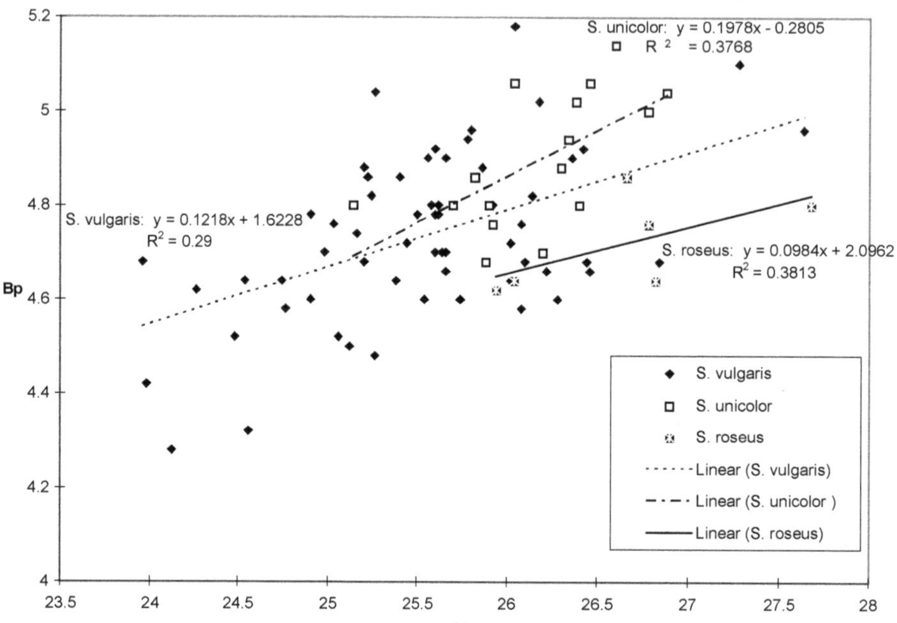

Figure 11.7. Scattergram of European *Sturnus* femora. Greatest length (GL) plotted against proximal breadth (Bp).

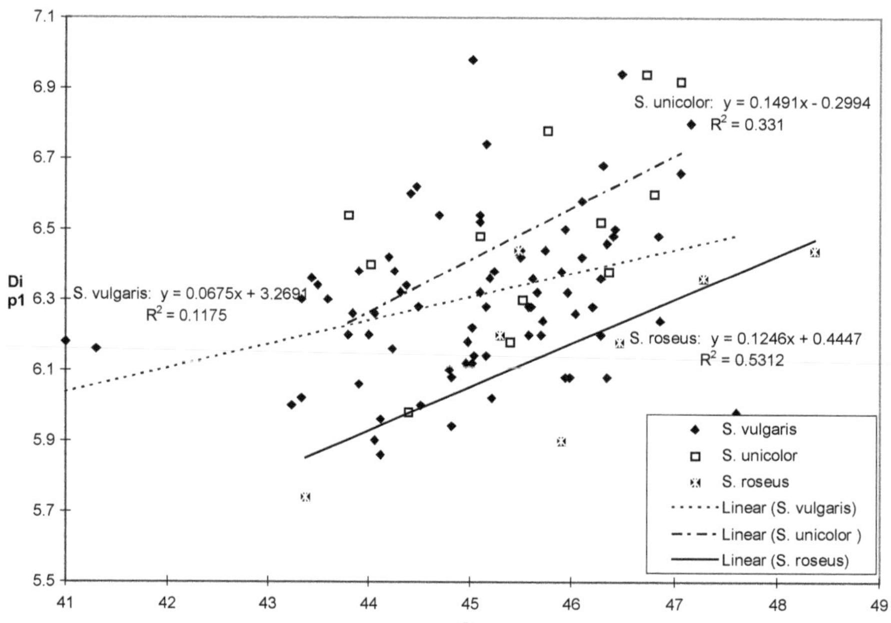

Figure 11.8. Scattergram of European *Sturnus* tibiotarsi. Greatest length (GL) plotted against proximal diagonal depth No. 1 (Dip1).

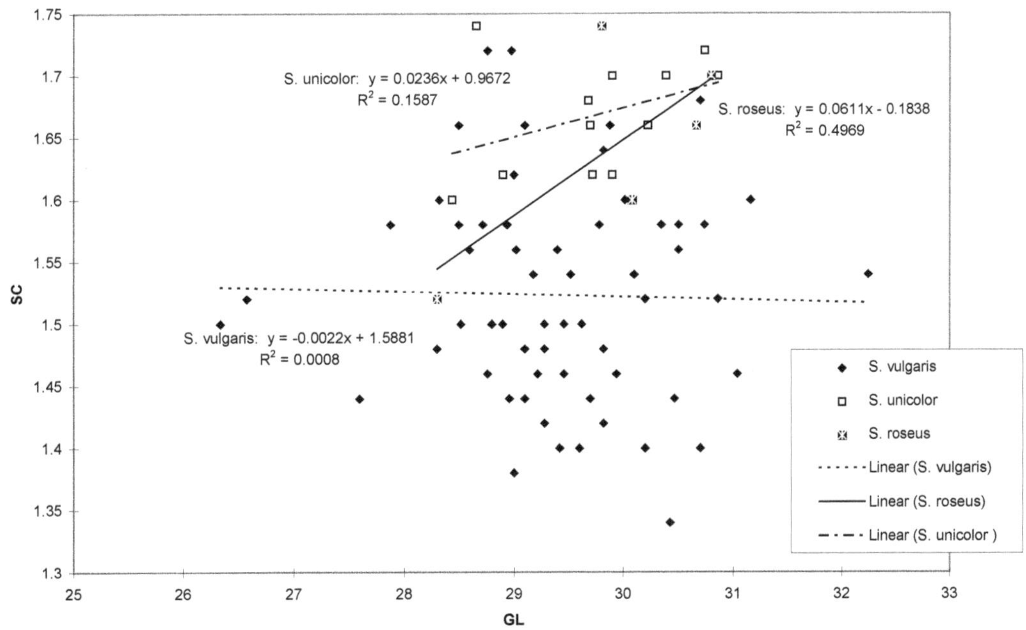

Figure 11.9. Scattergram of European *Sturnus* tarsometatrasi. Greatest length (GL) plotted against shaft breadth (SC).

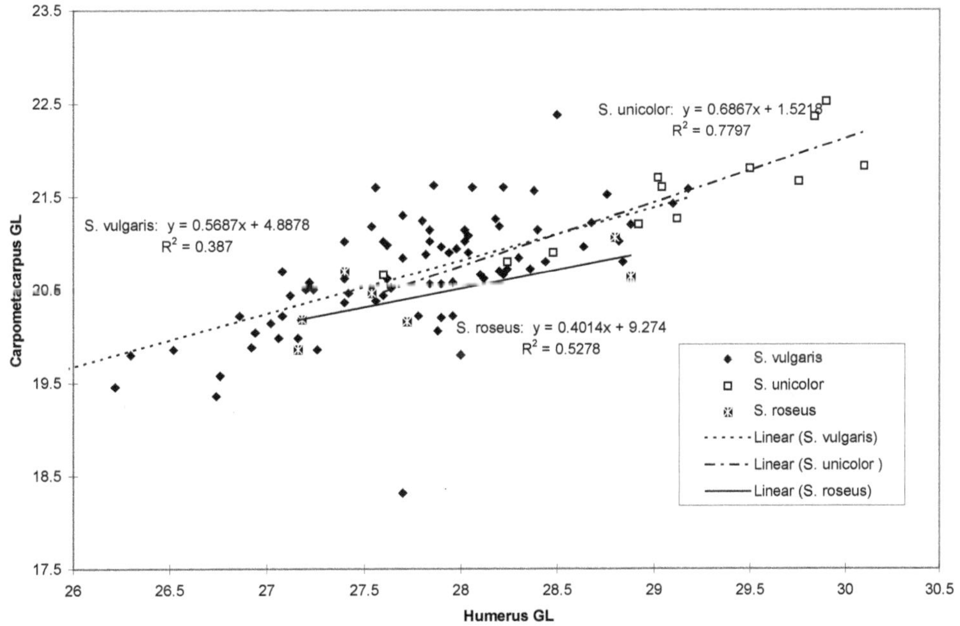

Figure 11.10. Scattergram of European *Sturnus* humeri to carpometacarpi ratios. Humerus greatest length (GL) plotted against carpometacarpus greatest length (GL).

Table 11.1. Mensural data of the modern *Sturnus* samples.

Sturnus samples	*Sturnus vulgaris*					*Sturnus unicolor*					*Sturnus roseus*				
Measurements	N	Min	Max	Mean	St. dev.	N	Min	Max	Mean	St. dev.	N	Min	Max	Mean	St. dev.
Hum GL	78	24.98	29.18	27.74	0.7	12	27.6	30.1	29.13	0.75	7	27.16	28.88	27.81	0.73
Hum Bp	80	8.3	9.58	9.01	0.25	13	8.9	9.82	9.36	0.25	7	8.12	9.02	8.56	0.28
Hum Bd	81	6.88	7.86	7.41	0.21	11	7.34	8.2	7.75	0.27	7	7.1	7.52	7.24	0.14
Hum SC	80	2.14	2.84	2.66	0.11	12	2.54	2.9	2.7	0.12	7	2.28	2.76	2.58	0.17
Uln GL	64	30.42	34.5	32.98	0.79	12	33.1	35.44	34.51	0.75	7	32.22	34.06	32.96	0.75
Uln Bp	64	4.18	5.1	4.68	0.19	12	4.5	5.24	4.82	0.25	7	4.3	4.86	4.62	0.2
Uln Dip	23	8.36	5.88	5.65	0.17	10	5.48	6.2	5.83	0.25	6	5.38	6.02	5.62	0.24
Uln Did	64	4.0	4.52	4.32	0.11	12	4.32	4.8	4.5	0.16	7	4.06	4.4	4.23	0.11
Uln SC	-	-	-	-	-	2	1.98	2.04	2.01	0.04	-	-	-	-	-
Rad GL	65	26.78	30.96	29.23	0.78	12	29.1	31.46	30.34	0.73	6	28.5	30.16	29.16	0.72
Rad Bp	65	1.82	2.74	2.08	0.17	12	1.06	2.36	2.03	0.33	6	1.94	2.4	2.1	0.16
Rad Bd	65	2.12	2.84	2.57	0.15	12	2.48	2.78	2.62	0.09	6	2.5	2.7	2.6	0.06
Cpt GL	83	19.36	28.32	20.78	1.02	12	20.66	22.52	21.52	0.58	7	19.86	21.06	20.44	0.4
Cpt Bp	83	4.0	5.62	5.26	0.21	12	5.24	6.08	5.58	0.26	7	4.92	5.4	5.24	0.19
Cpt Did	82	4.28	4.96	4.68	0.17	12	4.6	5.2	4.94	0.2	7	4.32	4.8	4.59	0.19
Fem GL	64	23.96	27.64	25.54	0.72	17	25.14	26.98	26.22	0.47	6	25.94	27.68	26.65	0.63
Fem Bp	65	4.28	5.18	4.73	0.18	17	4.68	5.92	4.96	0.28	7	4.62	5.1	4.77	0.17
Fem Bd	66	4.18	4.9	4.58	0.17	17	4.58	5.12	4.78	0.14	7	4.32	4.76	4.47	0.15
Fem SC	45	1.66	2.29	1.95	0.11	16	1.9	2.26	2.01	0.11	5	1.86	2.6	2.02	0.11
Tib GL	78	41.0	47.6	45.07	1.2	12	43.8	47.06	45.6	1.1	7	43.38	48.36	46.02	1.58
Tib Dip1	61	5.92	7.64	6.59	0.31	13	5.98	7.22	6.73	0.31	5	6.1	6.44	6.32	0.16
Tib Dip2	81	5.86	6.98	6.3	0.22	13	5.98	6.94	6.51	0.27	7	5.74	6.44	6.18	0.27
Tib Bd	80	3.58	4.3	3.38	0.15	12	3.84	4.32	4.05	0.15	7	3.72	4.14	3.95	0.15
Tib Dd	79	3.5	4.16	3.86	0.13	12	3.8	4.24	4.04	0.16	7	3.64	5.02	4.05	0.46
Tmt GL	62	26.34	32.24	29.43	1.02	12	28.44	30.86	29.76	0.77	7	28.3	31.58	30.32	1.07
Tmt Bp	63	3.94	4.72	4.2	0.15	12	4.24	4.72	4.53	0.15	6	4.06	4.58	4.33	0.2
Tmt Bd	64	2.92	4.32	3.23	0.2	12	3.24	3.8	3.39	0.17	7	3.14	6.12	3.71	1.07
Tmt SC	62	1.34	1.72	1.52	0.09	12	1.6	1.74	1.67	0.05	5	1.52	1.74	1.64	0.09
Cor GL	64	24.86	28.44	26.76	0.72	16	26.8	29.14	27.93	0.69	7	25.1	26.48	25.67	0.54
Cor Lm	64	23.74	27.3	25.65	0.71	16	25.84	27.66	26.81	0.62	7	24.22	25.54	24.73	0.51
Cor BF	64	5.02	6.98	5.68	0.34	16	5.12	6.4	5.7	0.32	7	5.06	5.96	5.4	0.3
Cor SC	40	1.34	1.66	1.5	0.06	16	1.48	1.7	1.59	0.07	4	1.28	1.48	1.42	0.09
Scap GL	53	29.8	34.44	32.21	0.99	13	31.38	34.74	33.19	1.1	5	27.02	32.9	30.55	2.23
Scap Dic	62	4.88	6.92	5.69	0.3	14	5.66	6.34	5.92	0.23	7	4.08	5.98	5.39	0.39

Table 11.2. *t*- tests between the three modern European *Sturnus* species.

	Sturnus vulgaris vs. *S. unicolor*	*S. vulgaris* vs. *S. roseus*	*S. unicolor* vs. *S. roseus*
Hum GL	***	NS	**
Hum Bp	***	***	***
Hum Bd	***	*	***
Hum SC	NS	NS	NS
Uln GL	***	NS	***
Uln Bp	*	NS	NS
Uln Dip	*	NS	NS
Uln Did	***	NS	**
Rad GL	***	NS	**
Rad Bp	NS	NS	NS
Rad Bd	NS	NS	NS
Cpt GL	*	NS	***
Cpt Bp	***	NS	**
Cpt Did	***	NS	**
Fem GL	***	***	NS
Fem Bp	***	NS	NS
Fem Bd	***	NS	***
Fem SC	*	NS	NS
Tib GL	NS	NS	NS
Tib Dip1	NS	*	*
Tib Dip2	**	NS	*
Tib Bd	**	NS	NS
Tib Dd	***	**	NS
Tmt GL	NS	*	NS
Tmt Bp	***	NS	*
Tmt Bd	**	**	NS
Tmt SC	***	**	NS
Cor GL	***	***	***
Cor Lm	***	**	***
Cor BF	NS	*	*
Cor SC	***	*	***
Scap GL	**	**	***
Scap Dic	**	*	***

Key: *** $p< 0.001$, ** $p< 0.01$, * $p< 0.05$, ns Not significant at the 5% level.

11.7. Morphometric Comparison of *Sturnus vulgaris* samples

The most reliable samples of *S. vulgaris* measured are those from Britain and Scandinavia (Table 11.3). This is because they most accurately represent the breeding populations of the areas where the birds were collected. The other two large samples, from Poland and Belgium, may include a large number of birds which were collected during the winter and thus could include migrants. The Scandinavian and British samples are therefore the only ones which have been *t*-tested (Table 11.4). All the samples have been plotted in the scattergrams for completeness (Figures 11.11 – 11.20).

There are no clear differences in the humerus measurements of the populations sampled (Figure 11.11; Table 11.4). There are slight differences in the means (Table 11.3) and in the regression lines in Figure 11.11, although these are significant at the 5% level (Table 11.4).

The ulnae of Scandinavian birds have small proximal breadths (Bp) in relation to their greatest lengths (GL) (Figure 11.12; Table 11.4) when compared to the populations from Britain and Belgium, and the British starlings have narrower proximal ulnae than the Belgian sample.

The Scandinavian birds have relatively thin radii for their length, when compared to the British and Belgian samples, in that their distal ends are smaller in relation to their length (Figure 11.13; Table 11.4).

The carpometacarpi of British birds are shorter on average than those from Poland, Belgium and particularly Scandinavia (Figure 11.14; Table 11.3 and 11.4). This is the most significant difference between the Scandinavian and British populations (Table 11.4).

The Belgian sample shows a tendency for their coracoids to be larger than those from Britain (Figure 11.15) and the greatest length of the Scandinavian coracoids are significantly longer than are those of the British starlings (Table 11.4).

The Belgian scapulae have a greater articular dimension in comparison to those from Britain and Scandinavia (Figure 11.16; Table 11.3).

Much overlap is exhibited by the dimensions of the femora and tibiotarsi of the different *S. vulgaris* populations (Figures 11.17 and 11.18). No significant differences were found between any of the dimensions of these bones from Britain and Scandinavia. (Table 11.4).

The Scandinavian tarsometatarsi are on average narrower for their length when compared to those of both the British (Figure 11.19, Table 11.4) and Belgian starlings. This is particularly apparent of the shaft breadth (SC) (Table 11.4).

Figure 11.20 shows the relative humeral to carpometacarpal lengths. There appears to be much overlap between the populations but the Scandinavian line is steeper than the others. This indicates that their carpometacarpus lengths are longer in relation to their humerus lengths which in turn may be explained by the Scandinavian population's greater migratory habit (see later).

In conclusion, while in general there is a great deal of overlap between the various samples, there are some metric-shape trends. Certain breadth dimensions, the ulnae and radii, of the Scandinavian birds, are more gracile than those from other samples. There also appear to be signs of relative robusticity in the pectoral girdle bones from Belgium. Finally, the tarsometatarsi from Scandinavia are relatively gracile compared to other populations. The most different population of *S. vulgaris* is that from Scandinavia. This leads one to consider wether the morphological differences are attributable to differences in migratory behaviour between the Scandinavian and the more southern populations. This can relatively easily be conceived for the skeletal elements of the wing (ulna, radius and particularly the carpometacarpus lengths) due to the work by Winkler and Leisler (1992) on the ecomorphology of migratory birds. Birds with greater migratory distances tend to have relatively longer distal wing bones compared to their proximal wing bones. The relatively gracile legs may be explained by a smaller size for the Scandinavian birds. Small size is another correlate of migratory behaviour.

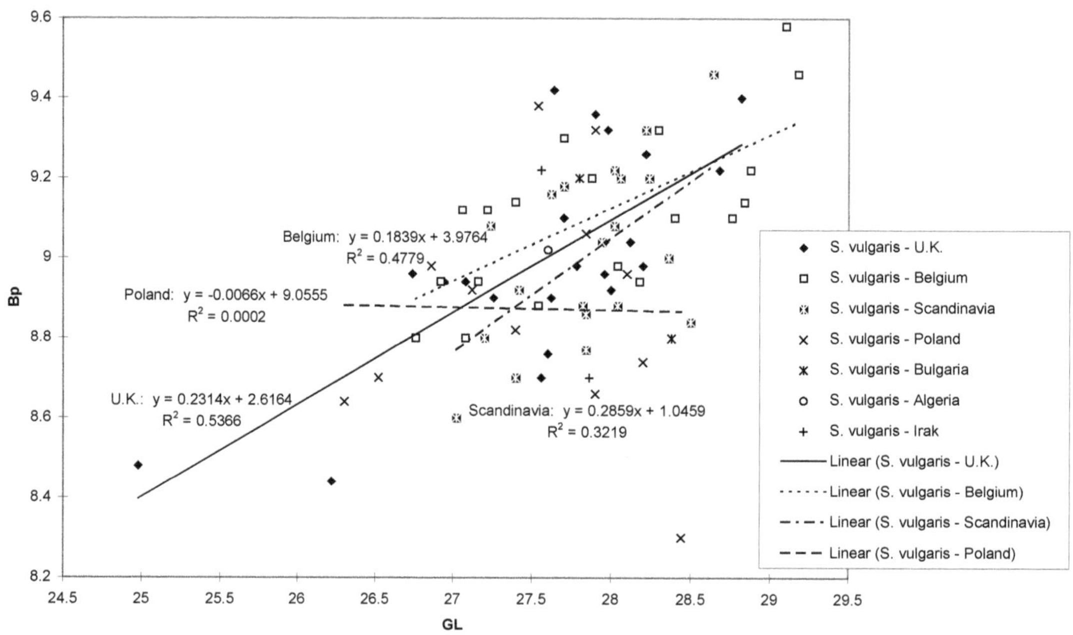

Figure 11.11. Scattergram of European *Sturnus vulgaris* humeri. Greatest length (GL) plotted against proximal breadth (Bp).

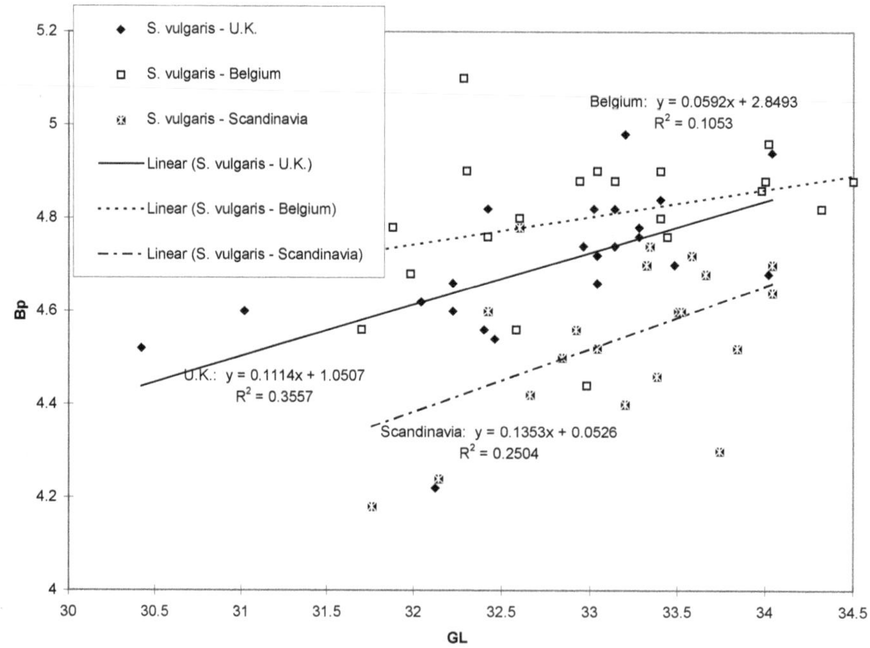

Figure 11.12. Scattergram of European *Sturnus vulgaris* ulnae. Greatest length (GL) plotted against proximal breadth (Bp).

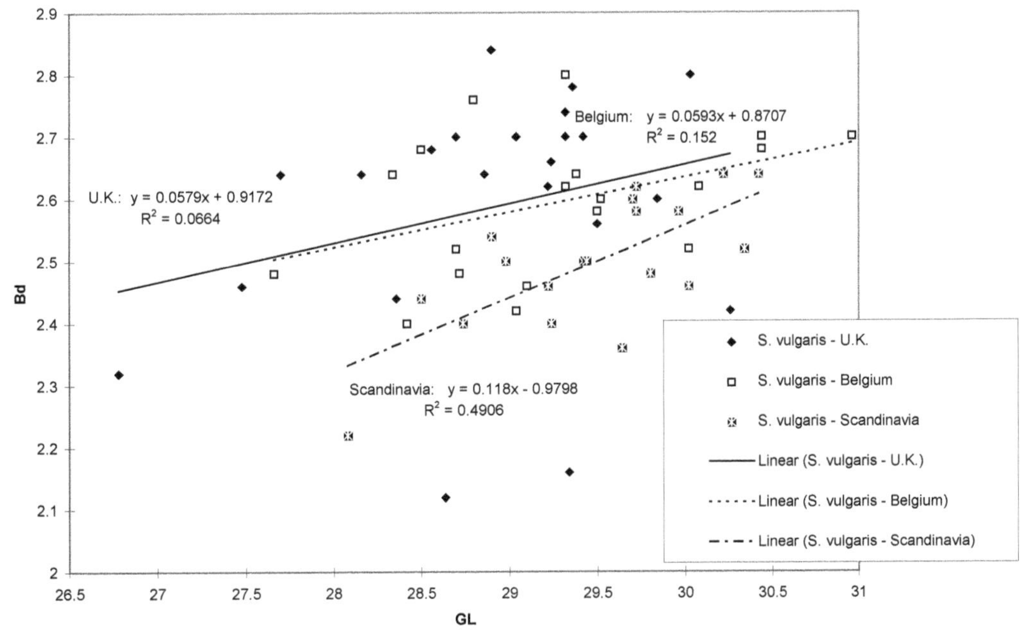

Figure 11.13. Scattergram of European *Sturnus vulgaris* radii. Greatest length (GL) plotted against distal breadth (Bd).

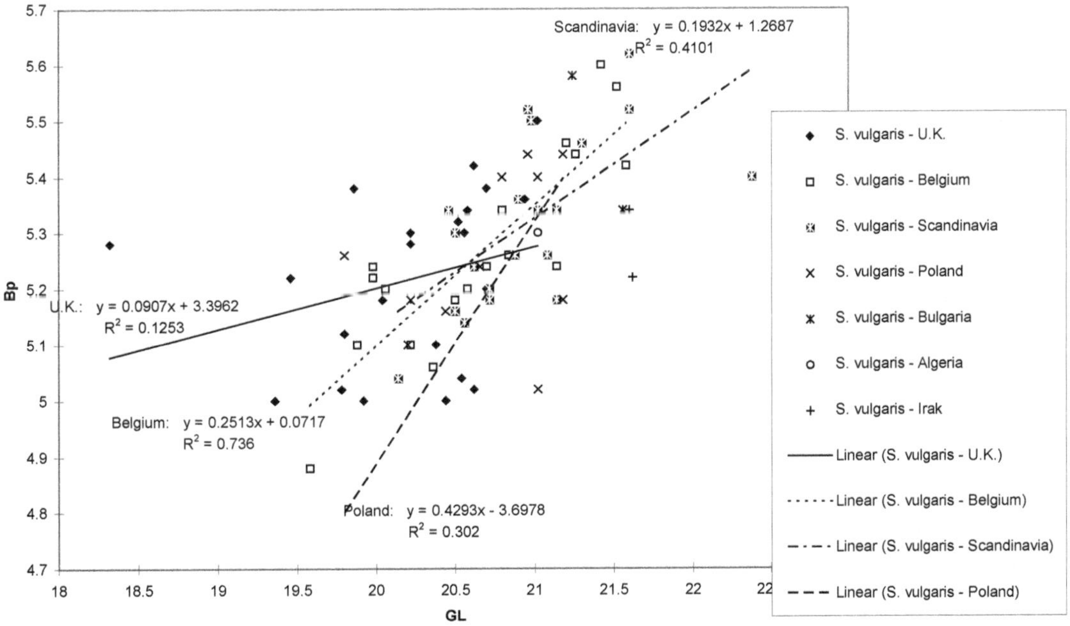

Figure 11.14. Scattergram of European *Sturnus vulgaris* carpometacarpi. Greatest length (GL) plotted against proximal breadth (Bp).

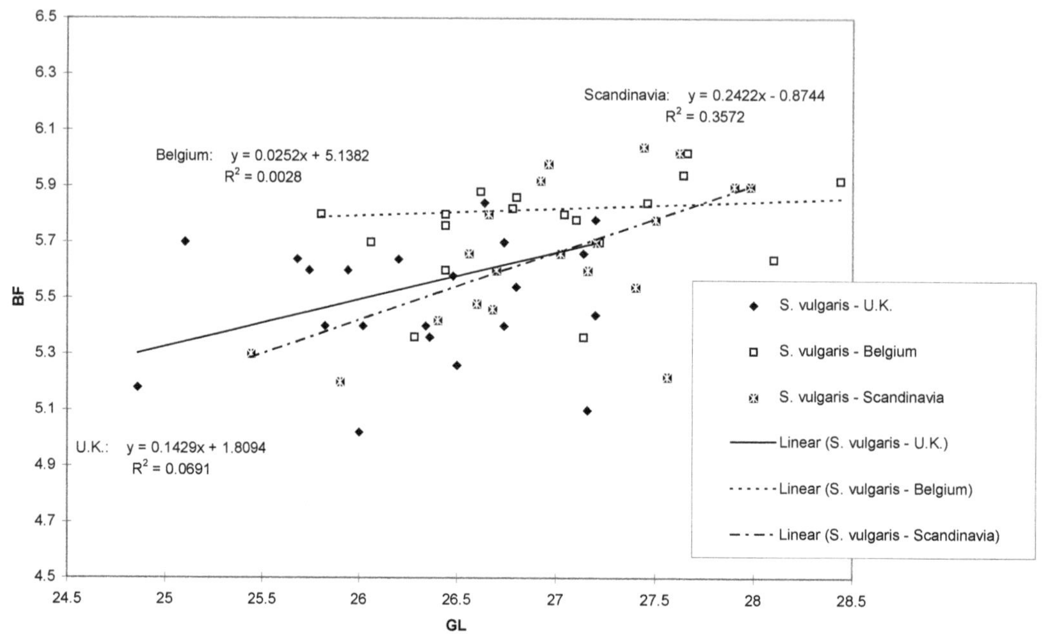

Figure 11.15. Scattergram of European *Sturnus vulgaris* coracoids. Greatest length (GL) plotted against the breadth of the facies sternalis (BF).

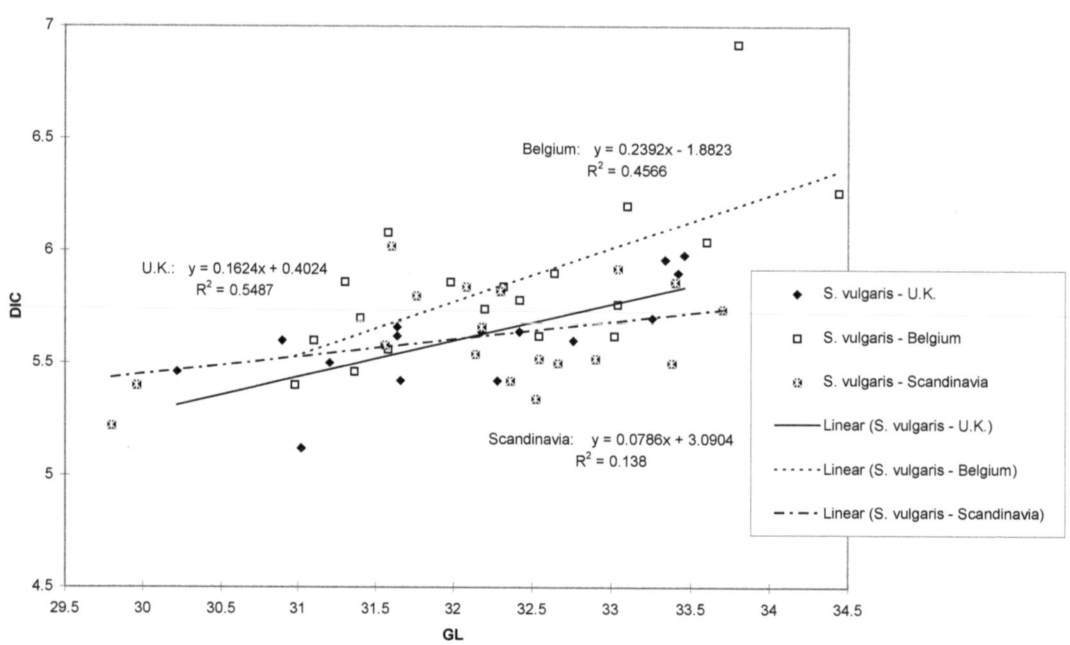

Figure 11.16. Scattergram of European *Sturnus vulgaris* scapulae. Greatest length (GL) plotted against proximal breadth (Dic).

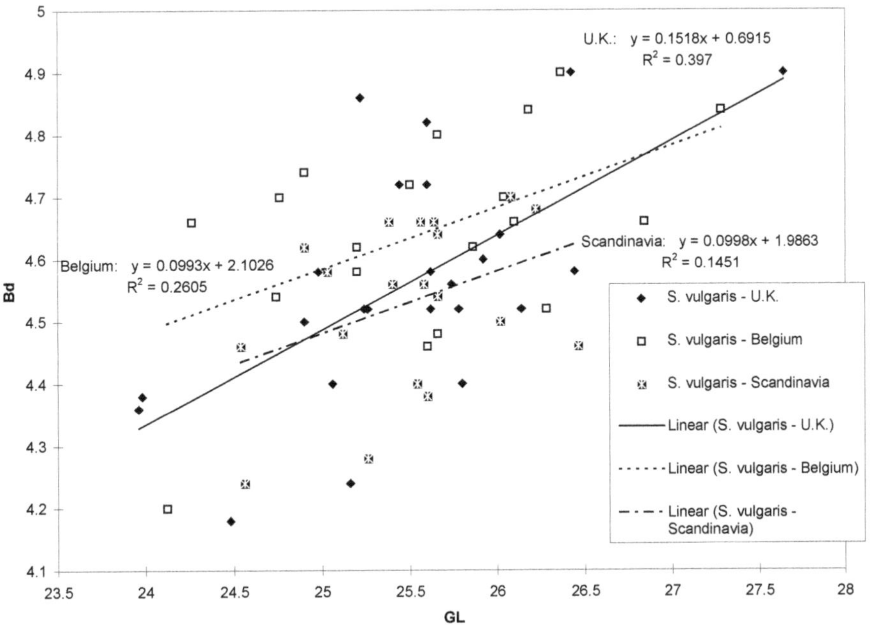

Figure 11.17. Scattergram of European *Sturnus vulgaris* femora. Greatest length (GL) plotted against distal breadth (Bd).

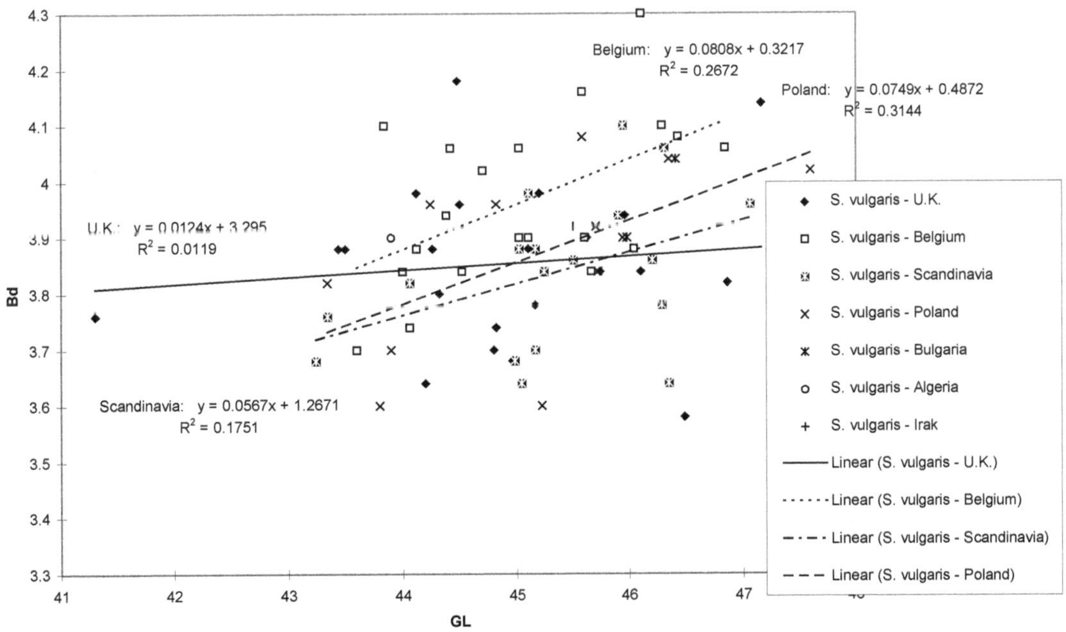

Figure 11.18. Scattergram of European *Sturnus vulgaris* tibiotarsi. Greatest length (GL) plotted against distal breadth (Bd).

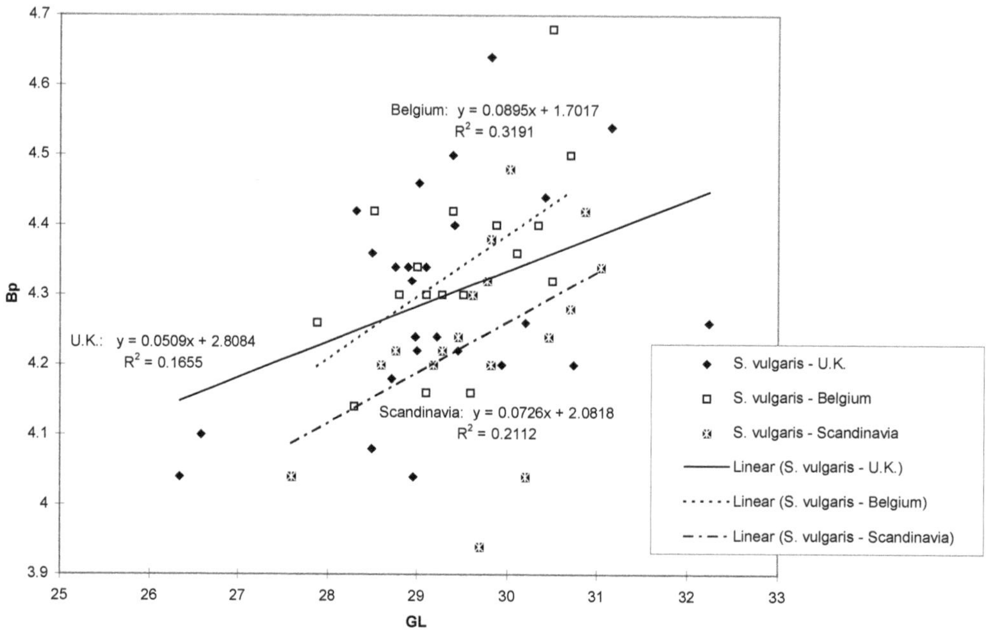

Figure 11.19. Scattergram of European *Sturnus vulgaris* tarsometatarsi. Greatest length (GL) plotted against proximal breadth (Bp).

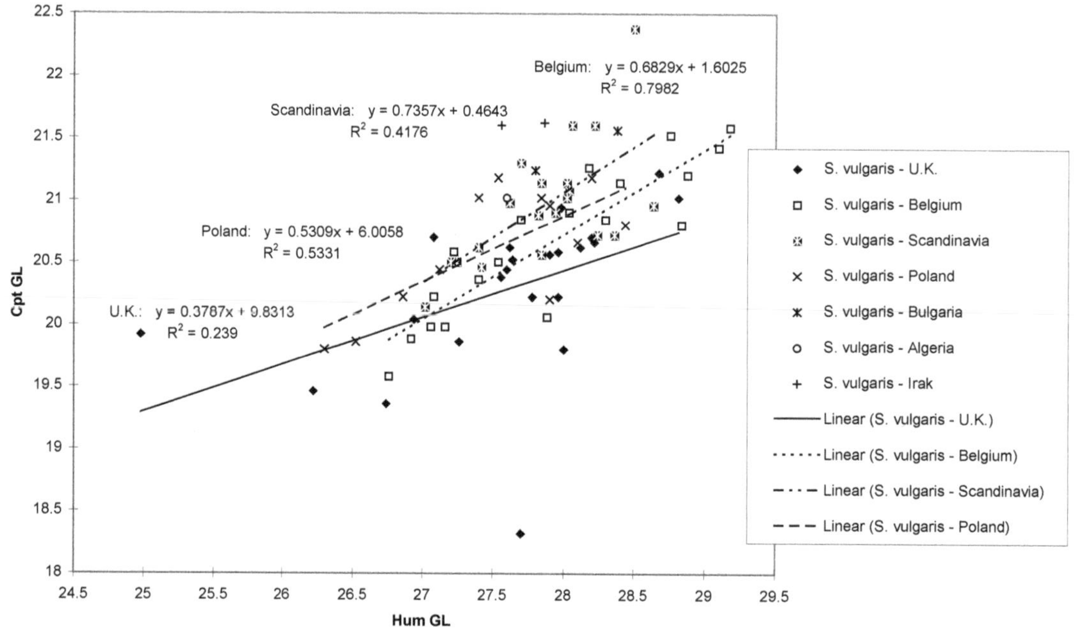

Figure 11.20. Scattergram of European *Sturnus* humeri and carpometacarpi. Humeral greatest length (GL) plotted against carpometacarpal greatest lengt (GL).

Table 11.3. Mensural data of the modern *Sturnus vulgaris* samples.

***Sturnus vulgaris* samples**	**Britain**					**Belgium**					**Scandinavia**				
Measurements	**N**	**Min**	**Max**	**Mean**	**St. dev.**	**N**	**Min**	**Max**	**Mean**	**St. dev.**	**N**	**Min**	**Max**	**Mean**	**St. dev.**
Hum GL	22	24.98	28.82	27.59	0.84	19	26.76	29.18	27.92	0.79	20	27.02	28.64	27.86	0.43
Hum Bp	23	8.44	9.42	9.0	0.26	20	8.8	9.58	9.1	0.21	20	8.6	9.46	9.01	0.22
Hum Bd	24	7.04	7.86	7.44	0.23	20	6.88	7.76	7.43	0.24	20	6.96	7.72	7.37	0.18
Hum SC	24	2.14	2.84	2.63	0.16	19	2.5	2.82	2.68	0.09	20	2.4	2.84	2.67	0.09
Uln GL	23	30.42	34.04	32.75	0.85	20	31.7	34.5	33.05	0.83	20	31.76	34.04	33.18	0.61
Uln Bp	24	4.42	4.98	4.7	0.16	20	4.44	5.1	4.81	0.15	20	4.18	4.78	4.54	0.16
Uln Dip	23	5.36	5.88	5.65	0.17	-	-	-	-	-	-	-	-	-	-
Uln Did	24	4.0	4.46	4.3	0.12	20	4.22	4.52	4.33	0.09	20	4.04	4.52	4.32	0.12
Uln SC	-	-	-	-	-	-	-	-	-	-	-	-	-	-	-
Rad GL	25	26.78	30.26	28.99	0.82	20	27.66	30.96	29.29	0.82	20	28.08	30.42	29.47	0.5
Rad Bp	25	1.86	2.74	2.1	0.21	20	1.88	2.5	2.11	0.14	20	1.82	2.22	2.02	0.1
Rad Bd	25	2.12	2.84	2.59	0.18	20	2.4	2.84	2.61	0.12	20	2.22	2.64	2.5	0.1
Cpt GL	24	19.36	21.22	20.37	1.66	20	19.58	21.58	20.67	0.59	20	20.14	22.38	20.96	0.49
Cpt Bp	25	5.0	5.5	5.24	0.16	20	5.22	5.22	5.27	0.17	20	5.04	5.62	5.32	0.15
Cpt Did	25	4.28	4.94	4.6	0.19	20	4.58	4.58	4.69	0.16	20	4.36	4.96	4.7	0.16
Fem GL	24	23.96	27.84	25.5	0.79	20	24.12	27.28	25.63	0.82	19	24.54	26.46	25.48	0.5
Fem Bp	25	4.34	5.04	4.74	0.18	20	4.28	5.18	4.76	0.21	19	4.32	4.9	4.68	0.13
Fem Bd	25	4.18	4.9	4.56	0.19	20	4.2	4.9	4.65	0.16	20	4.24	4.7	4.54	0.13
Fem SC	25	1.66	2.26	1.96	0.11	-	-	-	-	-	20	1.7	2.1	1.94	0.1
Tib GL	23	41.0	47.16	44.74	0.15	20	43.6	46.84	45.07	0.96	18	43.24	47.06	45.33	1.0
Tib Dip1	8	6.7	6.84	6.62	0.16	20	5.92	7.64	6.65	0.43	19	6.04	7.26	6.57	0.26
Tib Dip2	24	5.86	6.94	6.35	0.25	20	5.9	6.98	6.34	0.25	20	6	6.68	6.3	0.17
Tib Bd	23	3.58	4.18	3.85	0.14	20	3.7	4.3	3.97	0.15	20	3.64	4.1	3.8	0.14
Tib Dd	23	3.58	4.04	3.83	0.12	20	3.74	4.16	3.91	0.13	19	3.5	4.04	3.83	0.13
Tmt GL	25	26.34	32.24	29.23	1.23	18	27.88	30.7	29.48	0.81	18	27.6	31.04	29.68	0.83
Tmt Bp	25	4.04	4.64	4.3	0.15	19	4.14	4.72	4.36	0.15	18	3.94	4.48	4.24	0.13
Tmt Bd	25	3.06	3.52	3.23	0.12	19	2.92	4.32	3.28	0.3	19	2.92	3.5	3.18	0.13
Tmt SC	25	1.34	1.72	1.54	0.1	17	1.4	1.68	1.53	0.08	19	1.34	1.6	1.47	0.07
Cor GL	24	24.86	27.88	26.42	0.71	19	25.8	28.44	26.95	0.69	20	25.44	27.98	26.98	0.62
Cor Lm	24	23.74	26.74	25.35	0.72	19	24.86	27.3	25.87	0.66	20	24.24	26.98	25.9	0.64
Cor BF	24	5.02	6.98	5.58	0.39	19	5.36	6.96	5.82	0.33	20	5.2	6.04	5.66	0.25
Cor SC	1	-	-	1.44	-	19	1.34	1.66	1.49	0.07	20	1.42	1.64	1.52	0.06
Scap GL	16	30.22	33.46	32.06	1.0	19	30.98	34.44	32.34	0.98	18	29.8	33.7	32.21	1.01
Scap Dic	22	4.88	6.04	5.61	0.28	19	5.4	6.92	5.85	0.35	20	5.22	6.02	5.63	0.21

Table 11.3. (Cont.). Mensural data of the modern *Sturnus vulgaris* samples.

***Sturnus vulgaris* samples**

	Poland					Bulgaria	Irak	Algeria
Measurements	**N**	**Min**	**Max**	**Mean**	**St. dev.**			
Hum GL	12	26.3	28.44	27.51	0.68	27.8, 28.38	27.56, 27.86	27.6
Hum Bp	12	8.3	9.38	8.87	0.3	8.8, 9.2	8.7, 9.22	9.02
Hum Bd	10	7.12	7.62	7.36	0.18	7.32, 7.52	7.48, 7.52	7.42
Hum SC	12	2.52	2.82	2.67	0.8	2.66, 2.84	2.62, 2.72	2.7
Cpt GL	12	19.8	21.18	20.61	0.5	21.24, 21.56	21.6, 21.62	21.02
Cpt Bp	12	4.0	5.02	5.15	0.39	5.34, 5.58	5.22, 5.34	5.3
Cpt Did	12	4.56	4.92	4.71	0.12	4.6, 4.8	4.72, 4.92	4.76
Tib GL	12	43.34	47.6	45.18	1.22	45.98, 46.4	45.5	43.9
Tib Dip1	9	6.2	6.7	6.47	0.17	6.5, 4.76	6.68	6.5
Tib Dip2	12	5.98	6.46	6.14	0.13	6.08, 6.48	6.44	6.38
Tib Bd	12	3.6	4.08	3.87	0.16	3.9, 4.04	3.92	3.9
Tib Dd	12	3.64	3.98	3.84	0.1	3.92, 4.12	3.8	3.84

Table 11.4. *t*- tests between the Scandinavian vs. British samples of modern *Sturnus vulgaris*.

	Scandinavian vs. British samples
Hum GL	ns
Hum Bp	ns
Hum Bd	ns
Hum SC	ns
Uln GL	ns
Uln Bp	**
Uln Dip	-
Uln Did	ns
Rad GL	ns
Rad Bp	ns
Rad Bd	*
Cpt GL	***
Cpt Bp	ns
Cpt Did	ns
Fem GL	ns
Fem Bp	ns
Fem Bd	ns
Fem SC	ns
Tib GL	ns
Tib Dip1	ns
Tib Dip2	ns
Tib Bd	ns
Tib Dd	ns
Tmt GL	ns
Tmt Bp	ns
Tmt Bd	ns
Tmt SC	**
Cor GL	*
Cor Lm	ns
Cor BF	ns
Cor SC	ns
Scap GL	ns
Scap Dic	ns

Key: *** $p< 0.001$, ** $p< 0.01$, * $p< 0.05$, ns Not significant at the 5% level.

11.8. Sexual Dimorphism in the genus *Sturnus*

Sexual dimorphism apparently exists in the three extant European *Sturnus* species (Cramp and Perrins, 1994), although no statistically significant difference was found between the skeletal dimensions of the males and females of the Scandinavian sample. The Scandinavian sample is the one which is likely to be both the least heterogeneous while possesing the most sexed individuals which are approximately equally represented. It would appear therefore that the sexual dimorphism in *Sturnus* is not so large as to create further problems to the identification of the species.

11.9. Morphometric comparison of the fossil samples

Fossils previously identified as *Sturnus*, as well as ones which were originally assigned to the genus *Turdus*, were examined in collections to verify the identifications. Most early (Pre-Devensian) avifaunal assemblages from Britain were included, as well as a large number of Devensian ones. Sites outside Britain studied were those in the collections of the Natural History Museum.

It has not been possible to verify the identity of the proximal carpometacarpus, two distal tarsometatarsi and the fragmentary humerus from the West Runton Freshwater Bed identified by Harrison (1979a) as *Sturnus vulgaris / S. unicolor*, due to their degree of fragmentation. The West Runton Freshwater Bed is Cromerian *sensu stricto* in age and belongs to substages Ib - IIb according to West (1980). It has been suggested that this deposit correlates with either with Oxygen Isotope 15 or 13 (Funnell, 1995) and corresponds approximately in age to 600 Kyr. This would have been, had it been confirmed, the oldest record of the genus in Europe. Similarly, the specimens identified by Harrison as *Sturnus vulgaris* both at Westbury-sub-Mendip, Somerset (Harrison, unpublished notes) and Boxgrove in West Sussex (Harrison and Stewart, 1999) are too fragmentary to be confidently identified. These two sites are similar in age and belong to a period before the Anglian glaciation but post-dating the Cromerian *sensu stricto*. The fossils from these three early British sites are the most important records which are refuted because of their ambiguous morphology and fragmentary nature, and are now referred to *Sturnus / Turdus*. The specimens from these three sites are listed in Table 11.5.

The earliest reliable records of the genus in Europe are therefore those from the Hyaena Stratum in Tornewton Cave, as well as Bacon Hole and Minchin Hole, all of which are believed to correlate with Oxygen Isotope Stage 5 (Sutcliffe *et al.*, 1987; Currant, 1996). This does not mean that the genus was not present in Europe prior to this time as there are few deposits older than this which would be likely to yield evidence for the genus. Most vertebrate fossil sites older than Oxygen Isotope Stage 5 in Britain are open air sites which do not regularly contain many bird remains other than water birds, particularly ducks, and it is only recently that adequate recovery methods have been deployed on these sites. Open air sites only reveal significant quantities of small bird fossils when small diameter meshed sieves are used (Harrison and Stewart, 1999).

The record of the genus *Sturnus* in Britain is relatively unbroken from the oldest confirmed record in Oxygen Isotope Stage 5 to the present day (Table 11.6).

Table 11.5. Material previously identified as *Sturnus* now considered to be indeterminate.

Site	Skeletal elements	Reference
West Runton, Norfolk	Prox. R. carpometacarpus, Dist. L. and R. tarsometatarsi and Dist. L Humerus fragment from West Runton Fresh Water Bed.	Harrison, 1979a
Westbury-sub-Mendip, Somerset	Distal R. tarsometatarsus, Dist. R. Humerus and Dist. R. Ulna from Pink Breccia. Dist. R. Tarsometatarsus from Bed 10.	Harrison, unpublished notes with specimens (Natural History Museum).
Boxgrove, West Sussex	Right coracoid from Sample 86 - 36, Quarry 2, Area GTP3, Unit 4C, two right coracoids from Sample 87 - 251, Quarry 2, Area GTP 3, SQ 43, Unit 4C. Right coracoid from Quarry 2, Area GTP 17, Unit 4C.	Harrison and Stewart, 1999.

Table 11.6. Provenance of *Sturnus* fossils.

Sites	Age	References
London Wall, London	Roman? (43 – 410 AD?)	Harrison, 1980a.
Port Eynon, Gower	Early Holocene (9000 - 6000 BP).	Harrison, 1987a, 1987b.
Chudleigh Gorge, Devon	Late Pleistocene or Early Holocene	Harrison, 1987a.
Pin Hole Cave, Derbyshire	Devensian	Bramwell, 1984.
Bacon Hole, Gower	Oxygen Isotope Stage 5 (probably 5 c - 5a (125 -81 Kyr))	Stringer *et al*, 1986.
Minchin Hole, Gower	Oxygen Isotope Stage 5 (probably 5 c - 5a (125-81 Kyr))	Stringer *et al*, 1986.
Tornewton Cave, Devon Hyaena Stratum Reindeer Stratum	Oxygen Isotope Stage 5 Early Devensian	Sutcliffe and Kowalski, 1976; Harrison, 1980.
Torbryan Caves, Devon	Unknown	
Devil's Tower, Gibraltar	Weichselian	Currant pers. comm.
Tabun, Mount Carmel Caves, Israel	OIS 3 – 5 (dependent on dating technique)	Grün *et al*, 1991; Mercier *et al*, 1995.

Holocene

London Wall

Very little information regarding the excavation at London Wall is available. A single specimen of *Sturnus*, a humerus, is present in the Natural History Museum collection which was originally identified as *Turdus musicus*, the old name for the song thrush and the redwing (*T. philomelos* and *T. iliacus*). Despite the lack of information about the excavation there would appear to be no reason to doubt that it is Romano-British in age in its broadest sense. This would date the humerus to between 43 AD and 410 AD.

Despite the lack of provenance detail and tiny sample, the bone is of interest because of its diminutive size, although within the range of modern British *Sturnus vulgaris* (Figure 11.21).

Table 11.7. *Sturnus* sp. material from London Wall with measurements.

Specimen No.	Skeletal element	Measurements			
		GL	Bp	Bd	SC
A1031	R. humerus	26.4	8.52	7.38	2.64

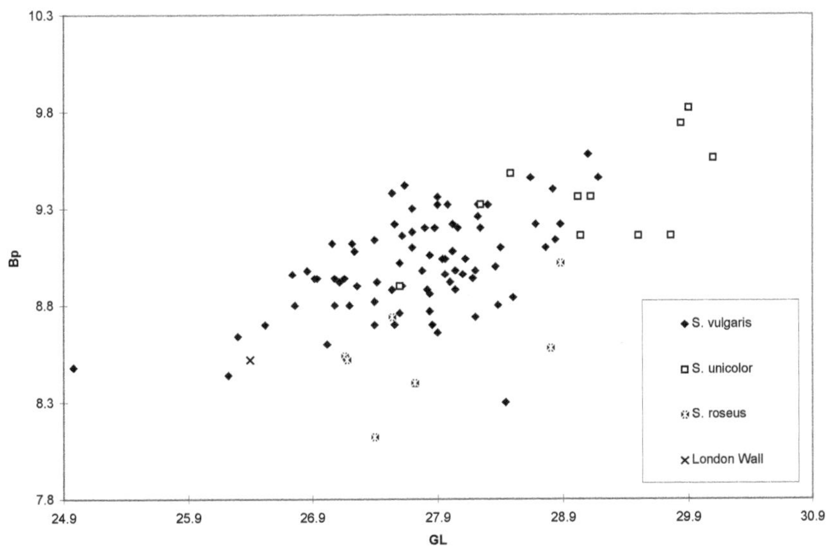

Figure 11.21. *Sturnus* sp. humerus from London Wall plotted with modern.

Port Eynon Cave, Gower, South Wales

Port Eynon is a sea-cliff cave on the southern side of the Gower Peninsula in South Wales. The fossils came from deposits scattered around the cave floor and the fauna has been reported to be Early Holocene (ca. 6000 - 9000 bp) (Harrison, 1987b). However, it is unclear why the early Holocene date has been suggested for the fauna. Pleistocene mammals such as mammoth and woolly rhino were said to have been recovered from the cave in the earliest excavations in the nineteenth century, and lion by the Rev. Cook in the nineteen thirties (Allen and Rutter, 1948). These three records may be irrelevant because the lion has since been reidentified as a bear and the other two species may in fact have come from another cave in the vicinity (Currant, pers. comm.). Currant and Jaccobi (1997) believe the mammalian fauna to be of Early Devensian (O.I.S. 4) in age, although it is possible that some of the birds are Holocene in age (Currant, pers. comm.). The *Gallus* bone described by Harrison led him to consider whether his *Gallus europaeus* (Harrison, 1978a) survived into the Early Holocene (Harrison, 1987b). If the bone does represent *Gallus* then it may be more likely that it represents a domesticated *G. gallus* dating from the Iron Age or later. This is because this species is believed not to have been introduced by man into Britain until the Iron Age. Therefore, there are suggestions that both earlier (Pleistocene) and later (post-Iron Age) material are part of the Port Eynon Cave assemblage. The assemblage is best considered to be a mixture of Devensian and Holocene material until demonstrated otherwise.

The *Sturnus* humerus is relatively short and robust in comparison to modern *S. vulgaris* (Figure 11.22). The carpometacarpus, meanwhile, is of a similar length to modern British *S. vulgaris,* although considerably more robust (Figure 11.23). The restricted number of fossils from Port Eynon limit what can be deduced from their dimensions.

Table 11.8. *Sturnus* sp. material from Port Eynon Cave with their measurements.

Specimen No.	Skeletal element	Measurements			
		GL	Bp	Bd	SC
A5799	L. humerus	26.4	8.96	7.1	2.8
		GL	Bp	Bd	
A5776	R. carpometacarpus	20.92	5.8	5.04	

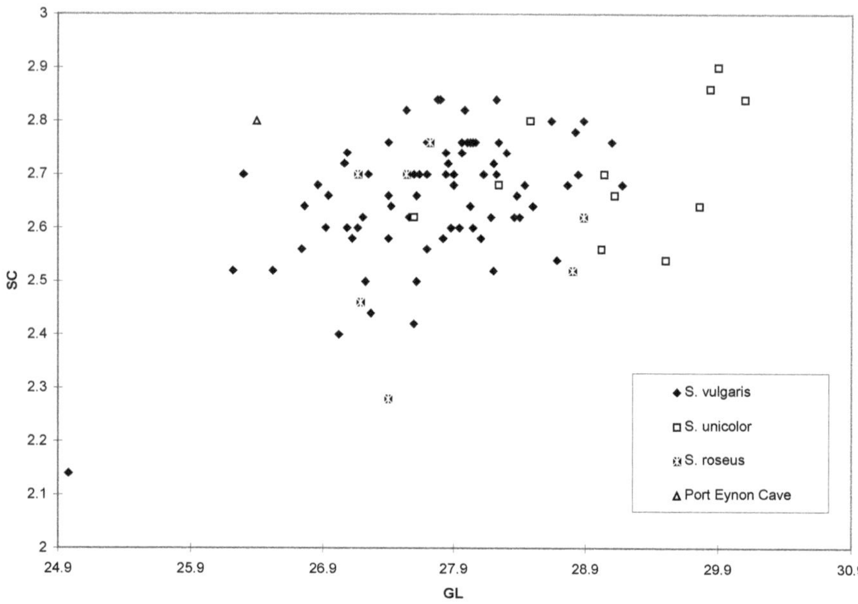

Figure 11.22. ***Sturnus*** **sp. humerus from Port Eynon Cave plotted with modern European** ***Sturnus*** **samples. Greatest length (GL) against shaft breadth (SC).**

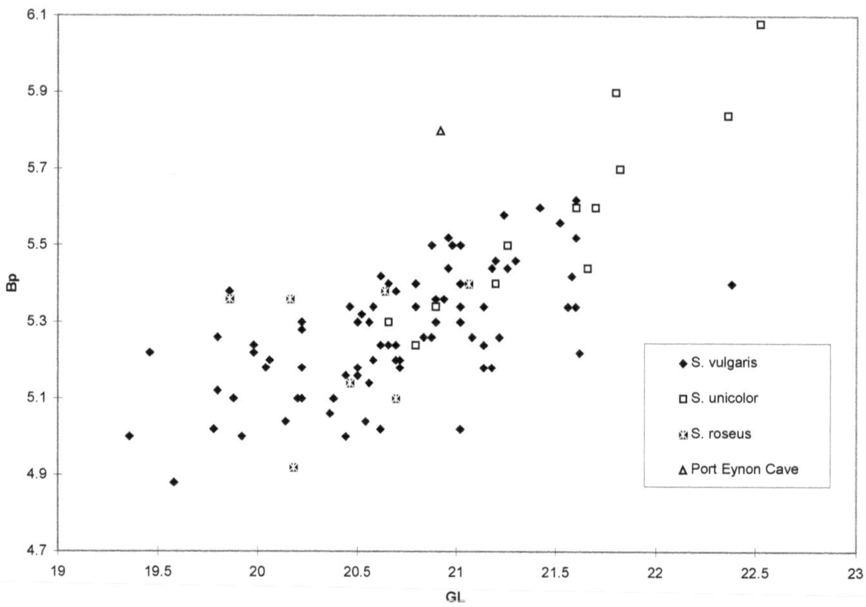

Figure 11.23. ***Sturnus*** **sp. carpometacarpus from Port Eynon Cave plotted with modern European** ***Sturnus*** **samples. Greatest length (GL) against proximal breadth (Bp).**

Devensian

Pin Hole Cave, Derbyshire

Pin Hole Cave is described in detail in Chapter 10 on *Lagopus*. The fossils from Pin Hole are likely to be a mixture of Late Pleistocene and Holocene in age.

Table 11.9. ***Sturnus*** **sp. material from Pin Hole Cave with their measurements.**

Specimen No.	Skeletal element	Measurements			
		GL	Bp	Bd	SC
PH(F) 12376 - 12380					
	L. humerus	27.0	8.75	7.0	2.7
	R. humerus	27.1	8.7	7.1	2.7
PH(F) 9064					
	R. humerus	27.92	8.68+	7.22	2.84
PH(F) 8302 - 8306					
	L. humerus	27.55	9.1	7.5	2.65

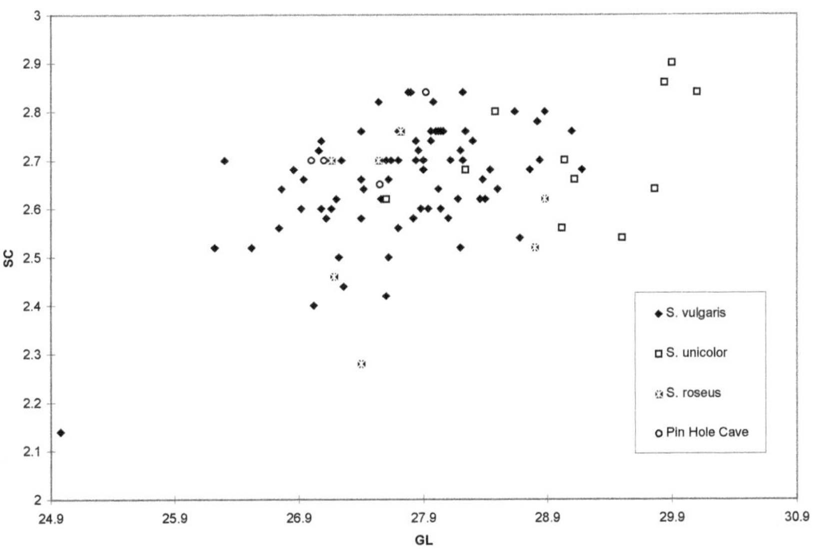

Figure 11.24. *Sturnus* sp. humerus from Pin Hole Cave plotted with modern European *Sturnus* samples. Greatest length (GL) against shaft breadth (SC).

The *Sturnus* humeri from Pin Hole Cave are within the range of morphology of *Sturnus vulgaris* and *S. roseus* today, and smaller than *S. unicolor* (Figure 11.24).

Chudleigh Fissure, Chudleigh Gorge, Devon

A.S. Kennard excavated a fissure deposit in Chudleigh Gorge, Devon, and the fossils were housed in the Natural History Museum in 1919. Unfortunately, the whereabouts of this fissure and details of the strata excavated by Kennard are now missing (Currant, pers comm.). Despite this major drawback, the fossil starlings from this site have proved interesting.

The birds identified by Harrison (1980a, 1987a) would appear to indicate that later Holocene material was part of the faunal assemblage. For instance, the presence of the domestic fowl *Gallus gallus*, first encountered during the Iron Age in Britain, is indicative of a mixed assemblage. Nevertheless, there is also an admixture of Pleistocene fossils as indicated by bird species such as the ptarmigan *Lagopus mutus*, which has not been found in lowland Britain since the Late Glacial (Chapter 9).

The bird remains were originally identified by E.T. Newton (1923), although at first his identifications were published by Bell (1915, 1922). Subsequent to this, Harrison re-examined the remains during his revision of many of the British Pleistocene bird fossils (Harrison, 1980a, 1987a). It should be noted that some of the *Sturnus* bones were identified as *Turdus* by Harrison. The carpometacarpi with larger proximal dimensions (A3497, Figure 11.26) were identified as *Turdus merula* by Harrison, the ones with smaller proximal dimensions as *S. vulgaris*, and all the humeri were identified as *T. iliacus* (A3518). The bones were all re-identified using the characters described in Chapter 7 as *Sturnus*.

Table 11.10. *Sturnus* sp. material from Chudleigh Fissure with their measurements.

Specimen No.	Skeletal element	Measurements			
		GL	Bp	Bd	SC
A3518					
	L. humerus	25.32	-	7.0	2.48
	R. humerus	25.88	8.44	6.98	2.58
	Proximal R. humerus	-	8.6	-	2.5
		GL	Bp	Bd	
A3497					
	R. carpometacarpus	20.76	5.7	4.54	
	R. carpometacarpus	20.8	5.9	4.82	
	R. carpometacarpus	20.32	5.6	4.92	
	L. carpometacarpus	20.3	5.88	4.86	
A5118					
	R. carpometacarpus	20.5	5.12	4.62	
	R. carpometacarpus	20.78	5.5	4.74	
	R. carpometacarpus	20.28	5.14	-	
	R. carpometacarpus	20.36	5.1	4.8	
	R. carpometacarpus	19.92	5.14	4.58	

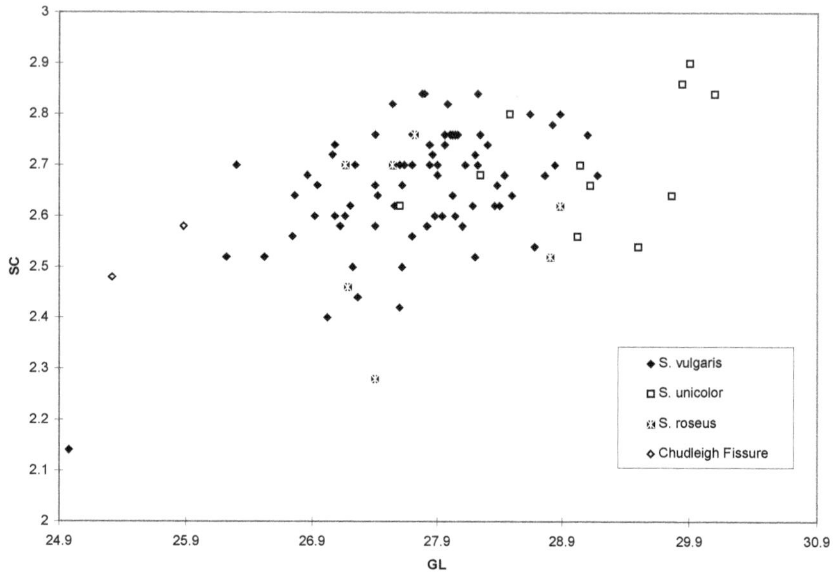

Figure 11.25. *Sturnus* sp. humeri from Chudleigh Fissure plotted with modern European *Sturnus* samples. Greatest length (GL) against proximal breadth (Bp).

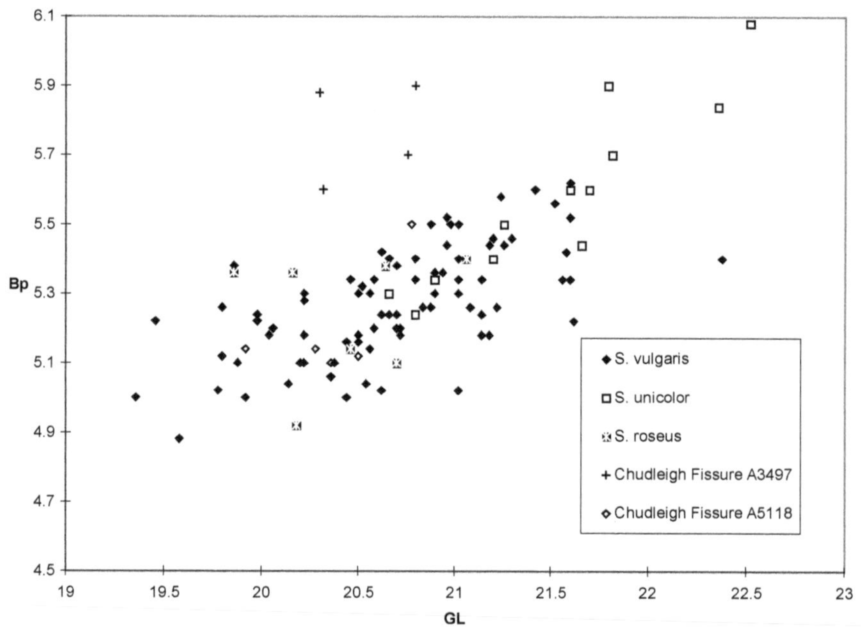

Figure 11.26. *Sturnus* sp. carpometacarpi from Chudleigh Fissure plotted with modern European *Sturnus* samples. Greatest length (GL) against proximal breadth (Bp).

The *Sturnus* humeri from Chudleigh Fissure are short when compared to all *Sturnus* species in Europe today (Figure 11.25). They are also relatively robust in the shaft as seen by the similar dimensions of this part of the bone to the longer modern starlings. The carpometacarpi, on the other hand, are more in line with the modern starlings in length, although their proximal breadths appear to split the bones into two size categories. There are those that are similar in size and shape to modern *Sturnus vulgaris* and *S. roseus*, and those which are considerably broader proximally (Figure 11.26). It is quite possible that two species are present in the assemblage, although it is difficult to determine what they are. An alternative explanation might lie in the chronologically mixed nature of the assemblage and that a single species is represented by different adaptive morphologies at different times. It seems likely that the small humeri correspond to the birds with the more gracile carpometacarpi, while the humeri of the birds with carpometacarpi which are broad proximally are not represented in the assemblage.

Oxygen Isotope Stage 5

Tornewton Cave, Devon

There are two reliable sources of material from Tornewton Cave. First, there is the material collected by A. Sutcliffe from 1944 to the early 1960's (Sutcliffe and Zeuner, 1962; Sutcliffe and Kowalski, 1976). Second is that recently excavated by a team from the British Museum (Roberts, 1996). In the first excavations starlings were recovered from the Reindeer and the Hyaena Strata. However, during the second excavation, the Reindeer Stratum was no longer available because it had been worked out and thus only the Hyaena Stratum yielded *Sturnus* fossils. The two collections shall therefore be treated separately. In addition to this there are collections of starling fossils from Torbryan Valley in the Natural History Museum and Torquay Museum which are unprovenanced, although they may have emanated from Tornewton Cave.

The stratigraphy in Tornewton is very complex and the ages of the various strata have been recently reinterpreted (Currant, 1996). The problems lie in the different biostratigraphic paradigms and in particular the different authors' beliefs regarding the association of different faunal elements. Currant believes that the combination of animals such as *Hippopotamus amphibius* and *Microtus oeconomus* in the Hyaena Stratum is impossible due to their respective warm and cold adapted tolerances. He considers the Hyaena Stratum of a later date than that suggested by Sutcliffe and Kowalski (1976) and that the hippo remains were erroneously associated with the deposit. He would therefore put the deposit in Oxygen Isotope Stage 5c or 5a, rather than 5e (Ipswichian *sensu stricto*). There is no question, however, that the deposits are from a warm part of oxygen isotope stage 5, and this will be assumed here.

The Hyaena Stratum was named due to the dominance of the remains of spotted hyaena (*Crocuta crocuta*) within it, as well as the evidence for prolonged use of the cave as a den by the animals during the period of the deposition. The evidence for this consists of large amounts of coprolitic material, bones and teeth of hyaenas of all ages, as well as digested and gnawed bones of a range of mammals including themselves. Indeed it is very noticeable that the condition of bones belonging to animals likely to have been eaten by hyaenas differs markedly from that of smaller animals which probably entered the cave either of their own accord or as prey of owls and/or diurnal birds of prey. Hence the small mammals and birds are in excellent condition, in contrast to the bones of larger mammals.

The other taxa recorded in the Hyaena Stratum by Sutcliffe consist of wolf (*Canis lupus*), fox (*Vulpes vulpes*), lion (*Panthera leo*), bear (*Ursus* sp.), rhinoceros (*Stephanorhinus hemitoechus*), fallow deer (*Dama dama*), red deer (*Cervus elaphus*), large bovid (*Bos / Bison*) and hippopotamus (*Hippopotamus amphibius*) among the large mammals and water vole (*Arvicola* sp.), hare (*Lepus* sp.), grass vole (*Microtus agrestis*), northern vole (*M. oeconomus*) and wood mouse (*Apodemus sylvaticus*) among the small mammals (Sutcliffe and Kowalski, 1976). The herptiles recorded from the deposit include natterjack toad (*Bufo calamita*), common frog (*Rana temporaria*) and slow worm (*Anguis fragilis*) (Gleed-Owen, 1996). The birds meanwhile include a small falcon (*Falco* sp.), a small dabbling duck (*Anas* sp.), a swallow (*Hirundo* sp.) and a large lark (Alaudidae) in addition to the starling remains (Stewart, 1996 and in prep). The important components of this fauna, which give indications about the age of the deposit and the environmental conditions at the time of deposition of the Hyaena Stratum are the rhino, the fallow deer, the hippo, the northern vole and the wood mouse among the mammals and the natterjack toad among the reptiles and amphibians. The hippo (if correctly assigned to the deposit) might be seen to indicate an Ipswichian age because they are generally considered to have only reached Britain in the later Pleistocene during that interglacial (Gascoyne, 1981; Stuart, 1986, 1991; Currant, 1989). The rhino is indicative of a Middle to Late Pleistocene temperate period. At one time its occurrence would probably have been taken to signify that the deposit was of Ipswichian age although recent finds attributed to Early Devensian (O.I.S. 5c or 5a) deposits mean that this is no longer certain. These late finds of *D. hemitoechus* were made in Bacon Hole, W. Glamorgan and were accompanied by *Palaeoloxodon antiquus* (Stringer *et al.*, 1986). Therefore, a temperate O.I.S. 5 age is likely based on these species. Similarly the occurrence of fallow deer and wood mouse indicate that temperate conditions prevailed during the deposition of the stratum (Currant, 1989). The natterjack toad may also suggest relatively warm conditions because today they only occupy areas of well drained soils and low vegetation cover outside of Iberia, although work on this is still at an early stage (Gleed-Owen, 1997). The apparent contradiction of the occurrence of *Microtus oeconomus* may have been overstated by Currant (1996) because they have clearly been found in unquestionable Ipswichian deposits with hippopotamus such as Swanton Morley in Norfolk (Coxon *et al.*, 1981) and recently by myself in Joint Mitnor Cave in Devon (cf. Sutcliffe, 1960). The name northern vole is also a misnomer because their modern geographic range includes parts of the Netherlands in addition to their occurrence further north (Macdonald and Barrett, 1993). It is therefore tempting to ignore Currant (1996) and to believe Sutcliffe and Kowalski (1976) by considering the Hyaena Stratum to be Ipswichian *sensu stricto* in age (i.e. with hippo). However, because of the occurence of *Stephanorhinus hemitoechus* and *Palaeoloxodon antiquus* in deposits dated radiometrically to later in Oxygen Isotope Stage 5 (5c or 5a) (Stringer *et al.*, 1986), together with abundant *M. oeconomus,* and the paucity of faunal remains from that period of time, it remains a possibility that hippo sites in Britain are not all from O.I.S. 5e, but include ones from 5a and / or 5c (Sutcliffe, pers comm.). For now it is safest to consider the Hyaena Stratum as belonging to O.I.S. 5 in its broadest sense, although a temperate part of that stage is highly likely.

The material which came from the Reindeer Stratum, if it is contemporary with the deposition of that stratum, would be of a Devensian age. There is no controversy about this stratum although an exact age is not yet available. However, there may be a reworked element to the fauna due to the presence of *Clethrionomys glareolus* (Sutcliffe and Kowalski, 1976), a vole usually associated with temperate woodlands (Currant, 1989), and mole (*Talpa europaea*) (Stuart, 1977). Other taxa forming part of this assemblage are wolf, fox, bear, stoat (*Mustela erminea*), hyaena, Man (*Homo* sp.), horse (*Equus caballus*), undetermined rhino, reindeer, hare (*Lepus* sp.) and a large bovid. All these species are consistent with a steppe-tundra habitat. Reindeer and bovids were the most common and may have represented the remains of human activity (Sutcliffe and Zeuner, 1962). The bird species found in the deposit include ptarmigan (*Lagopus mutus*) and red / willow grouse (*Lagopus lagopus*) (see Chapter 9), little bustard (*Tetrax tetrax*) and carrion crow / rook (*Corvus corone / frugilegus*) (Stewart, 1996). The bird species, and particularly the two *Lagopus* species, are also consistent with the taxa found in steppe-tundra faunas throughout Europe during the last glacial period (Tyrberg, 1991a).

Material from the Sutcliffe excavation

In his original description of the bird fossils from the cave based on the Sutcliffe excavations Harrison (1980b) mentioned the presence of carpometacarpi, ulnae, humeri, coracoids and femora as being attributable to the starling *Sturnus vulgaris*. He did, however, describe the material as being "on average slightly larger than those of Recent specimens". He also said two of the humeri were atypical in lacking the anconal fossa near the proximal end of the shaft. These remarks should therefore be confirmed and the phenomena considered futher.

Harrison (1980b) listed the following Natural History Museum specimen numbers as belonging to *Sturnus*: A4105, A4130, A4139, A4140, A4144 - 6. Each of these numbers refers to more than one specimen as can be gauged from Table 11.11. However, not all the specimens could be confirmed as *Sturnus,* due either to their condition or belonging to skeletal elements which have not yielded diagnostic morphological features (Chapter 7). Therefore, A4139 (5 coracoids), A4145 (12 ulnae) and A4146 (4 femora) have been excluded from the present analysis. Similarly, not all material listed under the other numbers could be confidently identified. Therefore, six of the seventeen humeri of A4140 and one of the seven carpometacarpi are excluded from the analysis. All that could be confirmed as being referable to *Sturnus* are listed with their measurements in Table 11.11.

The two humeri described by Harrison as atypical in lacking an anconal fossa were examined and were found to be the two most mineralised specimens. The region on the specimens where their anconal fossa (or second pneumatic fossa) should be is obscured by cemented cave sediment. However, one of these two humeri belongs to *Turdus* based on the shape of the crista pectoralis (see Chapter 7), indicating that this phenomenon was one of preservation quality rather than anatomy i.e. it is not likely that the same atypical state occurred in two different taxa. Nonetheless, the second fossa is one of the last areas on the humeri of both *Sturnus* and *Turdus* to form and is very poorly developed in immature individuals. Sadly, the preservation of the specimens prohibits assessment of the maturity of the bone. Therefore there is no clear evidence for absence of this fossa the Tornewton Cave humeri.

The measurements taken on the humeri when compared to samples of the extant European *Sturnus* species demonstrate that the fossils from Tornewton Cave are on the whole larger than both those of modern *Sturnus vulgaris* and *S. roseus*. They may also differ from those of *S. unicolor* in being somewhat more robust in the shaft on average (Figure 11.27). Figure 11.27 also shows that one specimen falls in the extreme small part of the range for modern European *Sturnus*, rivalling even some of the smallest *S. vulgaris* and *S. roseus*. The small size of the specimen in relation to the larger specimens and the range of sizes that are accomodated in extant species makes it unlikely that it is conspecific with the larger specimens. With this in mind the exact provenance of individual specimens becomes interesting. However, this does not easily explain the pattern as large starlings are present in both the Hyaena and Reindeer Strata while the small specimen is from the Reindeer Stratum. The range of sizes present in the Reindeer Stratum raises the possibility that two sympatric congeners were present at the time of its deposition. The alternative explanation would be that there has been reworking of the large specimens from the Hyaena into the Reindeer Stratum. However, it is not unreasonable to suggest that a large species of *Sturnus* was present in Devon during Oxygen-isotope 5a or 5c which survived later into the Devensian, when it was joined by a smaller species. The small specimen's proximal end is unfortunately damaged so it is not plotted on Figure 11.28. Figure 11.28, however, shows that in proximal breadth, the next smallest specimen is rather small in relation to the larger specimens and extant taxa. This may suggest that two species of *Sturnus* were living sympatrically during the time of deposition of the Reindeer Stratum.

Table 11.11. *Sturnus* sp. material from Sutcliffe excavations with their measurements.

Specimen No.	Skeletal element	Measurements			
		GL	Bp	Bd	SC
A4140					
Reindeer Stratum	L. humerus	28.7	9.76	7.86	2.88
	R. humerus	29.56	9.54	7.9	3.0
	R. humerus	30.1	10.0	8.1	2.88
	R. humerus	26.58	-	7.4	2.36
Hyaena Stratum	L. humerus	29.42	9.6	-	2.82
	R. humerus	28.3	9.2	-	2.82
	R. humerus	30.22	9.68+	8.04	2.9
	Prox. L. humerus	-	9.78	-	2.8
Stratum?	Prox. R. humerus	-	9.66	-	3.06
A4105					
Reindeer Stratum	R. humerus	29.52	-	8.15	3.24
A4144					
Reindeer Stratum	R. humerus	27.98	8.51	6.87	2.68
		GL	Bp	Bd	
A4153					
Reindeer Stratum	L. carpometacarpus	20.65	5.5	4.8	
	L. carpometacarpus	22.0	5.75	5.2	
	L. carpometacarpus	20.0	5.2	-	
	R. carpometacarpus	20.65	5.3	5.0	
	R. carpometacarpus	20.75	5.65	4.95	
	Prox. R. carpometacarpus	-	5.6	-	

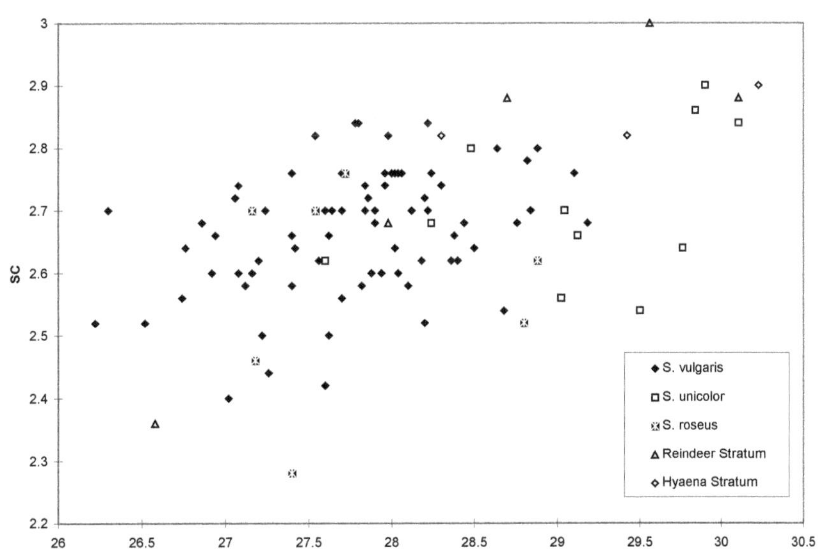

Figure 11.27. *Sturnus* sp. humeri from Tornewton Cave plotted with modern European *Sturnus* samples. Greatest length (GL) against shaft breadth (SC).

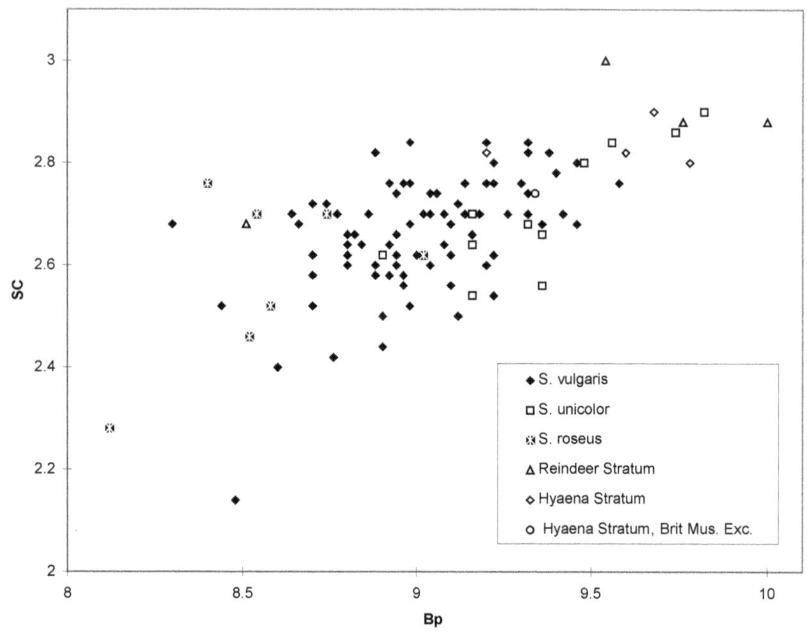

Figure 11.28. *Sturnus* sp. humeri from Tornewton Cave plotted with modern European *Sturnus* samples. Proximal breadth (Bp) against shaft breadth (SC).

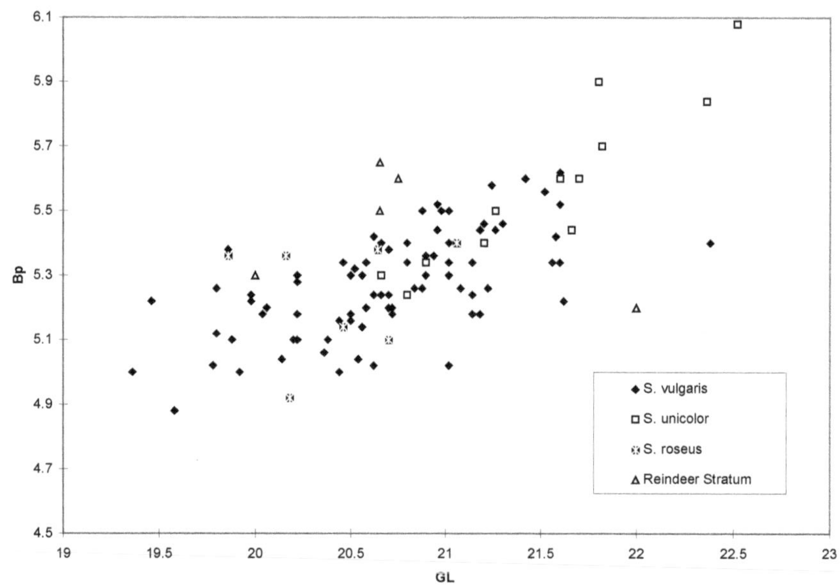

Figure 11.29. *Sturnus* sp. carpometacarpi from Tornewton Cave plotted with modern European *Sturnus* samples. Greatest length (GL) against proximal breadth (Bp).

The carpometacarpi from Tornewton Cave, which all came from the Reindeer Stratum, are mostly robust, as a result of being relatively short in relation to the humeri and mostly having higher proximal breadths than modern *Sturnus* spp. (Figure 11.29). There is one carpometacarpus, however, that is long and thin. These findings provide evidence that the *Sturnus* species in the reindeer Stratum are not simply isometrically similar to the local species in Devon today. The next challenge is to associate the different fossils with each other: which of the carpometacarpi, the gracile one or the robust ones belong respectively to the large and small humeri? Since articular breadth needs to be compatible for articulation, it would appear that the more robust carpometacarpi belong to the larger humeri and the gracile ones to the smaller humeri.

Therefore, Tornewton Cave may have revealed the presence of two species of *Sturnus*, one present in both the Hyaena Stratum and the Reindeer Stratum which was large with relatively short carpometacarpi, and another, present only in the Reindeer Stratum, which was small with relatively long carpometacarpi. The association is, however, tentative at present and needs to be tested with further material.

Material from British Museum excavation

The material excavated by the British Museum between 1989 and 1992 was the result of a more thorough sampling programe (ie. use of fine meshed sieves) than hitherto, although due to the limited extent of remaining deposits, produced far fewer measurable *Sturnus* fossils than the Sutcliffe excavation. The material which could be measured and confirmed as *Sturnus* is listed in Table 11.12.

The proximal humerus whose measurements are plotted in Figure 11. 29 is at the upper end of the size range of *S. vulgaris* in Devon today and in the middle of the range of

Table 11.12. *Sturnus* sp. material from British Museum excavation with their measurements.

Specimen No.	Skeletal element	Measurements			
		GL	Bp	Bd	SC
Hyaena Stratum TN92-252	Prox. R. humerus	-	9.34	-	2.74
		GL	Bp	Bd	
Hyaena Stratum TN92-252	Prox. R. carpometacarpus	-	5.58	-	
	Prox. R. carpometacarpus	-	4.96	-	

S. unicolor. It is undoubtedly a further example of the large starlings found in the Sutcliffe excavation although not as large as most of those specimens. The two carpometacarpi, both proximal ends, are incomplete and therefore it is difficult to consider how robust they are. Nevertheless one of the two is similar in size to carpometacarpi found previously in Tornewton Cave's Reindeer Stratum (Table 11.12). The other carpometacarpus, however, is smaller.

Torbryan Caves

This material is part of the collection of James Lyon Widger who was active in excavating various caves in Torbryan Valley between 1865 and 1890 (Measurements are listed in Table 11.13). Unfortunately, this collection lacks provenance data and it is impossible to know for sure from which of the caves the fossils came. Walker and Sutcliffe (1968) list the caves in the valley known to Widger which were Tornewton Cave, The Old Grotto, Three Holes Cave, Broken Cavern and an unnamed cave. It appears that he did not excavate The Old Grotto but the present fossils could have emanated from any of the others. It is tempting, considering the lack of starlings found in either Broken Cavern or Three Holes Cave during the British Museum excavation (Stewart, 1996 and in prep), to conclude that they were not from these caves, but that would be dangerous. No matter which cave or caves this material came from it is interesting that some of it was misidentified as blackbird (*Turdus merula*). This error is no doubt due to their being large (one of these is plotted in Figure 11.30), presumably the same starling taxon as from the more recent excavations. In addition, however, there were some correctly identified *Sturnus* humeri which were smaller (Figure 11.30). The lack of provenance information prevents further interpretation of the starlings excavated by Widger.

Table 11.13. *Sturnus* sp. material from the Widger collection with their measurements.

Specimen No.	Skeletal element	Measurements			
		GL	Bp	Bd	SC
A1134					
	L. humerus	27.6	8.8	7.5	2.8
	R. humerus	27.0	8.55	7.35	2.6
	R. humerus	27.45	8.55	7.1	2.65
A5150					
	L. humerus	29.44	9.82	7.9	2.9
	R. humerus	30.12	-	7.84	2.62
	Prox. L. humerus	-	9.44	-	2.54

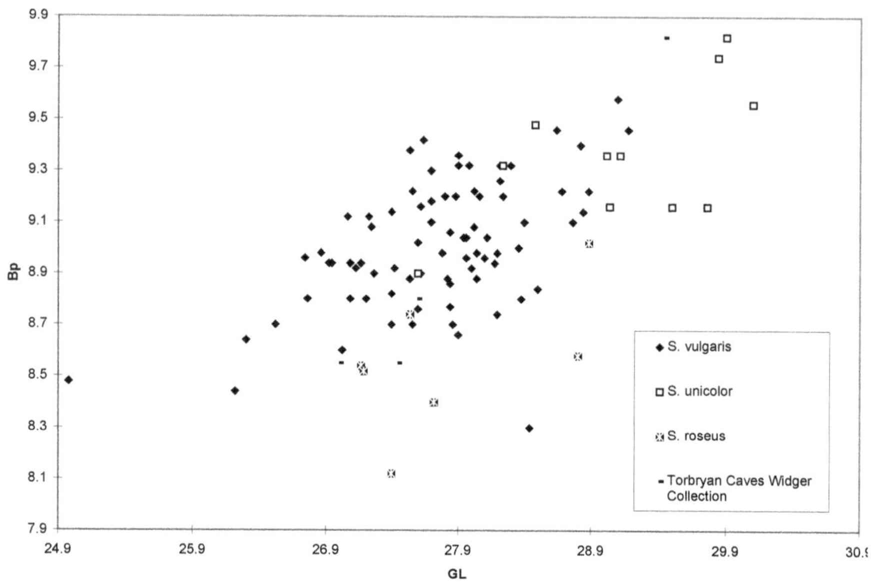

Figure 11.30. *Sturnus* sp. humeri from Torbryan Valley plotted with modern European *Sturnus* samples. Greatest length (GL) against proximal breadth (Bp).

Bacon Hole, West Glamorgan

The Bacon Hole succession has proved important in the development of our knowledge of the detailed history of the Late Pleistocene. In particular, it has helped elucidate the complexity of the Ipswichian / Devensian transition which until the advent of the oxygen isotope record had been assumed to be simple and gradual. Bacon Hole was the first British site to reveal that after the climatic optimum of O.I.S. 5 (5e) there was a cooler wetter temperate phase some time between 81 and 125 Kyr., probably equivalent to either O.I.S. 5a or 5c, at the begining of the Devensian (Stringer *et al.*, 1986).

A representative transverse section of some of the deposits outside the cave can be seen in Figure 11.31, and Table 11.14 gives the lateral equivalence of deposits ouside and inside the cave. The lowest deposit in the sequence is a coarse sand of aolean origin. The fauna associated with this deposit includes a large form of northern vole (*Microtus oeconomus*) and a small horse (*Equus* sp.) both of which would be consistent with a correlation of this deposit with Oxygen Isotope Stage 6 (Sutcliffe *et al.*, 1987; Bates *et al.*, 1998). The climate at this time was cold and dry.

The next three units are the Sandy Breccio-Conglomerate, the Sandy Cave Earth and the Shelly Sand whose molluscan and vertebrate fauna are indicative of a warm temperate interglacial environment. Amongst the rodents are the wood mouse (*Apodemus sylvaticus*), the bank vole (*Clethrionomys glareolus*), the water vole (*Arvicola* sp.) and field vole (*Microtus agrestris*) which is similar to the modern British fauna. The discovery of immature Cory's Shearwater (*Calonectris diomeda*) in these deposits 5° N of its present breeding range is significant as it implies higher temperatures than the present day (Harrison, 1977a). As well as the relatively high temperature the bird's occurrence together with the molluscs indicate close proximity to the comtemporary sea level. These deposits have therefore been attributed to the Ipswichian interglacial or O.I.S. 5e and correlated with the "Patella Beach" in Minchin Hole (Sutcliffe *et al.*, 1987).

The next unit is the Grey Clays, Silts and Sands which appears to represent a gradually deteriorating climate with a decrease of warm indicators such as *A. sylvaticus* and *C. glareolus*, while *M. oeconomus* appears to be replacing *M. agrestis* implying damper conditons. Significantly, the northern vole in this deposit is smaller than in the Coarse Sand deposit and this has been attributed to a warmer climate in accordance with Bergmann's Rule (Stuart, 1982 and Sutcliffe *et al.*, 1987). Also the only find of *Mammuthus primigenius* from the site came from this deposit.

The starling fossils were found in the next two deposits: the Upper Sands (Unit H) and the Upper Cave Earth (Unit I). These deposits also appear to indicate colder conditions than earlier deposits, although still temperate. Particularly significant in this regard are the occurences of *Palaeoloxodon antiquus* and *Stephanorhinus hemitoechus* in the Upper Cave Earth both of which are Middle and Late Pleistocene warm stage species. It is these deposits which were radiometrically dated as late O.I.S.5 coinciding with either O.I.S.5a or 5c (> 81 Kyrs.) (Stringer *et al.*, 1986).

The fossils from Bacon Hole confirmed as belonging to the genus *Sturnus* came from Unit 3b in Area I, Unit 3a in Area II, Unit I and H in Area IV and Unit K and W in Area VIII (Table 11.14). Some of these deposits are lateral equivalents, while others were laid down in relatively quick succession. Therefore, it is legitimate to treat the fossils from these various deposits as a cohesive sample.

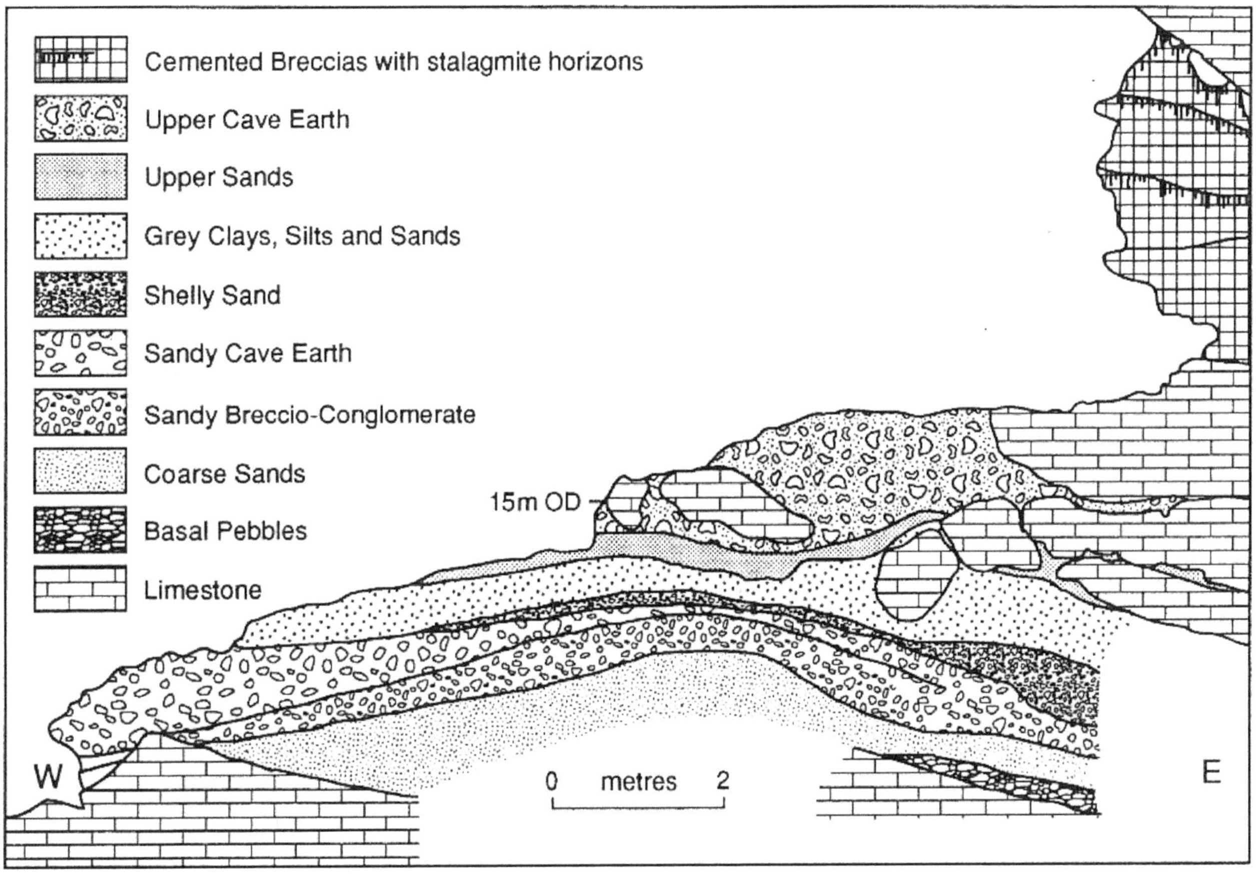

Figure 11. 31. Representative transverse (E-W) section of platform deposits of Bacon Hole (After Sutcliffe *et al.* 1987).

Table 11.14. Lateral equivalence of units inside and outside of Bacon Hole.

Deposits	Area I	Area II	Area IV	Area VIII
Upper Sands			Unit H	
Upper Cave Earth		Unit 3a	Unit I	Unit K
Cemented breccia	Unit 3b		Unit J	Unit W

The *Sturnus* humeri from Bacon Hole are larger on average than any representatives of that genus today (Figure 11.32 and 11.33). Figure 11.33 has more humeri plotted than Figure 11.32 because some specimens are incomplete. Figure 11.33 therefore more accurately reflects the range of size and proportions present in the population.

The carpometacarpi are slightly displaced from modern populations of *Sturnus* species in the robusticity of their proximal ends (Figure 11.34). While their length is on the whole relatively great there is possibly more overlap with *S. unicolor* and even larger *S. vulgaris,* than is seen with the humeri in Figures 11.32 - 33. It appears that the robusticity of the bones more consistently reflects these birds' size than does the length of individual bones alone.

Table 11.15. *Sturnus* sp. material from Bacon Hole with their measurements.

Specimen No.	Skeletal element	Measurements			
		GL	Bp	Bd	SC
Area II, Unit 3a					
	Prox. L. humerus	-	9.46	-	-
Area IV, Unit I					
Field No. 225	L. humerus	29.50	9.82	7.9	2.88
Area IV, Unit H					
Field No. 219	Prox. L. humerus	-	-	-	2.82
	Prox. L. humerus	-	9.05	-	-
Area VIII, Unit K					
	Prox L. humerus	-	9.07	-	2.78
Area VIII, Unit W					
Field No. 224	L. humerus	30.05	9.69	7.92	2.99
	R. humerus	30.06	9.86	8.27	3.1
Field No. 177	Prox. R. humerus	-	9.54	-	3.03
	Prox. L humerus	-	9.85	-	3.16
Field No. ???	L humerus	30.83	10.28	8.09	3.18
		GL	Bp	Bd	
Area I, Unit 3b					
	L. carpometacarpus	21.81	5.95	5	
	L. carpometacarpus	21.66	5.87	-	
Area IV, Unit I					
Field No. 236	L. carpometacarpus	21.27	5.64	5.03	
Area IV, Unit H					
Field No. 219	L. carpometacarpus	21.14	5.7	5.08	
	R. carpometacarpus	20.3	5.3	4.45	
Field No. 248	Prox. L. carpometacarpus	-	5.82	-	

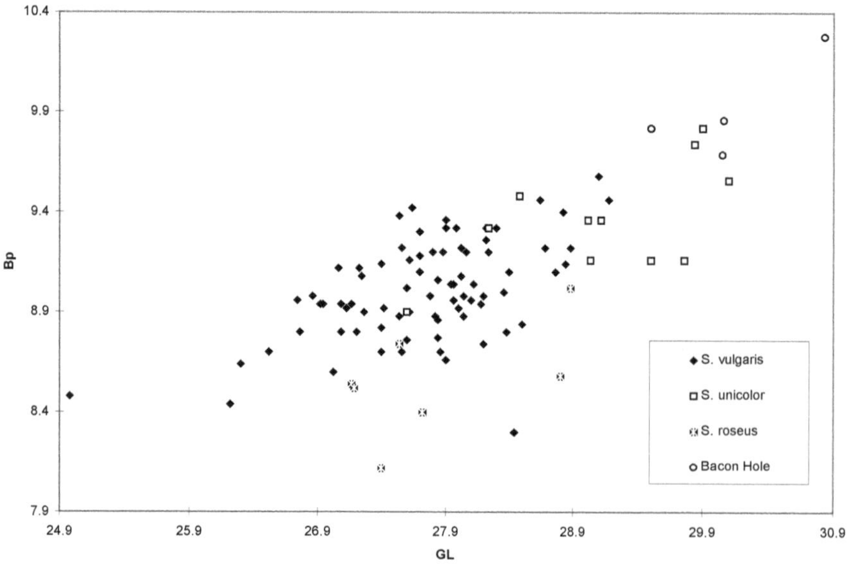

Figure 11.32. *Sturnus* sp. humeri from Bacon Hole plotted with modern European *Sturnus* samples. Greatest length (GL) against proximal breadth (Bp).

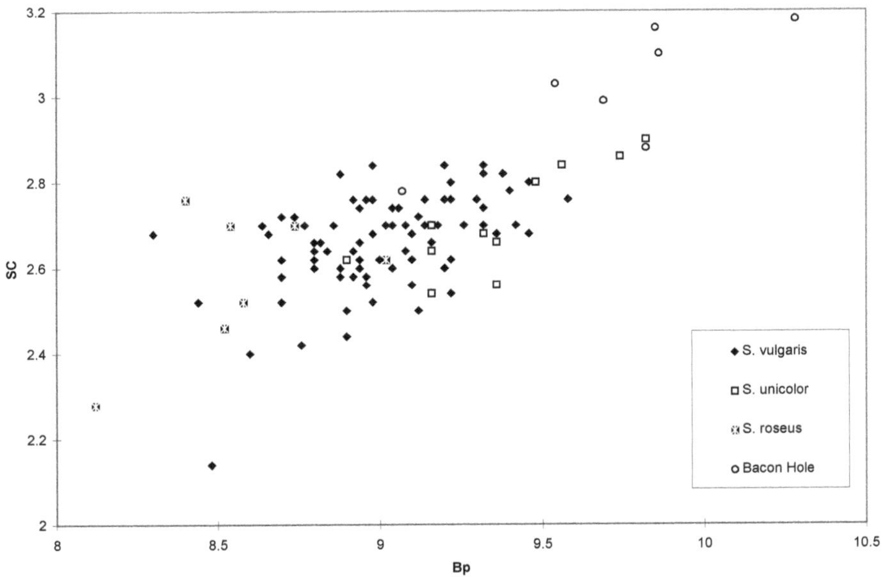

Figure 11.33. *Sturnus* sp. humeri from Bacon Hole plotted with modern European *Sturnus* samples. Proximal breadth (Bp) against shaft breadth (SC).

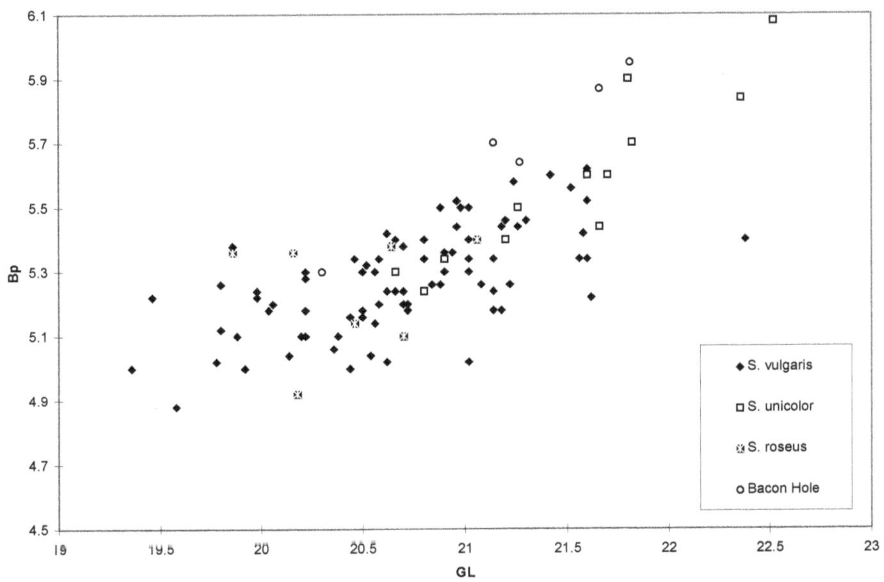

Figure 11.34. *Sturnus* sp. carpometacarpi from Bacon Hole plotted with modern European *Sturnus* samples. Greatest length (GL) against proximal breadth (Bp).

Minchin Hole, West Glamorgan

Minchin Hole, like Bacon Hole, is on the Gower Peninsula and has similar deposits, some of which are thought to be of equivalent ages (Sutcliffe *et al.*, 1987). The main difference between the two caves is that Minchin Hole appears to have had a more direct connection with the sea, reflected by a greater degree of marine influenced sediments such as the "Inner Beach", the "Patella Beach" and the "Neritoides Beach".

The oldest fill of the cave is the "Inner Beach", which is a thick deposit of bedded sand with marine molluscs. This deposit is thought to be an inter-tidal beach which sets the sea level three meters above the modern. The surface of the "Inner Beach" is covered by a thin white deposit which is incipient stalagmite and is followed by the "Lower Red Cave Earth". The latter deposit contains large *Microtus oeconomus* which may correlate it with the Coarse Sands of Bacon Hole and give an age of O.I.S. 6. The deposit also apparently represents a lowering of sea level. Next in the sequence is the "Patella Beach", which was deposited by a renewed high sea level and which lies unconformably on the previous deposits. It is a storm beach and includes marine molluscan taxa such as *Patella vulgata*, *Nucella lapillus* and *Littorina* species. The sea level represented is two meters higher than today. On top of this deposit is a broken block of flowstone with terrestrial molluscs which was dated by Uranium Series to between 107 and 127 Kyrs. This signifies that the Patella beach cannot be much older than 127 Kyrs old and probably represents OIS 5e. Then follows the deposit from which many of the *Sturnus* fossils derive, the "Neritoides Beach" or "Neritoides Deposit". This deposit consists of many small littoral gastropods and contains many small vertebrate remains. It is thought to be a regressive phase of the "Patella Beach" high sea level and therefore a later part of OIS 5. The mammalian species present in the "Neritoides Deposit" include *Apodemus sylvaticus*, *Clethrionomys glareolus*, *Microtus agrestis*, *Crocuta crocuta*, *Panthera leo*, *Dama dama* and *Sus scrofa*. Next comes the Earthy Breccia Series, thought to represent rock-falls inside the cave or material from the cliff slope. The fauna of this deposit appears to be cooler in character, although still temperate, than the "Neritoides Deposit", with small *Microtus oeconomus* and *Stephanorhinus hemitoechus*. After this are a series of deposits of limited interest and largely worked out by excavators before Sutcliffe.

The significance of the cave rests in the two different high sea level stands which may correspond to Oxygen Isotope Stage 7 (Inner Beach) and part of OIS 5 for the "Patella Beach" (Sutcliffe *et al.*, 1987).

Sturnus fossils were found in unit 2b, unit 4b ("Neritoides Deposit"), the "Pink Bone Horizon" and in the "Bouldery Brown Cave Earth" of the "Hyaena Tunnel". While these deposits are not all mentioned in Sutcliffe *et al.* (1987), they are apparently more or less contemporary with the "Neritoides Deposit" and of late OIS 5 age (Currant, pers. comm.) and can thus be lumped together as a single sample.

The humeri belonging to *Sturnus* at Minchin Hole are large and of similar proportions to *S. unicolor*, while larger on average than those of *S. vulgaris* and *S. roseus* (Figure 11.35). The single carpometacarpus from the cave, like the humerus, is large and of the same order of size and proportion as *S. unicolor* (Figure 11.36). The samples are, however, very small.

The single tibiotarsus again falls in the large size range of modern European *Sturnus* (Figure 11.37).

Table 11.16. *Sturnus* sp. material from Minchin Hole with their measurements.

Specimen No.	Skeletal element	Measurements				
		GL	Bp	Bd	SC	
"Pink Bone Horizon"	L. humerus	29.68	9.76	7.92	2.82	
Unit 2b	L. humerus	29.4	9.4	-	2.7	
"Hyaena Tunnel" Field No. 172	L. humerus	29.16	-	7.76	2.76	
"Hyaena Tunnel" Field No. 211	L. humerus	29.84	9.82	8.02	2.96	
		GL	Bp	Bd		
Unit 4b "Neritoides Deposit"	L. carpometacarpus	21.8	5.64	4.04		
	Distal L. carpometacarpus	-	-	4.9		
		GL	Dip (V.D.D.)	Dip (J.R.S.)	Bd	Dd
"Hyaena Tunnel" Field No. 211	Prox. R. tibiotarsus	-	6.76	7.06	-	-
Unit 4b "Neritoides Deposit"	R. tibiotarsus	46.96	-	7.12	4.2	4.26
	Prox. L. tibiotarsus	-	6.96	-	-	-

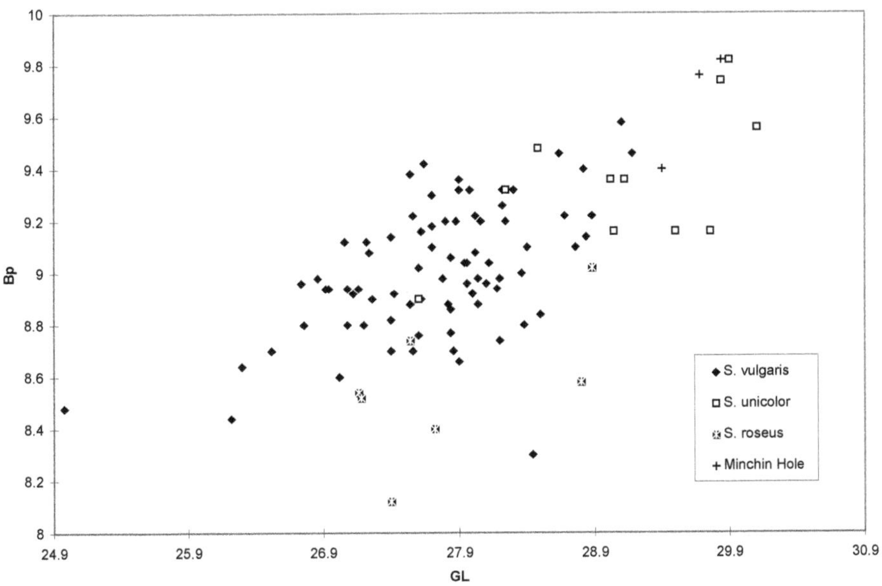

Figure 11.35. *Sturnus* sp. humeri from Minchin Hole plotted with modern European *Sturnus* samples. Greatest length (GL) against proximal breadth (Bp).

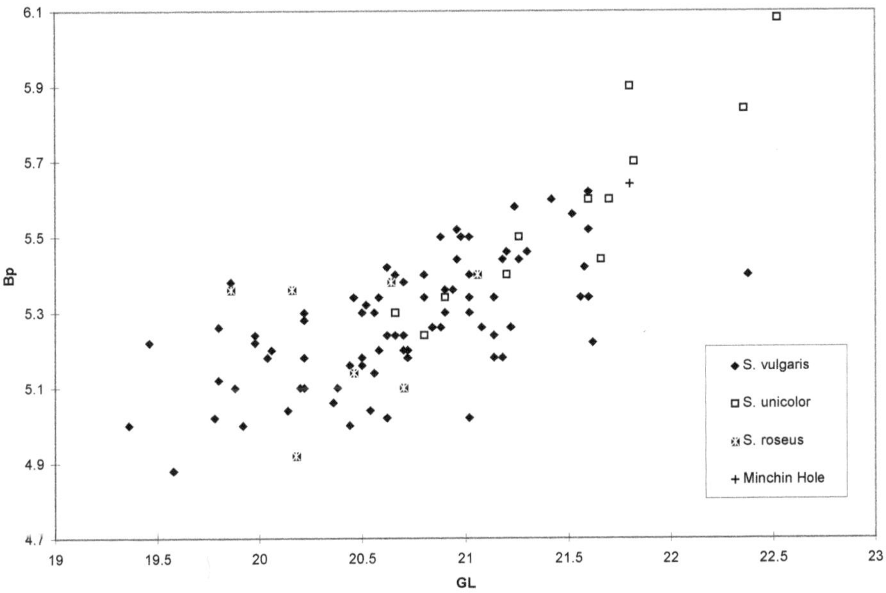

Figure 11.36. *Sturnus* sp. carpometacarpi from Minchin Hole plotted with modern European *Sturnus* samples. Greatest length (GL) against proximal breadth (Bp).

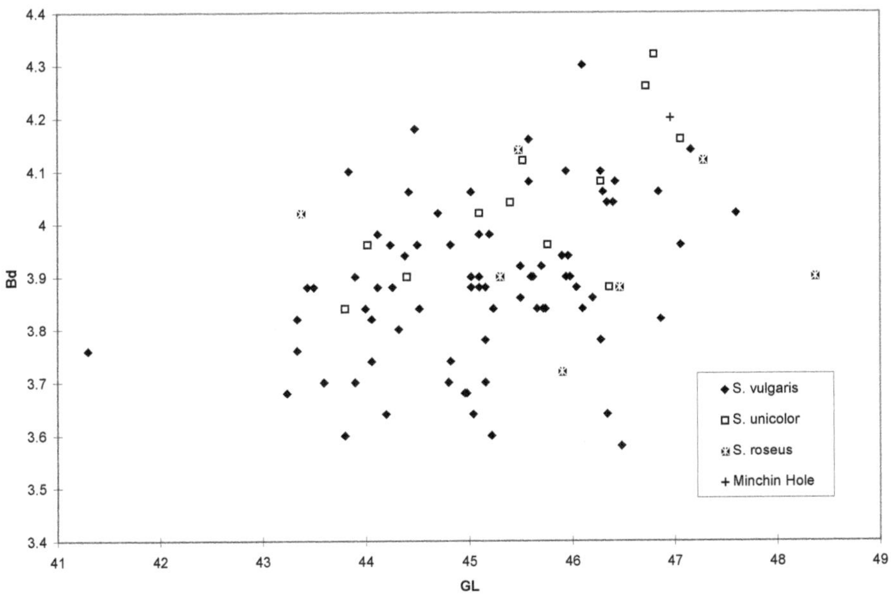

Figure 11.37. *Sturnus* sp. tibiotarsi from Minchin Hole plotted with modern European *Sturnus* samples. Greatest length (GL) against distal breadth (Bd).

Weischelian – Gibraltar

Devil's Tower, Gibraltar

The *Sturnus* fossils from Devil's Tower were excavated by D.A.E. Garrod (Garrod *et al.*, 1928). The age of the fossils from this cave, as is the case with many older excavations, is difficult to gauge very accurately. However, the *Sturnus* fossils appear to have been present together with animals which are undoubtedly Pleistocene such as elephant and hyaena (D.M.A. Bate in Garrod *et al.*, 1928). In addition the presence of Mousterian artifacts from these deposits would eliminate the most recent part of the Pleistocene, the last 25 000 years (Davies, pers. comm.). The assemblage is probably early to middle Weischelian, although an earlier date cannot be ruled out.

The three different *Sturnus* skeletal elements represented at Devil's Tower, the humeri, the carpometacarpus and the tibiotarsus, are in the general size range of *Sturnus vulgaris* and *S. roseus* (Figures 11.38 - 11.41). Although all the bones except one (the carpometacarpus) are within the length range of *S. unicolor*, they do have an unusual range of breadth distributions. They are either gracile, like one humerus and a tibiotarsus, or stout like the carpometacarpus. This forces one to consider the possibility that more than one species of *Sturnus* is represented at Devil's Tower.

Table 11.17. *Sturnus* sp. material from Devil's Tower with their measurements.

Specimen No.	Skeletal element	Measurements				
		GL	Bp	Bd	SC	
A1835	L. humerus	27.74	9.14	-	2.68	
	L. humerus	28.08	-	7.56	2.86	
A1834	R. humerus	29.16	8.74	6.52	2.34	
		GL	Bp	Bd		
A7440	L. carpometacarpus	20.42	5.62	4.44		
		GL	Dip1	Dip2	Bd	Dd
A6786	R. tibiotarsus	46.64	-	6.36	3.6	3.74

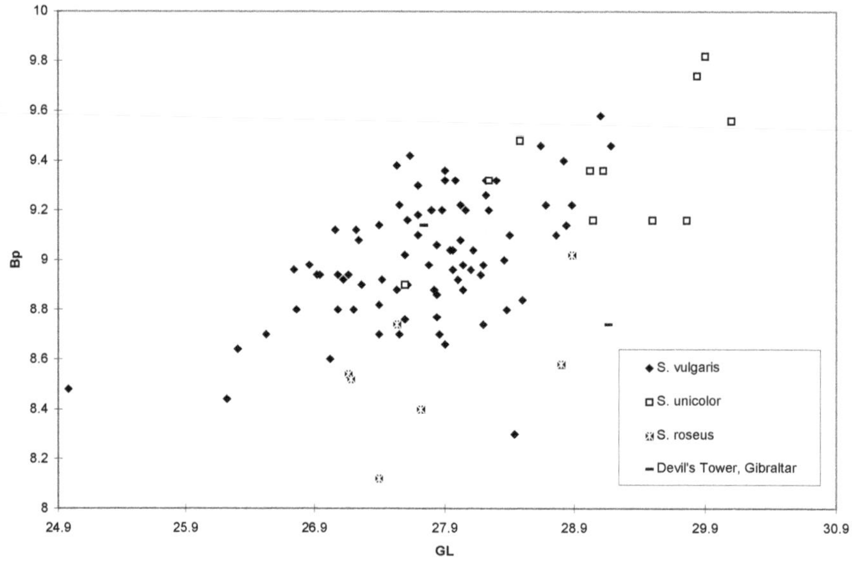

Figure 11.38. *Sturnus* sp. humeri from Devil's Tower plotted with modern European *Sturnus* samples. Greatest length (GL) against proximal breadth (Bp).

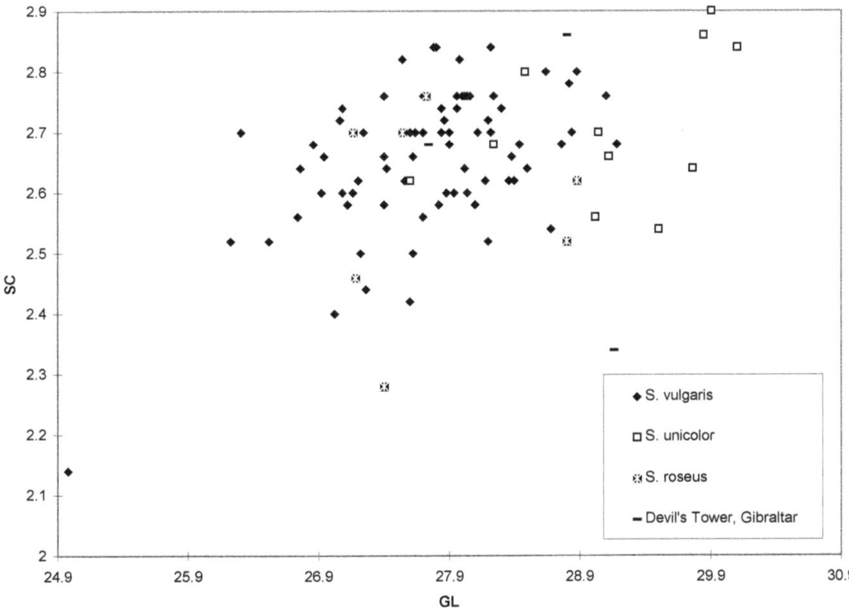

Figure 11.39. *Sturnus* sp. humeri from Devil's Tower plotted with modern European *Sturnus* samples. Greatest length (GL) against shaft breadth (SC).

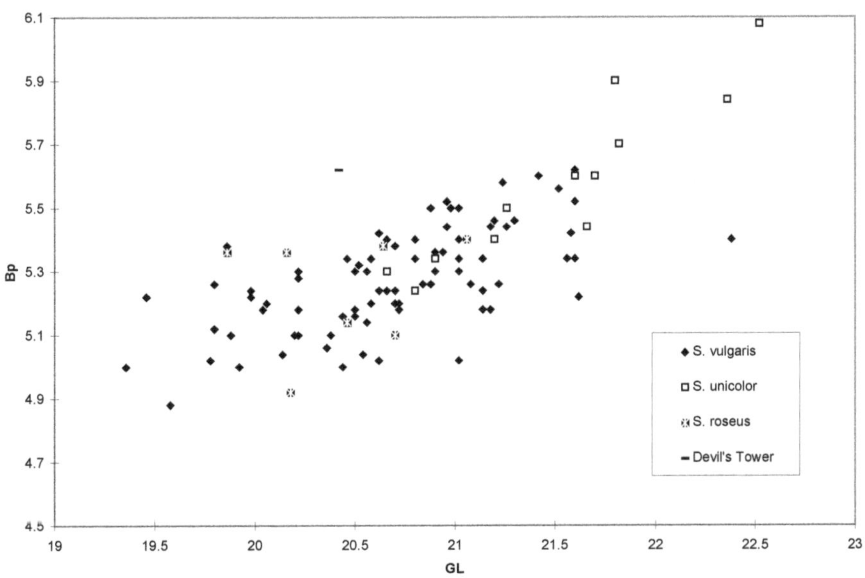

Figure 11.40. *Sturnus* sp. carpometacarpi from Devil's Tower plotted with modern European *Sturnus* samples. Greatest length (GL) against proximal breadth (Bp).

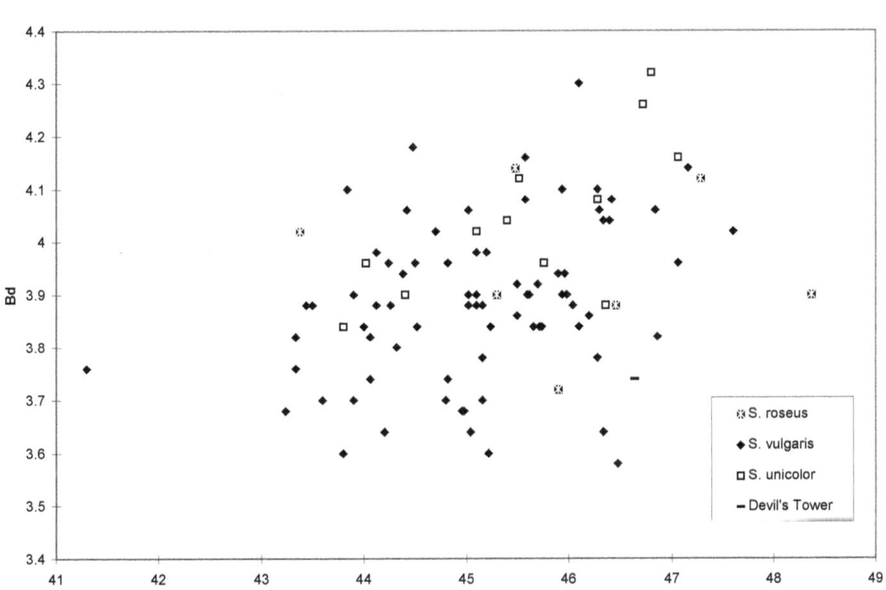

Figure 11.41. *Sturnus* sp. tibiotarsi from Devil's Tower plotted with modern European *Sturnus* samples. Greatest length (GL) against distal breadth (Bd).

Table 11.18. Oxygen Isotope Ages of Layers at Tabun with *Sturnus* fossils according to different authors.

Author	Layer C	Layer D	Layer Eb
Garrod and Bate, 1937.	Lower Levalloiso-Mousterian		Upper Acheulean (Micoquan)
Jelineck, 1982.	Stage 3	Stage 4/5	Stage 5
Bar-Yosef, 1989.	Stage 3/4	Stage 5	Stage 5/6
Grün *et al.*, 1991.	Stage 5	Stage 6	Stage 6
Mercier *et al.*, 1995.	Stage 7	Stage 8	Stage 9

Pleistocene, Israel

Tabun, Mount Carmel Caves, Israel

The Tabun Cave is located on the western edge of Mount Carmel at the entrance to Wadi Mughara, about 20 km south of Haifa in Israel (once Palestine). The *Sturnus* fossils from Tabun were collected By D. Garrod between 1929 and 1934. The specimens listed in Table 11.19 derive from Layers C, D and Eb which were initially described as Late Levalloiso-Mousterian (Upper Pleistocene) in age (Garrod and Bate, 1937). Recent attempts to date these deposits by Electron Spin Resonance (Grün *et al.*, 1991) and Thermoluminecence dating (Mercier *et al.*, 1995) have, however, produced conflicting results, although there is consensus that these deposits are older than previously believed (Table 11.18). In fact the respective deposits became older with each subsequent attempt at dating. It is therefore perhaps best to consider the starlings in relation to the contemporary mammalian faunas. Hence, the birds will be considered in relation to the environment indicated by those faunas. It should be borne in mind, however, that the situation of the cave signifies that a wide range of habitats fell into its potential catchment (Garrod and Bate, 1937).

The mammalian faunas of Tabun are characterised by a combination of persistent species such as gazelle (*Gazella gazella*) and fallow deer (*Dama dama mesopotamica*) as well as an admixture of a number of various other components which are less persistent. According to Bate in Garrod and Bate (1937) the microfauna includes a number of extinct taxa at lower levels which disappear and are replaced by modern species by Tabun B (dated to either OIS 3 or 5 (Grün *et al.*, 1991; Mercier *et al.*, 1995)). The gazelle and fallow deer are the commonest elements and because they live in very different habitats, dry desert and more moist woodlands respectively, they were used by Bate to track the changing climate and environment through the different levels. The levels from which the starlings derive, C and D, are characterised by a high gazelle to deer ratios while Eb is the reverse. This appear to imply that the starlings at Tabun may have been restricted to dry phases in climate unless one of the humeri in Table 11.19 comes from Layer Eb.

The six *Sturnus* humeri in various states of preservation accord approximately in size with extant *S. vulgaris* and are slightly larger than *S. roseus*, the two species which are found in the Levant today (Figure 11.42). Furthermore it is noticeable that the Tabun *Sturnus* humeri are relatively stout compared to the modern starlings plotted in Figure 11.42. There is no indication, bearing in mind the limited sample size, that the starlings at Tabun differed in relation to the layers they came from.

Table 11.19. *Sturnus* sp. material fromTabun Cave with their measurements.

Specimen No.	Skeletal element	Measurements			
		GL	Bp	Bd	SC
A4773					
Layer Eb or D??	L. humerus	27.44	8.9	7.18	2.78
Layer D	R. humerus	27.6	8.8	7.5	2.76
Layer D	Prox. L. humerus	-	9.1	-	2.84
A4774					
Layer D	R. humerus	27.5	8.64	7.02	2.6
A4719					
Layer C	R. humerus	-	9.06	-	2.82
A4798					
Layer D	R. humerus	27.36	8.98	7.2	2.68

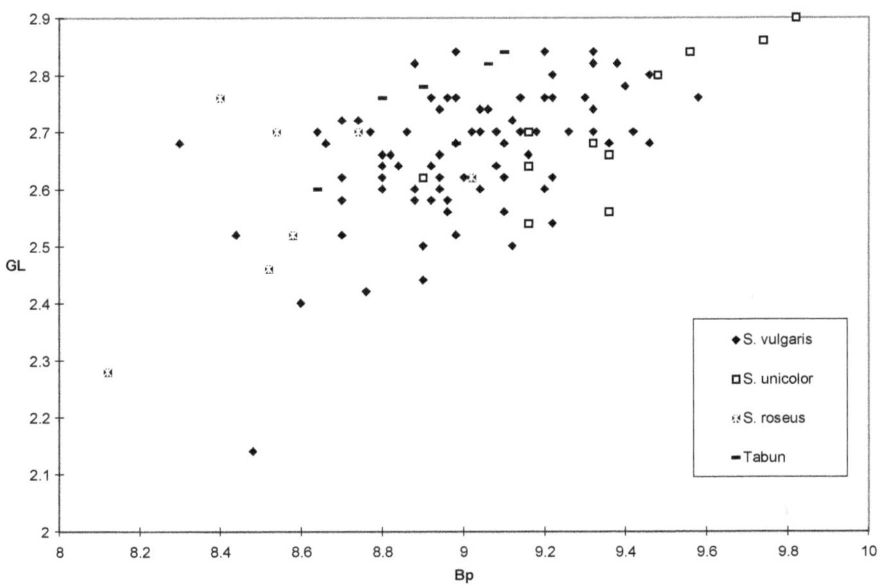

Figure 11.42. *Sturnus* sp. humeri from Tabun Cave plotted with modern European *Sturnus* samples. Greatest length (GL) against proximal breadth (Bp).

11.10. Conclusions

11.10.1. What is the pattern of size and shape in modern members of the genus *Sturnus* both inter- and intraspecifically and what does the pattern signify?

The comparison of the three European *Sturnus* species has revealed that *S. unicolor* is on average larger than both *S. roseus* and *S. vulgaris*. Additionally, the humeri of *S. roseus* are narrower proximally than the similarly sized *S. vulgaris*. *S. roseus* also has relatively long and thin femora and relatively long tarsometatarsi, overlapping with *S. unicolor*. The significance of these differences is difficult to interpret. The large size of *S. unicolor* may be explicable by their sedentary life style. This is a characteristic of relatively non-migratory bird species identified by Winkler and Leisler (1992) (Chapter 2). The metric-shape differences seen between *S. roseus* and *S. vulgaris* are, however, more difficult to explain. *S. roseus* is an obligate migrant, which may explain the different proximal humeral dimension, although it is not clear how. Meanwhile the different leg proportions of *S. roseus* may well reflect a greater use of the ground when feeding or alternatively a different or lesser use of trees for roosting (Chapter 2). This cannot, however, be confirmed as no comparative study of the use of ground or trees for roosting exists in the literature (e.g. Cramp and Perrins, 1994; Feare, 1984).

Of the three European *Sturnus* species it has only been possible to collect sufficient data for *S. vulgaris* to elucidate any intraspecific geographical patterns in morphology. The populations of *S. vulgaris* are in general very similar. There are problems with these samples, however, as many of them cannot be confirmed as comprising individuals which bred in the respective areas from which they were collected. Certain populations of *S. vulgaris* are migratory and move to areas where sedentary individuals live. The Scandinavian sample can be assumed to be mostly migratory and is therefore a homogeneous sample, although most of the more southern European samples, with the exception of the British sample, probably include both the local sedentary individuals, as well as visiting migratory birds. The British sample is the best of the southern samples because most individuals died between April and June when they breed and after the wintering, migratory birds have left. This means that the samples, other than the Scandinavian and British samples are, potentially heterogeneous. This in turn makes interpretation of the samples' relative mensural data problematic. However, any differences seen between the Scandinavian and British populations are probably meaningful. Table 11.4 gives *t*-tests for the dimensions of these two populations. The results show that the carpometacarpus greatest length, in particular, differs between the two samples. The Scandinavian carpometacarpi are significantly longer than those from Britain. This results in a different distal to proximal anterior limb bone proportion. Furthermore, this difference is most likely to be due to the relative migratory behaviour of the two populations sampled and not due to relative openness of habitat because starlings are generally found in more open areas in both areas. Birds are known to respond to habitat openness by having relatively longer wings than closely related birds in dense woodland (Chapter 2). This is because of the practicalities of flight in dense woodland compared to open habitats. In any case, the difference between the British and Scandinavian populations is in relative proportions of the carpometacarpus to humerus length rather than in total wing length. The results appear to agree with Winkler and Leisler (1992) who found a positive relationship between migratory distance and relative distal wing bone lengths (carpometacarpus to humerus length ratios) in different bird species.

Table 11.20. Ratios of mean humerus to carpometacarpus lengths of modern and fossil *Sturnus* populations.

Sample	**N**	**Ratio**
Modern:		
Sturnus vulgaris		
Britain	Hum: 22, Cpt: 24	1.36
Scandinavia	Hum: 19, Cpt: 20	1.33
Sturnus unicolor	Hum: 12, Cpt: 12	1.35
Sturnus roseus	Hum: 7, Cpt: 7	1.36
Fossil:		
Tornewton Cave Reindeer Stratum	Hum: 5, Cpt: 5	1.4
Tornewton Cave Reindeer and Hyaena Stratum	Hum: 9, Cpt: 5	1.41
Bacon Hole	Hum: 4, Cpt: 5	1.42
Minchin Hole	Hum: 4, Cpt: 1	1.35
Chudleigh Fissure - Small *Sturnus**	Hum: 2, Cpt: 5	1.26

*Note: The Chudleigh Fissue Starlings appear to fall into two categories. The figures given here are for the small *Sturnus*. The small humeri are presumed to be associated with the gracile carpometacarpi.

11.10.2. What is the pattern of size and shape in fossil *Sturnus* and what does the pattern signify?

The specific identity of the fossil *Sturnus* material from the various localities is difficult to resolve because of the overlap in morphology among extant species. It would be dangerous to interpret their identity based on either their gross size or shape because such characters can vary greatly within modern species and are likely to have varied even more through time. Therefore, it is perhaps best to concentrate on ecomorphological aspects of the starling fossils rather than their taxonomic affinities.

The most impressive set of *Sturnus* fossils, in terms of different relative measurements, are those from Oxygen Isotope Stage 5 from the south-west of Britain (Tornewton Cave in Devon, and Bacon and Minchin Holes in South Wales). These fossil starlings are significantly larger than the local members of that genus today (Figures 11.21 to 11.31), and it is likely that they represent a distinct population living during Stage 5 in the south-west of Britain. Furthermore, they appear to differ from all other populations of that species measured in Europe in the present study, as well as from the other species encountered in the area today. *S. unicolor* is the closest in size to the fossils because it too is relatively large, although it appears to be differently proportioned to them. The humerus of all the OIS 5 birds are statistically more robust, particularly the midshaft breadth, than the similarly lengthed *S. unicolor* (for humerus shaft breadth SC: *t*-test $p = 0.000777$).

Due to the differences in relative bone dimensions of the two populations of *S. vulgaris* above (i.e. from Scandinavia and Britain), a study of the ratio of carpometacarpus to humerus length of the OIS 5 starlings was devised (Table 11.20). Ratios have been calculated for each modern and fossil population. Table 11.20 shows these ratios and reveals that the relationship between these ratios and the relative migratory behaviour of the different *Sturnus* populations sampled does not always follow the predictions of Winkler and Leisler (1992) (Chapter 2). Specifically, the more migratory populations and species should possess lower values compared to those which are more sedentary. Even though the samples which represent migratory individuals (*S. vulgaris* from Scandinavia) are statistically different from the sedentary populations (*S. vulgaris* Britain) (carpometacarpus GL to humerus GL ratio *t*-Test $p = 0.007797$) their ratios of the means are very similar (Table 11.20.). This is probably because the ratios represent calculations which are averaging the values derived from individual samples masking the true trend. This was necessary because the fossils with which the modern data was to be compared consisted of isolated bones and not associated skeletons like the modern material. Although caution should be exercised when interpreting the ratios, the relative order of these ratios shall be considered below.

S. roseus, a migrant, in Table 11.20. appears to have a ratio more indicative of sedentism than modern sedentary birds and *S. unicolor* which ought to have a ratio indicative of a particularly sedentary bird is equal to that of British *S. vulgaris*. This is puzzling, although it may be a product of small sample size. Nevertheless, all the fossils from south-west Britain proved to have the shortest carpometacarpi in relation to their humeri (Table 11.20). It is also interesting to note that the relatively small starlings from Chudleigh Fissure may have the proportions of a highly migratory bird. However, unlike the ratio for modern populations the ratios of the fossils were calculated from unassociated humeri and carpometacarpi.

This investigation proved a little unsatisfactory because of the narrowness of the difference between modern populations with known migratory or sedentary behaviour. However, these narrow differences can be statistically significant as seen from the result of the comparison of the British and Scandinavian populations of *S. vulgaris*. These statistical significances follow that predicted by Winkler and Leisler (1992) and may therefore be biologically significant.

11.10.3. Can the present study of *Sturnus* in the Quaternary of Europe shed any light on the historical biogeography of the genus *Sturnus*?

In the light of the present analysis of *Sturnus* fossils it is appropriate to reconsider the historical biogeography of the genus. As stated by Grande (1985), fossils can be very informative in further elucidating the origins of taxa, although only if the fossils are correctly assigned to the taxon and if the dating of the fossils is reliable. As shall be seen it is not necessary that the identity of the fossils be totally clear at a specific level to derive an understanding of pattern and process.

The oldest unquestionable remains we have of the genus *Sturnus* in Western Europe are the O.I.S. 5 fossils from the south-west of Britain. They occur in the Hyaena Stratum in Tornewton Cave, as well as similar deposits in Minchin Hole and Bacon Hole in South Wales. These starlings are large with a humerus to carpometacarpus length ratio which probably indicates that they had a sedentary habit. It is therefore tempting, given the prediction made by Feare (1984), to interpret these fossils as evidence of a former wider distribution of *S. unicolor*. This is because, like the fossils, *S. unicolor* is a large starling which is relatively sedentary. However, the fossils do appear to have different proportions to *S. unicolor*, calling into question the hypothesis that they might belong to that species, although this cannot be ruled out. Alternatively, the fossils might represent a large, differently proportioned population of *S. vulgaris* or even *S. roseus*. Finally, the large starlings might represent an extinct species which shared a common ancestor or was ancestral to some of the modern members of the genus *Sturnus*. There is a more important conclusion no matter what the identity of the O.I.S. 5 fossils. That is that the genus *Sturnus* may have a tendency to give rise to larger, more sedentary populations along the western and south-western edge of the taxon's range. Hence, today there is the large relatively sedentary species (*S. unicolor*) and large sedentary subspecies (*S. vulgaris zetanticus* and *S. v. faroensis*) all distributed along the western edge of the genus' range. *S. v. granti*, also with a south-western distribution (the Azores), although no larger than *S. v. vulgaris*, is nevertheless sedentary. Certainly the trend holds true for *S. v. vulgaris* regarding their degree of migratory behaviour; the more westward their breeding location, the more sedentary they are (Feare, 1984). The reason proposed for the difference in behaviour from South-West to North-East is the degree of climatic continentality, which correlates with more intense winter ground frost and therefore reduced the possibilities for feeding on insects in the ground. Therefore, the further North-East the population, the greater the likelihood that it migrates. The fossils from O.I.S. 5 are found in deposits whose faunas are interglacial and warm in character. This correlation between equable climates and westerly distributed large, relatively sedentary members of the genus appears to be a recurring pattern. Given that these patterns occur at both the species and subspecies level they do not help elucidate the level of taxonomic distinctness of the fossils. The changes in size and metric-shape ratios which presumably relate to greater or lesser degrees of migratory behaviour are adaptive not related to taxonomy. However, it is possible that time is a necessary determinant of the degree to which these adaptations have been acquired. Therefore, there may be a gradual spectrum from populations of migratory *S. vulgaris vulgaris* in the North and East (e.g. Scandinavia in this study), to populations of sedentary *S. v. vulgaris* in the West (e.g. Britain and Belgium in this study), to sedentary subspecies in the West (*S. v. granti*), to large sedentary subspecies (*S. v. zetanticus* and *S. v. faroensis*), to large relatively sedentary species (*S. unicolor*). If this sequence has any merit it may be that the OIS 5 *Sturnus* are more different than any of the extant starlings (species and subspecies) are from each other, raising the possibility that the fossils are a distinct species. However, because there were no comparative skeletons of the three western subspecies mentioned, their osteological metric-shape cannot be considered.

The next set of fossil starlings in Britain probably dates from the Devensian and is from the Reindeer Stratum in Tornewton Cave. Given the model proposed above, a prediction can be made that these birds, being present in a probable cold environment, would be relatively small with proportional characters of the wing indicating a migratory habit. The starling fossils from the Reindeer Stratum may, as already mentioned, represent two species. There are large specimens which appear to correspond to birds of the same size found in the OIS 5 deposits in western Britain, and maybe a smaller species. This scenario could be explained by the survival of the warm adapted sedentary species into the later Devensian, assuming that it is not reworked, followed by the arrival of a more cold adapted species from the east which may or may not have been a resident breeding bird. It is tempting, given the model, to suggest that it is a winter visitor from the North or East. There is also a possibility that these two forms, assuming they are distinct species, represent *S. unicolor* and *S. vulgaris* which came into contact further north at this period.

The fossils from Chudleigh Gorge, which are poorly dated, cannot be adequately considered because there is little information regarding their contemporary environment. However, if the possibility of sympatry between the larger and smaller starlings from that cave were confirmed it may signify that the large O.I.S. 5 starlings survived into later in the Devensian than the period represented by the Reindeer Stratum at Tornewton Cave. This, in turn, might require that the large O.I.S. 5 birds cannot be *Sturnus vulgaris*, assuming the small birds are *S. vulgaris*. However, if in this scenario one substitutes two different sized subspecies for two species coming together during the winter, this conclusion could not be made.

Port Eynon Cave, whose fauna is poorly dated, may provide further evidence of two forms of starling. This conclusion cannot be made with any conviction because the fauna could well be a mixture of Devensian and Holocene material.

The sample of *Sturnus* humeri from Pin Hole Cave in Derbyshire is consistent in morphology with *S. vulgaris* from Britain today. This may be because they are Holocene in age, as the cave has a mixed fauna.

Great caution should also be taken in integrating the Holocene record available from Britain into this model because only one specimen has been seen, a Roman occurrence from London Wall. Bearing in mind the dangers imposed by the very small sample size it is interesting that the humerus falls at the lower end of the variation of modern British birds. No conclusion can be drawn from this but if a larger sample showed that there was a statistically significant difference this could be explicable by the populations being more migratory than modern populations. The following is speculative but if starlings at the time were different to the modern *S. vulgaris*, this might add credence to the suggestion that *S. vulgaris* have only recently increased in numbers and sedentariness in western Europe due to climatic amelioration (Berthold, 1968). This requires testing with further data and further interpretation because the Roman period was one of relative warm temperatures.

The Tabun starlings appear to be morphologically similar to *S. vulgaris* and may well represent that species, although it is possible that they are *S. roseus*. Regardless of their identity it is important to note the presence of starlings in the east of the Western Palaearctic at this time, which may be between O.I.S. 5 and 9. The Tabun specimens could be the oldest starlings confirmed by this study.

The birds from Gibraltar are of interest because they might indicate the starlings present in Iberia during the last glacial period. In particular, they may confirm the minimum age of *S. unicolor* in the peninsula. This is particularly important because of the potential that the Iberian peninsula formed a refugium for starlings during this cold stage. In fact Feare (1984) suggested that this is where the ancestors of *S. unicolor* retreated, evolving into this distinct species during an unspecified cold stage from which it never re-emerged. Unfortunately, the small number of specimens available for study is ambiguous. There are no unquestionable specimens consistent in size with *S. unicolor*, although there are small specimens which are below the modern species' size range. This does not necessarily signify that the specimens represent *S. vulgaris*. It is quite possible that *S. unicolor* was differently adapted during glacial periods and may have been smaller and even less sedentary. This is the important contribution that these fossils make, that there is no evidence for a large sedentary starling in Iberia during the last glaciation. This could provide a maximum date after which the Iberian starlings (*S. unicolor*) aquired their present form, if not a maximum age for the origination of the *S. unicolor* species. This should of course be tempered by the fact that the sample is as yet small and only from a single site.

To conclude, it can be stated that fossils are very important in helping test the model produced by Feare (1984) of *Sturnus* historical biogeography. Furthermore, the fossils add perspective that could not even be predicted from modern biogeographic patterns.

11.10.4. Are there any extinct *Sturnus* species in the fossil record?

It has not been possible to confirm with certainty the presence of extant taxa at any of the sites described. It is equally difficult to know whether the fossils represent extinct taxa. Phylogenetic recontructions are therefore not possible because it has not been possible to completely separate the modern species osteologically, implying that their common ancestors will be equally difficult to recognise in the fossil record. Furthermore, no fossil *Sturnus* appear so different that they must have been divergent sister-groups, now extinct.

The existence at Chudleigh Fissure and in the Reindeer Stratum of Tornewton Cave of possible sympatric *Sturnus* species does not really help the elucidation of whether either of them represents an extinct species. This because it is always possible that both represent extant species.

The problem can be addressed from a more tangential perspective. Consideration was given to whether certain of the fossil samples represented isolated populations which by analogy may be argued to be distinct species. This method is implicit in much of the work done on island faunas in areas such as the Pacific islands (For a review see Millberg and Tyrberg, 1993). In such cases workers often name new species on the basis of characters such as size, and presumably feel safe to do so because of the theoretical perspective of the biological species concept which highlights allopatry as a mechanism. The large starlings in western Britain during Oxygen Isotope Stage 5 may well be an extinct species. They certainly differ in certain respects from both *Sturnus vulgaris* and *S. unicolor* to a greater extent than each does from the other. Furthermore, they may have been relatively sedentary, making them even more prone to genetic isolation and hence to divergence, forming a distinct species. This can be posed as an hypothesis with regard to these fossil starlings, which requires further testing, perhaps with more fossils, including an analysis of other skeletal elements if they can be retrieved from the sites in question.

Section IV

Discussion and Conclusions

Chapter 12

Discussion and Conclusion

At the beginning of this thesis the general aims of the study were outlined. In the broadest sense, it was designed to gain an insight into the taxonomic and adaptive ecomorphological significance of Quaternary bird fossils. This was to be achieved by studying a limited number of taxa in great detail across both time and space, including cranes of the genus *Grus*, ptarmigan and grouse of the genus *Lagopus*, ravens *Corvus corax* and starlings of the genus *Sturnus*. The central aim was to assess the degree to which the two perspectives of the thesis, the taxonomic and ecomorphological (adaptive), were compatible forms of knowledge. In other words, are the sizes and shapes of bones of closely related birds more indicative of the birds' taxonomic affinities or of their adaptations to their modes of life or local environments?

12.1. What kinds of morphological variation are present within species today and in the past, and what are their significance?

Variation today

Species in all four taxonomic groupings studied are polytypic, and this has been shown to be reflected to a greater or lesser extent in their osteology. The common crane *Grus grus* is the exception in that no material of the other Western Palaearctic subspecies of common crane *G. g. lifordi* was available for comparison with *G. g. grus*.

The two members of the ptarmigan and grouse genus, *L. lagopus* and *L. mutus*, both vary morphologically between subspecies. In *L. mutus*, *L. m. hyperboreas* from Spitzbergen is the largest population studied here, followed by the samples of *L. m. millaisi* from Scotland, *L. m. helveticus* from the French Alps and *L. m. mutus* from Scandinavia. In *L. lagopus*, the largest specimen examined was an individual of *L. l. major* from Kazakhstan, and the smallest was *L. l. brevirostris* from the Altai Mountains in Siberia. Intermediate to these in size are *L. l. scoticus* and *L. l. lagopus*, the former being the larger of the two. Additionally, *L. l. scoticus* has significantly longer leg bones than *L. lagopus*. The differences in size may reflect variations in food quality available, with larger subspecies living in areas with better quality vegetation. It may be that climatic adaptation (following Bergmann's rule) is also a factor, particularly in the case of *L. mutus* where the largest individual examined came from the furthest north i.e. Spitzbergen. The longer leg dimensions of *L. l. scoticus* may relate to shorter periods of snow cover in Scotland than in mainland Europe. The Scottish birds lack of the white winter plumage of the continental willow grouse which is used as camouflage during periods of snow cover. Therefore, it may be that the longer leg length is another adaptation to lower snow precipitation.

The raven *C. corax* varies greatly across its modern range and appears to conform to the ecogeographic rules of Bergmann and Allen. With regard to the former rule, there is a general cline from north to south, with the populations of larger individuals coming from the north and the smaller ones from the south. The Polish sample breaks that trend and their bone dimensions are small on average relative to their geographic position in Europe. Allen's rule is demonstrated by the longer legs of the Iberian subspecies *C. c. hispanus* although no cline is present in this feature.

In the starling genus it is the common starling *S. vulgaris* that can be studied in relation to intraspecific variation although only one subspecies was available. The most distinct population examined comes from Scandinavia. These starlings have more gracile tarsometatarsi, ulnae, radii and possibly carpometacarpi than other populations measured. Additionally, the carpometacarpi of the Scandinavian population are significantly longer than those from Britain. It seems likely that the differences of the wing elements reflect the difference in migratory behaviour of the populations (Winkler and Leisler, 1992). Many Scandinavian birds move from their breeding areas north east as far as Britain, while British breeding birds are sedentary (Cramp and Perrins, 1994). The difference in the tarsometatarsal breadth has not been explained but may be due to a pleiotropic effect (Cock, 1969).

Variation in the past

The existence of anomalously sized and proportioned fossil skeletal remains was confirmed for the four taxa. Thus, contrary to Harrison (1980a, 1988a), bird species can be shown to have changed in size (as well as shape) during the Quaternary. The different taxa pinpoint several problems with the investigation of such matters. The two *Lagopus* species, whose size probably decreased at the end of the Pleistocene, do so in a relatively subtle way because the size decrease was not an isometric one i.e. the smaller body size was reflected in the decrease in tarsometatarsal shaft breadth and proximal humeral breadth. The common crane *Grus grus* may have decreased in size since the Roman times, although this change was disguised because all the larger individuals were allocated to an extinct species *Grus primigenia*. Alternatively, it may be that British and other more southern-breeding crane populations that are now extinct formed a different, larger subspecies to those living today in Scandinavia which formed the modern comparative sample. *Corvus corax* also appears to have varied in size through time, with fossils found which differed from local modern populations. As with the large cranes, the distinct dimensions of the ravens from the "Rissian" of La Fage and other Middle Pleistocene sites, which were smaller, have resulted in the specimens being attributed to an extinct species. Again, the naming of a distinct species has had the effect of hiding what was probably an adaptation to different environmental conditions by an extant species in the past, i.e. of hiding intraspecific variation.

The results of the analysis of the *Sturnus* fossils are perhaps the most difficult to interpret. The large *Sturnus* fossils found in OIS 5 in the south-west of Britain may represent size change in one of the modern species such as *S. vulgaris*, as implied by Harrison (1980b); distribution change in *S. unicolor*; or else an extinct species. The last explanation has some merit due to the zoogeographic patterns seen among starlings in Europe today. Isolated populations of large starlings today are considered to be either subspecies, *S. vulgaris zetlandicus* and *S. v. faroensis* in the Shetlands and Faroes, respectively, or species *S. unicolor* in Iberia and other parts of the western Mediterranean (Feare, 1984). In addition to being distincively large compared to modern British starlings the OIS 5 birds probably had carpometacarpus to humerus ratios which may be indicative of sedentism as predicted by Winkler and Leisler (1992). The degree of morphological difference seen between the OIS 5 starlings in Britain and the two modern western European sister species (*S. vulgaris* and *S. unicolor*) may exceed that between the extant species. This may suggest that a distinct species was present although this is not certain and therefore the OIS 5 starlings remain unnamed.

12.2. What kinds of morphological variation are present between species in the taxa studied ?

The most striking similarity between the four genera studied was that the osteological distinction of neospecies within the genera appears to be restricted to size and metric-shape. These differences, however, are probably not reliable for species diagnosis as they can vary considerably within species today and may have varied to an even greater extent during the past, particularly given the dramatic climatic oscillations which have been documented for the Quaternary. In other words, there is a distinct possibility that neospecies in a species-rich genus may have undergone complete or partial shifts in size or metric-shape through time, from the morphospace occupied by one modern species to that of another. This would give the false impression that a specimen belongs to a certain neospecies when it belonged to another.

The osteology of *Grus* has long been considered by some as being uniform in all characters other than size (Lydekker, 1891a). Those non-metric traits that have been identified as species specific by authors such as Northcote and Mourer-Chauviré (1985, 1988) have been shown here to be less reliable when a larger range of specimens is compared (Chapter 9).

In *Lagopus*, the same situation exists as no non-metric characters exist to distinguish the two species *L. lagopus* and *L. mutus* (Kraft, 1972 and Chapter 10). Only in the case of the tarsometatarsus can the two species be at all reliably identified and only because the tarsometatarsi of *L. mutus* are consistently smaller than those of *L. lagopus*. Other skeletal elements are more problematic, and many can not be reliably identified, even to their family within the order Gallifomes. The size difference between the two *Lagopus* species is thought to be reliable as a diagnostic character because it is unlikely that the two would differ in their size order, i.e. *L. lagopus* has probably always been the larger of the two species. However, the use of this character does rely on both species being present in sufficient quantities in an assemblage to be clearly bimodal.

In the genus *Corvus* the same lack of morphological differentiation is seen among the species. This has been previously noted by Mourer-Chauviré (1975a) when she described the new taxa *C. antecorax* and *C. pliocaenus janossyi*. The diagnoses of these two taxa were that they were intermediate in size between various other members of the genus *Corvus*. No discrete non-metric characters were given as diagnostic of these extinct corvids.

Lastly, the three Palaearctic species of *Sturnus* studied here also appear to be uniform in their post-cranial osteology. Indeed it proved difficult to distinguish many of the bones of *Sturnus* from those of members of the thrush genus *Turdus* (see Chapter 7). It is quite possible given the work by Beecher (1978) that this is not the case with the species' crania. However, only the post cranium has received attention here because Quaternary avian crania are extremely rare, and if found are generally poorly preserved.

It would be wrong to conclude from the study of the four present taxa that distinct non-metric traits do not exist in the post-cranial bones of birds below the genus level. However, it is my belief that such characters are uncommon and therefore workers who identify species, particularly within species-rich genera, should describe the characters as recommended by Stewart and Hernàndez Carrasquilla (1997). Awareness that this problem exists is not apparently widespread, judging from the species-level identifications in the lists given in site reports by many authors (see Tyrberg, 1998). Furthermore, it is difficult to demonstrate the magnitude of the problem; indeed it is difficult to prove that there is a general absence of distinguishing characters below the genus level. Therefore, the burden of proof should lie with workers who assert that certain problematic taxa can be identified and not with those who advocate caution. It is only once the diagnostic characters are described that they can be tested in the manner described for the difference between the genus *Turdus* and *Sturnus* in Chapter 7.

All four taxa have, in addition, demonstrated that variation over and above that seen in the modern species is present during the Quaternary. In the case of *Lagopus* it is possible to attribute many of these distinct fossils to either *L. lagopus* or *L. mutus* because there are only two possible modern candidates. This is, however, only true of their tarsometatarsi and possibly their humeri. No evidence has been found for an extra small species living in the past as claimed by E.T. Newton (1924a). If this were true it could cause considerable problems with identification of small fossils to either the small species or *L. mutus*. Newton's belief probably arose because a number of the *L. mutus* specimens found at Merlin's Cave were smaller that the British subspecies, *L. m. millaisi*, which was the only subspecies represented in Newton's reference collection. It is possible that Newton may have been further influenced by the presence in North America of a smaller third species in the genus i.e. the white-tailed ptarmigan *L. leucurus*.

In *Grus*, the confidence with which a species is identified varies according to whether one accepts that there is a possibility that the Asian sarus crane *G. antigone* may have been resident in Europe in the past. The fact that cranes are highly sexually dimorphic and most of the Holocene *Grus* fossils are either the size of the modern *G. grus* males (which are larger than the females) or larger appears to indicate that *G. grus*, as a whole species, was larger in the past. The alternative explanation, due to the fact that this pattern is not well marked in the Danish Holocene fossils, is that there was a larger population of *G. grus* living and breeding in Britain and possibly other more southern parts of Europe.

In the raven *Corvus corax*, size appears to exclude any other specific identity for the fossils. This is because the raven is the largest corvid - indeed the largest passerine - in the Western Palaearctic. *C. corone*, the nearest sized congener, is considerably smaller. Most importantly there is no size overlap between the raven and any other members of the family.

The starlings *Sturnus* meanwhile present the greatest difficulty in the attribution of fossils to species. The overlap between the three modern species in Europe signifies that no confidence can exist when attempting to identify individual fossils to species. Therefore, none of the *Sturnus* fossils have been identified to species.

The presence of anomalously sized and proportioned Quaternary fossils in the four taxa studied has implications to the identification of the fossils to species. In instances where species are identified in the fossil record on the basis of characters such as size and metric shape, they may be correct, although it depends on the circumstances. These kinds of characters could be argued to be reliable for more recent fossils such as those of the Holocene, although with older material there is an increased possibility that such characters have changed and thus may appear to be reliable when they are not. The present study of *Grus* fossils in Britain together with studies of other taxa in the Holocene (Northcote, 1979b, 1981a, 1982, 1983) and ones showing changes in modern birds such as sparrows (Johnston and Selander, 1964) and Darwin's finches (Grant, 1986) would highlight a need for caution. This is another reason that none of the *Sturnus* fossils described in this thesis has been ascribed to species level. The assumptions used in the identification of *Corvus corax* and the two *Lagopus* species are that in *Lagopus* the order of size has not changed, i.e. that *L. lagopus* has always been the larger of the two, and in the case of *C. corax*, that species are unlikely to change more than a certain amount over time. In *Grus*, the situation is different again because historically there has been a consideration that a species whose modern geographical range does not include the Western Palaearctic, *G. antigone*, may have been present in the past. This was discounted here due to the very significant habitat requirements of the extant sarus crane which were not available in Europe during the Quaternary (Johnsgard, 1983). If there is a need to consider the possible presence of modern species whose ranges are far outside the area in which bird fossils are studied, many analyses of Quaternary birds could be flawed. One situation where this definitely becomes a problem is in more recent times when movement of taxa have taken place by humans. An example of this was discovered recently by the present author when a parrot was identified from post-medieval Norwich (Albarella *et al.*, 1997).

Still further problems to the identification of fossils arise if extinct species are believed to have been present. This further increases the number of possible candidates when identification is attempted (see next section).

12.3. When can extinct Quaternary bird taxa justifiably be named, and to what taxonomic level should the fossils be referred?

A variety of factors appear to influence authors when they name new bird species in the Quaternary fossil record. These include both intrinsic factors, such as the size and/or the shape of the fossils, and extrinsic factors, such as whether the new species is accompanied by a sympatric congeneric neospecies, the relative age of the fossils, or whether the fossil animals had conceivably formed an allopatric population (e.g. on an island). The degree to which intraspecific variation is acknowledged as a potential influence on birds' morphology will also determine what decisions are made. The other extrinsic factor influencing the authors of new species is the species concept he or she prefers. The latter shall be dealt with in the next section.

The first category includes the characters invoked in the "Diagnosis" of new taxa. These diagnoses include statements about the size and shape of the bones of the new taxon in relation to those of closely related taxa, usually extant ones. It is rarely stated, but apparently believed by many workers, that size differences alone are not in themselves adequate for the purposes of defining a new species. This is because of two related reasons. First, size differences have been observed within many modern species across their geographic ranges today. This means that any size difference observed in the fossil record could either be sub-specific or specific, and without a means of telling which, one must be conservative and assume that it is sub-specific. Second, size is considered to be an adaptive character and hence subject to convergent evolution. However, definitions of species and subspecies are to some extent arbitrary in the sense that they depend entirely on a taxonomists philosophical leaning (see next section). If an individual believes in certain species concepts he or she might be inclined to elevate the subspecies to new species. Given that some subspecies are defined on the basis of size this has great significance to how new species are described. Moreover, size difference is clearly a factor in allowing similar species to co-exist through character displacement. Certainly, if two different sized birds within one genus were present at a point in time at a specific place then it is not unreasonable to suggest that they represent different species. (This is assuming that they are not conspecific geographical variants that have come together during migration). If that genus today only includes one species it would seem that a different species had existed in the past in addition to the present one. Which one of the two fossils of different size represents the modern taxon is more problematic. It is tempting to allocate the birds which conform in size with the modern species to that species. However, size change in time due both to environmental effects and the precence of other species (character displacement) may have taken place. In such a situation it would be incorrect to name a new species as it would not be clear what material to base the taxon on. Therefore, in such a circumstance it would be possible to say that an additional species within the genus had been present although the allocation of names to the fossil material is not possible.

Two further problems arise in this situation. First, there could theoretically have been a reduction in sex-related size difference within a species. This is difficult to address adequately but it is clear that sexual dimorphism of this nature and magnitude is usually confined to specific taxonomic groups of birds such as diurnal birds of prey, cranes and certain galliformes, due to very particular selective mechanisms. If the birds analysed do not belong to such taxa then it could be argued that neither fossil bird represents the extant species.

Metric shape variation might be considered as being more important in terms of taxonomic significance. This is because metric shape differences are more complex than simpler size variation, so that to evolve a different size is more easily achieved than differences in metric shape. However, metric shape change (or variation) is or can also be adaptive in origin and hence be acquired independently. Metric shape variation is seen to occur among subspecies as well as species within genera. Therefore, metric shape differences may often be no more informative about the taxonomic affinities of birds than isometric ones.

There is an overriding sense that non-metric, so called discrete, characters are the most reliable indicators of the taxonomic affinities of fossil remains. This is as true of palaeornithology as other areas of vertebrate palaeontology. The most prominent comment by a Quaternary worker with regard to birds is that by Jánossy (1987): "Since the beginning of the paleornithological investigations, many research workers have attributed to recent species the bird remains of Upper Pliocene and Lower Pleistocene. Actually these fossil forms show often very weak but constant differences comparatively to recent ones, and belong to extinct species." The precise meaning of the "very weak but constant differences" may be more clear from another prase in the same paper where the differences are described as "a smallest constant morphological difference". This would appear to refer to non-metric characters, which although subtle can be used to distinguish extinct taxa. A similar argument has been made in palaeoanthropology when Tattersall (1986; 1992) suggested that the number of fossil hominids was being grossly underestimated. He made the point that the differences seen between many closely related primates today are often limited to soft tissue and behaviour. This he takes to signify that the often greater hard-tissue variation seen in the hominid fossil record can in many instances be regarded as indicative of greater numbers of species than considered at that time. While this may be true in birds, and indeed primates, it ignores the fact that subtle non-metric differences will also be variable across a species' range just as metric characters are. It also ignores the fact that only subtle differences exist, which may imply that it is unrealistic to take the identification of the bones to species level in the first place.

The second category is not generally given explicitly as part of the justification for the naming of extinct taxa. These are the extrinsic factors such as the presence of congeners, the age of the fossils and the geographic locality of the find. It is clear, however, from the literature that such factors influence authors of species. Among the taxa studied here, the diagnosis of the large species of crane named *G. primigenia* was that it was larger than the sarus crane *G. antigone*, with distinct dimensions of the distal tibiotarsus which formed the holotype (Milne-Edwards, 1867 – 71). Later work used non-metric morphological characters to distinguish *G. primigenia* from other large cranes like *G. antigone* and *G. melitensis*. It is interesting that in this case the degree of size difference from the smaller *G. grus* generally appeared to exclude *G. grus* as a possible candidate. Nevertheless, apparent sympatry between the large crane and a smaller species believed to be *G. grus* was used to further justify the extinct taxon. In the case of *G. melitensis*, not dealt with in this volume, the isolated nature of the fossil population as well as their extreme size may have helped justify the taxon.

In *Lagopus*, the different proportions of their tarsometatarsi were given as diagnostic of distinct subspecies during the Rissian. However, these metrical characters also distinguished later populations, therefore implying that the age of the material from La Fage was also an influence in the decision. However, age should not be used to diagnose taxa because in that case, any analysis of evolutionary rates is affected by circularity. The *Lagopus* fossils from La Fage provide an interesting contrast to those of the raven. Here the smaller raven fossils were accorded a full species level distinction and named *C. antecorax*. It may be that the existence of the two species of *Lagopus* in the Rissian, as now, may have helped in the more conservative taxonomic assignment of those fossils. It may also be that the magnitude of the metrical difference in the ravens was considered greater than that in the *Lagopus* fossils. This is because Mourer-Chauviré described *C. antecorax* as intermediate between *C. corax* and *C. corone* / *C. frugilegus*. However, this greater magnitude of the difference may have been overstated, as seen from the current work. *C. antecorax* has been shown to differ from modern ravens to a similar extent as extant subspecies differ from each other (Chapter 10).

In starlings, the anomalous population in the Pleistocene from south-western Britain was not named (Harrison, 1980b). This is reasonable given the problems of distinguishing *Sturnus* species in Europe today. While it is possible to infer on theoretical grounds (the Biological Species Concept) that the fossil population in south-west Britain constituted a distinct species, it is not reasonable to do so, because a species should not be named on the basis of size or limb-bone length ratio differences alone. However, it could be argued that there is more justification for naming the south-western British starlings as a new taxon than the new taxa named by others in the other groups dealt with in this volume.

12.4. How do the different species concepts affect the interpretation of the four taxa?

As discussed in chapter 2, detailing the theoretical background to this thesis, the species concepts to which the palaeontologists adhere are perhaps the most influencial in the taxonomic descisions they make. These concepts fall into two broad categories, the pattern-based concepts and the process-based concepts. In this section an attempt shall be made to consider the different taxa studied in the light of the different concepts. It is felt that this is useful as most palaeontologists do not explicitely express their stance on this issue despite the clear influence it will have on their interpretation. The following discussion will detail each taxon in turn although a few concepts are best summed up here as they differ not in the implications for each taxon. The biggest problem possed to the process related concepts is that opperationally they end up being pattern related concepts because process is even less visible in the fossil record than among living organisms.

The evolutionary species concept might tend to have the same effect as the biological species concept because it is mostly a modification of the latter to better include a temporal dimension. The phenetic species concept is reliant on multivariate statistical analysis and since this has not been attempted here because of the lack of variables available for most of the fossils its effect on the conclusions of this thesis shall not be discussed. Finally, the recognition concept is difficult to apply because the characters that are generally considered meaningfull from this perspective in birds are unlikely to enter the fossil record.

Grus (Cranes)

The central question regarding *Grus* in the Quaternary is whether the large crane remains represent an extinct species *Grus primigenia* (Milne-Edwards, 1867 – 71). The species was named at an early date, when an unsophisticated typological species concept was generally followed. Such a concept regarded species as groups of individuals that conformed morphologically to a type. The types were the specimens upon which the species was defined. The trend at the time was for splitting. Although the validity of *G. primigenia* has been questioned on a regular basis since its description and naming, it has not been dismissed for long.

The Biological Species Concept may imply that the fossils considered to belong to *G. primigenia* were a distinct form of *Grus grus* rather than a distinct species because the concept better accommodates polytypy within species. The Ecological Species Concept may have falsely given credence to the size of the large specimens, implying that they belonged to a distinct congeneric species *G. primigenia*, sympatric with the smaller *G. grus*. However, the evidence for the occurrence of two sympatric *Grus* species in the Quaternary of northern Europe is complicated by the high degree of sexual dimorphism in *Grus grus*. The Phylogenetic Species Concept, meanwhile, would attach importance to non-metric traits over characters such as size because the concept highlights characteristics which are acquired and not adaptive. However, because the present study was unable to demonstrate the reliability of the characters previously described it suggests that the large *Grus* fossils belong to *G. grus*.

Lagopus (Red / willow grouse and ptarmigan)

Only new subspecies have been named for the *Lagopus* fossils examined during the present study (Mourer-Chauviré, 1975a). They were named on the basis of their differently-shaped (more robust) tarsometarsi at the site of La Fage, which is late Middle Pleistocene in age. The decision not to name new species on the basis of the *Lagopus* fossils at this site, as mentioned earlier, is because two sizes of population could be defined. Therefore, the smaller and larger clusters were allocated to the small and large extant species *L. mutus* and *L. lagopus* respectively. It could be argued from this that an aspect of the Ecological Species Concept influenced this decision. This is because the separation of two species into two sizes is justified on ecological grounds, each species is filling a slightly different niche. The Phylogenetic Species Concept would agree with the assignments because the tarsometatarsal breadth is likely to be adaptive and therefore only subspecific. The Biological Species Concept would probably only support the use of a subspecies for the La Fage population and any later ones. The lack of extreme allopatry, as might be observed if the fossils represented isolated island populations, is one reason for this. The other is that the Biological Species Concept favours the concept of polytypy among widely distributed species.

***Corvus corax* and *C. antecorax* (Ravens)**

The species *C. antecorax* was named on the basis that the fossils were intermediate in size between *C. corax* and *C. praecorax*. It seems that the species concept being followed was essentially a typological one. The BSC in the case of the ravens would suggest that *C. antecorax* should be subsumed into *C. corax* due to the polytypic nature of *C. corax* today. The Ecological Species Concept does not effect the interpretation of the fossils attributed to *C. antecorax* because there are no fossils of *C. corax* together with it.

The Phylogenetic Species Concept, if Mourer-Chauviré (1993) is correct in regarding *C. antecorax* as solely ancestral to *C. corax*, would cause one to regard *Corvus antecorax* as conspecific with *C. corax*. This is because *C. antecorax* is not a cladogenetic ancestor of *C. corax* and merely a chronospecies, whose relationship to *C. corax* is anagenic. Cladists (adherants to the Phylogenetic Species Concept) regard such species as invalid, as speciation should entail a cladogenetic event. This does, however, depend heavily on whether the Phylogenetic Species Concept would split the modern subspecies of *C. corax* into separate species. I suspect that, in this instance, the species would remain intact despite its clear polytypic character.

***Sturnus* (Starlings)**

The *Sturnus* fossils are more difficult to assess in relation to the different species concepts because of the greater number of similarly-sized species within the genus than in the other taxa. The Biological Species Concept, however, might suggest that the large starlings in the south-west of Britain during OIS 5 constituted an isolated allopatric population. This is further influenced by the suggestion that these birds formed a relatively sedentary population. The possible existence of a sympatric smaller species accompanying the large starlings may imply that two species were present at the time and that one, possibly the larger one, was an extinct member of the genus *Sturnus*. The Ecological Species Concept would help bolster the view that the two sized starlings were different species because it views species as ecological entities which often divide up niches and assume different sizes. The pattern-based species concepts such as the typological and the phylogenetic species concepts may imply that the large differently proportioned starling fossils belonged to a separate species.

Finally, there is an unexpected perspective that the consideration of the Recognition Concept of species gave to the study of fossil birds. That is a theoretical explanation for the difficulty of distinguishing closely related bird species: the Specific Mate Recognition System (see Chapter 2). The explanation lies in the fact that the RC only considers characters useful in mate recognition as relevant to the identification of species. In birds these characters lie mostly in their plumage, song and behaviour, characters which will not enter the fossil record. Therefore, any differences that are found between species in their osteology, such as size and metric-shape, are likely to be analogous of those species rather than diagnostic of them. In other words the characters may aid the correct identification of a bone although the characters in themselves are not what make them distinct species. The differences, often present, are ones of size and metric-shape which are labile adaptive characters and likely to be acquired independently by different species, especially closely related congeners. This explanation is particularly satisfying given that it fits the authors view that birds are conservative in their "bauplan". The "bauplan" could be thought of as an osteological "type", and variations on the theme of the type would mostly be expressed in size and metric shape, both below and above the species level. Therefore, natural selection is transposing adaptive differences between populations both below and above the species level. Furthermore, these adaptations are nothing to do with the inateness of those species. If this is correct it raises some interesting problems for the whole field of avian palaeontology.

This perspective would certainly explain many of the arguments which have taken place regarding the taxonomic significance of Quaternary birds fossils since the beginnings of their study (Chapter 1). Nevertheless, it is not surprising that the debate has never been opened up as such a fundamental problem, because if fully acknowledged, it would be regarded by some as heralding a "depressing" situation in Quaternary avian taxonomy. However, if taxonomy is not the sole aim of the avian palaeontologist, a similar approach to that taken by Trinkaus (1981), who studied Neanderthal adaptations, could be taken. The study of adaptive form would then be highlighted, and hence there would still be much fertile ground in the study of Quaternary bird fossils. With a greater understanding of patterns of adaptive morphology an insight into patterns of species diversity (if not actual taxonomy) during the fossil record might be gained.

12.5. Can evolutionary rates be gauged in the fossil record of the four taxa?

In general the difficulty in recognising ancestral species in the fossil record indicates that species-level evolutionary rates can not generally be gauged. This hampers any assessment of when modern species of birds generally evolved. Hence, fossils are not very useful to help with the debate started by Brodkorb (1960) and Wetmore (1959) earlier this century, and revived recently by molecular biological work such as that by Klicka and Zink (1997). However, the rates of adaptive change within species, and hence subspecies differentiation, may be possible to detect.

In *Grus*, it is possible that the common crane has become smaller since about Roman times due to large scale wetland drainage. Assuming this to be true, younger samples (post-Roman) from Britain may elucidate when, or the rate at which, this took place. This would signify that cranes have undergone a change in morphology as well as geographic breeding range in response to human interference with the environment. Alternatively, it is possible that the British population of *G. grus*, and maybe others in Europe, formed a distinct, larger subspecies, meaning that no actual size reduction took place. Instead the larger, more southern population became locally extinct. This signifies that there are no contemporary British cranes to compare with the fossils. The fossils have instead been compared with modern Scandinavian breeding birds. Specimens from Britain, intermediate in age between the present and the Roman period, may indicate which of these hypotheses is correct.

The two *Lagopus* species appear to have reduced their size, in a complex manner, between the end of the Pleistocene (the Late Glacial) and the present day. However, the lack of early Holocene fossils has prevented the investigation of whether this took place rapidly, shortly after the end of the Pleistocene, or at a more recent time. It seems likely that the Late Glacial / Holocene boundary is the time when it took place because it is also probably the time during which the geographic range of both species reduced dramatically. In Europe, the two species retreated northwards with the climate, leaving isolated populations on the higher ground of the Alps and the Pyreneese, in the case of *L. mutus*, and in Britain in the case of *L. lagopus* and *L. mutus*. These isolated populations have subsequently differentiated into subspecies. The present day subspecies of the two *Lagopus* species, therefore, probably arose during the last ten thousand years. If the red grouse were once again considered to be as separate species rather than merely a distinct subspecies, then the species would probably have evolved during the last 10 kyr.

The timing of splitting of the two *Lagopus* species is more difficult to ascertain. The oldest indisputable record of *L. mutus* is late Middle Pleistocene ("Rissian") in age, at La Fage. Mourer-Chauviré (1993) has suggested that this may have been shortly after their evolution from *L. lagopus*. Certainly all older records of *Lagopus*, including those from the Pliocene, are more akin to *L. lagopus* in size. While it could be argued that size is not a reliable character to gauge whether *L. lagopus* gave rise to *L. mutus*, there is no reason to doubt that the first time two differently-sized members of *Lagopus* arose was in the late Middle Pleistocene. The Lower Pleistocene record of *L. mutus* at Westbury-sub-Mendip in Britain (Andrews, 1990) has not been confirmed by the present study. As to the rate at which the split took place, this is probably related to the mode with which it took place. If a climate-driven (glacial / interglacial) allopatric speciation mode is postulated, the interval could be a single interglacial. This may imply that approximately 10 000 years would suffice for isolation of a population of *L. lagopus* sized birds in, for instance, an upland region in Southern Europe, for differentiation to take place. However, if no isolation is required for speciation to take place, i.e. sympatric speciation, a rate cannot be estimated.

The timing of the differentiation of ravens from other species in the genus *Corvus* cannot be clarified with present data. Furthermore, given that both Mourer-Chauviré (1975a) and the present work found no morphological details of the post-cranium upon which to base the identification of individual species within the genus, future work is unlikely to help. All differences in the genus are of size and metric-shape. Nevertheless, it may be noteworthy that it is only comparatively late, during the "Rissian", that raven-sized *Corvus* fossils first appear. All earlier finds of the genus *Corvus* are smaller and nearer in size to the carrion crow *C. corone*, such as *C. praecorax* and *C. pliocaenus*. This need not, however, imply that the timing of the split between raven and carrion crow occurred during the time between these early *Corvus* fossils and the first appearance of the raven-sized *Corvus* fossils.

Subspecific differentiation within *C. corax* into distinct populations may have also taken place since the last glaciation. Ravens from the last glaciation in Britain do not differ from present day local populations. However, there is an indication, albeit based on limited fossil data, that the birds in Spain (Iberia) differed from those there today. This is demonstrated by the shorter legs of ravens from Spain in the last glacial in contrast to the relatively long legs of the extant subspecies there today, *C. c. hispanus*. The exact age of the Spanish fossils would give a maximum length of time during which the leg lengthening may have taken place. A greater number of raven fossils from the last glaciation would help confirm whether this conclusion is correct, and ideally, Late Glacial and earlier Holocene fossils would help tie down the rate at which the change took place.

The rate of the differentiation of *Sturnus* species can not be gauged due to the problems inherent in the identification of *Sturnus* fossils to species. The rate at which subspecies of *S. vulgaris* have differentiated may be further elucidated. Work by Fear (1984) on *S. vulgaris* subspecies suggested that *S. vulgaris* moved into Europe from the East in different waves. The first wave left various populations which subsequently became the isolated sedentary subspecies on the islands, such as the Azores (*S. v. granti*), the Faroes (*S. v. faroensis*) and the Shetlands (*S. v. zetlandicus*). The second wave formed the nominate, widely distributed subspecies in Europe. The timing of this movement was linked to glacial-interglacial climate variation (Feare, 1984). It is difficult, however, to pinpoint which particular cycle this involved, although it is most likely that all the subspecific differentiation of form has occurred since the last glaciation. This is because the morphology of these subspecies is adaptive and hence linked to the environmental conditions which they are living in. These conditions are likely to have been different during the last Glaciation. However, this does not help solve when these populations first arrived.

The anomalously large *Sturnus* fossils from OIS 5 in western Britain may also possibly be the result of a single interglacial. If, as postulated in chapter 11, these large starlings represent an isolated population, which may be a distinct species, then this must have taken place during OIS 5. Precise dating of fossils in relation to the climatic oscillations of the stage will help clarify the interpretation of these fossils. It would also be useful to have continental *Sturnus* fossils from the same period to confirm any morphological distinctiveness. This would help confirm the isolated nature of the south-western British population.

12.6. Can speciation modes be detected in the fossil record of the four taxa?

Speciation modes, or indeed speciation, can only rarely be observed directly in the fossil record. However, the patterns of morphology and taxonomic occurrence, both today and in the past, may imply that certain processes are more likely to have taken place than others. The two main alternative modes of species origination are allopatric and sympatric speciation.

To date, allopatric speciation is widely accepted as the most probable mode of speciation within vertebrates (Grant and Grant, 1989). However, instances where a large number of niches in a heterogeneous environment exist, that are occupied by a large number of related taxa, has regularly led to a consideration of sympatric speciation (Grant and Grant, 1989; Bush, 1994). Furthermore, Maynard-Smith (1966) has provided a theoretical mechanism for this kind of speciation and hence given further credence to this alternative model.

Due to the order of events that have been proposed for the different speciation modes, certain predictions can be made regarding the fossil record. For instance, if evolution in sympatry has taken place, one would not expect there to be any geographical patterns of occurrence of species which would suggest that isolated populations were involved. In other words, particularly among closely related species, one would not expect there to be geographical barriers to gene flow if sympatric speciation were the norm.

At this juncture, it is has to be pointed out that the taxa chosen in this study do not easily lend themselves to a study of sympatric speciation. Specifically, the criteria on which they were chosen, i.e. species-poor genera to facilitate the identification of the fossils, are exactly the opposite to the taxa which have given rise to a consideration of sympatric speciation. Nevertheless, the genus *Lagopus* may perhaps be the result of a sympatric speciation event. Today, where *Lagopus mutus* and *L. lagopus* are found together, their breeding locations are separated by altitude. For the past, this is difficult to ascertain, because their fossil remains are usually found in areas and latitudes where they are unknown today. The altitudes at which the fossils occur may be more to do with avian predators spanning a wide range of altitudes than the two *Lagopus* species overlapping to a greater extent in the past. However, both species certainly were more widespread in the past and the diminution of their respective ranges to the refugial higher grounds and more northern areas where they are found today, presumably resulted in the allopatric subspecies currently known. Species which are associated with cold climates have their refugia in the north during interglacials. This is important to note because refugia in Europe with respect to glacial / interglacial cycles are generally thought of as Mediterranean, and present during the glacial episodes. Despite this, the event which led to the separation of *L. lagopus* from *L. mutus* may well have been sympatric. It is feasible that a population became split in one area along altitudinal breeding lines and hence speciated sympatrically. (It should be noted that not everyone would call this sympatric and would consider altitude to be a 'barrier'.) However, it is equally possible that the speciation occurred in allopatry and the altitudinal division occurred when the populations met and came into competition. If allopatric speciation occurred, the splitting of populations presumably took place during an interglacial when the widespread *Lagopus* population of the previous glaciation became separated into two. This may have been during the Holsteinian complex prior to the Saalian / "Rissian", where the two species first turn up, at for instance La Fage.

Of the other taxa studied here, *Grus* and *C. corax* are of little use in elucidating speciation modes because they effectively constitute single species in this study. The only thing to be said is that both the common crane *G. grus* and the raven *C. corax* are polytypic. The very existence of geographic subspecies favours a belief that allopatric speciation is an important mechanism in their formation.

The genus *Sturnus,* on the other hand, provides a far more complex and interesting scenario. If only the *Sturnus vulgaris* / *S. unicolor* superspecies is considered, allopatric speciation would seem to have been the mode of differentiation of the ancestor into the two sister species. This is primarily because the distribution of the modern breeding ranges of *S. unicolor* is mainly in the south-western edge of the European landmass, mostly in Iberia, and that of *S. vulgaris* is to the north. The Pyrenees seem to have posed a barrier to the overlap between *S. unicolor* to the south and *S. vulgaris*, because the two species' breeding distributions overlapped only recently. There is a far greater overlap between the taxa during the winter although this does not cause a problem to the degree of allopatry as it is the breeding range which is significant. However, there is a problem if one considers Iberia to be a potential refuge for the northern species during last glacial, as is generally believed for the northern temperate species (Bilton *et al.*, 1998) i.e. this is only really tenable if *S. unicolor* split from *S. vulgaris* since the last glaciation (*ca.* 10,000 yrs. BP). This in turn is unlikely because of the existence of the more southern subspecies of common starling on the Azores, *S. vulgaris granti.* It is difficult to see how this island race would have reached there and only become a subspecies, while the mainland isolate had differentiated to the point of full species. This implies an age greater than the last glaciation for the split between *S. vulgaris* and *S. unicolor*. Therefore, the proposition that *S. vulgaris* had used an eastern refuge (Feare, 1984) is attractive. Further to this, the fact that the more northern and eastern populations of *S. vulgaris* today are more migratory is interesting, particularly from the point of view that the more western and southern populations of the genus *Sturnus* are more sedentary. This points to a relationship between geography and migratory behaviour. It appears to suggest that the more migratory starlings of the East, where the genus originated, periodically produce sedentary populations in the warmer regions of Europe in the West and South. This, in turn, is probably related to the fact that migratory behaviour has been shown to be genetically determined and that climate can select for different migratory routes, as is shown in the case of blackcaps (Berthold, *et al.*, 1992). Presumably degrees of migratory behaviour will also be selected for by climate. Therefore, if there is a link between climate and migratory behaviour which may lead to isolated sedentary populations in the south-west of Europe, this is a mechanism whereby the genus *Sturnus* is producing new species and subspecies. The final perspective on this is provided by the *Sturnus* fossils found in the south-west of Britain during Oxygen Isotope Stage 5. These fossils indicate that these starlings were relatively large, which may signify that they were relatively sedentary and, in turn, may have been an extinct species of starling. This is speculative, although it is no worse a deduction in relation to investigating the possible occurrence of extinct species than if size alone were used as a justification.

The most important perspective of the starling study is that it has highlighted a relationship between relative migratory/sedentary behaviour, climate change and speciation. This may signify that herein lies a novel version of the allopatric speciation mode which accommodates migratory birds.

12.7. What next?

Future work on Quaternary birds in Europe would benefit from a further concentration on individual taxa. The approach followed during the present study was to gain a far greater knowledge of variation within the taxa analysed than previously. This included both fossil and modern skeletal material and hence incorporated a greater geographic and temporal perspective of the taxa than previously achieved. In turn, this allowed for zoogeographic patterning to be considered within a more realistic framework than if merely modern geographical data were used. The approach has produced some unexpected results, particularly with respect to possible scenarios of subspecies, and maybe species, origination. This will undoubtedly lead to some interesting results if other taxa are similarly treated. The example of *Sturnus* shows that even taxa which prove difficult to disentangle taxonomically can yield interesting patterns.

The study of ecomorphology is underrepresented in vertebrate palaeontology as a whole compared to taxonomic treatments, and in birds, site faunal reports predominate. The study of the influence of the environment on form has many consequences, and given the dramatic changes that took place during the Quaternary will undoubtedly lead to unexpected discoveries. The present work on the ecomorphology of migration in *Sturnus*, while not wholly satisfactory, will undoubtedly provide some interesting results with further data. More work is planned on *Sturnus*, and further taxa are being considered. The potential for examining relative migratory behaviour in the fossil record is exciting and novel. This is particularly so because of the link which has been drawn between relative migratory behaviour - genetically determined - and climate (Berthold *et al.*, 1992). This is clearly important in the Quaternary, and although the importance of Quaternary climate to migratory birds has previously been recognised, no mechanism has put forward a link between this and population differentiation and speciation.

Bibliography

Abbot , I.L; Abbot, L.K. and Grant, P.R. 1977. Comparative ecology of Galapagos ground finches (*Geospiza* Gould): evaluation of the importance of floristic diversity and interspecific competition. *Ecological Monographs*, 47: 151 – 184.

Abrams, P.A. 1987. Alternative models of character displacement and niche shift. I. Adaptive shifts in resource use when there is competition for nutritionally nonsubstitutable resources. *Evolution*, 41(3): 651 - 661.

Albarella, U. and Davis, S. 1996. Mammals and birds from Launceston castle, Cornwall: decline in status and the rise of agriculture. *Circaea*, 12(1): 1 - 156.

Albarella, U.; Beech, M. and Mulville, J. 1997. *The Saxon, Medieval and Post-Medieval Mammal and Bird Bones Excavated 1989 – 1991 from Castle Mall, Norwich, Norfolk*. Ancient Monuments Laboratory, Report 72/97.

Albrecht, G.H. and Miller, J.M.A. 1993. Geographic variation in primates. A review with implications for interpreting fossils. In: W.H. Kimbel and L.B. Martin (Eds.). *Species, Species Concepts and Primate Evolution*. Plenum Press, New York. 123 - 161.

Aldrich, J.W. 1984. *Ecogeographic variation in size and proportions of Song Sparrows (*Melospiza melodia*)*. Ornithol. Monogr. No. 35.

Allen, E.E. and Rutter, J.G. 1948. *A Survey of the Gower Caves with an Account of Recent Excavations*. Part 2. Welsh Guides, Swansea.

Amadon, D. 1943. Genera of starlings and their relationships. *American Museum Novitates*, 1247: 1 - 16.

Amadon, D. 1956. Remarks on the Starlings, Family Sturnidae. *American Museum Novitates*, 1803: 1 - 41.

Amadon, D. 1962. Family Sturnidae, Starlings. In: E. Mayr and J.C. Greenway (Eds.). *Peters Check-list of Birds of the World*. Vol. 15. Museum of Comparative Zoology, Cambridge.

Amadon, D. 1963. Comparison of fossil and recent species: some difficulties. *The Condor*, 65(5): 407 - 409.

Andrews, C.W. 1920. Remains of the great auk and ptarmigan in the channel islands. *Ann. Mag. Nat. Hist.* 6(2): 166.

Andrews, P. 1990. *Owls, Caves and Fossils: predation, preservation and accumulation of small mammal bones in caves, with analysis of the Pleistocene cave faunas from Westbury-sub-Mendip, Somerset, UK*. Natural History Museum, London.

Archibald, G.W. 1976a. *The Unison Call of Cranes as a Usefull Taxonomic Tool*. Ph.D. dissertation. Ithaca, New York, Cornell University.

Archibald, G.W. 1976b. Crane taxonomy as revealed by the unison call. *Proc. Int. Crane Workshop*, 1: 225 - 251.

Armstrong, A.L. 1926. Excavations in the Pin Hole Cave, Creswell Crags, Derbyshire. *Reports of the British Association for the Advancement of Science*, 392.

Armstrong, A.L. 1928. Excavations in the Pin Hole Cave, Creswell Crags, Derbyshire. Proceedings of the Prehistoric Society of East Anglia, 6: 330 – 334.

Armstrong, A.L. 1929. Excavations in the Pin Hole Cave, Creswell Crags, Derbyshire. *Transactions of the Hunter Archaeological Society*, III: 116 – 122.

Arnold-Bemrose, H.H. and Newton, E.T. 1905. On an Ossiferous Cavern of Pleistocene age at Hoe Grange Quarry, Longcliffe, near Brassington (Derbyshire). *Quart. Journ. Geol. Soc.*, Vol.LXI: 43 - 63.

Ashley, J. F. 1941. A study of the structure of the humerus in the Corvidae. Condor, 43: 184 - 195.

Atkinson, T.C.; Briffa, K.R. and Coope, G.R. 1987. Seasonal temperatures in Britain during the past 22,000 years, reconstructed using beetle remains. *Nature*, 325: 587 – 592.

Baales, M. 1992. Accumulation of bones of *Lagopus* in Late Pleistocene sediments. Are they caused by man or animals? *Cranium* 9(1): 17 - 22.

Bacher, A. 1967. *Vergleichend morphologische Untersuchungen an Einzelknochen des postkranial Skeletts in Mitteleuropa Vorkommender Schwane und Ganse*. Dissertation. Institute fur Palaeoanatomie, Munchen.

Ballman, P. 1973. Die Fossielen Vogel Aus Dem Jung Pleistozän der Hohle Marie-Jeanne bei Hastiere, Belgien. *Le Gerfaut*, 63: 3-16.

Ballman, P. 1978. Knochenfunde von Vogeln aus der Abeit Sankt Peters zu Gent, Belgien (Vll Bis XVll Jahrundert). *Le Gerfaut*, 68: 551-576.

Bar-Yosef, O. 1989. Geochronology of the levantine middle Palaeolithic. In: P. Mellars and C. Stringer (Eds.). *The Human Revolution*. Edingburgh University Press, Edingburgh. 589 – 610

Barret-Hamilton, G.E.H. 1889. Notes on bird bones from Irish caves. *The Irish Naturalist*. January, 17.

Bate, D.M.A. 1916. On a small collection of vertebrate remains from the Har Dalam Cavern, Malta, with note on a new species of *Cygnus*. *Proc. Zool. Soc. London*, 421 – 430.

Bate, D.M.A. 1928. The animal remains. In: D.A.E. Garrod, L.H.D. Buxton, G. Elliot Smith and D.M.A. Bate. Excavation of a Mousterian Rock-shelter at Devil's Tower, Gibraltar. *J. Royal. Anth. Inst. Gr. Br. and Ire.*, 58: 33 - 115.

Bates, M.R. 1998. Pleistocene sequences at Norton Farm, Chichester, West Sussex. In: J.M. Murton, C.A. Whiteman, M.R. Bates, D.R. Bridgeland, A.J. Long, M.B. Roberts and M.P. Walker (Eds.). *Quaternary of Kent and Sussex: Field Guide*. Quaternary Research Association, London.

Bates, M.R., Bates, C.R., Gibbard, P.L., Macphail, R.I., Owen, F., Parfitt, S.A., Preece, R.C., Roberts, M.B., Robinson, J.E., Whittaker, J.E. and Wilkinson, K.N.

2000 Late Middle Pleistocene deposits at Norton Farm on the West Sussex coastal plain, southern England. *Journal of Quaternary Science*, 15, 61 - 89.

Baumel, J.J. (Ed.) 1979. *Nomina Anatomica Avium*. Academic Press, London.

Beecher, W.J. 1953. A phylogeny of the Oscines. *The Auk*, 70: 270 - 333.

Beecher, W.J. 1978. Feeding adaptions and evolution in the starlings. *Bulletin of the Chicago Academy of Sciences*, 11(8): 269 - 298.

Behre, K.E. 1989. Biostratigraphy for the Last Glacial Period in Europe. *Quaternary Science Reviews*, 8: 25 – 44.

Bell, A. 1915. Pleistocene and later bird fauna of Great Britain and Ireland. *Zool. (London)*, 4(19): 401 - 412.

Bell, A. 1922. Pleistocene and later bird fauna of Great Britain and Ireland. *Naturalist (Hull)*, 251 - 253.

Berger, A.J. 1957. On the anatomy and relationships of *Fregilupus varius*, an extinct starling from the mascarene Islands. *Bull. Amer. Mus. Nat. Hist.*, 113: 225 - 272.

Bergmann, C. 1847. Über die verhältnisse der Wärmeökonomie der Thiere zu ihrer Grösse. *Göttinger Studien*, 1: 595 – 708.

Berthold, P. 1968. Die Massenvermehrung des Stars *Sturnus vulgaris* in fortpflanzungs - physiologischer Sicht. *J. Ornith.*, 109: 141 - 147.

Berthold, P. 1988. Evolutionary aspects of migratory behavior in European warblers. *J. evol. Biol.*, 1: 195 - 209.

Berthold, P. and Terrill, S.B. 1991. Recent advances in studies of bird migration. *Ann. Rev. Ecol. Syst.*, 22: 357 - 378.

Berthold, P.; Helbig, A.J.; Mohr, G. and Querner, U. 1992. Rapid microevolution of migratory behaviour in a wild bird species. *Nature*, 360: 668 - 670.

Bickart, K.J. 1984. A field experiment in avian taphonomy. *Journal of Paleontology*, 4(4): 525 - 535.

Bilton, D.T.; Mirol, P.M.; Mascheretti, S.; Fredga, K.; Zima, J. and Searle, J.B. 1998. Mediterranean Europe as and area of endemism for small mammals rather than a source for nothwards postglacial coloniazation. *Proc. R. Soc. Lond.* B, 265: 1219 – 1226.

Bishop, M.J. 1974. A preliminary report on the Middle Pleistocene mammal bearing deposits of Westbury-Sub-Mendip, Somerset. *Proc. Univ. Bristol Spelaeol. Soc.* 13, 301 – 318.

Bishop, M.J. 1975. Earliest record of Man's presence in Britain. *Nature*, 253: 95 - 97

Bishop, M.J. 1982. *The Mamal Fauna of the Early Middle Pleistocene Cavern Infill Site of Westbury-sub-Mendip, Somerset.* Special Papers in Palaeontology No. 28. Palaeontological Association, London.

Bjordal, H. 1988. Metrical and mechanical properties of some skeletal elements from the house sparrow, *Passer domesticus*, a contribution to the understanding of zooarchaeological problems. *Ossa*, 13: 49 - 59.

Björklund, M. 1992. Selection of bill size proportions in the common rosefinch (*Carpodacus erythrinus*). *The Auk*, 109(3): 637 - 642.

Blytt, A.1876. *Essay on the Immigration of the Norwegian Flora During Alternating Rainy and Dry Periods*. Cammermeyer, Kristiana.

Boag, P.T. 1983. Galapagos evolution continues. *Nature*, 301: 12.

Boag, P.T. and Grant, P.R. 1981. Intense natural selection in a population of Darwin's Finches (*Geospizinae*) in the Galapagos. *Science*, 214: 82 – 85.

Bochenski, Z. 1974. *Ptaki mlodszego czwartorzedu Polski*. Warszswa-Krakow.

Bochenski, Z. 1981. Szczatki kopalne ptakow z Jaskini Mamutowej. *Folia Quaternaria*, Krakow, 54: 1 - 24.

Bochenski, Z. 1984. Middle Pleistocene remains of birds from Kozi Grzbiet in the Swietokrzyskie Mts. (Holy Cross Mts. -Central Poland). *Acta zool. Cracov.*, 27(10):177-186.

Bochenski, Z. 1985. Osteological differentiation in willow grouse. *Fortshritte der Zoologie*, Stuttgart, 30: 69 - 72.

Bochenski, Z. 1989. Problems of skeletal proportions in fossil bird research. *Forschritte der Zoologie*, 35: 445 - 450.

Bochenski, Z. 1991. Pliocene grouse of the genus *Lagopus* from Poland. *Acta zool. Cracov.*, 34(2): 563 - 577.

Bochenski, Z. 1993. Catalogue of fossil and subfossil birds of Poland. *Acta zool. Cracov.*, 36(2): 329 - 460.

Bochenski, Z. and Bochenski, Z. Jun. 1992. Correlation between the wing length of birds and measurements of their bones. *Belg. J. Zool.*, 122 (1): 123 - 132.

Bochenski, Z. jun. 1990. Fossil birds from Dziadowa Skala Cave, Central Poland. *Acta zool. Cracov.*, 33(8): 133 - 147.

Bochenski, Z.M. 1994. The comparative osteology of grebes (*Aves: Podicepediformes*) and its systematic implications. *Acta zool. Cracov.*, 37(1): 191 - 346.

Bochenski, Z., Bochenski, Z.M. and Stewart, J.R. (Eds.). 2002. Proceedings of the 4th. Meeting of the ICAZ Bird Working Group, 11th to the 15th September 2001, Krakow, Poland. *Acta Zoologica Cracoviensia* Vol. 45 (special issue).

Bochenski, Z.M.; Tomek, T.; Boev, Z. and Mitev, I. 1993. Patterns of bird bone fragmentation in pellets of Tawny owl (*Strix aluco*) and the eagle owl (*Bubo bubo*) and their taphonomic implications. *Acta zool. Cracov.*, 36(2): 313 - 328.

Bochenski, Z.M. and Tomek, T. 1994. Fossil and subfossil bird remains from five Austrian Caves. *Acta zool. Cracov.*, 37(1): 347 - 358.

Bochenski, Z.M. and Tomek, T. 1995. How many comparative skeletons do we need to identify a bird bone? *Courier Forschungsinstitut Senckenberg*, 181: 357 - 361.

Bochenski, Z.M.; Korovin, V.A.; Nekrasov, A.E. and Tomek, T. 1997. Fragmentation of bird bones in food remains of imperial eagle (*Aquila heliaca*).

International Journal of Osteoarchaeology, 7: 165 - 171.

Bock, W.J. 1960. The palatine process of the premaxilla in the Passeres. *Bull. Mus. Comp. Zool.*, 122: 361 - 488.

Bock, W.J. 1962. The pneumatic fossa of the humerus in the passeres. *The Auk*, 79: 425 - 443.

Bock, W.J. 1966. An approach to the functional analysis of bill shape. *The Auk*, 83: 10 – 51.

Boessneck, J. 1987. Early bronze age bones of the Siberian White Crane (*Grus leucogeranus*) in South East Turkey. *Ököl. Vogel*, 9: 127 - 129.

Boisseau, S. and Yalden D.W. 1998. The former status of the crane *Grus grus* in Britain. *Ibis*, 140: 482 – 500.

Boyce, M.S. 1979. Seasonality and patterns of natural selection for life histories. *Am. Nat.*, 114: 564 – 583.

Bramwell, D. 1956. The eyrie cave on Ossom's Crag. *Peakland Archaeological Society Newsletter*, 13: 10 - 11.

Bramwell, D. 1957. Report on the bird bones from the Pleistocene levels of Sun Hole, Cheddar. *Proc. Univ. of Bristol Speleol. Soc.*, 8: 39.

Bramwell, D. 1958a. Report on work at Ossom's Eyrie Cave, season 1957. *Peakland Archaeological Society Newsletter*, 14: 8 - 15.

Bramwell, D. 1958b. Final report on Ossom's Eyrie Cave. *Peakland Archaeological Society Newsletter*, 15: 8.

Bramwell, D. 1960a. Some research into bird distribution during the Late Glacial and Post-Glacial Periods. *Bird Reports of the Merseyside Naturalists' Association*, 1959-60: 51 - 58.

Bramwell, D. 1960b. report on a collection of bird bones from the 1929 excavations at Soldiers Hole, Cheddar. *Proceedings of the Somerset Archaeology and Natural History*, 104: 87 – 90.

Bramwell, D. 1976. The vertebrate fauna at Wetton Mill Rock Shelter. In: J.H. Kelley (Ed.). *The Excavation of Wetton Mill Rock Shelter, Manifold Valley, Staffs.* City Museum and Art Gallerey, Stoke-on-Trent, Staffs.

Bramwell, D. 1977. Bird and vole bones from Buckquoy. In: A. Ritchie. Pictish and Vicking Age Farmsteads at Buckquoy, Orkney. *Proc. Soc. Antiq. Scot.*, 108: 209 – 211.

Bramwell, D. 1978. The fossil birds of Derbyshire. In: R.A. Frost. *Birds of Derbyshire.* Buxton: Moorland Publishing Company. 160 - 163.

Bramwell, D. 1984. The birds of Britain: when did they arrive? In: D.D. Gilbertson and R.D.S. Jenkinson (Eds.). *In the Shadow of Extinction: a Quaternary Archaeology and Palaeoecology of the Lake, Fissures and smaller Caves at Cresswell Crags SSSI.* University of Sheffield - John Collis - Publisher.

Bramwell, D. and Yalden, D.W. 1988. Birds from the Mesolithic of Demen's Dale, Derbyshire. *Naturalist*, 113: 141 – 146.

Bramwell, D.; Yalden, D.W. and Yalden, P.E. 1987. Black grouse as the prey of the golden eagle at an archaeological site. *Journal of Archaeological Science*, 14: 195 - 200.

Bramwell, D.; Yalden, D.W. and Yalden, P.E. 1990. Ossom's Eyrie Cave: an archaeological contribution to the recent history of vertebrates in Britain. *Zoological Journal of the Linneann Society*, 98: 1 - 25.

Bramwell, D.; Scott, K.; Stuart, A.J. and Cook, J. 1987. Ossom's Cave, Staffordshire: A study of its vertebrate remains and Late Pleistocene environments. *Staffordshire Archaeological Studies No. 4, Museum Archaeological Society report (new series)*, 25 – 59.

Bridgland, D.R. 1988. The Pleistocene fluvial stratigraphy and palaegeography of Essex. *Proceedings of the Geological Association*, 99: 291 – 314.

Bridgland, D.R. 1994. *Quaternary of the Thames.* Chapman and Hall, London.

Bridgland, D.R. 1995. The Quaternary sequence of the eastern Thames basin: problems of correlation. In: D.R. Bridgland, P. Allen and B.A. Haggart (Eds.). *The Quaternary of the Lower Reaches of the Thames.* Field Guide. Quaternary Research Association, London.

Bridgland, D.R.; Allen, P.; Currant, A.C.; Gibbard, P.L.; Lister, A.M.; Preece, R.C.; Robinson, J.E.; Stuart, A.J. and Sutcliffe, A.J. 1988. Report of Geologist's Field Meeting in north-east Essex, May 22nd - 24th, 1987. *Proceeedings of the Geologist's Association*, 99: 315 - 334.

Bridgeland, D. and Allen, P. 1996. A revised model for terrace formation and its significance for the early Middle Pleistocene terrace aggredation of north-east Essex, England. In: C.Turner (Ed.). *The Early Middle Pleistocene in Europe*. Balkema, Rotterdam. 121 - 134.

Brodkorb, P. 1959. The Pleistocene avifauna of Arredondo, Florida. *Florida State Mus. Bull.*, 4: 269 – 291.

Brodkorb, P. 1960. How many species of birds have existed? *Florida State Mus. Bull.*, 5: 41 – 53.

Brodkorb, P. 1962. Catalogue of Fossil Birds: Part 1 (Archaeopterygiformes through Ardeiformes). *Bulletin of the Florida State Museum Biological Sciences*, 17(4): 179 - 293.

Brodkorb, P. 1964. Catalogue of Fossil Birds: Part 2 (Anseriformes through Galliformes). *Bulletin of the Florida State Museum Biological Sciences*, 8(3): 195 - 335.

Brodkorb, P. 1967. Catalogue of Fossil Birds: Part 3 (Ralliformes, Ichthyornithiformes, Charadriiformes). *Bulletin of the Florida State Museum Biological Sciences*, 11(3): 99 - 220.

Brodkorb, P. 1970. Catalogue of Fossil Birds: Part 4 (Columbiformes through Piciformes). *Bulletin of the Florida State Museum Biological Sciences,* 15(4): 163 - 266.

Brodkorb, P. 1971. Origins and evolution in birds. In: D.S. Farner and J.R. King (Eds.). *Avian Biology.* Academic Press, New York. Vol. 1. 19 - 55.

Brodkorb, P. 1978. Catalogue of Fossil Birds: Part 5 (Passeriformes). *Bulletin of the Florida State Museum Biological Sciences*, 23(3): 139 - 228.

Brothwell, D. 1993. Avian osteopathology and its evaluation. In: A. Morales Muniz (Ed.). Archaeornithology: Birds and the Archaeological Record. Proceedings of the First Meeting of the ICAZ Bird Working Group. Madrid. Octobre 1992. *Archaeofauna*, 2: 33 - 43.

Brown, W.L., Jr. 1978. Punctuated equilibrium excused: the original examples fail to support it. *Biological Journal of the Linnean Society*, 31, 383 – 404.

Brown W.L., Jr. and Wilson, E.O.1956. Character displacement. *Systematic Zoology*, 5: 49 - 64.

Buckland, W. 1822. An account of an assemblage of fossil teeth and bones discovered in a cave at Kirkdale. *Phil. Trans. R. Soc. Lond.*, 122: 171 – 236.

Buckland, W. 1823. *Reliquiae Diluvianae*. John Murray, London.

Bulleid, A. and Grey, St. G. 1917. *The Glastonbury Lake Village*. The Glastonbury Antiq. Soc. Folk Press, London.

Bullough, W.S. 1942. On the external morphology of the British and Continental races of Starling (*Sturnus vulgaris* L.). *Ibis*, 6: 225 - 239.

Burness, G.P. and Montevecchi, W.A. 1992. Oceanographic-related variation in the bone size of extinct great auks. *Polar Biol.*, 11: 545 - 551.

Bush, G.L. 1994. Sympatric speciation in animals: new wine in old bottles. *TREE*, 9(8): 285 – 288.

Cain, A.J. 1953. Geography, ecology and coexistence in relation to the biological definition of the species. *Evolution*, 7: 76 - 83.

Campbell, K.E., Jr. and Marcus, L. 1992. The relationship of hindlimb bone dimentions to body weight in birds. In: K.E. Campbell, Jr. (Ed.). *Papers in Avian Paleontology Honouring Pierce Brodkorb*. The Natural History Museum of Los Angeles County.

Cassoli, P.F. 1972. Lo Pteroclide (Aves, Pteroclidae) fossile nei livelli del Palaeolitico superiore e medio nel Pleistocene dell'Italia meridionale. *Quaternaria*, XVI: 225 - 245.

Cassoli, P.F. and Aldo, G. S. 1985. L'alca impenne *Alca impennis* del Pleistocene d'Italia. *Atti 3 Convegno Italiano di Ornithologia*, 251 - 254.

Cassoli, P.F. and Tagliacozzo, A. 1991(1994). Considerazioni paleontologiche, paleoecologiche sui macromammiferi gli uccelli dei livelli del Pleistocene superior del Riparo di Fumane (VR) (Scavi 1988 - 91). *Boll. Mus. civ. St. nat. Verona*, 18: 349 - 445.

Chaline, J. 1972. *Les Rongeurs du Pléistocène Moyen et Supérieur de France*. Cahier de Paléontologie, Edition du C.N.R.S., Paris.

Chaline, J. 1975. Les rongeurs, l'age et la chronolgie climatique du remplissage de l'Aven I de La Fage (Correze). *Nouv. Arch. Mus. Hist. nat. Lyon*, 13: 113 - 117.

Clark, J.G.D. 1954. *Excavations at Star Carr*. Cambridge University Press, Cambridge.

Clot, A.; Chaline, J.; Jammot, D.; Mourer-Chauvire, C. and Rage, J.C. 1976. Les poches ossifière du Pléistocène moyen et inférieur de Montoussé (Hautes-Pyrénées). *Bul. Soc. Hist. Nat. Toulouse*, 112 (1-2): 146 – 161.

Clutton-Brock, J.; Dennis-Bryan, K.; Armitage, P.L. and Jewel, P.A. 1990. Osteology of soay sheep. *Bull. Br. Mus. nat. Hist. (Zool.)*, 56(1): 1 - 56.

Cock, A.G. 1969. Genetical studies on growth and form in the fowl. 2. The complexity of change in skeletal proportions produced by selection. *Genet. Res.*, 14: 237 - 247.

Cohen, A. and Serjeantson, D. A. 1986. *Manual for the Identification of Bird Bones from Archaeological sites*. Alen Cohen, 2 Powys Lane, London N14.

Cooper J.H. & Voous K., 1999. Birds bones and biogeography: Iberian Azure-winged Magpies come in from the cold. *British Birds* 92: 659-665.

Cooper J.H., 2000. First fossil record of Azure-winged Magpie *Cyanopica cyanus* in Europe. *Ibis* 142: 150-151.

Cooper J.H., 1999. Late Pleistocene avifaunas of Gibraltar and their palaeoenvironmental significance. University of London. Unpublished PhD Dissertation.

Cooper, J.H. In press. A preliminary report on the Pleistocene avifauna of Ibex Cave, Gibraltar. In: Finlayson, J. C. (Ed). *Gibraltar during the Quaternary: the southernmost part of Europe in the last two million years*. Gibraltar Heritage Publications, Gibraltar.

Cooper, R.A. and Ni, Y. 1986. Taxonomy, phylogeny and variability of *Pseudograptus* Beavis. *Palaeontology*, 29: 313 - 363.

Cordy, J-M. 1991. Palaeoecology of the Late Glacial and early Postglacial of Belgium and neighbouring areas. In: N. Barton, A.J. Roberts and D.A. Roe (eds). *The Late Glacial in north-west Europe: Human adptation and environmental change at the end of the Pleistocene* Council for British archaeology, London. 40 – 47.

Cowles, G.S. 1981. Identification of bird bones. In: C. L. Mathews. A roman cess-pit with skeletons. *Bedfordshire Archaeological Journal*, 15: 63 - 73.

Coxon, P.; Hall, A.R.; Lister, A. and Stuart, A.J. 1981. New evidence on the vertebrate fauna, stratigraphy and palaeobotany of the interglacial deposits at Swanton Morley, Norfolk. *Geol. Mag.*, 117 (6): 525 - 546.

Cracraft, J. 1983. Species concepts and speciation analysis. *Current Ornithology*, 1: 159 - 187.

Cramp, S. 1977. *Handbook of the birds of Europe, the Middle East and North Africa: the birds of the Western Palaearctic. Volume 1. Ostrich to Ducks*. Oxford University Press, Oxford.

Cramp, S. 1980. *Handbook of the birds of Europe, the Middle East and North Africa: the birds of the Western Palaearctic. Volume 2. Hawks to Bustards*. Oxford University Press, Oxford.

Cramp, S. and Perrins, C. 1994. *Handbook of the birds of Europe, the Middle East and North Africa: the birds*

of the Western Palaearctic. Volume 8. Crows to Finches. Oxford University Press, Oxford.
Cuisin, J. 1981. L'identification des crânes de passereaux. *L'oiseau et la Rev. Fr. d'Orn.*, 51(1): 17 - 31.
Cuisin, J. 1982. L'identification des crânes de passereaux II. *L'oiseau et la Rev. Fr. d'Orn.*, 52(1): 15 - 19.
Cuisin, J. 1983. L'identification des crânes de passereaux III. *L'oiseau et la Rev. Fr. d'Orn.*, 51(1): 177 - 179.
Cuisin, J. 1989. *L'Identification des Crânes de Passereaux (Passeriformes:Aves)*. Ph.D. Dissertation, University of Boulogne, France.
Cunliffe, B. 1984. *Danebury: and Iron Age Hillfort in Hampshire. Volume 2, The excavations, 1969 – 1978: the finds*. Council for British Archaeology, Research Report 52.
Currant, A.P. 1989. The Quaternary origins of the modern British mammal fauna. *Biol. Journal of the Linnean Society*, 38: 23 - 30.
Currant, A.P. 1991. A Late Glacial Interstadial mammal fauna from Gough's Cave, Somerset, England. In: N. Barton, A.J. Roberts and D.A. Roe (Eds.). The Late Glacial in North-West Europe: Human Adaption and Environmental Change at the End of the Pleistocene. CBA Research Report 77, CBA.
Currant, A.P. 1996. Tornewton Cave and the palaeontological succession. In: D.J. Charman, R.M. Newnham and D.G. Croot (Eds.). *The Quaternary of Devon and East Cornwall: Field Guide*. Quaternary Research Association, London.
Currant, A. and Jacobi, R. 1997. Vertebrate faunas of the British Late Pleistocene and the chronology of human settlement. *Quaternary Newsletter*, 82: 1 – 8.
Darwin, C. 1959. *On the Origin of Species by Means of Natural Selection*. John Murray, London.
Davis, S.J. 1981.The effects of temperature change and domestication on the body size of Late Pleistocene to Holocene mammals of Israel. *Paleobiology*, 7(1): 101 - 114.
Davis, S. and Stewart, J.R. 2002. The mammals and birds from the Gruta Caldeirão, Portugal. *Revista Portuguesa de Arqueologia*, 5(2): 29 – 98.
Davison, G.W. 1985. Avian spurs. *J. Zool. London (A)*, 206: 353 - 366.
Dayan, T.; Simberloff, D.; Tchernov, E. and Yom-Tov, Y. 1991. Calibrating the paleothermometer: climate, communities, and the evolution of size. *Palaeobiology*, 17(2): 189 - 199.
Delacour, J.and Vaurie, C. 1957. A classification of the Oscines (Aves). *Contr. in Science*, No. 16, 6pp.
Delpech, F. 1967. Recherches Palaeontologiques Concernant Quelques Gisements du Magdalenien VI: Station de la Gare de Couze (Dordogne), du Morin (Gironde) et de Duruthy (Lande). *Avhandl. Bordeaux*.
Delpech, F. 1975. *Les Faunes du Paleolithique Superieur dans le Sud-Ouest de la France*. Thèse de doctorat d'état ès Sciences Naturelles. Université de Bordeaux I.
Dement'ev, G.P. and N.A. Gladkov (Eds.) 1967. *Birds of the Soviet Union*. Volume IV. Israel Program for Scientific Translation, Jerusalem.
Dennison, M.D and A.J. Barker. 1991. Morphometric variability in continental and Atlantic Island populations of chaffinches (*Fringilla coelebs*). *Evolution*, 45(1): 29 - 39.
Depéret, C. 1897. Les animaux pliocène du Poussillon. *Mémoire de la Société géologique de France, Paléontologie*, 9.
Diamond A.W. and Hamond, A.C. 1980. The distribution of forest passerine birds and the Quaternary climatic change in tropical Africa. *J. Zool. London*. 191, 379 – 402.
Dobney, K.; Jacques, D.; Pathinarak, T.; Searle, J.; Stewart, J. and von den Driesche, A. In prep. Extinct, extant, resident or migratory?: Evidence for the status of *Grus* sp. in Britain during the Holocene. Journal to be descided.
Dozhansky, Th. 1937. *Genetics and the Origin of Species*. Columbia University Press, New York.
Eastham, A. 1967. The avifauna of Gorham's Cave, Gibraltar. *Bulletin of the Institute of Archaeology*, 1: 37 - 42.
Eldredge, N. and Cracraft, J. 1980. *Phyletic Patterns and the Evolutionary Process*. Columbia University Press, New York.
Eldredge, N. and Gould, S.J. 1972. Punctuated equilibria: an alternative to phyletic gradualism. In: T.J.M Schopf (Ed.). *Models in Paleobiology*. Freeman, Cooper and Co., San Francisco.
Ellias, S.A. 1994. *Quaternary Insects and their Environments*. Smithsonian Institution Press, Washington.
Emslie, S.D. 1985. Late Pleistocene vertebrates from Gunnison County, Colorado. *Journal of Palaeontology*, 60(1): 170 – 176.
Emslie, S.D. 1986. The late Pleistocene (Rancholabrean) avifauna of Little Box Elder Cave, Wyoming. *Contributions to Geology, University of Wyoming*, 23(2): 63 – 82.
Emslie, S. 1992. Two late Blancan avifaunas from Florida and the extinction of wetland birds in the Plio-Pleistocene. *Nat. Hist. Mus. Los Angeles Co., Sci. Ser.*, 36: 249 - 269.
Emslie, S.D. 1996. A fossil scrub-jay supports a recent systematic decision. *The Condor*, 98: 675 – 680.
Emslie, S.D. and Heaton, T. H. 1987. The Late Pleistocene avifauna of Crystal Ball Cave, Utah. *Journal of the Arizona-Nevada Academy of Science*, 21: 53 – 60.
Erbersdobler, K. 1968.*Vergleichend morphologische Untersuchungen an Einzelknochen des postcranial Skeletts in Mitteleuropa vorkommender mittelgrosser Huhnervogel*. Ludwig-Maximilians-Universitat, Munchen.
Ereshefsky, M. (Ed.). 1992. *The Units of Evolution: Essays on the Nature of Species*. Massachusetts Institute of Technology, Cambridge, Massachusetts.

Ericson, P.G.P. 1986. A multivariate analysis of microevolution in the humerus of the Eider (*Somateria mollisima* L.). In: L.-K. Königsson (Ed.). Nordic Late Quaternary Biology and Ecology. *Striae*, 24: 107 - 109.

Ericson, P.G.P. 1987a. *Osteology of the eider (*Somateria mollisima *L.). A study of Sexual, Geographic and Temporal Morphometric Variation in the Eider Skeleton.* Studies 5. Statens Historiska Museum, Stockholm.

Ericson, P.G.P. 1987b. Interpretation of archaeological bird remains: a taphonomic approach. *Journal of Archaeological Science*, 14: 65 - 75.

Ericson, P.G.P. and Hernández Carrasquilla, F. 1997. Subspecific affinity of prehistoric Baltic cormorants (Aves: Phalacrocoracidae). *Ardea*, 85: 1 - 7.

Feare, C. 1984. *The Starling.* Oxford University Press, Oxford.

Fenchel, T. 1975. Character displacement and coexistence in mud snails (Hydrobiidae). *Oecologia (Berl.)*, 20: 19 - 32.

Fick, O.K.W. 1974. *Vergleichend Morphologische Untersuchungen an Einzelknochen europaischer Taubenarten.* Ludwig-Maximilians-Universitat, Munchen.

Fitzgerald, G.R. 1978. Pleistocene grebes of the Old Crow Basin, Yukon Territory, Canada. *Canadian Journal of Earth Science*, 15(2): 1887 - 1892.

Fitzgerald, G.R. 1980. Pleistocene loons of the Old Crow Basin, Yukon Territory, Canada. *Canadian Journal of Earth Science*, 17: 1593 - 1598.

Fitzgerald, G.R. 1991. Pleistocene ducks of the Old Crow Basin, Yukon Territory, Canada. *Canadian Journal of Earth Science*, 28: 1561 - 1571.

Fleicher, R.F. and Johnston, R.F 1982. Natural selection on body size and proportions in house sparrows. *Nature*, 198: 747 - 749.

Florit, X. Mourer-Chauviré, C. and Alcover, J.A. 1989. Els ocells pleistocenics d'es Pouas, Eivissa. Nota preliminar. *Butll. Inst. Catalana Hist. Nat.*, 56(Sec. geol. 5): 35 - 46.

Foley, R.A. 1991. How many species of hominid should there be? *Journal of Human Evolution*, 20: 413 – 427.

Fraguglione, D. 1982. La Bécasse fossile (*Scolopax rusticola* L.). Son origine, son expansion géographique au cours des millénaires. *La Mordoré*, 143: 3 – 47.

Fraser, F.C. and King, J.E. 1954. Faunal Remains. In: J.G.D. Clark (Ed.). *Excavations at Star Carr.* Cambridge University Press, Cambridge.

Fretwell, S. 1969. Ecotypic variation in the nonbreeding season in migratory populations: a study of tarsal length in some Fringillidae. *Evolution*, 23: 406 - 420.

Fischer, K. and Stephan, B. 1974. Eine pleistozäne Avifauna aus der Ghar Dalam-Höhle, Malta. *Z. geol. Wis. Berl.*, 2: 515 - 523.

Funnell, B.M. 1995. Global sea-level and the (pen-)insularity of late Cenozoic Britain. In: R.C. Preece (Ed.). *Island Britain: A Quaternary Perspective.* Geological Society Special Publication No. 96, 3 – 13.

Gardarsson, A. and Moss, R. 1970. Selection of food by icelandic ptarmigan in relation to its availability and nutritive value. In: A. Watson (Ed.). *Animal exploitations in relation to their food resources.* Oxford. 47 – 71.

Garrod, D.A.E.; Buxton, L.H.D.; Elliot Smith, G. and Bate, D.M.A. 1928. Excavation of a Mousterian Rock-shelter at Devil's Tower, Gibraltar. *J. Royal. Anth. Inst. Gr. Br. and Ire.*, 58: 33 - 115.

Garrod, D. and Bate, D. 1937. *The Stone Age of Mount Carmel*, Vol. 1. Oxford University Press, Oxford.

Gascoyne, M.; Currant, A.P. and Lord, T.C. 1981. Ipswichian fauna of Victoria Cave and the marine palaeoclimatic record. *Nature,* 297: 652 - 654.

Gee, H. 1993. The distinction between postcranial bones of *Bos primigenius* Bojanus, 1827 and *Bison priscus* Bojanus, 1827 from the British Pleistocene and the taxanomic status of *Bos* and *Bison. Journal of Quaternary Science*, 8(1): 79 - 92.

Geist, V. 1987. Bergman's rule is invalid. *Can. J. Jool.*, 65: 1035 - 1038.

Geist, V. 1991. Taxonomy: On an objective definition of subspecies, taxa as legal entities, and its application to *Rangifer tarandus* Lin. 1758. In: C.E. Butler and S.P. Mahoney (Eds.). *Proceedings 4th. North American Caribou Workshop*. St. John's, Newfoundland.

Geist, V. 1992. Endangered species and the law. *Nature*, 357: 274 - 276.

Gleed-Owen, C.P. 1996. Amphibian and reptile remains. In: D.J. Charman, R.M. Newnham and D.G. Croot (Eds.). *The Quaternary of Devon and East Cornwall: Field Guide.* Quaternary Research Association, London.

Gleed-Owen, C.P. 1997. The Devensian Late-Glacial arrival of natterjack toad, *Bufo calamita*, in Britain and its implications for colonisation routes and land-bridges. *Quaternary Newsletter*, 81: 18 – 24.

Gibbard, P.L. 1994. *Pleistocene History of the Lower Thames Valley.* Cambridge University Press, Cambridge.

Gibbard, P.L. 1995. Palaeogeographical evolution of the Lower Thames. In: D.R. Bridgland, P. Allen and B.A. Haggart. (Eds.). *The Quaternary of the Lower Reaches of the Thames.* Field Guide, Quaternary Research Association, London.

Gibbard, P.L.; West, R.G.; Zagwijn, W.H.; Balson, P.S.; Burger, A.W.; Funnell, B.M.; Jeffery, D.H.; de Jong, J.; van Kolfschoten, T.; Lister, A.M.; Meijer, T.; Norton, P.E.P.; Preece, R.C.; Rose, J.; Stuart, A.J.; Whiteman, C.A. and Zalasiewicz, J.A. 1991. Early and early middle Pleistocene correlations in the southern North Sea basin. *Quaternary Science Reviews*, 10: 23 - 52.

Gilbert, B.M., Martin, L.D. and Savage, H.G. 1981. *Avian Osteology*. B. Miles Gilbert, 709 Kearney, Laramie, Wyoming 82070.

Godwin, H. 1940. Pollen analysis and forest history of England and Wales. *New Phytologist*, 39: 370 – 400.

Godwin, H. 1978. *Fenland: its Ancient Past and Uncertain Future*. Cambridge University Press, Cambridge.

Gould, S.J. and Eldredge, N. 1977. Punctuated equlibrium: the tempo and mode of evolution reconsidered. *Palaeobiology*, 3: 115 - 151.

Gould, S.J. and Eldredge, N. 1993. Punctuated equilibrium comes of age. *Nature*, 366: 223 - 227.

Grande, L. 1985. The use of paleontology in systematics and biogeography, and a time control refinement for historical bigeography. *Paleobiology*, 11(2): 234 - 243.

Grant, P.R. 1971. Variation in the tarsus length of birds in island and mainland regions. *Evolution*, 25: 599 - 614.

Grant, P.R. 1972. Convergent and divergent character displacement. *Biological Journal of the Linnean Society*, 4: 39 - 68.

Grant, P.R. 1986. *Ecology and Evolution of Darwin's Finches*. Princetown University Press, Princetown.

Grant, P.R. and Grant, B.R. 1989. Sympatric speciation and Darwin's Finches. In: D. Otte and J.A. Endler (Eds.). *Speciation and its Consequences*. Sinauer Associates, Inc. Publishers Sunderland, Massachusets.

Green, C.P.; Coope, G.R.; Currant, A.C.; Holyoak, D.T.; Ivanovich, M.; Jones, R.L.; Keen, D.H.; McGregor, D.F.M. and Robinson, J.E. 1984. Evidence for two temperate episodes in late Pleistocene deposits at Marsworth, U.K. *Nature*, 309: 778 – 781.

Grieve, S. 1885. *The Great Auk or Garefowl* Alca impennis, *its History, Archaeology and Remains.* Thomas C. Jack, London.

Gruber, A. 1990. *Vergleichend morphologische Untersuchungen an Einzelknochen in Ägypten Vorkommender Ciconiidae*. Ludwig-Maximilians-Universität, München.

Grün, R.; Stringer, C.B. and Schwarcz, H.P. 1991. ESR dating of teeth from Garrod's Tabun cave collection. *Journal of Human Evolution*, 20: 231 - 248.

Guiot, J.; Pons, A.; De Beaulieu, J.L.; Reille, M. 1989. A 140,000 year continental climate reconstruction from two European pollen records. *Nature*, 338: 309 – 313.

Gurney, J.H. 1921. *Early Annals of Ornithology*. H.F. and G. Witherby, London.

Guthrie, R.D. 1984. Alaskan megabucks, megabulls, and megarams; the issue of Pleistocene gigantism. In: H.H. Genoways and M.R. Dawson (Eds.). *Contributions in Quaternary Vertebrate Paleontology: A Volume in memorial to John E. Guilday*. Carnegie Museum of Natural History Special Publication No. 8, Pittsburgh.

Guthrie, R.D. 1990. *Frozen Fauna of the Mammoth Steppe: the Story of Blue Babe*. The University of Chicago Press, London.

Hale, W.G. 1984. Evolutionary trends in wading birds. *Biol. Journal of the Linnean Society*, 23: 323 - 332.

Hamilton, T.H. 1961. The adaptive significance of intra-specific trends of variation in wing length and body size among bird species. *Evolution*, 15: 180 – 195.

Hamilton-Dyer, S. 1993. The animal Bone. In: J. D. Zienkiewicz (Ed.). Excavations in the Scamnum Tribunorum at Caerleon: The Legionary Museum Site 1983 -5. *Britania*, XXIV: 132 - 136.

Harrison, C.J.O. 1969. Additional information on the carpometacarpal process as a taxonomic character. *Bulletin of the British Ornithological Club*, 89: 27 - 29.

Harrison, C.J.O. 1976. The tendinal canals of the hypotarsi as a taxonomic character of water birds. *Bulletin of the British Ornithological Club*, 96(3): 82-88.

Harrison, C.J.O. 1977a. Non-passerine birds of the Ipswichian interglacial from Gower caves. *Trans. British Cave Research Assoc.*, 4(4): 441 - 442.

Harrison, C.J.O. 1977b. A new Guillemot, *Cepphus* species, from the Lower Pleistocene. *Suffolk nat. Hist.*, 17: 238 - 239.

Harrison, C.J.O. 1978a. A new jungle-fowl from the Pleistocene of Europe. *J. Archaeological Science*, 5: 373 - 376.

Harrison, C.J.O. 1978b. A longtailed-duck *Clangula hyemalis*, from the Norwich Crag of Suffolk. *Suffolk nat. Hist.*, 18: 358 - 359.

Harrison, C.J.O. 1979a. Birds of the Cromer Forest Bed Series of the East Anglian Pleistocene. *Trans. Norfolk Nat. Soc.*, 24: 177 - 286.

Harrison, C.J.O. 1979b. Pleistocene birds from Swanscombe, Kent. *London Naturalist*, 58: 6 - 8.

Harrison. C.J.O. 1980a. A re-examination of British Devensian and Earlier Holocene Bird bones in the British Museum (Natural History). *J. Archaeological Science*, 7: 53 - 68.

Harrison, C.J.O. 1980b. Pleistocene bird remains from Tornewton Cave and the Brixham Windmill Hill Cave in south Devon. *Bull. Brit. nat. Hist. (Geol.)*, 33 (2): 91 - 100.

Harrison, C.J.O. 1980c. Additional birds from the Lower Pleistocene of Olduvai, Tanzania: and potential evidence of Pleistocene bird migration. *Ibis*, 122: 530 – 532.

Harrison, C.J.O. 1985. The Pleistocene Birds of South-Eastern England. *Bull. geol. Soc. Norfolk*, 35: 53 - 69.

Harrison, C.J.O. 1986. Bird remains from Gough's Cave Cheddar, Somerset. *Proc. Univ. Bristol Spelaeol. Soc.*, 17 (3): 305 - 310.

Harrison, C.J.O. 1987a. Pleistocene and Prehistoric birds of South-Western Britain. *Proc. Univ. Bristol Spelaeol. Soc.*, 18(1): 81 - 104.

Harrison, C.J.O. 1987b. The Early Holocene avifauna of Port Eynon Cave, Gower, South Wales. *Gower*, 38: 60 - 65.

Harrison, C.J.O. 1987c. A re-examination of Star Carr birds. *Naturalist*, 112: 141.

Harrison, C.J.O. 1988a. *The History of the Birds of Britain*. Collins, London.

Harrison, C.J.O. 1988b. Bird bones from Soldier's Hole, Cheddar, Somerset. *Proc. Univ. Bristol Spelaeol. Soc.*, 18(2): 258 - 264.

Harrison, C.J.O. 1989a. Bird remains from Gough's Old Cave, Cheddar, Somerset. *Proc. Univ. Bristol Spelaeol. Soc.*, 18(3): 409 - 411.

Harrison, C.J.O. 1989b. Bird bones Chelm's Combe Shelter, Cheddar, Somerset. *Proc. Univ. Bristol Spelaeol. Soc.*, 18(3): 412 - 414.

Harrison, C.J.O. and Cowles, G.S. 1977. The extinct Large cranes of the North-West Palaearctic. *J. Archaeological Science*, 4: 25 - 27.

Harrison, C.J.O. and Stewart, J.R. 1999. The birds remains. In: M.B. Roberts and S.A. Parfitt (Eds.). *The Middle Pleistocene Site at ARC Eartham Quarry, Boxgrove, West Sussex, U.K.* English Heritage Monograph Series 16, London.

Harrison, C.J.O. and Walker, C.A. 1977. A re-examination of the fossil birds from the Upper Pleistocene in the London Basin. *The London Naturalist*, 56: 6 - 8.

Hedges, R.E.M., R.A. Housley, C. Bronk Ramsey and G.J. Van Klinken. 1994. Radiocarbon dates from the Oxford AMS system: *Archaeometry* Datelist 18. *Archaeometry,* 36(2): 337 - 374.

Helbig, A.J.; Knox, A.G.; Parkin, D.T.; Sangster, G. and Collinson, M. 2002. Guidelines for assigning species rank. *Ibis*, 144: 518 – 525.

Hennig, W. 1965. Phylogenetic systematics. *Annual Review of Entomology*, 10: 97 - 116.

Hernández Carrasquilla, F. 1995. Cueva de Nerja: Las Aves de la campañas de 1980 Y 1982. In: M. Pellicer Catalán and A. Morales Muniz (Eds.). Fauna de la Cueva de Nerja I, Salas de la Mina de la Torca, campañas 1980 - 82. *Trabajos Sobres la Cueva de Nerja*, 5: 221 - 293.

Hernández, E.C., Martín, M and Rando, J.C. 1993. Estudio osteológico comparado de dos suespecies de *Corvus corax* (Aves: Passeriformes). In: A. Morales Muniz (Ed.). Archaeornothology: Birds and The Archaeological Record. Proceedings of the First Meeting of the ICAZ Bird Working Group. Madrid. Octobre 1992. *Archaeofauna*, 2: 1 - 13.

Hewer, T.E. 1925. First report on excavations in the Wye Valley. *Proc. Univ. Bristol Spelaeol. Soc.*, 2(2): 147 – 158.

Hewitt, G. 1996. Some genetic consequences of ice ages, and their role in divergence and speciation. *Biol. J. Linn. Soc.*, 58: 247 – 276.

Hewitt, G. 1999. Post-glacial re-colonisation of European biota. *Biol. J. Linn. Soc.*, 68: 87 – 112.

Hewitt, G. 2000. The genetic legacy of the Quaternary ice ages. *Nature*, 405: 907 – 913.

Hinton, M.A.C. 1926. *Monograph of Voles and Lemmings (Microtinae) Living and Extinct.* London, British Museum (Natural History).

Hoch, E. 1974. The spotted crake, *Porzana porzana* (L.) (Rallidae, Aves), from a Bavarian Eemian interglacial lake deposit. *Mitt. Bayer. Staatsssamml. Palaont. hist. Geol.* 14: 11 - 20.

Housley, R. A. 1991. AMS dates from the Late Glacial and early Postglacial in north-west Europe: a review. In: N. Barton, A.J. Roberts and D.A. Roe (Eds.). *The Late Glacial in north-west Europe: human adaption and environmental change at the end of the Pleistocene.* CBA Research Report, 77: 25 - 39.

Howard, H. 1929. The avifauna of Emeryville Shellmound. University of California, *Publications in Zoology*, 32: 301 - 294, pls. 1 - 4.

Howard, R. and Moore, A. 1991. *A Complete Checklist of birds of the World.* Academic Press, London.

Hufthammer, A.K. 1982. *Geirfuglens utbredelse og morfologiske variasjon I Skandinavia.* Unpubl. thesis. Universitet I Bergen, Bergen, Norway. pp. 60.

Huntley, B. and Birks, H.J.B. 1983. *An Atlas of Past and Present Pollen Maps for Europe: 0 – 13000 Years Ago.* Cambridge university Press, Cambridge.

Huxley, J. 1942. *Evolution the Modern Synthesis.* George Allen and Unwin Ltd, London.

Ingold, J.L.; Guttman, S.I. and Osborne, D.R. 1987a. Biochemical systematics and the evolution of the cranes (Aves: Gruidae). In: G.W. Archibald and R.F. Pasquier (Eds.). *Proceedings of the 1983 Internatinal Crane Workshop.* Baraboo, Wisconsin, International Crane Foundation. 575 - 584.

Ingold, J.L.; Guttman, S.I. and Osborne, D.R. 1987b. Biochemical systematics of crowned cranes. In: G.W. Archibald and R.F. Pasquier (Eds.). *Proceedings of the 1983 Internatinal Crane Workshop.* Baraboo, Wisconsin, International Crane Foundation. 317 – 322.

Ingold, J.L.; Vaughn, J.C.; Guttman, S.I. and Maxson, L.R. 1989. Phylogeny of the cranes (Aves: Gruidae) as deduced from DNA-DNA hybridization and albumin micro-complement fixation analysis. *The Auk*, 106: 595 - 602.

Jacobi, R.M.; Rowe, P.J.; Gilmour, M.A.; Grün, R. and Atkinson, T.C. 1998. Radiometric dating of the Middle Palaeolithic tool industry and associated fauna of Pin Hole Cave, Creswell crags, England. *J. Quaternary Sci.*, 13: 29 – 42.

James, F.C. 1970. Geographic size variation in birds and its relationship to climate. *Ecology*, 51: 365 – 390.

James, F. C. 1983. Environmental component of morphological differentiation in birds. *Science*, 221: 184 - 186.

Jánossy, D. 1972. Die mittelpleistozane Vogelfauna der Stránská Skála. *Anthropos (Brno)*, 20 (N.S.12): 35 - 64.

Jánossy, D. 1974a. Upper Pliocene and Lower Pleistocene Bird remains from Poland. *Acta zool. Cracov.*, XIX(21): 531 - 564.

Jánossy, D. 1974b. Die mittelpleistozäne Vögelfauna von Hundsheim (Niedeosterreich). *Sitzunsberichten der Osterr. Akademie de Wissenscxhaften, Mathem.-naturw. Kl.*, Abt. I, 182 (6 – 8): 211 – 257.

Jánossy, D. 1976a. Plio-Pleistocene bird remains of the Carpathian Basin. I. Galliformes. 1. Tetraonidae. *Aquila*, 82: 13 - 36.

Jánossy, D. 1976b. Plio-Pleistocene bird remains of the Carpathian Basin. II. Galliformes. 2. Phasianidae. *Aquila*, 83: 29 - 42.

Jánossy, D. 1977. Plio-Pleistocene bird remains of the Carpathian Basin. III. Striformes, Falcinoformes, Caprimulgiformes, Apodiformes. *Aquila*, 84: 9 - 36.

Jánossy, D. 1979a. Plio-Pleistocene bird remains of the Carpathian Basin. IV. Anseriformes, Gruiformes, Charadriiformes and Passeriformes. *Aquila*, 85: 11 - 39.

Jánossy, D. 1979b. Plio-Pleistocene bird remains of the Carpathian Basin.V. Podicepiformes, Ciconiiformes, Otidiformes, Columbiformes, Piciformes. *Aquila*, 86: 19 - 33.

Jánossy, D. 1980. Plio-Pleistocene bird remains of the Carpathian Basin.VI. Systematical and geographical catalogue. *Aquila*, 87: 9 - 22.

Jánossy, D. 1983. Humeri of Central European Passeriformes. *Fragmenta Mineralogica et Paleontologica*, 11: 85 - 112.

Jánossy, D. 1987. Taxonomical status of Upper Pliocene and Lower Pleistocene bird remains. In: C. Mourer-Chauvire (Ed.). L'Evolution des oiseaux d'apres le temoignage des fossiles. *Document des Lab. Geol. Lyon*, 99: 189 - 192.

Jánossy, D. 1992. Lower Pleistocene bird remains from Beremend, (S-Hungary, Loc. 15. And 16). *Aquila*, 99: 9 – 25.

Jehl, J.R. 1966. Subspecies of Recent and fossil birds. *The Auk*, 83(2): 306 - 307.

Jelinek, A. 1982. The Tabun Cave and Palaeolithic Man in the the Levant. *Science*, 216, 1369 - 1375.

Jenkinson, R.D.S. 1984. *Creswell Crags: Late Pleistocene Sites in the East Midlands*. Oxford: BAR Series 122.

Johnsgard, P.A. 1983. *Cranes of the World*. Croom Helm, London.

Johnson, J.G. 1982. Occurrence of phyletic gradualism and punctuated equilibria through geological time. *Journal of Paleontology*, 56(6): 1329 – 1331.

Johnston, R.F. 1992. Evolution in the rock dove: skeletal morphology. *The Auk*, 109(3): 530 – 542.

Johnston, R.F. and Selander, R.K. 1964. House Sparrows: Rapid evolution of races in north America. *Science* 144: 548 - 550.

Jones R.L. and Keen, D.H. 1993. *Pleistocene Environments in the British Isles*. Chapman and Hall, London.

Jollie, M. 1958. Comments on the phylogeny and skull of the Passeriformes. *The Auk*, 75: 26 - 35.

Kellner, M. 1986. *Vergleichend Morphologische Untersuchungen an Einzelknochen des Postkranialen Skeletts in Europa Vorkommender Ardeidae*, Ludwig-Maximilians-Universität, München.

Kelly, J.H. 1976. *The Excavation of Wetton Mills Rock Shelter, Manifold Valley, Staffs*. City Museum and Art Gallery, Stoke on Trent.

Kendeigh, S. C. 1969. Tolerance of cold and Bergamnn's Rule. *The Auk*, 86: 13 - 25.

Kessler, E. and Moldovai, M. 1993. Studii biometrice si Morfologice asupra scheletului Membrelor la Corvidae (*Aves*) (Patea I). *Studia Univ. Babes-Bolyai, Biologia*, XXXVIII, 1-2: 69 - 80.

Kimbel, W.H. and Martin, L.B. (Eds.). 1993. *Species, Species Concepts and Primate Evolution*. Plenum Press, New York.

Klein, R.G. 1986. Carnivore size and Quaternary climatic change in southern Africa. *Quaternary Research*, 26: 153 – 170.

Klein, R.G. and Scott, K. 1989. Glacial/Interglacial size variation in fossil spotted hyenas (*Crocuta crocuta*) from Britain. *Quaternary Research*, 32: 88 - 95.

Klicka, J. and Zink, R.M. 1997. The importance of recent ice ages in speciation: a failed paradigm. *Science*, 277: 1666 – 1669.

Knox, A.G.; Collinson, M.; Helbig, A.J.; Parkin, D.T. and Sangster, G. Taxonomic recommendations for British birds. *Ibis*, 144: 707 – 710.

Kolfschoten, T. van. 1990. The evolution of the mammal fauna in the Netherlands and the middle Rhine area (West Germany) during the Late Middle Pleistocene. *Mededelingen Rijks Geologische Dienst*, 43: 1 - 69.

Kolstrup, E. 1991. Palaeoenvironmental developments during the Late Glacial of the Weischelian. . In: N. Barton, A.J. Roberts and D.A. Roe (Eds.). *The Late Glacial in North-West Europe: Human Adaptation and Environmental Change at the End of the Pleistocene*. CBA Research Report No. 77. 1 - 6.

Kraft, E. 1972. *Vergleichend morphologische Untersuchungen an Einzelknochen nord und mitteleuropaischer kleinerer Huhnervogel*. Ludwig-Maximilians-Universität, München.

Krajewski, C. 1989. Phylogenetic relationships among cranes (Gruiformes: Gruidae) based on DNA hybridization. *The Auk*, 106: 603 - 618.

Kurtén, B. 1968. *Pleistocene Mammals of Europe*. London, Weidenfeld and Nicolson.

Kutzbach, J.E. and Guetter, P.J. 1986. The influence of changing orbital parameters and surface boundary conditions on climatic simulations for the past 18 000 years. *Journal of Atmospheric Sciences*, 43: 1726 – 1759.

Lack, D. 1944. Ecological aspects of species formation in Passerine birds. *Ibis*, 1944: 260 - 286.

Lack, D. 1947. *Darwin's Finches*. Cambridge University Press, Cambridge.

Lack, D. 1971. *Ecological isolation in Birds*. Blackwell Scientific Publications, Oxford.

Lamb, H.H. 1977. The late Quaternary history of the climate of the British Isles. In: F.W. Shotton (Ed.). *British Quaternary Studies: Recent Advances*. Clarendon Press, Oxford.

Lambrecht, K. 1914. Morphologie des Mittelhandknochens-Os metacarpi-der Vogel. *Aquila*, 21: 53 - 84.

Lambrecht, K. 1933. *Handbuch der Palaornithologie*. Borntrager, Berlin.

Langer, G. 1980. *Vergleichend Morphologische Untersuchungen an Einzelknochen in Mitteleuropa vorkommender mittelgrosse Eulenarten*. Ludwig-Maximilians-Universitat, Munchen.

Lawlor, L. R. and Maynard Smith, J. 1976. The coevolution and stability of competing species. *The American Naturalist.*, 110(971): 79 - 99.

Leisler, H. and Winkler B. 1985. Morphological aspects of habitat selection in birds. In: M.L. Cody (Ed.). *Habitat Selection in Birds*. London: Academic Press. 415 – 434

Linsted, S.L. and Boyce, M.S. 1985. Seasonality, fasting endurance, and body size in mammals. *Am. Nat.*, 123: 115 – 124.

Lister, A.M. 1984. Evolutionary case histories from the fossil record. *Nature*, 309: 114 - 115.

Lister, A.M. 1987. Giant deer and giant red deer from Kent's Cavern, and the status of *Strongyloceros spelaeus* Owen. *Trans. Proc. Torquay nat. Hist. Soc.* 19: 189 - 198.

Lister, A.M. 1992. Mammalian fossils and Quaternary biostratigraphy. *Quaternary Science Review*, 11: 329 - 344.

Lister, A.M. 1993. Patterns of evolution in Quaternary mammal lineages. In: D.R. Lees and D. Edwards (Eds.). Evolutionary patterns and processes. *Linnean Society Symposium Series*,14: 71 - 93.

Lister, A.M. 1996. The morphological distinction between bones and teeth of fallow deer (*Dama dama*) and red deer (*Cervus elaphus*). *International Journal of Osteoarchaeology*, 6: 119 - 143.

Lister, A.M. 1998. The age of Early Pleistocene mammal faunas from the 'Weybourne Crag' and Cromer Forest-bed Formation (Norfolk, England). *Mededelingen Nederlands Instituut voor Toegepaste Geowetenschappen TNO*, 60: 271 – 280.

Lister, A. M. and A.V. Sher. 1995. Ice cores and mammoth extinction. *Nature*, 378: 23 - 24.

Livingston, S.D. 1989. The taphonomic interpretation of avian skeletal part frequencies. *Journal of Archaeological Science*, 16: 537 - 547.

Lorch, R. 1992. *Vergleichend Morphologische Untersuchungen Skeletts von* Pelecanus onocrotalus, Pelecanus crispus *und* Pelecanus rufescens. Ludwig-Maximilians-Universitat, Munchen.

Losos, J.B., Warheit, K.I. and Schoener, T.W. 1997. Adaptive differentiation following experimental island colonisation in *Anolis* lizards. *Nature*, 387: 70 - 73.

Louchart, A. 2002. Les oisaux de Pleistocene de corse et quelque localites sardes, ecologie, evolution, biogeographie et extinction. *Docum. Lab. Geol. Lyon*, 155: 1 - 287.

Louchart, A. 2004. An extinct large thrush (Aves: Turdidae) from the Late Quaternary of Mediterranean Europe. *N. Jb. Geol. Paläont. Abh.*, 233(2): 275 – 296.

Lowe, J.J.; Coope, G.R.; Harkness, D.D.; Sheldrick, C.; and Walker, M.J.C. 1995. Direct comparison of UK temperatures and Greenland snow accumulation rates, 15 - 12,000 years ago. *Journal of Quaternary Science*, 10(2): 175 - 180.

Lydekker, R. 1890. On the remains of some large extinct birds from the cavern deposits of Malta. *Proceedings of the Zoological Society of London*, 28: 403 - 411.

Lydekker, R. 1891a. *Catalogue of the Fossil Birds in the British Museum (Natural History)*. British Museum Publ., London.

Lydekker, R. 1891b. On British fossil birds. *The Ibis*, 3(6): 381 - 410.

Lydekker, R. 1891c. On Pleistocene bird-remains from Sardinian and Corsican Islands. *Proc. zool. Soc. London*. 467 - 476.

MacDonald, K.C. 1992. The domestic chicken (*Gallus gallus*) in sub-Saharan Africa: a background to its introduction and its osteological differentiation from indigenous fowls (Numidinae and *Francolius* sp.). *Journal of Archaeological Science*, 19: 303-318.

Macdonald, D. and Barrett, P. 1993. *Mammals of Britain and Europe*. Harper Collins Publishers, London.

Malmquist, M.G. 1985. Character dislacement and biogeography of the pygmy shrew in Northern Europe. *Ecology*, 66(2): 372 - 377.

Martin, L.B. and Andrews, P. 1993. Species recognition in the Middle Miocene Hominoids. In: W.H. Kimbel and L.B. Martin (Eds.). *Species, Species Concepts and Primate Evolution*. Plenum Press, New York.

Masters, J.C. and Rayner, R.J. 1996. The recognition concept and the fossil record: putting the genetics back into phylogenetic species. *South African Journal of Science*, 92: 225 - 231.

Matthews, C. L. 1981. A roman cess-pit with skeletons. *Bedfordshire Archaeological Journal*, 15: 63 - 73.

Mayhew, D.F. 1975. *The Quaternary History of some British Rodents and Lagomorphs*. Ph.D. Thesis, University of Cambridge.

Maynard-Smith, J. 1966. Sympatric speciation. *Am. Natur.*, 100: 637 – 650.

Mayr, E. 1942. *Systematics and the Origin of Species*. Columbia University Press, New York.

Mayr, E. 1956. Geographical character gradients and climatic adaption. *Evolution*, 10: 105 – 108.

Mayr, E. 1957. Species concepts and definitions. In: E. Mayr (Ed.). *The Species Problem*. American Association for the Advancement of Science Publication No. 50. 1 – 22.

Mayr, E. 1963. *Animal Species and Evolution*. Harvard University Press, Cambridge, MA.

Mayr, E. 1969. Bird speciation in the tropics. *Biol. J. Linn. Soc.*, 1: 1 – 17.

Mayr, E. 1982. *The Growth of Biological Thought*. Belknap Press, Harvard University.

McNab, B.K. 1971. On the ecological significance of Bergmann's Rule. *Ecology*, 52: 845 - 854.

McNeill, J. 1979. Purposefull phenetics. *Systematic Zoology*. 28: 465 - 482.

Meijer, T. and Preece, R.C. 1996. Malacological evidence relating to the stratigraphical position of the Cromerian. In: C. Turner (Ed.). *The Early Middle Pleistocene in Europe*. Balkema, Rotterdam.

Mein, C.D. and Archibold, G.W. 1996. *Status Survey and Conservation Action Plan. The Cranes*. IUCN/SSC Crane Specialist group. IUCN, Gland, Switzerland.

Mein, P. 1990. Updating of MN zones. In: E.H. Lindsay, V. Fahlbusch, and P. Mein (Eds.). *European Neogene Mammal Chronology*. Plenum Press, New York. 73 - 90.

Mello, J.M. 1875. On some bone caves in Creswell Crags. *Quaterly Journal of the Geological Society of London*, 31: 679 – 691.

Mello, J.M. 1876. On the bone caves of Creswell. *Quaterly Journal of the Geological Society of London*, 32: 240 – 259.

Mercier, N.; Valladas, H.; Reyss, J.-L.; Jelinek, A.; Meignen, L. and Joron, J.-L. 1995. TL dates of burnt flint from Jelinek's excavation at Tabun and their implications. *Journal of Archaeological Science*, 22(4): 495 - 510.

Millberg, P. and Tyrberg, T. 1993. Naive birds and noble savages - a review of man-caused prehistoric extinctions of island birds. *Ecography*, 16: 229 - 250.

Milne-Edwards, A. 1867 - 71. *Recherches Anatomiques pour Servir a la Histoire des Oiseaux Fossiles de la France*. Masson, Paris.

Milne-Edwards, A. 1875. Observation sur les oiseaux dont les ossements ont ete trouves dans les cavernes du Sud-Ouest de la France. *Materiaux pour L'Histoire Primitive et Naturelles de l'Homme*, Paris, 2(6): 473 - 503.

Mitchell, G.F.; Penny, L.F.; Shotton, F.W. and West, R.G. 1973. *A Correlation of Quaternary Deposits in the British Isles*. Geological Society of London Special Publication No. 4. Geological Society, London.

Mlikovsky, J. 1982a. Zur systematischen Stellung von *Buccephala angustipes* Jánossy, 1965 und *Somateria gravipes* Harrison, 1979 (Aves: Anseriformes) aus dem Pleistozan Europas. *Z. geol. Wiss. Berlin*, 10(11): 1463 - 1475.

Mlikovsky, J. 1982b. Taxonomische Identität der *Anas submajor* Jánossy, 1979 (Aves: Anseriformes) aus dem Oberpliozän Ungarns. *Vest. cs. Spolec. Zool.*, 46: 199 - 202.

Mönkkönen, M. 1995. Do migrant birds have more pointed wings?: a comparative study. *Evolutionary Ecology*, 9: 520 - 528.

Moss, R. 1967. Food selection by red grouse (*Lagopus lagopus scoticus* (Lath.)) in relation to chemical composition. In: K. Petrusewicz (Ed.). *Secondary Productivity of Terrestrial Ecosystems*. Warsaw.

Moss, R. 1968. Food selection and nutrition in ptarmigan (*Lagopus mutus*). *Symp. zool. Soc. Lond.*, 21: 207 - 216.

Moss, W.W. and Hendrickson, J.A. Jr. 1973. Numerical taxonomy. *Annual Review of Entomology*, 18: 227 – 258.

Morales Muniz, A. 1993. Ornithoarchaeology: the various aspects of the classification of bird remains from archaeological sites. In: A. Morales Muniz (Ed.). Archaeornothology: Birds and The Archaeological Record. Proceedings of the First Meeting of the ICAZ Bird Working Group. Madrid. Octobre 1992. *Archaeofauna*, 2: 1 - 13.

Moreno, E. 1985. Clave osteologica para la identification de los paserifomes ibericos. *Ardeola*, 32(2): 295-378.

Moreno, E. 1986. Clave osteologica para la identification de los paserifomes ibericos. *Ardeola*, 33(1-2): 69-130.

Moreno, E. 1987. Clave osteologica para la identification de los paserifomes ibericos. *Ardeola*, 34(2): 243-374.

Mourer-Chauviré, C. 1975a. Les Oiseaux du Pléistocène Moyen et Supérieur de France. *Documents de Laboratoire de Géologie de la Faculté de Lyon* No. 64, 2 Fasc., 624 pp.

Mourer-Chauviré, C. 1975b. Les oiseaux (*Aves*) du gisement Pléistocène Moyen des Abimes de la Fage a Noailles (Corrèze). *Nouv. Arch. Mus. Hist. nat. Lyon.* Fasc. 13: 89 - 112.

Mourer-Chauviré, C. 1975c. Faunes d' oiseaux du Pleistocene de France: systematique, evolution et adaption, interpretation paleoclimatique. *Géogios*, 8(5): 333 - 352.

Mourer-Chauviré, C. 1979. Les Oiseaux de la fin des temps glaciaires en France. La disparition des espèces froides. In: D. de Sonneville-Bordes (Ed.). *Colloques internationaux C.N.R.S.* No. 271.

Mourer-Chauviré, C. 1983. Les oiseaux dans les habitats palaeoloithiques: gibier des hommes ou proies des rapaces? In: C. Grigson and J. Clutton-Brock (Eds.). *Animals in Archaeology, Tome II. Shell Middens, Fishes and Birds*. BAR International Series 183, Oxford. 111 - 124.

Mourer-Chauviré, C. 1993. The Pleistocene avifaunas of Europe. In: A. Morales Muniz (Ed.). Archaeornithology: Birds and The Archaeological Record. Proceedings of the First Meeting of the ICAZ Bird Working Group. Madrid. Octobre 1992. *Archaeofauna*, 2: 53 - 66.

Mourer-Chauviré, C. 2004. Les oiseaux du Pliocène supérieure de Saint-Vallier (Drôme, France). *Geobios* 37: S97 –S114.

Mourer-Chauviré, C. and Antunes, M.T. 1991. Presence du grand pingouin, *Pinguinis impennis* (Aves, Charadriiformes) dans le Pleistocene du Portugal. *Géogios*, 24(2): 201 - 205.

Mourer-Chauvire, C.; Adrover, R. and Pons, J. 1975. Presence de *Grus antigone* (L.) dans l' ''Avence De Na Corna'' a Majorque (Espagne). *Nouv. Arch. Mus. Hist. nat. Lyon. Fasc.*, 13, suppl. : 45 - 50.

Mourer-Chauviré, C., Moya, S. and Adrover, R. 1977. Les oiseaux des gisements quaternaires de Majorque. *Nouv. Arch. Mus. Hist. nat. Lyon, Suppl.*, 15: 61 - 64.

Mourer-Chauviré, C., Alcover, J.A. Moya, S., Pons, J. 1980. Une nouvelles forme insulaire d'effraie geante, *Tyto balearica* n. sp., (Aves, Strigiformes), du plio-pleistocene des Baleares. *Géogios*, 13(5): 803 - 811.

Mourer-Chauviré, C.; Philippe, M.; Quinif, Y.; Chaline, J.; Debard, E.; Guérin, C. and Hugueney, M. 2003. Position of the palaeontological site Aven I des Abîme de la Fage, at Noailles (Corrèze, France), in the European Pleistocene chronology. *Boreas*, 32: 521 – 531.

Nadachowski, A. 1993. The species concept and Quaternary mammals. *Quaternary International*, 19: 9 - 11.

Nelson, G. and Platnick, N. 1981. *Systematics and Biogeography*. Columbia University Press, New York.

Newton, A. 1901. On some Cranes' bones found in Norfolk. *Trans. Norf. Norw. nat. Hist. Soc.* 7: 158 159.

Newton, E.T. 1883. On the remains of a red-throated diver, *Colymbus septentrionalis*, Linn., from the "Mundesley River Bed". *The Geological Magazine*, 10: 96 - 100.

Newton, E.T. 1894. The vertebrate fauna collected by Mr. Lewis Abbott from the Fissure near Ightham, Kent. *Quaterly Journal of the Geological Society of London*, 50: 188 – 211.

Newton, E.T. 1916 - 17. Notes on bones found in the Creag Nan Uamh Cave, Inchnadamff, Assynt, Sutherland. *Proc. Royal Soc. Edingburgh*, 37: 344 - 349.

Newton, E.T. 1922a. List of avian species identified from Aveline's Hole, Burrington. *Proceedings of the Bristol Spelaeological Society*, 1: 119 - 121.

Newton, E.T. 1922b. Fossil bird-remiains collected by Dr. Forsyth-Major in Sardinia, Corsica and Greece. *Proc. Zool. Soc. London*, 229 - 232.

Newton, E.T. 1923. Pleistocene bird remains from Chudleigh. *Naturalist*, 40: 264 - 265.

Newton, E.T. 1924a. Note on bird's bones from Merlin's Cave. *Proceedings of the University of Bristol Speleological Society*. 2: 159 - 161.

Newton, E.T. 1924b. Note on additional species of birds from Aveline' Hole. *Proceedings of the University of Bristol Speleological Society*. 2:121.

Northcote, E.M. 1979a. Determination of age and sex of long bones of mute swan *Cygnus olor*. *Ibis*, 121:74-80.

Northcote, E.M. 1979b. *Comparative and Historical Studies of European Quaternary Swans and other Aquatic Birds*. Unpublished PhD Thesis, Lucy Cavendish College, Cambridge University.

Northcote, E.M. 1980a. Some Cambridgeshire Neolithic to Bronze Age birds and their presence or absence in England in the Late-Glacial and Early Holocene. *Journal of Archaeological Science* 7: 379 - 383.

Northcote, E.M. 1980a. Sexual dimorphism of the long bones of whooper swans *Cygnus cygnus cygnus*. *Ibis*, 122: 369-372.

Northcote, E.M. 1981b. Size differences between limb bones of recent and subfossil mute swans *Cygnus olor*. *Journal of Archaeological Science*, 8: 89-98.

Northcote, E.M. 1981b. Difference in weight and habit of Whooper *Cygnus cygnus cygnus* and Mute *C. olor* Swans in relation to differences in their long bones. *Bull. Br. Ornithol. Club*., 101: 266 - 267.

Northcote, E.M. 1982a. The extinct maltese crane *Grus melitensis*. *Ibis*, 124: 76-80.

Northcote, E.M. 1982b. Size, form and habit of the extinct Maltese swan *Cygnus falconeri*. *Ibis*, 124: 148 - 158.

Northcote, E.M. 1982c. Sympatry of common cranes, *Grus grus* with larger cranes in the last c. 125 000 years. *Bull. Brit. Orn. Club*, 102: 141 - 142.

Northcote, E.M. 1983. Morphology of the mute swan, *Cygnus olor*, in relation to domestication. In: C. Grigson and J. Clutton-Brock (Eds.): *Animals in Archaeology. Tome II. Shell middens, Fishes and Birds*. BAR International Series 183. Oxford. 173 - 179.

Northcote, E.M. 1984. Crane *Grus* fossils from the Maltese Pleistocene. *Palaeontology*, 27: 733 - 735.

Northcote, E.M. 1988. An extinct 'swan goose' from the Pleistocene of Malta. *Palaeontology*, 31 (30: 725 - 740.

Northcote, E.M. and Mourer-Chauviré, C. 1985. The distinction between the extinct Pleistocene European crane *Grus primigenia*, and the extant Asian sarus crane, *G. antigone*. *Geobios*, 18: 877 - 881.

Northcote, E.M and Mourer-Chauviré,C. 1988. The extinct crane *Grus primigenia* Milne-Edwards in Majorca (Spain). *Geobios*, 21(2): 201 - 208.

Oliver, J.S. and Graham, R.W. 1994. A catastrophic kill of ice-trapped coots: time averaged versus scavenger-specific disarticulation patterns. *Paleobiology*, 20(2): 229 - 244.

Olsen, S.J. 1979. Osteology for the Archaeologist. *Papers of the Peabody Museum of Archaeology and Ethnology*, 56, 3, 4 and 5. Harvard.

Olson, S.L. 1977. A great auk, *Pinguinis*, from the Pliocene of North Carolina (Aves: Alcidae). *Proc. Biol. Soc. Wash.* 90(3): 690 - 697.

Olson, S.L. and James, H.F. 1982. Fossil birds from Hawaiian Islands: evidence for wholesale extinction by man before western contact. *Science*, 217: 633 - 635.

Olson, S.L. 1985. The fossil record of birds. In: *Avian Biology Vol. VII*, D.S. Farner and K.C. Parks (Eds.). Academic Press, New York. 79 - 238.

Olson, S.L. and Rasmussen, P.C. 2001. Miocene and Pliocene birds from the Lee Creek Mine, North Carolina. *Smithsonian Contrib. Palaeobiol.*, 90: 235 - 265.

Osborne, P.J. 1976. Evidence from the insects of climatic variation during the Flandrian period: a preliminary note. *World Archaeology*, 8: 150 - 158.

Otto, C. 1981. *Vergleichend morphologische Untersuchungen an Einzelknochen in Zentraleeuropa vorkommender mittelgrossen Accipitridae: 1, Schadel, Brustbein, Schultergurtel und Vorderextremitat*. Ludwig-Maximilians-Universitat, Munchen.

Owen, R. 1839. On the bone of an unknown struthious bird from New Zealand. *Proceedings of the Zoological Society of London*, 1939, 169 – 170.

Owen, R. 1846. *A History of British Fossil Mammals and Birds*. Van Voorst, London.

Owen, R. 1865. Description of the skeleton of the Great Auk (*Alca impennis*). *Transactions of the Zoological Society of London*, V:4, 317 - 335.

Parker, A.J. 1988. The birds of Roman Britain. *Oxford Journal of Archaeology*, 7(2): 197 - 226.

Parry, S. 1996. The avifaunal remains. In: B. Conway, J. McNabb and N. Ashton (Eds.). *Excavations at Barnfield Pit, Swanscombe, 1968 - 72.* British Museum Occasional Paper 94. 137 - 143.

Patterson, H.E.H. 1981. The continuing search for the unknown and unknowable: a critique of contemporary ideas on speciation. *South African Journal of Science*, 77: 133 – 119.

Patterson, H.E.H. 1985. The recognition concept of species. In: E.S. Vbra (Ed.). *Species and Speciation.* Transvaal Museum Monographs, 4: 21 – 29. 21 - 29.

Payne, R.B. 1984. *Sexual Selection, Lek and Arena Behavior, and Sexual Size Dimorphism in Birds.* Ornithological Monograph No. 33. The American Ornithologists' Union, Washington, D.C.

Peach, B.N. and Horne, J. 1916 – 17. The bone caves in the Valley of Allt nan Uamh Cave (Burn of the Caves), Inchnadamff, Assynt, Sutherland. *Proc. Royal Soc. Edingburgh*, 37:344 - 349.

Persson, L. 1985. Asymetrical competition: are larger animals competitively superior? *The American Naturalist*, 126(2): 261 - 266.

Peters, J.L. 1934. *Check-List of Birds of the World, Vol. 2*. Harvard Univ. Press, Cambridge.

Phillips, L. 1974. Vegetational history of the Ipswichian/Eemian interglacial in Britain and Continental Europe. *New Phytologist*, 73: 589 –604.

Popper, K.R. 1968. *The Logic of Scientific Theory.* Harper Torch Books, New York.

Potapova, O.R. 1986. The Pleistocene ptarmigans G. *Lagopus* from the northern Urals. USSR Academy of Sciences. *Proceedings of the Zoological Institute*, Leningrad 147: 46 - 58.

Price, T. and Langen, T. 1992. Evolution of correlated characters. *TREE*, 7(9): 307 - 310.

Rand, A.L. 1948. Glaciation, an isolating factor in speciation. *Evolution*, 2: 314 - 321.

Randi, E.; Alkon, P.U. and Meriggi, A. 1992. A new model of *Alectoris* Evolution based on Biochemical analysis. *Gibier Faune Sauvage*, 9: 661 - 666.

Ratcliffe, L.M. and Grant, P.R. 1983. Species recognition in Darwin's finches (*Geospiza* Gould). I. Discrimination by morphological cues. *Animal Behaviour*, 31: 1139 – 1153.

Regalia, E. 1902. Sette uccelli pliocenici del Pisano e del Valdarno superiore. *Palaeontographia italica*, VIII: 219 – 238.

Reichstein, H. von and Pieper, H. 1986. *Untersuchungen an Skelettresten von Vogeln aus Haithabu (Aausgrabung 1966 - 1969).* Karl Wachholtz Verlag: Neumunster.

Rensch, B. 1936. Studien über klimatische Parallelität der Merkmalsausprägung bei Vögeln und Säugern. *Arch. f. Naturgesch.* N.F., 5: 317 - 363.

Reynolds, S.J. 1996. Establishing body-size indicators in a study of female spruce grouse *Dendragapus canadensis. J. Zool., Lond.*, 240: 113 - 122.

Rich, P.V. 1980. Preliminary report on the fossil avian remains from the late Tertiary sediments at Langebaanweg (Cape Province), South Africa. *South African Journal of Science*, 76: 166 - 170.

Ricklef, R.E. 1980. Geographical variation in clutch size among passerine birds: Ashmole's hypothesis. *Auk*, 97: 38 – 49.

Roberts, M.B. 1986. Excavation of a Lower Palaeolithic site at Amey's Eartham Pit, Boxgrove, West Sussex; A preliminary report. *Proceedings of the Prehistoric Society*, 52: 215 - 245.

Roberts, M.B.; Stringer, C.B. and Parfitt, S.A. 1993. A hominid tibia from Middle Pleistocene sediments at Boxgrove, UK. *Nature*, 369, 311 - 313.

Roberts, M.B.; Gamble, C.S. and Bridgland, D.R. 1995. The earliest occupation of Europe: the British Isles. In: W. Roebroeks and T. van Kolfschoten (Eds.). *The Earliest Occupation of Europe*. University of Leiden Press, Leiden.

Roberts, A. 1996. Evidence for Late Pleistocene and Early Holocene activity and environmental change from the Torbryan Valley, South Devon. In: D.J. Charman, R.M. Newnham and D.G. Croot (Eds.). *The Quaternary of Devon and East Cornwall: Field Guide.* Quaternary Research Association, London.

Rohlf, F.J. and Marcus L.F. 1993. A revolution in morphometrics. *TREE*, 8(4) 129 - 132.

Rose, J. 1987. The status of the Wolstonian glaciation in the British Quaternary. *Quaternary Newsletter*. 53: 1 - 9.

Rosen, D.E. 1979. Fishes from the uplands and intermontane basins of Guatemala: revisionary studies and comparative geography. *Bull. Am. Mus. Nat. Hist.*, 162: 267 – 376.

Rosenzweig, M.L. 1997. Tempo and mode of speciation. *Science*, 277: 1622 – 1623.

Röttger, U. 1987. Schmelzbandbreiten an Molaren von Schermäusen (*Arvicola* Lacepede, 1799). *Bonn. Zool. Beitr.*, 38(2): 95 – 105.

Saint-Périer, R. de 1920. La grotte des Harpons a Lespugne (Haute-Garonne). *L' Anthropologie*, 30(3 - 4): 209 - 234.

Saint-Périer, R. de 1927. La grotte de Gouërris a Lespugne. *L' Anthropologie*, 37: 263 - 276.

Salomonson, F. 1955. The evolutionary significance of bird migration. *Dansk Biologiske Meddelelser*, 22(6): 1 - 62.

Sánchez Marco, A. 1987a. Aves fosiles de Atapuerca. In: E. Aguirre; E. Carbonell and J.M. Bermudez de Castro (Eds.): *El hombre fosil de Iberas y El Pleistoceno de la Sierra de Atapuerca. I.* Junta de Castilla y Leon, Consejeria de Cultura y Bienestar Social. 67 - 73.

Sánchez Marco, A. 1987b. Saalian S.L. avian findings in the Northern Iberian Plateau.. In: Mourer-Chauvire (Ed.): L'Evolution des Oiseaux d'apres le

Temoignage des Fossiles. *Docum. Lab. Geol. Lyon*, no. 99. 193 – 196
Sánchez Marco, A. 1990. Aves de los yacimientos mesopleistocenios de Torralba y Ambrona (Soria, España). *Actos de Paleontologia*, XX: 349 – 356.
Sánchez Marco, A. 1992. Birds of the stratigraphic unit TG-11 at la Galeria (Sierra de Atapuerca, Burgos, Spain). *Evolución humana en Europa y los yacimientos de la sierra de Atapuerca*, 1: 137 - 146.
Sánchez Marco, A. 1996. The presence of the Waldrapp *Geronticus eremita* (Plataleidae) in the Plio-Pleistocene boundary in Spain. *Ibis*, 138(3): 560 - 561.
Sangster, G.; Collinson, J.M.; Helbig, A.J.; Knox, A.G. and Parkin, D.T. 2002. Taxonomic recommendations for British birds: second report. *Ibis*, 146: 153 – 157.
Schmidt-Burger, P. 1982. *Vergleichend morphologische Untersuchungen an Einzelknochen in Zentraleuropa vorkommender mittelgrossen Accipitridae: II, Beken und Hinterextremität.* Ludwig-Maximilians-Universitat, Munchen.
Scholander, P.F. 1956. Climatic rules. *Evolution*, 10: 339 - 340.
Schluter, D. 1986. Character divergence between distantly related taxa? Finches and bees in the Galapagos. *American Naturalist*, 127: 95 - 102.
Schluter, D. 1988. Character divergence and the adaptive divergence of finches on islands and continents. *American Naturalist*, 131, 799 - 824.
Scott, K.M. 1985. Allometric trends and locomotor adaptions in the Bovidae. *Bulletin of the American Museum of Natural History*, 179(2): 197 - 288.
Searcy, W.J. 1980. Otimum body size at different abient temperatures: an energetic explanation of Bergmann's Rule. *J. theor. Biol.*, 83: 579 - 593.
Selander, R.K. 1965. Avian speciation in the Quaternary. In: H.E. Wright, Jr. and D.G. Frey (Eds.). *The Quaternary of the United States*. Priceton: Princeton University Press. 527 – 542.
Selander, R.K. 1966. Sexual dimorphism and differential niche utilization in birds. *Condor*, 68: 113 – 151.
Serebrovsky, P.V. 1940. New species of birds from the Binagady Beds. *Doklady Akad. Nauk SSSR* , 27: 7: 766 - 768.
Sernander, R. 1908. On the evidence of Post-glacial changes of climate furnished by the peat mosses of northern Europe. *Geologiska Föreningen i Stockholm Förhandlingar* 30: 365 – 478.
Serjeantson, D. 1991. The birds. In: B. Cunliffe and C. Poole (Eds.). *Danebury: An Iron Age Hillfort in Hampshire. Vol. 5. The Excavation 1979 - 1988: The Finds*. Research Report No. 73. Council for British Archaeology.
Serjeantson, D. 1997. Subsistence and symbol: The interpretation of bird remains in archaeology. *International Journal of Osteoarchaeology*, 7(4): 255 – 259.
Shackleton, N.J. and Opdyke, N.D. 1973. Oxygen isotope and palaeomagnetic stratigraphy of equatorial Pacific core V28 - 238: oxygen isotope temperatures and ice volumes on a 10^5 and 10^6 year scale. *Quaternary Research*, 3: 39 - 55.
Shuffeldt, R.W. 1888. On the skeleton in the genus *Sturnella* with osteological notes upon other North American Icteridae and the corvidae. *J. Anat. Physiol.*, 22: 309 - 348.
Sibley, C.J. and Ahlquist, J.E. 1972. *A Comparative Study of the Egg White Proteins of Non-Passerine Birds.* Peabody museum of Natural History Yale University Bulletin No. 39.
Simpson, G.G. 1961. *Principles of Animal Taxonomy.* Columbia University Press, New York.
Slotow, R. and Goodfriend, W. 1996. Ecogeographic variation in body size and shape of cape sparrows (*Passer melanurus*) in southern Africa. *J. Zool., Lond.*, 283: 279 - 286.
Smith, A.B. 1994. *Systematics and the Fossil Record: Documenting Evolutionary Patterns.* Blackwell Scientific Publications, Oxford.
Sneath, P.H.A. and Sokal, R.R. 1973. *Numerical Taxonomy*. W.H. Freeman, San Francisco.
Snow, D.W. 1954. Trends in geographical variation in Palearctic members of the genus *Parus*. *Evolution*, 8: 19 – 28.
Snow, D.W. 1996. Should the biological be supersede by the phylogenetic species concept? *Bull. Brit. Ornith. Club*, 117(2): 110 - 121.
Soergel, E. 1955. Uber einige Vogelreste (Seeadler, Kranishe) aus dem Neolitikum von Erhenstein bei Ulm. *Jahres. Ver. vaterl. Naturk. Wurtt.*, 110: 121 - 124.
Soergel, E. 1961. Die Vogelreste. I Schuldt, E. (red). Hohen Viecheln. Ein mittelsteinzeitlicher Wohnplatz in Mecklenburg: 64-69. *Deutsch. Akad Wiss. Berlin Schr. Sekt. Voru. Frugesch. Berlin.*
Sokal, R.R. and Crovello, T.J. 1970. The biological species concept: a critical evaluation. *American Naturalist*, 104: 127 – 153.
Solti, B. 1981. Vergleigend osteologische Untersuchungen an Skelettsystem der Falkenarten *Falco cherrug* Gray und *Falco peregrinus* Tunstall. *Vertebr. Hung.*, 20: 75 - 125.
Southwell, T. 1901. On the breeding of the Crane in East Anglia. *Trans. Norf. Norw. nat. Hist. Soc.* 7: 160 - 170.
Steadman, D. 1981. Vertebrate fossils in lava tubes in the Galapagos Islands. *Proceedings of the 8th International Congress in Speleology*, 549 - 550.
Steadman, D. 1985. Holocene vertebrate fossils from Isla Florena, Galapagos. *Smithsonian Contributions to Zoology*, No. 413.
Stewart, J.R. 1992. *The Turdidae of Pin Hole Cave, Derbyshire*. Unpublished MSc. Dissertation, City of London, North London and Thames Polytechnics.
Stewart, J.R. 1996. Quaternary birds from Torbryan Valley. In: D.J. Charman, R.M. Newnham and D.G. Croot (Eds.). *The Quaternary of Devon and East Cornwall: Field Guide*. Quaternary Research Association, London, England.

Stewart, J.R. 1999a. *The Evolution of Quaternary Birds in the Western Palaearctic: aspects of taxonomy and ecomorphology*. University of London. Unpublished PhD Dissertation.

Stewart, J.R. 1999b. Intraspecific variation in modern and Quaternary European *Lagopus*. *Smithsonian Contributions to Paleobiology*, 89:159 – 168.

Stewart, J.R. 1999c. The avifauna. In: N.M. Ashton, S.G. Lewis and S. Parfitt.(Eds.) *Excavations at the Lower Palaeolithic Site at Barnham 1989 - 94*. British Museum Press.

Stewart, J.R. 2001. Wetland birds in the archaeological and recent palaeontological record of Britain and Europe. Pp. 141 – 148. In: B. Coles and D.E. Bull (Eds.). *Heritage Management of Wetlands*. Europae Archaeologiae Consilium.

Stewart, J.R. 2002a. The evidence for the timing of speciation of modern continental birds and the taxonomic ambiguity of the Quaternary fossil record. Pp. 261 – 282. *Proceeding of the 5th Symposium of the Society of Avian Paleontology and Evolution.* In: Z. Zhou and F. Zhang (eds.). China Science Press, Beijing.

Stewart, J.R. 2002b. Sea Birds from coastal and non-coastal archaeological and "natural" Pleistocene Deposits or not all unexpected deposition is of human origin. *Acta Zoologica Cracoviensia,* Vol. 45 (special issue), Pp. 167 – 178.

Stewart, J.R. 2003. Comment on "Buffered tree population changes in a Quaternary refugium: Evolutionary implications". *Science*, 299: 825a.

Stewart, J.R. 2004. Wetland Birds in the Recent Fossil Record of Britain and North-West Europe. *British Birds*, 97(1): 33 – 43.

Stewart, J.R. 2005. The use of modern geographical ranges in the identification of archaeological bird remains. In: Gisela Grupe and Joris Peters (Eds.). Feathers, grit and symbolisms. Birds and archaeology in the Old and New Worlds. Proceedings of the ICAZ Bird Remains Working Group meeting in Munich 2004. *Documenta Archaeobiologiae* 3: 43 – 54.

Stewart, J.R. In press a. The bird remains. In: M.B. Roberts and S.A. Parfitt (Eds.). *The Boxgrove Hominid Site*. English Heritage Monograph Series, London.

Stewart, J.R. In press b. The avifauna of the deposits surrounding the West Runton elephant. *Quaternary International*.

Stewart, J.R. In prep. The birds. In: A. Roberts, A. Currant and N. Barton (Eds.) *The History of a Valley*. Torbryan. British Museum Press, London.

Stewart, J.R. and Beech, M. 2006. The Miocene birds of Abu Dhabi with comments on the likelihood that modern species lived in the Neogene. *Journal of Historical Biology* 18(2): 103–113.

Stewart, J.R. and Hernández Carrasquilla, F. 1997. The identification of Extant European Birds remains: a review of the literature. *International Journal of Osteoarchaeology*, 7: 364 – 371.

Stewart, J.R. and Lister, A.M. 2001. Cryptic northern refugia and the origins of modern biota. *Trends in Ecology and Evolution*, 16(11): 608 - 613.

Storer, R.W. 1966. Sexual dimorphism and food habits in habits in three North Americam accipiters. *Auk*, 83: 423 – 436.

Storer, R.W. 1992. Intraspecific variation and the identification of Pliocene and Pleistocene grebes. *Nat. Hist. Mus. Los Angeles Co., Sci. Ser.*, 36: 419 - 422.

Stringer, C.B. 1975. A preliminary report on new excavations at Bacon Hole Cave. *Gower,* 26: 32 - 37.

Stringer, C.B. 1977. Evidence of climatic change and human occupation during the last interglacial at Bacon Hole, Gower. *Gower*, 28: 36 - 44.

Stringer, C.B.; Currant, A.P.; Schwartcz, H.P. and Collcutt, S.N. 1986. Age of Pleistocene faunas from Bacon Hole, Wales. *Nature*, 320: 59 - 62.

Stringer, C.B.; Andrews, P. and Currant, A. 1996. Palaeoclimatic significance of mammalian faunas from Westbury Cave, Somerset, England. In: C. Turner (Ed.). *The early Middle Pleistocene in Europe.* Balkema, Rotterdam.

Stuart, A.J. 1976. The history of the mammal fauna during the Ipswichian/Last Interglacial in England. *Philosophical Transactions of the Royal Society of London B*, 276: 221 – 250.

Stuart, A.J. 1977. The vertebrates of the last cold stage in Britain and Ireland. *Philosophical Transactions of the Royal Society of London B*, 280: 295 – 312.

Stuart, A.J. 1982. *Pleistocene Vertebrates in the British Isles*. Longman, London.

Stuart, A.J. 1983. Pleistocene bone caves in Britain and Ireland. *Studies in Speleology*, 4: 8 - 36.

Stuart, A.J. 1986. Pleistocene occurances of hippopotamus in Britain. *Quärtarpalaontologie,* 6: 209 - 218.

Stuart, A.J. 1991. Mammalian extinctions in the Late Pleistocene of Northern Eurasia and North America. *Biol. Rev.*, 66: 453 - 562.

Stuart, A.J. and West, R.G. 1976. Late Cromerian fauna and flora at Ostend, Norfolk. *Geological Magazine*, 113: 469 – 473.

Sutcliffe, A.J. 1960. Joint Mitnor Cave, Buckfastleigh. *Trans. Proc. Torquay Nat. Hist. Soc.*, 13: 1 – 26.

Sutcliffe, A.J. 1976. The British glacial – interglacial sequence a reply to Professor R.G. West. *Quatenary Newsletter*, 17: 1 - 3.

Sutcliffe, A.J. 1975. A hazard in the interpretation of glacial/integlacial sequences. *Quaternary Newsletter*, 17: 1 - 3.

Sutcliffe, A.J. and Kowalski, K. 1976. Pleistocene Rodents of the British Isles. *Bulletin of the Biritish Isles (Natural History). Geology*. 27(2): 31 – 147.

Sutcliffe, A.J. and Zeuner, F. 1962. Excavations in the Torbryan Caves, Devonshire. 1. Tornewton Cave. *Proc. Devon archaeol. Explor. Soc.*, Torquay, 5: 127 - 145.

Sutcliffe, A.J.; Currant, A.P. and Stringer, C.B. 1987. Evidence of sea-level change from coastal caves with raised beach deposits, terrestrial faunas and dated stalagmites. *Progress in Oceanography*, 18: 243 - 271.

Sutherland, S. 1986. *The Archaeozoological Study of Bird Bones Past and Present with Particular Reference to some Late Pleistocene Avifaunas in Britain.* Unpublished PhD Thesis, University of Sheffield.

Tattersall, I. 1986. Species recognition in human Paleontology. *Journal of Human Evolution*, 15: 165 – 175.

Tattersall, I. 1991. What was the human revolution. *Journal of Human Evolution*, 20: 77 – 83.

Tchernov, E. 1962. Palaeolithic avifauna in Palestine. *Bull. Res. Counc. of Israel*, 11B3: 95 - 131.

Temple, S.A. 1972. Systematics and evolution of the North American Merlins. *Auk*, 89: 325 – 338.

Thorne A.G. and Wolpoff, M.H. 1992. The multiregional evolution of humans. *Scientific American*, April 1992: 76 - 83.

Tordoff, H.B. 1954. A systematic study of the avian familly Fringillidae, based on the structure of the skull. *Misc. Publ. Mus. Zool. Univ. Michigan*, 81: 1 - 63.

Tratman, E.K. 1960. Gough's Old Cave, Cheddar, Somerset. *Proc. Univ. Bristol Spelaeol. Soc.*, 9(1): 7 - 21.

Tyrberg, T. 1985. *Grus primigenia* - Europas utdöda jättetrana. *Fauna och Flora*, 80: 169 - 176.

Tyrberg, T. 1991a. Arctic, Montane and Steppe birds as Glacial relicts in West Palearctic. *Orn. Verh.*, 25: 29 - 49.

Tyrberg, T. 1991b. Crossbill (Genus *Loxia*) evolution in the West Palearctic - a look at the fossil evidence. *Ornis Svecica*, 1: 3 - 10.

Tyrberg, T. 1995. Palaeobiogeography of the genus *Lagopus* in the West Palearctic. *Courier Forschungsinstitut Senckenberg*, 181: 275 - 291.

Tyrberg, T. 1998. Pleistocene birds of the Palaearctic. A Catalogue. *Publ. Nutall Ornith. Club*, 27: 1 - 720.

Tyrberg, T. 1999. Seabirds and late Pleistocene Environments in the Northeast Atlantic and Mediterranean. *Smithsonian Contributions to Paleobiology*, 89:139 – 157.

Tyrberg, T. 2002. Avian species turnover and species longevity in the Pleistocene of the Palaearctic. Pp. 281 – 289. *Proceeding of the 5th Symposium of the Society of Avian Paleontology and Evolution.* In: Z. Zhou and F. Zhang (eds.). China Science Press, Beijing.

Turner, A. 1993. Species and speciation. Evolution and the fossil record. *Quaternary International*, 19: 5 - 8.

Turner, A. and Chamberlain, A.T. 1989. Speciation, morphological change and the status of of African *Homo erectus*. *Journal of Human Evolution*, 18: 115 - 130.

Turner, C. and West, R.G. 1968. The subdivision and zonation of interglacial periods. *Eiszeitalter un Gegenwart*, 19: 93 – 101.

Turner, C. (Ed.). 1996. *The Early Middle Pleistocene in Europe*. Balkema, Rotterdam.

Tzedakis, P.C.; Lawson, I.T.; Frogley, M.R.; Hewitt, G.M.; Preece, R.C. 2002. Buffered tree population changes in a Quaternary Refugium: evolutionary implications. *Science*, 297: 2044 –2047.

Valen, L. van. 1976. Ecological species, multispecies and oaks. *Taxon*, 25: 233 - 239.

Vaurie, C. 1962. *The birds of the Palaearctic fauna. Passeriformes*. Witherby, London.

Vbra, E.S. 1980. Evolution, species and fossils: how does life evolve? *S. Afr. J. Sci.*, 76, 61 – 84.

Vbra, E.S. 1983. Evolutionary pattern and process in the sister-group Alcelaphini-Aepycerontini (Mammalia: Bovidae). In: N. Eldredge and S.M. Stanley (Eds.). *Living Fossils*. Springer-Verlag, New York. 62 – 79.

Von den Driesche, A. 1976. *A Guide to the Measurement of Animal Bones from Archaeological Sites*. Peabody Museum Bulletin 1. Peabody Museum of Archaeology and Ethnology, Harvard University.

Von den Driesch, A. 1999. The cranes, *Grus grus*, in prehistoric Europe and its relation to the Pleistocene crane, *Grus primigenia*. In Norbert Benecke (Ed.). *The Holocene History of the European Vertebrate Fauna Modern Aspects of Research.* Workshop, 6th to 9th April 1998, Berlin. Verlag Marie Leidorf GmbH, Rahden/Westf.

Von den Driesch, A and Boessneck, J. 1969. Die Fauna des Cabezo Redondo bei Villena (Prov. Alicante). *Studien uber Fruhe Tierknochenfunde von der Iberischen Halbinsel,* 1: 43 - 89.

Voous, K.H. 1960. *Atlas of European Birds*. Nelson, Amsterdam.

Walkinshaw, L. 1973. *Cranes of the World.* Winchester Press, New York.

Walker, C.A.; Wragg, G.M. and Harrison, C.J.O. 1990. A new shearwater from the Pleistocene of the Canary Islands and its bearing on the evolution of certain *Puffinus* shearwaters. *Historical Biology*, 3: 203 - 224.

Walker, H.H. and Sutcliffe, A.J. 1968. James Lyon Widger, 1823 - 1892, and the Torbryan Caves. *Transactions of the Devonshire Association*, 99, 49 - 110.

Warheit, K.I. 1992. The role of morphometrics and cladistics in the taxonomy of fossils: a palaeornithological example. *Syst. Biol.*, 41(3): 345 - 369.

Weesie, D.M. 1988. The Quaternary Avifauna of Crete, Greece. *Palaeovertebrata*, 18 (1): 1-94.

Weiner, J. 1994. *The Beak of the Finch.* Vintage, London.

West, R.G. 1956. The Quaternary deposits at Hoxne, Suffolk. *Philosophical Proceedings of the Royal Society of London.* B239: 265 – 356.

West, R.G. 1976. The British Glacial-Interglacial sequence. *Quaternary Newsletter*, 18: 1.

West, R.G. 1977. *Pleistocene Geology and Biology: with Especial Reference to the British Isles*. Second edition. Longman, London.

West, R.G. 1979. Further on the Flandrian. *Boreas*, 8: 426.

West, R.G. 1980. Pleistocene forest history in East Anglia. *New Phytologist*, 85: 571 - 622.

Wetmore, A. 1931. The avifauna of the Pleistocene of Florida. *Smithsonian Misc. Coll.*, 85: 1 – 41.

Wetmore, A. 1951. Recent additions to our knowledge of prehistoric birds, 1933 – 1949. *X Intern. Ornithol. Congr. Proc.*, 51 – 74.

Wetmore, A. 1957. The classification of the Oscine Passeriformes. *Condor*, 59(3): 207 - 209.

Wetmore, A. 1959. Birds of the Pleistocene of North America. *Smithsonian Misc. Coll.*, 138: 1 – 24.

Wiens, J.A. and Rotenberry, J.T. 1980. Patterns of morphology and ecology in grassland and shrubsteppe bird populations. *Ecological Monographs*, 50(3): 287 - 308.

Wheeler, A. 1978. Why were there no fish remains at Star Carr? *Journal of Archaeological Science*, 5: 85 – 89.

Willey, E.O. 1978. The evolutionary species concept reconsidered. *Syst. Zool.*, 27: 17 - 26.

Williams, B.J. 1987. Rates of evolution: is there a conflict between neo-darwinian evolutionary theory and the fossils record? *American Journal of Physical Antrhropology*, 73: 99 – 109.

Willis, K.J.; Bennett, K.D. and Walker, D. 2004. The evolutionary legacy of the Ice Ages. *Philosophical Transactions of the Royal Society. Biological Sciences*, 359 (1442): 155 - 303.

Willis, K.J. and Whittaker, R.J. 2000. The refugial debate. *Science*, 287: 1406 – 1407.

Willis, K.J.; Rudner, E. and Sümegi, P. 2000. The full-glacial forest of Central and Southeastern Europe. *Quaternary Research*, 53: 203 – 213.

Winkler, H. and Leisler, B. 1992. On the ecomorphology of migrants. *Ibis,* 134, Suppl. 1: 21 - 28.

Wiszniowska, T. 1991. Middle Pleistocene bird remains from the Ferdynandovian interglacial near Belchatow (Central Poland). *Folia Quaternaria*, 61/62: 125 - 130.

Whitlock, R. 1953. *Rare and Extinct Birds of Britain.* Phoenix House Ltd., London.

Witherby, H.F.; Jourdain, F.C.R.; Ticehurst, N.F. and Tucker, B.W. 1941. *The Handbook of British Birds*, Vol. 5, H.F. and G. Witherby , London.

Woelffle, E. 1967. *Vergleichend morphologische Untersuchungen an Einzelknochen des postcranialen Skeletts in Mitteleuropa vorkommender Enten, Halbgänse und Säger.* Ludwig-Maximilians-Universitat, Munchen.

Wood, D.S. 1979. The phenetic relationships within the Family Gruidae. *Wilson Bull.*, 91: 384 - 399.

Woolfenden, G.E. 1961. Postcranial osteology of the waterfowl. *Bulletin of the State of Florida State Museum, Biological Sciences*, 6(1): 1 - 129.

Zink, R.M. 1997. Species concepts. *Bull. British Onithologists' Club*, 117(2): 97 - 109.

Zink, R.M.; Klicka, J. and Barber, B.R. 2004. The tempo of avian diversification during the Quaternary. In: K.J. Willis; K.D. Bennett and D. Walker (Eds.). The Evolutionary Legacy of the Ice Ages. *Philosophical Transactions of the Royal Society; Biological Sciences*, 359 (1442): 215 – 220.

Zink, R.M. and Remsen, J.V. Jr. 1986. Evolutionary processes and patterns of geographic variation in birds. *Curr. Ornithol*, 4:1 - 69.

www.ingramcontent.com/pod-product-compliance
Ingram Content Group UK Ltd.
Pitfield, Milton Keynes, MK11 3LW, UK
UKHW061845041026
15591UKWH00015B/926

9 781407 300894